The International Karakoram Project

THE INTERNATIONAL

KARAKORAM

PROJECT

Volume 2

Proceedings of the International Conference
held at
The Royal Geographical Society
London

EDITED BY K. J. MILLER

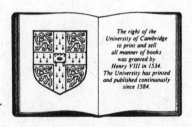

*The right of the
University of Cambridge
to print and sell
all manner of books
was granted by
Henry VIII in 1534.
The University has printed
and published continuously
since 1584.*

Cambridge University Press

Cambridge

London New York New Rochelle
Melbourne Sydney

CAMBRIDGE UNIVERSITY PRESS
Cambridge, New York, Melbourne, Madrid, Cape Town, Singapore,
São Paulo, Delhi, Dubai, Tokyo

Cambridge University Press
The Edinburgh Building, Cambridge CB2 8RU, UK

Published in the United States of America by Cambridge University Press, New York

www.cambridge.org
Information on this title: www.cambridge.org/9780521129763

First published 1984
This digitally printed version 2009

A catalogue record for this publication is available from the British Library

ISBN 978-0-521-26340-5 Hardback
ISBN 978-0-521-12976-3 Paperback

These proceedings are dedicated to the memory of
J. F. BISHOP
who died on Kurkun (4730 m) on 14 July 1980
during the course of
The International Karakoram Project

Contents of Volume 2

KARAKORAM RESEARCH CELL

Contents of Volume 1

Preface

These volumes detail a comprehensive series of inter-related studies carried out by seventy-three scientists from China, Pakistan and Britain who visited the Karakoram mountains in the summer of 1980. These men and women were members of the Royal Geographical Society Project to the world's highest mountain range as part of the Society's 150th anniversary celebrations.

The proceedings are divided into two volumes; the first volume being reserved for papers presented and discussed, in conference, in Islamabad, Pakistan prior to the field work. The second volume is reserved for papers based on the field work conducted during the course of the International Karakoram Project, or IKP for short, and read at a conference at the Royal Geographical Society, London.

The two volumes present a comprehensive cover of the earth sciences, ranging in scale from surveys to measure the effects of impacts between the Eurasian and Indian tectonic plates, to electron microscope studies on weathering processes. The six general subject areas are: geology, glaciology, geomorphology, seismology and topography (survey). The sixth subject is concerned with the indigenous population namely social, economic and health aspects of villagers, the architectural form and structural stability of their homes and the order of priorities they have to accept when combating the greatest number of natural hazards known to man which include earthquakes, drought, fire, floods, famine, landslides, avalanches and disease.

These proceedings therefore bring together a collection of papers to provide a datuum for future scientific research in the Karakoram. The IKP also proved to be a catalyst for the founding of the Karakoram Research Cell (KRC) by the University Grants Commission of Pakistan. The KRC will be responsible for co-ordinating the research work conducted in the Karakoram by future researchers; see Volume 2.

The varied scientific programmes undertaken were supported by many national and international bodies, listed elsewhere, but a special mention must be made of Academia Sinica, Beijing and the Ministry of Science and Technology, Islamabad who joined the Royal Geographical Society in sponsoring the venture. These three together with The Overseas Development Administration of the UK Government and the British Council initially made the project a reality and finally an international success in scientific collaboration.

K. J. Miller
Leader of the
International Karakoram Project

Sheffield 1982.

Other publications related to the International Karakoram Project: Continents in Collision, 212 pp, K.J. Miller, ISBN 0 540 010669. George Philip and Son Ltd. London.

Preface to Volume 2

This volume presents the completed work of the 73 scientists associated with the International Karakoram Project during the summer of 1980 together with their friends and colleagues in home laboratories who assisted in the interpretation of the collected data. The 34 papers are divided into several groups, including:

Geology

Glaciology

Survey

Earthquakes

Housing and Natural Hazards

Geomorphology

but even a superficial glance of the papers will show that many of them could be placed into alternative categories such was the multidisciplinary nature, both of the individual studies and the expertise of the investigators.

The text is supported by some 80 Tables of data, 49 graphs, 108 photographs, 71 maps and 52 figures illustrating the range and depth of the research undertaken by the Chinese, Pakistan, Swiss and British scientists. Undoubtedly this volume (along with volume 1) will constitute essential reading for scientists, naturalists and mountaineers visiting the Karakoram, and who may wish to extend the breadth and depth of man's knowledge of an area which only a little over a century ago was a blank on the map.

The first part of the proceedings, covering the opening session of the conference presents an appraisal of the IKP immediately after an introduction by the President of the Royal Geographical Society, Professor M J Wise. There follows a review of the history of the Karakoram exploration by Dorothy Middleton, an account of modern climbs and new techniques employed by today's mountaineers, written by Paul Nunn, and finally a historical note by Michael Vyvyan, one of the first men to attempt Rakaposhi; the mountain that provided the backcloth of the IKP research studies.

The two Geology papers are complimentary, the first rationalising and so simplifying the geology formed by continental collision, the second paper concentrating on the complexity and inconsistencies of the known geological patterns. Both papers indicate the immense research potential of the area.

The glaciology section presents the first ever results of precise ice-depth soundings of Karakoram-Himalayan glaciers, together with suggestions for improvements of this new technique using impulse-radar equipment. Transverse and longitudinal profiles of the Hispar and Ghulkin glaciers are presented.

The survey papers give the results of the 1980 programme and compares the data with the 1913 survey linking India to Russia in order to detect the location and magnitude of movements between the continental masses. An alternative method to the use of co-ordinates for deformation analysis is also presented and further results ten years hence are now awaited with interest to compare with the 1913 and 1980 studies.

The section on earthquakes reports the first ever micro-earthquake study ever carried out in the Karakoram. Most of the 371 recordings are of shallow and intermediate depth earthquakes that are in contrast to the deeper shocks to the north in China and west in the Hindu Kush of Afghanistan.

A brief summary of the eight papers in the Housing and Natural Hazards section is impossible within these few lines; sufficient to report that for the first time a socio-economic cultural study has been made of the human pattern of life within the context of the rapid geomorphological changes of the area so as to provide order to what has previously been a chaotic and grossly misunderstood - misinterpreted scene only equalled by the surrounding landscape.

The volume draws to a close with nine papers reporting on the geomorphology of the Hunza · region, e.g., Glacier positions, scree slope stability, weathering processes, sediment classification, water chemistry etc. all of which set a basis for continuing studies. This section more than any other gives valuable and substantial reference information in the form of photographs that clearly indicate the reasons why more catastrophic events occur in the Karakoram than elsewhere in the world.

Thus the 600 plus pages represent only one beginning of an integrated search to unravel the natural mysteries and solve the human problems associated with the most chaotic landscape on the surface of our earth. To help achieve such objectives, future researchers are directed towards the final paper in these proceedings which outlines the aims of the Karakoram Research Cell.

In this respect it is important to appreciate the philosophy that guided the sponsors of the IKP namely that the Karakoram mountains of Pakistan represent a laboratory for world science and that all who go to this area should acknowledge that the great forces that distort the surface of our earth can by equalled by man's energy and ability to work together to the benefit of all.

K. J. Miller

Sheffield, 1983

Acknowledgements

The authors of these proceedings and all members of the IKP acknowledge the assistance given, in terms of time, effort and finance, by persons far too numerous to list here. Without the help given by these countless men and women, and those listed below, the IKP would never have completed the numerous studies reported in these volumes. We are forever in their debt.

IKP SPONSORS

Government of Pakistan
Academia Sinica, Beijing
Royal Geographical Society

INTERNATIONAL PROJECT COMMITTEE

Dr John B. Auden (Chairman)
Dr G. Colin L. Bertram
Professor Eric H. Brown
Sir Douglas Busk
Dr Andrew Goudie (Deputy Leader)
George Greenfield
Lt. Col. David Hall
Brigadier George Hardy
John Hemming (Director RGS)
Lord Hunt of Llanfair Waterdine
    (President RGS)
Professor Keith J. Miller (Leader)
Mr Michael Ward
Nigel de N. Winser (Deputy Leader
    and Secretary)

PAKISTAN LIAISON

David Latter, British Council
Representative, Islamabad

DISTINGUISHED VISITORS

H. E. The British Ambassador
    Oliver Forster
Major General Mushtaq Ahmad Gill
Lord and Lady Hunt
Lord Shackleton

CONSULTANTS

GEOMORPHOLOGICAL RESEARCH

Professor E. H. Brown
Professor K. J. Gregory
Professor N. J. Stevens
Dr R. J. Price
Dr J. M. Grove
Dr D. Q. Bowen

RADAR ICE-DEPTH SOUNDING RESEARCH

Sir Vivian Fuchs
Dr S. Evans
Dr D. Drewry
Dr G. de Q. Robin
Dr C. W. M. Swithinbank
Dr C. M. Doake

cont'd...

SURVEY EQUIPMENT

Aga Geotronics
British Aluminium
Casio Electronics
Decca Surveys
Hunting Surveys
Mapping and Charting Establishment RE

School of Military Survey
SLD Sitelink Services
Survey & General Instruments
Tellurometer (UK) Ltd
Wild Heerbrugg

OTHER SCIENTIFIC EQUIPMENT

Avon   Rubber Company
Chloride Batteries (Pakistan) Ltd

Hewlett Packard Ltd (Geneva)
Perex Ltd

TRANSPORT

Anchor Line Ship Management Ltd
Associated Tyre Specialists
Automobile Association (Stockport)
British Airways
Brown Jenkinson (Liverpool) Ltd
Burmah Castrol Ltd
Caltex Oil (Pakistan) Ltd
Fairey Engineering Ltd
Land Rover Ltd

Lifting Gear Hire Ltd
Lloyds Industries
Manor National Group Motors
Michelin Tyre Company Ltd
Pakistan National Shipping Corp
Pakistan State Oil Co Ltd
Protofram Ltd
Trailvan Ltd
Unipart Ltd

FOODSTUFFS

British Food Export Council
Coca Cola Corporation
Colmans Foods
Drinkmaster Ltd
Frank Cooper
H J Heinz
The Honey Bureau & Gales Honey
John West Foods
Nabisco Ltd

Raven Food
Ryvita
Sabatani & Taylor Associated
    for Schwartz Spices
St Ivel Ltd
Tate & Lyle
Weetabix Ltd
Whitworths Holdings
Unilever Export Ltd

MEDICAL SUPPLIES

Abbott Laboratories Ltd
A D International Ltd
Allen & Hanburys Ltd
Astra Chemicals Ltd
Bayer (UK) Ltd
Beecham Research Laboratories
Bencard
Boehringer Ingleheim Ltd
Boots Company Ltd
Calmic Medical Division of the
    Wellcome Foundation
Ciba Laboratories
Davis & Geck
Dista Products
Dome Laboratories
Downs Surgical Ltd
Duncan Flockhart & Co Ltd

Merck Sharp & Dohme Ltd
Miles Laboratories
Montedison Pharmaceuticals Ltd
Nicholas Laboratories
Novo Laboratories
Lakeland Plastics
Parke Davis & Co Ltd
Pfizer Ltd
Pharmacia (GB) Ltd
Pharmax Ltd
A H Robins Co Ltd
Roche Products Ltd
Roussel
Reckitt & Colman Pharmaceuticals
Scholl (UK) Ltd
Serle Laboratories
R Rycroft & Sons Ltd

Eli Lily & Co Ltd
Fair Laboratories
Farley Health Products
Geigy Pharmaceuticals
Glaxo Laboratories
Hoechst Pharmaceuticals
Imperial Chemical Industries
Janseen Pharmaceuticals Ltd
Johnson & Johnson Ltd
Kirby Warwick Ltd
Laboratories for Applied Biology
Lederle Laboratories
Leo Laboratories
May & Baker Ltd

Smith & Nephew Pharmaceuticals
Smith Kline & French Laboratories
Strentex Fabrics Ltd
Chas. Thackery Ltd
3M UK Ltd
Upjohn Ltd
Vernon Carus Ltd
Wander Pharmaceuticals Ltd
Wlm Warner Ltd
Warner-Lambert Ltd
WB Pharmaceuticals
The Wellcome Foundation
Winthrop Laboratories
Wyeth Laboratories

PHOTOGRAPHIC WORK

Classic Vases
C & J Clark Ltd
Geographical Magazine
Japanese Cameras (Minolta)
Kodak Ltd

Laptech Studies
National Geographic
Quest Vest
Three Arrows Films
Tenba Bags

INSURANCE SERVICES

Sedgewick International

Results of the
International Karakoram Project

14–18 September 1981

ORGANIZED BY

The Royal Geographical Society

IN COOPERATION WITH

Academia Sinica, Beijing
Ministry of Science and Technology, Islamabad
British Council, London

CONFERENCE COMMITTEE

Dr Ian Davis

Mr Marcus Francis

Dr Andrew Goudie (Deputy Chairman)

Dr James Jackson

Professor Keith J. Miller (Chairman)

Professor Rashid A. K. Tahirkheli

Mr Jonathan Walton

Mr Nigel de N. Winser (Secretary)

Mrs Shane Winser (nee Wesley-Smith)

Authors and
Members of the Expedition

Members of the IKP

	Profession/Team
Abbas, S. Ghazanfar	Deputy Director Geological Survey of Pakistan Geology Team
Ahmad, Shabbir	Lecturer Quaid-i-Azam University Seismology Team
Akbar, Khurshid	Assistant Geophysicist Geological Survey of Pakistan Seismology Team
Allen, John	Postgraduate research student Sheffield University Survey Team
Aman, Ashraf	Mountaineer Rawalpindi Tourist Industries Survey Team
Atkinson, Nigel	Surveyor Hunting Surveys and Consultants Ltd Survey Team
Awan, Abdul Razzaq	Surveyor Survey of Pakistan Survey Team
Bilham, Roger	Geophysicist Lamont-Doherty Geological Observatory Survey Team
Bishop, James	Civil Engineer Sir Alexander Gibb and Partners Survey Team
Brunsden, Denys	University Reader King's College, University of London Geomorphology Team
Charlesworth, Ronald	Freelance Photographer Three Arrows Limited Film Unit
Chen, Jianming	Engineer and Surveyor Institute of Glaciology and Cryopedology, Lanzhou Survey Team

Collins, David Nigel   University Lecturer
                            Manchester University
                            Geomorphology Team

Colvill, Alan John   Chartered Accountant/Land Surveyor
                            University of Colorado
                            Survey Team

Crompton, Thomas Oliver   University Lecturer
                            University College, University of London
                            Survey Team, Deputy Director

Davis, Ian Robert   Principal Lecturer
                            Oxford Polytechnic
                            Director, Housing and Natural Hazards
                                Team

Davison, Ian   Geologist
                            British National Oil Corporation
                            Seismology Team

Derbyshire, Edward   University Reader
                            University of Keele
                            Geomorphology Team

Dong, Zhi Bin   University Lecturer
                            Lanzhou University
                            Radar Ice-Depth Sounding Team

Durrani, Nasir Ali   Professor of Geology
                            University of Baluchistan, Quetta
                            Geology Team

Farooq, Mohamed   Surveyor
                            Survey of Pakistan
                            Survey Team

Ferguson, Robert Ian   University Senior Lecturer
                            Stirling University
                            Geomorphology Team

Ferrari, Ronald Leslie   University Lecturer
                            Cambridge University
                            Radar Ice-Depth Sounding Team

Francis, Marcus   Research Engineer
                            British Hydraulics Research Association,
                                Cranfield
                            Co-Director, Radar Ice-Depth Sounding
                                Team

Ghauri, Arif Ali Khan   Associate Professor
                            Peshawar University
                            Geology Team

Giles, David Peter Vaughan Lindsey	Doctor Medical Practice, Bude Director, Medical Team
Goudie, Andrew Shaw	University Lecturer Oxford University Deputy Leader and Director, Geomorphology Team
Holmes, Robert Edward	Photographer (Stills) Freelance Photographer Film Unit
Hughes, Richard Edward	Geotechnical Archaeologist Ove Arup & Partners Housing and Natural Hazards Team
Illi, Dieter	University Lecturer Zurich University Housing and Natural Hazards Team
Islam, Shaukat	University Lecturer Peshawar University Botanist
Israr-ud-Din	Associate Professor Peshawar University Housing and Natural Hazards Team
Jan, Qasim M.	Associate Professor Peshawar University Geology Team
Jackson, James Anthony	Research Fellow Queen's College, Cambridge University Co-Director, Seismology Team
Jones, David Keith Crozier	University Lecturer London School of Economics, University of London Geomorphology Team
Khan, Islam M.	University Lecturer Peshawar University Botanist
Khan, Muhammad Zakir	Major, Pakistan Army Liaison Officer
Khattak, Rehman	Electrical Engineer Pakistan Atomic Energy Commission Seismology Team
King, Geoffrey Charles Plume	Senior Research Assistant Cambridge University Director, Seismology Team

Li, Jijun   Associate Professor
                                   Lanzhou University
                                   Geomorphology Team

Lin, Ban Zuo   Engineer
                                   Institute of Geophysics, Beijing
                                   Seismology Team and Co-Leader of
                                        Chinese Team

Massil, Helen   Doctor
                                   Medical Practitioner
                                   Medical Team

Miller, Keith John   Professor of Mechanical Engineering
                                   Sheffield University
                                   Leader of Expedition

Moughtin, James Clifford   Professor of Planning
                                   Nottingham University
                                   Housing and Natural Hazards Team

Moughtin, Timothy   Assistant
                                   Housing and Natural Hazards Team

Muir Wood, Robert   Journalist/Geologist
                                   Writer for New Scientist

Musil, George Jiri   Postgraduate Research Student
                                   British Antarctic Survey
                                   Radar Ice-Depth Sounding Team

Nash, David Francis Tyris   University Lecturer
                                   Bristol University
                                   Housing and Natural Hazards Team
                                   (Mrs Nash also gave valuable service
                                        to the Project)

Nunn, Paul James   Mountaineer/Lecturer
                                   Sheffield Polytechnic
                                   Film Unit/Radar Ice-Depth Sounding Team

Oswald, Gordon Kenneth   Electronics Research Engineer
    Andrew   Cambridge Consultants Ltd
                                   Co-Director, Radar Ice-Depth Sounding Team

Perrott, Frances Alayne   Lecturer
                                   Oxford University
                                   Geomorphology Team

Rajab, Ali   Sirdar of Porters
                                   Survey Team

Rana, Javaid Akhtar   Major, Pakistan Army
                                   Liaison Officer

Redhead, Charles Stephen

Driver
Transport Team

Rehman, Mohammad Abdul

Geologist
Pakistan Atomic Energy Commission
Geology Team

Rendell, Helen

Lecturer
Sussex University
Geomorphology Team

Riley, Anthony

Freelance Cameraman
Three Arrows Limited
Film Unit

Said, Mohammed

Professor of Geography
Peshawar University
Geomorphology Team

Smith, Edward Whittaker

Lecturer
University of Manchester
    Institute of Science and Technology
Survey Team

D'Souza, Frances

Director
International Disaster Institute
Housing and Natural Hazards Team

Spence, Robin John

Lecturer
Cambridge University
Deputy Director, Housing and Natural
    Hazards Team
(Dr Spence had his research student,
    Andrew Coburn, assist for a short period)

Stoodley, Robert Arthur

Chairman and Managing Director
Manor National Group Motors Ltd
Transport Team Manager

Tahir, Iqbal

Major, Pakistan Army
Liaison Officer

Tahirkheli, Rashid Ahmad
    Khan

Professor of Geology
Peshawar University
Director, Geology Team and
    Leader of Pakistani Scientists

Walton, Jonathan Lancelot
    William

Land Surveyor
University College, London University
Director, Survey Team

Waters, Ronald Sidney

Professor of Geography
Sheffield University
Geomorphology Team

Wesley-Smith, Shane   Administrator
                               Royal Geographical Society
                               Administrative Team

Whalley, Brian   Lecturer
                               Queens' University, Belfast
                               Geomorphology Team

Winser, Nigel de Northop   Administrator
                               Royal Geographical Society
                               Deputy Leader and Director,
                                   Administrative Team

Xu, Shuying   Associate Professor
                               Lanzhou University
                               Geomorphology Team

Yielding, Graham   Postgraduate research student
                               Cambridge University
                               Seismology Team

Zafar, Hashmat   Hydrogeologist
                               Water and Power Development Authority
                               Geomorphology Team

Zhang, Xiangsong   Associate Professor
                               Institute of Glaciology and Cryopedology,
                                   Lanzhou
                               Radar Ice-Depth Sounding Team and
                                   Co-Leader of Chinese Team

OTHER AUTHORS OF CONFERENCE PAPERS WHO WERE NOT MEMBERS OF THE IKP

Coburn, A. W.   University of Cambridge, UK

Coward, M. P.   Leeds University, UK

Jacoby, G. C.   Lamont-Doherty, Geological Observatory of
                                   Columbia University, New York, USA

Jafree, S. A. R.   Quaid-i-Azam University, Islamabad, Pakistan

McGreevy, J. P.   Queens' University, Belfast, UK

Middleton, C.   Royal Geographical Society, London, UK

Plant, G. B.   Lamont-Doherty, Geological Observatory of
                                   Columbia University, New York, USA

Rex, D.   Leeds University, UK

Shu Pei Yi   Institute of Geophysics, Academia Sinica,
                                   Beijing, Peoples Republic of China

Tarney, J.   Leicester University, UK

Thirwall, M.   Leeds University, UK

Vyvyan, M.   Trinity College, Cambridge, UK

Windley, B. F.   Leicester University, UK

Wise, M. J.   Royal Geographical Society, London, UK

Conversion Units

To convert from	to	multiply by
inch	metre (m)	2.54×10^{-2}
foot	metre (m)	30.48×10^{-2}
mile	kilometre (km)	1.609
pound force	newton (N)	4.448
kilogram force	newton (N)	9.807
kilogram force/metre^2	pascal (Pa)	9.807
pound mass	kilogram mass (kg)	4.536×10^{-1}
pound force/in^2	pascal (Pa)	6.895×10^3
torr	pascal (Pa)	1.333×10^2
bar	pascal (Pa)	1×10^5
angstrom	metre (m)	1×10^{-10}
calorie	joule (J)	4.184
foot-pound	joule (J)	1.356
degree Celsius	kelvin (K)	$T_K = T_C + 273.15$

Multiplication factor	Prefix	Symbol
10^{-12}	pico	p
10^{-9}	nano	n
10^{-6}	micro	μ
10^{-3}	milli	m
10^3	kilo	k
10^6	mega	M
10^9	giga	G

Inauguration

International Karakoram Conference
Opening Remarks by The President of The Royal Geographical Society, Professor M.J. Wise CBE, MC, on Monday September 14th, 1981

Your Excellency, My Lords, Ladies and Gentlemen,

It is with very great pleasure that I welcome you all to this conference which will review in detail the scientific work achieved by members of the International Karakoram Expedition 1980. I am especially pleased to see present strong representations from the teams from Pakistan and the People's Republic of China who joined with British scientists to mount a very strong inter-disciplinary and international expedition. It was a very happy and entirely appropriate arrangement to mount this expedition in the year of the Society's 150th Anniversary and I am very glad to see with us Lord Hunt who was our President at that time and who took an active role in the planning of the expedition and who participated in it, and Lord Shackleton another Past President who also took an influential part in the Expedition's work in Pakistan. We also welcome representatives of many organisations, diplomatic, scientific, industrial, commercial, engineering, mountaineering and educational who in so many ways supported the idea of the expedition.

As the names inscribed around the walls of this hall remind us, the Society, throughout its history, has stimulated and sponsored expeditions. But compared with its earlier days the objectives and character of our expeditions have changed. It is no longer the opening up of new lands that concerns us: the emphasis, rather, is on cooperation and collaboration between British scientists and those of other countries in the interests both of advancing science and of applying scientific results to the benefit of the people in the regions under study. In this connection Lord Hunt has recently suggested the importance of conservation, or wise management of resources, as a prevailing theme.

There is another important aspect. The scientific disciplines whose members gather in our House, geography, geology, surveying, engineering, botany, zoology, archaeology, to name only some, have greatly refined their specialist skills and enhanced their techniques. Thus it becomes ever more important to bring them together in teams with well-defined objectives both to provide opportunities for cross-fertilisation between disciplines and to apply the wealth of the scientific resources available to the problems of a region or area.

The changed objectives can well be seen in the context of the record of major expeditions launched by the Society in the 1970's, Makran, South Turkhana, the Mato Grosso (jointly with the Royal Society), Mt. Mulu (Sarawak) and now to the Karakoram. The Karakoram was indeed well chosen for our 150th Anniversary Expedition with its dramatic landscape of formidable mountains, great glaciers, surging rivers, extremely high rates of erosion, its earthquakes and natural hazards to human life, the intensive occupance of the limited land in its deeply incised valleys.

At the outset of the conference we express our thanks to all the members of the expedition for the courage, energy and initiative which they displayed and for the spirit which welded the members and the constituent teams together into a happy and productive group. Even when backed by the forethought and care of experienced administrative members and by the scientific and medical skills, expeditions are still not without risk and we are conscious today of the absence of one of the leading members who failed to return.

The Society has already received preliminary reports on the work of the Expedition. The reports we shall hear this week will enable us to form a more exact measure of the work of the six main programmes. Of the general success of the International Karakoram Project we are in no doubt and I am sure that all members of this Conference will have been pleased by the recognition of that success in the award to Professor Keith Miller, the leader, of the Society's Founder's Medal for 1981.

The International Karakoram Project: an appraisal

K. J. Miller – Royal Geographical Society, London

INTRODUCTION

My key-note speech at the conference at Quaid-i-Azam University, which heralded the start of our Project in Pakistan, attempted to show how and why an international interdisciplinary expedition should address itself to the scientific and technological problems posed by the Karakoram. In that sense it looked forward to what we now know was a most successful venture.

But now it is necessary to look back and present an appraisal of what we did. Whilst this conference will give a full account of both our practical and academic achievements, and these will be shown not to be mutually exclusive, a far wider view is needed especially in these days when technology is viewed with suspicion by environmentalists and science research expenditure increases exponentially as more and more problems are unearthed while too many problems related to the suffering of mankind remain unsolved.

It follows therefore that it is doubly necessary to convince the public at large and our sponsors that we accomplished a worthwhile series of studies, and that such scholarly interchanges between nations, the hallmark of democracy, are to be encouraged since they are a springboard for inter-national collaboration and one means of achieving a better understanding between peoples that can not but help reduce world tensions. For this reason and for another reason, namely that all ventures should be accountable to those who give their support, in terms of services, goods, time, finance and encouragement, it beholds us to take a rational view of our work and to see in detail our achievements in the widest possible sense with the hope that those who supported this type of venture will continue to do so.

A FRAMEWORK OF ANALYSIS

I am going to argue that we succeeded in a wider context than just by our specific research activities alone. Although science and technology can do much to alleviate the problems of mankind we see today that political, economic, racial, social, cultural and religious barriers are as impenetrable as the Karakoram was to early travellers at the end of the last century and it is the relative lack of progress in these areas that is frequently the limiting factor.

It is alongside all these aspects that we will be judged if only through questions such as:-

"How much did it cost?"

"What long term benefits will accrue and to whom?"

"Did you aggravate the situation in Afghanistan?"

"Should this kind of exercise be repeated?"

"Where next, British exploration?"

I therefore intend to address this lecture to these wider apsects with the objectives of informing politicians, diplomats, educationalists and businessmen in industry and commerce that we did succeed and that others should follow our example.

EDUCATION

Scientists from four countries amalgamated their skills to give lectures to several scientific bodies in Pakistan. Forty technical papers were read at the International Conference in Islamabad whose theme was "Recent Technological Advances in Earth Sciences". Lectures on engineering design relevant to the nuclear industry and the construction industry were given in other centres of learning and several members gave seminars on how the people of Pakistan could best protect themselves against the ravages of floods, earthquakes and other natural catastrophes. Just as important was the interaction of scientists with colleagues from other disciplines from other countries. The ease of cross-breeding of ideas is one hallmark of the British educational system and this Project was a well planned exercise in that respect. At the end of the Project we had all lectured to and learned from our colleagues and finally dispersed to our home countries full of praise for the endeavours and abilities of our overseas friends.

Perhaps the culmination of this aspect of our work was the creation of the Karakoram Research Cell about which we will learn more later but let me say now that scientists who visit Pakistan in the future will have a recognised home where they can report, collect data from previous expeditions, enrol scientists from the indigenous population to participate in their own ventures and finally leave behind a record of achievements for the benefit of those who follow rather than return home with the spoils of their research without further communication with the host country.

This organisation is possibly the first step in an attempt to establish the Karakoram Range as an international science park for the benefit of collaborating scientists from all the countries of the world and this was uppermost in my mind when I was requested to prepare a memorandum on the formation of the Cell by our friend, Dr Afzal, Head of the Pakistan University Grants Commission.

However the work of the Cell cannot be achieved by one nation in isolation and so it is beholden to all the members of our Project and those institutions who helped us in Pakistan, China and the U.K. to assist the Cell to grow in strength and purpose of mission.

Finally, do not let us forget that the educational aspects of our work were approved and greatly assisted by the Pakistan University Grants Commission, by the Minister of Education, Muhammad Ali Khan, who helped inaugurate our first conference, and by the British Council in Pakistan, which, via the capable hands of David Latter, threw much of its manpower, skill, tact and experience in overseas educational work into the organization of the Project. Without this kind of assistance we would not have succeeded.

RELIGIOUS ASPECTS

Many religions were represented on the expedition but every member respected the different viewpoints of his colleagues. It was never a point of tolerating another religion or philosophy but rather an active participation to render assistance to those who, although having different faiths, had necessarily to conform to different patterns of behaviour. This was particularly true during the month of fasting, Ramadan, and the Eid celebration at its termination. Such behaviour was so interwoven into the fabric of the expedition that at its termination we all had a deeper understanding of one another. When one compares this attitude to that pertaining in some currently troubled parts of the earth, the benefits of international projects such as ours are clear to see.

MEDICAL CARE

The work of our doctors cannot but be applauded by everyone. More will be reported later about this work by our colleague David Giles but I am sure he will be too modest and will confine himself to purely medical matters. He and Dr. Helen Massil provided far more than a non-stop service. They overcame religious, cultural and social obstacles by exhibiting sympathy, love and affection to all those who came for help, and there were many. It was this breaking down of ethnic barriers that revealed a far greater depth of deprivation than I thought existed based on my own previous experience. The Aga Khan has already started good work in the Hunza Valley but far more needs to be done. A far worse situation exists for those not of the Ismaili faith. The final report of our doctors will be sent to all those organizations who may help alleviate the sufferings of these people whom we once thought to be the healthiest, longest living members of the human race.

Our sympathy for these people who suffer from the greatest collection of natural hazards known to man, including floods, avalanches, earthquakes and rockfalls, as well as disease, is heightened by the knowledge that many live below subsistence levels and are now suffering stress due to rapid social changes consequent to the building of the Karakoram Highway. The realisation that half the population (women) is seldom able (or does not wish) to seek medical aid is probably one of the most important features of this problem while the more obvious threat from the flies and mosquitoes, lack of adequate sanitation, mica filled streams and no provision for clean drinking water, daily appals the privileged visitors such as ourselves.

Now that we have clearly demonstrated these deficiencies we must gather all available assistance or surely we will have failed in one of our major objectives to bring some benefit to the peoples of Pakistan.

CULTURAL BENEFITS

I will not detail here what the British and Chinese members of the Project learned from our visit to Pakistan. That will be fed back into our own cultural environments but I will detail what impact we had on Pakistani society with the assistance of the British Council.

The Pakistan newspapers of different political leanings gave us a very good press. Four press conferences, two television programmes and several radio commentaries gave full coverage to our activities. A deeper understanding of our scientific endeavours was made available by demonstrations of all our equipment to the public; and since our return the layman in our own country has learned of our work through the writings of Robert Muir Wood in the popular weekly journal "The New Scientist". More detailed and precise technical information is obtainable in these two volumes of proceedings of our scientific work which will be easily available to those who, in their love of mountains, will want to know what we studied. Perhaps my book "Continents in Collision" which I have just finished writing will assist in that process of bringing science closer to the man in the street or should I say man on the track. No one can say that we have not had good coverage and once our film has been completed and edited we hope that our endeavours will have an even greater impact on the public.

In the context of cultural associations two episodes come to mind. The first was the Sports Day organised by Shane and Nigel. The entire male population of Hasanabad, Aliabad, Karimbad, Baltit and Ganesh must have been squeezed into our stadium. It was so magnificent an exercise in international relationships that instead of the seventy guests expected for dinner that evening an additional sixty persons stayed on to continue the celebrations until rain stopped play. The second memory was the constant reference to the BBC and the value of its overseas programmes. Transmissions from London are regarded as an essential part of the daily timetable and BBC impartiality and non-political bias to world affairs was greatly appreciated. It also reinforced my view that the English language is no longer the preserve of the English peoples but is now the tongue for world discussions.

DIPLOMATIC TRIUMPHS

Embassies today are sometimes like fortresses and all countries have their own problems in this respect. It was therefore with pleasure that we found ourselves not just aided but encouraged by all members of the British Embassy staff in Islamabad orchestrated by its popular and dynamic Ambassador, Oliver Forster. At the same time the Chinese Embassies in London and Islamabad did all that we asked of them as did the Chinese Government in Beijing (e.g., the granting of border visas to permit access into Sinkiang). The assistance of the Chinese Academy of Sciences was also a major factor in the successful outcome of our project.

A clear recognition of the status of our venture was accorded us by the President of Pakistan, General Zia-ul-Haq, who granted two private consultations and who inaugurated the international conference held at Quaid-i-Azam University, and this at a time when he had far greater problems to contend with. Let us not forget that it was he who gave a final blessing to the Project and financial assistance to the participating Pakistani university scientists. At this point we must record our thanks to two other gentlemen who gave us their support. Of course I refer to two past Presidents of the Royal Geographical Society, Lord Hunt and Lord Shackleton. What they did for us on the diplomatic front matched the unsparing efforts of Oliver Forster and David and Alison Latter. They also convinced me that there is more energy available in the Upper House than is commonly believed.

It was because of our diplomatic successes that we now see several of our Chinese and Pakistani colleagues working alongside us in our own establishments. I personally have acquired three more Chinese friends and next year more scientists from that country will come to work in my laboratories where they will join a different kind of international team but one kindled with the same kind of enthusiasm.

POLITICAL SUCCESSES

Perhaps success can sometimes be gauged by proving one's initial arguments and showing that those who did not have the ability to understand our motives were wrong. But such a victory may only be a transitory success and needs to be followed up.

Here I am referring to the false accusations made against the Project by the APN news agency of Moscow. Since our return I have met several Russian friends in Sheffield and at international conferences who reported that they knew nothing of the incident. The fact that they wished to hear all about our scientific achievements was the reaction I desired. Furthermore I am pleased to report that four of my colour photographs are now decorating the homes of some of my Russian friends. More will follow.

More important than this however is that we succeeded in carrying out our programmes of research when many senior politicians in many countries thought we were doomed to fail especially after the revolution in Iran and the take-over in Afghanistan.

In the final analysis it is to the credit of the Government of Pakistan that some ten Ministries and four other government bodies eventually gave us their blessing despite the inconvenience we must have caused. It was a courageous decision that was fully supported by the Chinese and also very handsomely by the British Government through the good offices of Neil Marten, the Minister for the Overseas Development Administration.

In time I am sure that our academic colleagues in Russia, Afghanistan and India will come to have similar proposals for research programmes in the Pamirs, Hindu Kush and the Himalaya and I look forward to the day when they too get the support we did. Sometimes I have been branded as an optimist, a charge I do not reject, and so you will forgive me in my hope that in the not too distant future all countries interested in mountain research (or any scientific endeavour) will be free to associate with one another to the benefit of all mankind.

SCIENTIFIC SUCCESSES

As I have said previously the extent of our success will be judged by others reading the publications that have been and are continuing to be derived from our work. But now once again, I will try to paint a wider canvas.

International interdisciplinary scientific studies bring their own rewards. All too frequently an investigation will concentrate on a single specialized study. For example, the social structure in villages or the surges of glaciers or the analysis of sediments in rivers or the frequency and strength of earthquakes or the geological problems concerned with the building of dams or the salinity of lakes or mapping by triangulation, or the engineering of roads and bridges.

When tackled separately useful results may emerge that will help a clearer understanding in that discipline. But far greater rewards come to those who can use their skills to assist other branches of science. I have dealt with some aspects of interdisciplinary benefits in my paper published in Volume I of these Proceedings, but to appreciate this point more fully, I would draw your attention to Figure 1. This shows that

DISCIPLINE	NUMBER OF MEMBERS
ANTHROPOLOGY	2
ARCHAEOLOGY	1
ARCHITECTURE	3
BOTANY	2
CIVIL ENGINEERING	6
ELECTRICAL ENGINEERING	6
GEOLOGY	7
GEOMORPHOLOGY	13
GEOPHYSICS	8
GLACIOLOGY	1
MECHANICAL ENGINEERING	2
MEDICINE	2
MOUNTAINEERING	2
PHOTOGRAPHY	3
SURVEYING	6
ADMINISTRATIVE TEAM	9
TOTAL	73

FIG. 1. THE DISTRIBUTION OF DISCIPLINES WITHIN THE PROJECT.
(Only the major interest of members are listed).

in the six teams, thirteen disciplines were represented. Thus the geomorphologists studying the deposition of sediments could relate to civil engineers concerned with the reduction of operating life of the Tarbela dam power station due to the 5,000,000 tons of sediment carried daily by the Indus in mid-summer. The glaciologist could appreciate the usefulness of electrical engineers designing radar ice depth sounding equipment and the geophysicists were made aware that slip line field solutions applied to the movement of land masses during earthquakes are essentially those derived by mechanical engineers who study the processing of hot metal. More examples of this type of interaction will become apparent as this conference progresses.

Figure 2 shows the degree of interaction between the six teams comprising

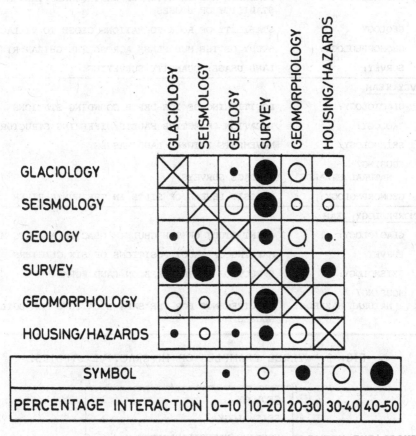

FIG. 2. THE VARIOUS DEGREES OF INTERACTION BETWEEN THE GROUPS.

the Project. The interactive estimates are my own with 50% being a maximum limit since higher levels than this would indicate a reversal of minor and major roles. My subjective analysis is based on the number of interactive studies, the period of interaction, the number of people involved, and the scientific level of interaction.

Examples of the interactions are given in Fig. 3 but it must be noted that during the first half of the expedition the glaciological programme being carried out on the Hispar Glacier was six to seven days march from base camp which limited interaction. This was compensated for during the latter half of the expedition, when the team worked with geomorphologists and surveyors on the Ghulkin Glacier. The same difficulty arose with the Housing and Natural Hazards Group who carried out only one study in the Hunza Valley but, as will be seen from Fig. 4, they were an inter-disciplinary team in their own right employing anthropologists, an archaeologist, civil engineers, a geologist and surveyors as well

HOUSING/NATURAL HAZARDS TEAM	
GLACIOLOGY:	GLACIER MELT WATER SURGES THAT DESTROY CROPS AND HOMES
SEISMOLOGY:	STRENGTH AND LOCATION OF EARTHQUAKES THAT AFFECT STABILITY OF HOUSES
GEOLOGY:	STABILITY OF ROCK FORMATIONS CLOSE TO VILLAGES
GEOMORPHOLOGY:	STUDY OF THE MUD FLASH ACROSS THE GHIZAR RIVER
SURVEY:	LAND USAGE; QUANTITY SURVEYING
SURVEY TEAM	
GLACIOLOGY:	POSITIONING OF ICE-DEPTH SOUNDING STATIONS
GEOLOGY:	LOCATION OF ACTIVE FAULTS/DIFFERING STRUCTURES
SEISMOLOGY:	MOVEMENTS BETWEEN LAND MASSES
HOUSING/ NATURAL HAZARDS	VILLAGE SURVEYS
GEOMORPHOLOGY:	DEPOSIT LEVELS OF SILTS IN THE HUNZA RIVER
GEOMORPHOLOGY TEAM	
GLACIOLOGY:	JOINT STUDY OF THE GHULKIN GLACIER AND ITS MORAINES
SURVEY:	MAPPING OF SNOUT POSITIONS OF SIX GLACIERS
SEISMOLOGY:	EFFECT OF EARTHQUAKES ON LAND FORMS
HOUSING/ NATURAL HAZARDS:	REQUIREMENTS FOR THE SAFE LOCATIONS FOR BUILDING HOUSES

FIG. 3.   THREE EXAMPLES OF INTERACTIVE STUDIES.

GEOLOGY	1
SEISMOLOGY	1
HOUSING/NATURAL HAZARDS	6
GLACIOLOGY	3
SURVEY	2
GEOMORPHOLOGY	2

FIG. 4.   OTHER DISCIPLINES AVAILABLE IN THE VARIOUS TEAMS.

as architects.

It was this general philosophy of interdisciplinary studies that caused Clive Warren, our visitor from the Overseas Development Administration, to remark "It is this kind of expedition that combines the talents of several consultants in different fields which would allow the ODA to arrive quickly and efficiently at optimum solutions taking all issues into account".

I would fully endorse such a sentiment and would urge the UK Government to be sympathetic to future ventures of this kind that can achieve all the benefits outlined in these papers and in other publications derived from the International Karakoram Project 1980.

To emphasize the success of the Project we need not go further than our lists of Sponsors. I was told prior to our project that many of the UK scientific research councils had never previously supported an RGS sponsored expedition. Having previously sat on two Science Research Council Committees I can appreciate both sides of this argument, but for the Karakoram Project we did our homework correctly. Proposals clearly stated in detail what scientific work was to be done, how it was to be done and its value to the community. A large number of consultants gave us advice in depth. Consequently we received grants from

The Science and Engineering Research Council
The Natural Environment Research Council
The Royal Society

It was because of the substantial funding received from these three organisations that we could feel confident in our research objectives, especially at a time when research funding was being restricted. Although significant additional financial contributions came from many other sources which have been acknowledged elsewhere, special thanks must be given to the British Council, Carnegie Trust, Mount Everest Foundation, National Geographic Society, the Royal Geographical Society, The Royal Institution of Chartered Surveyors, George Wimpey Ltd, The United Nations and finally the ODA of the UK Government, all of whom approved of the scientific research work we wished to undertake.

INDUSTRIAL COLLABORATION

In an increasingly competitive world it was encouraging to see many business and commercial organisations getting involved with our Project. For example, British Airways, Land Rover and George Wimpey were not just sponsors, but helped us in many different ways. They too wanted to know why our work was considered valuable and accountability was an important consideration in our dealings with these organisations. The one exception was Bob Stoodley's group, Manor National Group Motors. His tremendous assistance falls into a special category.

CONCLUSIONS

Figure 5 provides a brief list of the major scientific successes of the Project. These coupled with the statements above on cultural, social, religious, political, diplomatic, industrial and commercial successes should convince our sponsors that we did a good job. We have held ourselves to be accountable and in so doing I hope that the Royal Geographical Society will be given further support to continue the kind of operation we mounted in the summer of 1980.

And now ladies and gentlemen, we have four days to listen to reports on the individual projects undertaken during the course of the Project.

14

FIG. 5

THE MAJOR SUCCESSES OF THE PROJECT
(and a list of collaborative ventures)
(in no order of priority)

1. DETERMINATION OF THE CORRECT SURVEY STATIONS ON HACHINDAR, BURIHARAR AND SHUNUK.

2. DISCOVERY OF THE 1913 ATABAD SURVEY STATION.

3. POSITIONING OF KEY LOCATIONS BY SATELLITE RECEIVER EQUIPMENT TO ENABLE RAPID FUTURE ADVANCES BY THE SURVEY OF PAKISTAN (SOP).

4. POSITIONING OF MOST OF THE MAJOR SEISMIC STATIONS BY SATELLITE DOPPLER EQUIPMENT.

5. FIRST EVER 3-DIMENSIONAL RECORDING OF THE SHAPE, SIZE AND POSITION OF KARAKORAM GLACIERS (HISPAR AND GHULKIN); A FIRST STEP TOWARDS DOCUMENTING A NATURAL RESOURCE AND A HAZARD.

6. THE FIRST MULTI- AND INTER-DISCIPLINARY STUDY OF THE VULNERABILITY OF LOW-COST HOUSING AND SETTLEMENTS TO NATURAL HAZARDS.

7. DEVELOPMENT OF A NEW METHODOLOGY TO ASSESS NATURAL HAZARDS.

8. A DETAILED CRITICAL ASSESSMENT OF THE HEALTH AND MEDICAL CARE PROVIDED FOR THE PEOPLE OF THE KARAKORAM.

9. THE EXPOSURE OF PRIORITY NEEDS IN VILLAGES ESPECIALLY THE NEED FOR PREVENTION BEFORE A CATASTROPHE, RATHER THAN AID AFTER THE EVENT.

10. ESTABLISHMENT OF THE CLEAR NEED FOR EDUCATION ON:

    I. CLEAN WATER SUPPLIES

    II. ERADICATION OF POOR BUILDING STYLES

    III. INTEGRATION OF SOCIAL/MEDICAL SERVICES

    IV. DEFORESTATION CONTROL

11. THE NEED FOR OPERATIONAL RESEARCH PROCEDURES TO THE MAINTENANCE OF THE KARAKORAM HIGHWAY.

12. THE ESTABLISHMENT OF THE NEED TO HAVE AN INTER-DISCIPLINARY APPROACH BY GEOMORPHOLOGISTS, ARCHITECTS AND ENGINEERS TO THE DESIGN AND SAFE LOCATION OF NEW BUILDINGS.

13. A REPORT ON THE STRUCTURAL REPAIR WORK URGENTLY REQUIRED TO PRESERVE THE MOST HISTORIC BUILDING IN THE KARAKORAM, BALTIT FORT.

14. THE FIRST DETAILED SEISMOLOGICAL STUDY EVER CARRIED OUT IN THE KARAKORAM.

15. THE MOST THOROUGH SERIES OF GEOMORPHOLOGICAL INVESTIGATIONS EVER PERFORMED IN THE KARAKORAM THAT WILL ASSIST A MORE PRECISE DOCUMENTATION AND HENCE A WIDER APPRECIATION OF THE NATURAL HAZARDS FACED BY THE INDIGENOUS POPULATION.

FIG. 5. (CONTINUED)

16. A CONTINUATION OF GEOLOGICAL SURVEYS IN WHAT IS, AND WILL LONG REMAIN, A RELATIVELY UNEXPLORED AREA IN GEOLOGICAL TERMS.

17. THE FIRST DETAILED STUDY OF THE GLACIAL SEDIMENTOLOGY OF THE UPPER HUNZA REGION.

18. THE FIRST THERMOLUMINESCENCE DATINGS OF PLEISTOCENE AND RECENT SEDIMENTS IN THE KARAKORAM.

19. THE FIRST CORING OF QUATERNARY LAKE SEDIMENTS IN THE UPPER HUNZA VALLEY.

20. AN EXTENSION OF KNOWN QUATERNARY GLACIAL STRATIGRAPHY OF THE HUNZA VALLEY.

21. DETAILED MAPPING OF GLACIER SNOUT POSITIONS AND FLUCTUATIONS IN THE HUNZA AREA.

22. PROVISION OF DETAILED WATER CHEMISTRY DATA FOR THE HUNZA RIVER AND ITS TRIBUTARIES.

23. MEASUREMENT OF HIGH ALTITUDE ROCK TEMPERATURES.

COLLABORATIVE VENTURES

24. CLOSE COLLABORATION WITH SEVERAL GOVERNMENT AGENCIES IN PAKISTAN, NAMELY THE WATER AND POWER DEVELOPMENT AUTHORITY, THE SURVEY OF PAKISTAN, THE GEOLOGICAL SURVEY OF PAKISTAN, THE FRONTIER WORKS ORGANISATION OF THE PAKISTAN ARMY AND THE KOHISTAN DEVELOPMENT BOARD.

25. CLOSE COLLABORATION WITH, AND LECTURES TO, PAKISTAN UNIVERSITIES AT KARACHI, QUETTA, PUNJAB, ENGINEERING TECHNOLOGY OF LAHORE, PESHAWAR AND QUAID-I-AZAM ISLAMABAD.

26. LECTURES TO THE ENGINEERING DESIGN STAFF OF THE PAKISTAN ATOMIC ENERGY COMMISSION.

27. LECTURES TO THE METROPOLITAN PLANNING BOARD IN KARACHI.

28. SUCCESSFUL COLLABORATION WITH SEVERAL INSTITUTIONS IN CHINA INCLUDING THE INSTITUTE OF GLACIOLOGY AND CRYOPEDOLOGY, LANZHOU, LANZHOU UNIVERSITY, BEIJING UNIVERSITY, FOUR OTHER INSTITUTES OF ACADEMIA SINICA.

29. CONTINUING AND LONG-TERM CO-OPERATION WITH ALL PAKISTAN AND CHINESE COLLABORATORS.

30. A CONTINUATION OF ASSISTANCE BY SPONSORS, PARTICULARLY THE BRITISH COUNCIL, TO THOSE ASSOCIATED WITH THE PROJECT, E.G., LONG-TERM LIAISON WITH OVERSEAS UNIVERSITIES.

16

ACKNOWLEDGEMENTS

The scientific work of IKP80 was brought to a successful conclusion due to the outstanding efforts of the Directors of the programmes, namely: Andrew Goudie (Geomorphology; and deputy leader), Ian Davis and Robin Spence (Housing and Natural Hazards), James Jackson and Geoff King (Seismology), Gordon Oswald and Marcus Francis (Glaciology), Tahirkheli (Geology) and Jon Walton (Survey), together with their teams. Their efforts plus that of the logistics team, Nigel Winser and Shane Wesley-Smith (now Shane Winser), are gratefully acknowledged.

Karakoram history; early exploration

D. Middleton – The Royal Geographical Society, London

ABSTRACT

The Karakoram is the most heavily glaciated area outside the Polar regions and contains some of the world's highest peaks. It is also the most formidable of the parallel mountain ranges dividing the Indian sub-continent from Central Asia. Early exploration was directed to finding a way beyond the Karakoram to trading centres in Eastern Turkestan (now Sinkiang). Attention focused first on the route from Leh to Yarkand by the Karakoram Pass, later on the Gilgit road into Hunza. William Moorcroft (1765-1825), G. T. Vigne (1801-1863) and Thomas Thomson (1817-1878) were among the pioneers who opened up the Karakoram, and the first official maps were made by the Survey of India operating in Kashmir between 1857 and 1864. T. G. Montgomerie (1830-1878) was the first to see K2 and H. H. Godwin-Austen (1834-1923) the first actually to put it on the map with its surrounding peaks. The explorations of George Hayward (d. 1870) and the Intelligence work of Francis Younghusband (1863-1942) helped to build up a general picture of the region, and so paved the way for private enterprise and the scientific exploration of the twentieth century. Lord Conway, T. G. Longstaff, the Duke of the Abruzzi, Filippo De Filippi, the Duke of Spoleto, Giotto Danielli, Ardito Desio, Dr and Mrs Visser, Kenneth Mason and Eric Shipton were active in the Karakoram between 1908 and the outbreak of the Second World War.

INTRODUCTION

The Karakoram (Turkish for 'Black Mountains') is one of the most formidable upland regions on earth, the most heavily glaciated outside the Poles, and containing some of the world's highest peaks. The range, 200 miles or more from north-west to south-east, is the watershed between the drainage to the Indian Ocean and that into Central Asia. It lies between the lower Shyok and Indus rivers on the south and the Shaksgam tributary of the Yarkand river on the north. Tom Longstaff, most famous and experienced of Himalayan explorers, who was there in 1908, summarizes the lie of the land thus: 'A continuous series of vast glaciers lie in a trench along its whole extent. On the west the great Hispar Glacier connects at its head with the Biafo. After a relatively short distance comes the westward flowing Baltoro Glacier, near the head of which rises the Siachen Glacier flowing generally south-eastward. The great peaks align themselves in two axes on either side of these great glaciers. The highest, and northern range, starts north of Nagyr and follows through K2 and the Gasherbrums on to Teram Kangri. Thence the axis bends somewhat to the south-east, crossed by the Sasir Pass, to end in the great Nubra peaks. The southern and generally lower axis starts in the west with Rakaposhi, near Gilgit, and runs through Masherbrum and the Kondus peaks. It is crossed by the Saltoro Pass and continues south-east... ending in the spur which forms the sharp corner where the Nubra joins the Shyok.' (Longstaff, pp. 191-2).

The story of how all this became known and laid down on the map is closely bound up with the position of the British in India. They came there, of course, as merchants in the seventeenth century, establishing posts in Bombay, Madras and Calcutta. Calcutta became the headquarters of an increasingly powerful administrative machine with its own Army and Navy. By the early 1800s the Honourable East India Company domi-nated great tracts of the subcontinent and was moving steadily north towards the great lines of mountains which formed an undefined boundary between India and Central Asia. Trade and politics rather than geographi-cal curiosity were the spurs to exploration, for the Karakoram was regarded as one of the most effective barriers between the lands where the Company's writ ran and the possibly lucrative markets beyond. Along the ancient Silk Road into China lay cities whose names have rung romanti-cally in Western ears at least since Marco Polo trudged across the Pamirs and the Gobi Desert, and returned to tell the tale 700 years ago. Bukhara and Samarkand, Kashgar and Yarkand – these were the targets of the early travellers. Geographical exploration, of which the Karakoram Project was the most recent example, came later with the failure of commercial missions and with the necessity of fixing defensible frontiers.

To get to the promised lands of Marco Polo it was necessary to go not across but round the Karakoram. One way, it appeared, was through Kashmir to Leh in Ladakh over country which, though mountainous, was negotiable. The passes over the Pir Panjal and the Great Himalaya were certainly hard going but offered no insuperable barrier; beyond lay a more difficult stretch to the Karakoram Pass which even as lately as 1908 Longstaff found strewn with bones of men and animals. 'The track' he wrote, was 'worn by generations of caravans and marked the whole way by the bleaching bones of dead pack-horses and the occasional skeleton of a camel. As I rode I counted 100 skulls in half an hour'. (Longstaff, p. 185). Beyond this death-trap it was supposed that the road ran down-hill all the way to Yarkand, but when Thomas Thomson breasted the pass in 1848, the first European to do so, he saw before him the further obstacle of the Kun Lun. Another way into Eastern Turkestan (modern Sinkiang) was south of the range, down the Indus to Gilgit and from there through Hunza to Kashgar – easier going, but beset with turbulent and independent tribes whose local sport was robbing the traveller.

The exploration of and into Central Asia was never a straightforward business of pressing forward over unmapped territory and bringing back reports to the Royal Geographical Society. In Africa and in the Polar regions the aim was geographical enlightenment; in Australia and the Americas, living room; in Central Asia motives were so mixed up with the prospecting of trade routes, with security and with imperial expansion, as actually to inhibit the spread of knowledge. The nineteenth century explorers in the region were either in official employ, or dependent on official permission. The RGS got what it could from the published reports

GODFREY THOMAS VIGNE (1801 - 1863)

(By kind permission of H.D'O. Vigne Esq.)

of the Survey of India, and between the Society's founding in 1830 and the end of the century had presented twelve Gold Medals to explorers into the Himalaya and beyond.   But only one man - George Hayward - was openly acknowledged as their representative.

The story begins with William Moorcroft, veterinary surgeon in the employ of the East India Company.   John Keay, whose When Men and Mountains Meet has provided me with much of the material for this after- noon's talk, calls him 'the father of modern exploration in the western Himalaya and Central Asia' (Keay, 1977, p. 18), but stresses that his interests were almost purely commercial.   In 1812 he applied for permission to travel north to look for horses for breeding purposes at the stud farm in Bihar of which he was superintendent.   Rather unaccountably, he crossed the Himalaya into Tibet where - as Keay points out - there are no suitable horses.   He came back instead with goats whose wool was of the kind used in weaving the famous shawls of Kashmir.   He also came back with important geographical news - the correct identification of the sources and courses of both the Sutlej and the Indus, showing them to be uncon- nected with the Gangetic system.   In 1820 he was off again, floating across the Sutlej on the traditional inflated buffalo skin, on his way through Kashmir to Leh.   He hoped to gain the Karakoram Pass and thence to travel to Yarkand and - who knows - beyond to the fabled markets of China. Difficulties with local authorities in Ladakh kept him at Leh for two years and he got no further.   He did however make a reconnaissance into the Nubra Valley and elsewhere which began to lift the curtain on the approaches to and round the Karakoram Range.   He was later to visit Bukhara and to die in mysterious circumstances on his way back to India.

The next important name in the history of Karakoram exploration is that of Godfrey Thomas Vigne.   Vigne had no official connection with India and no commercial axe to grind.   His chief interest was in sport, and he rejoiced in the nickname of 'Ramrod' Vigne, his equally outdoor brother being known as 'Nimrod'.   He was an independent gentleman traveller, who had decided early in his career as a London barrister that his health was 'too delicate' to survive such a life.   He took to travel, arriving in Bombay by way of Persia in 1834 with no special programme in mind and, 'as the climate of Bombay did not tend much to the improvement of my health, I determined, not having had the slightest intention of the kind when I left England, to run up at once to the cool air of the Himalaya' (Vigne, Vol. 1, p. 23).   Simla gave him his first sight of the mountains and from there he embarked on some three years' extensive wanderings in Kashmir, Ladakh and Baltistan.   He crossed the Sutlej in 1835 riding, like Moorcroft, on the usual blown-up buffalo skin; his luggage included rods, guns and an easel, for he was also an accomplished artist.   Vigne was the first European to enter Baltistan, the small state which lies along the Indus on the western edge of the Karakoram, and he based himself on Skardu (which he called Iskardo).

From Skardu he made several attempts to penetrate or to circumvent the range before him.   In the fashion of the time, he was looking for a road into Eastern Turkestan.   He made one sally towards Gilgit, gaining a vantage point near Bunji from where he could look along the valley - the route taken today by the Karakoram Highway.   He was turned back by the local people, who had taken fright at the approach of strangers, and returned to Skardu.   He also conceived the bold idea of tackling the Hispar Glacier into Hunza and from there crossing the Pamirs to Khokand. The Hispar route had to wait for Martin Conway in 1892, with all the paraphernalia of a modern expedition, and Vigne was defeated at the

outset.   Intrigued by rumours of the long since disused Mustagh Pass
on the watershed, he made attempts also to penetrate the mountains up
the Shigar river from Skardu, but the Mustagh at 18000 feet (or some
5500 m) would have been far beyond his capacity even had he got so
far.   Younghusband was to force his way over, literally at peril of his
life in 1887.   Vigne also went east, finding his way up the Saltoro valley
in search of a pass to take him into Turkestan, but weather and local
opposition turned him back.   Vigne had neither the mountaineering skill
nor the equipment to make much headway into what he called 'the snowy
sierra...extending from Hunzeh to Nubra (Vigne, Vol. 2, p. 358), but he
understood its nature.   He had an eye for country, cultivated during many
days of outdoor sport; he was a Fellow of the Geological Society; above
all he had the artist's vision.   His is the first description of Nanga
Parbat, come upon as he travelled from Srinagar into Baltistan in 1835:
'Mountain seemed piled upon mountain, to sustain a most stupendous confusion of mist and glacier, glistening with the dazzling and reciprocated
brightness of snow and sunbeam...I feel sure that, had Mont Blanc, in
its wildest phase, been placed beside it, it would have been disregarded
for a time, on account of its comparative insignificance, and greater
tameness of configuration.' (Vigne, Vol. 2, p. 296).

Vigne's Travels in Kashmir, Ladakh, Iskardo... was published in 1842
just four years before the East India Company entered into a treaty with
the Maharajah of Kashmir and seven years before the annexation of the
Punjab.   Better knowledge of the mountain lands, observed tantalizingly
from the new summer capital at Simla, became imperative.   In 1847 the
first official initiative into the western Himalaya was taken when the
botanist, Thomas Thomson, and two companions travelled into Kashmir with
a roving commission which was to split the party three ways.   Thomson
is our man: from Leh he travelled into Baltistan hoping to get to Gilgit
from there and so to Yarkand.   The weather stopped him and he was the
first European to winter in Baltistan - not even now a comfortable sojourn
as one may gather from Dervia Murphy's experiences (Murphy, 1977).
Thomson passed the time in making notes about the flowers and rocks,
in recording the weather, and in collecting general information.   Like
Vigne, he heard about the Mustagh Pass, but he did not venture to look
for it.   Come the spring, he returned to Leh and from there tried again
for Yarkand, crossing the Sasir Pass, the Shyok river and the
Depsang plateau to stand at last on the Karakoram Pass.   Seeing the
further barrier of the Kun Lun ahead, he turned back.   The German
brothers Robert and Herman Schlagintweit were the first Europeans actually
to cross the Pass, during extensive Himalayan travels in the 1850s.

So much for the overture to the Karakoram drama; the first serious
assault on the region was to come with the arrival of the Kashmir Survey.
Since 1802 the Survey of India had been working on the Grand Trigono-
metrical Survey which, within 100 years, was to map the whole of the
subcontinent from Cape Comorin to the Himalaya.   It was a colossal
undertaking, perhaps the finest legacy left to India by the British.
Through war and peace, in the most impossible conditions of weather
and of terrain, the Survey pressed on regardless.   In mid-century, the
President of the RGS described its work in southern India as being 'much
hindered by the unhealthiness of the country and the number of tigers'
(JRGS, 1875).   The late Sir Clinton Lewis, speaking in this very hall
described his problems as a young man working in 10-ft high grass which
concealed herds of wild elephants (Geogrl J., 1955).   On the northern
frontiers there were no tigers and elephants, but during the Afghan wars
a cannon ball might well bowl over a carefully posed plane-table.   The

22

CAPTAIN F. E. YOUNGHUSBAND. KING'S DRAGOON GUARDS

(Reproduced by kind permission of the Illustrated London News
(May 17, 1890, 613)).

Mutiny in 1857, which sent shock waves through the land, did not stop the Survey of Kashmir which began in that year to make the first geographical inroads into the Karakoram. John Keay sums up what these early surveyors had to contend with: their 'mountaineering skills' he writes, 'amounted to strong lungs and a cool head and climbing equipment was what you might find in a garden shed, a spade, a pick and a hefty coil of rope,' (Keay, 1977, p. 189). As we have seen, the terrain, though unfamiliar, was not entirely unknown when Lt. T. G. Montgomerie of the Royal Engineers brought the Survey to Kashmir. Vigne had indicated the extent of the range; Thomson's account of crossing the Sasir Pass and on, uphill and down, to the Karakoram Pass, stressed its complexity. The Survey had now to put it on the map.

The curtain rises on Montgomerie gazing north from his perch on Haramukh in the Vale of Kashmir. 'I had the pleasure to see the various ranges of the Himalaya right up to the Karakoram' he noted. 'There was nothing remarkable in the first six or seven ridges...Beyond that came the snowy points of the Karakoram Range and behind them I saw two fine peaks standing very high above the general range...possibly 140 miles away' (quoted in Keay, 1977, pp. 192-3). Montgomerie called these two peaks K1 and K2 - the nearer (which he thought the higher) was in fact Masherbrum, the lower of the two. The further away and, as it turned out the loftier, kept its name of K2, the second highest mountain in the world - 8611 m. Montgomerie wanted to call it after H. H. Godwin-Austen who actually put it on the map, but the Survey early made a rule to use no proper names for peaks other than for Everest, and K2 it has remained. In 1860 Godwin-Austen was on the way from Skardu to carry the Survey up to the watershed, supposed to mark the frontier between Kashmir and Turkestan. Godwin-Austen knew of the Mustagh Pass and fixed it as his goal, travelling up the Shigar river and over the Skoro Pass, then onto the Panmah Glacier. He had to retreat in face of the appalling conditions, and to make his way instead onto the Baltoro Glacier which was to lead him to the mountain always associated with his name even if he was not allowed to be so commemorated - K2. Struggling up the Baltoro, Godwin-Austen came to what John Keay describes as 'the innermost sanctum of the western Himalaya, an amphitheatre of the greatest mountains on this planet. In the space of about fifteen miles the Baltoro holds in its icy embrace ten of the world's thirty highest peaks (Keay, 1977, p. 199). Impressed, no doubt, but with a job to do, Godwin-Austen lost no time in seeking a sight of K2, first spotted by Montgomerie from 140 miles away. Straight ahead of him was Gasherbrum, to the right Masherbrum and it was up the latter that he climbed. One thousand, two thousand feet and at last he had the view he was seeking - K2, the great mountain which turned out to stand just on the crest of the watershed, and on the boundary.

The Kashmir Survey laid down maps of Ladakh, Baltistan and the country to the south of them, but there were still unexplored stretches north, exasperating to the RGS which found themselves horribly inhibited so long as all travel was in official hands in India, the more so since it was becoming the exclusive business of the Intelligence Department. The Society therefore nominated the adventurous George Hayward as their 'Envoy in Central Asia.' Hayward's goal was the Pamirs where he hoped to reach the source of the Oxus, either by way of disturbed tribal territory through Gilgit or by the desolate upland route round the south-eastern end of the Karakoram. In 1869 he was on the latter way, bound for Yarkand, when he diverged from the track to do some useful reconnaissance work on the headwaters of the Yarkand river. He eventually got as far as Kashgar where he was arrested by the local authorities and had to return to base. An attempt in the following year to reach the Pamirs through Gilgit ended in his murder in Yasin. Members of the Karakoram Project, accompanied by Lord and Lady Hunt, visited his grave, near Gilgit Fort.

The commercial dream faded during the 1860s and 1870s. The hardships of the route to Yarkand and Kashgar by the Karakoram Pass proved simply not worth the meagre results. Robert Shaw, a tea planter, got through only to be put under house arrest and dismissed with nothing to show for it. Douglas Forsyth took two Government Missions into Eastern Turkestan, but the response was disappointing – the local people did not care for Indian tea, for one thing. One independent trader, Andrew Dalgleish, kept going until he was murdered on the Karokoram Pass in 1888. Dalgleish's description of the way from Leh to Yarkand – 'the worst trade route in the whole wide world' (quoted in Keay, 1977, p. 262) proved the right one. Failure of the trade missions did not however put an end to official interest in the western Himalaya as a whole. The Government of India, which took over from the Company after the Mutiny, was becoming increasingly aware of the political importance of the lands lying beyond its ill-defined northern frontier. There was a dangerous no-man's-land north of the Karakoram between the Russian and the British empires, and a St Petersburg-inspired invasion from the north over the Pamirs seemed a not impossible threat. The impractical nature of the Leh-Yarkand route combined with these fears to turn attention increasingly towards the Gilgit road. The founding of the Gilgit Agency and Francis Younghusband's journeys across the range and round the north through Hunza were all part of what came to be called 'The Great Game'. Played with gusto during the 1880s by Army Intelligence and the Survey, the 'Game' dominates the scene.

Younghusband was an army officer of the dedicated and adventurous type common in those days who was appointed in 1885 to intelligence work in Kashmir. A year later saw him returning from extended leave in Manchuria by way of the Gobi Desert and Yarkand. From Yarkand he broke new ground by crossing the Aghil Mountains into the upper valley of the Shaksgam river. He was the first white man to see the north face of K2. He then made his way – by guess and by God as the old mariners used to say – over the Mustagh Pass, long since abandoned by local caravans because of dangerous shifts in the glacier pattern. This, of course, was the pass Vigne and Thomson had heard tell of, and that Godwin-Austen had looked for on his way to K2. Younghusband had no experience of mountaineering but by sheer grit and resolution he tackled it – scrambling over with his porters, using knotted turbans for ropes, gripping sheer ice drops with their frozen feet. It seemed unlikely, at least, that the Russians would send an invading force by this route.

Younghusband's journeys were undertaken in collaboration with the civil power, represented by the Gilgit Agency. The story is told in full in John Keay's The Gilgit Game (Keay, 1979) and is concerned more with imperial politics than with exploration – but it yielded some useful geographical information on the Karakoram. In 1889 he was investigating Hunza raids on caravans using the Karakoram Pass, and looking for alternative ways over the range. This took him again across the Aghil Mountains into the Shaksgam Valley. Here a Survey of India expedition under Kenneth Mason was to be at work as much as forty years later, so complex is the terrain. Mason's work was to be checked and extended by Shipton in 1937 – thus do explorers climb on one another's shoulders.

MARTIN CONWAY (1856 - 1937)

(Reproduced by kind permission of the RGS)

One of Younghusband's most interesting adventures was to run into a Russian expedition along the Yarkand river, led by Captain Grombchevsky. This was a friendly meeting - joint photography and exchange of information - but in the following year Younghusband was arrested by a Cossack patrol in the Pamirs, and firmly escorted off what they insisted was Russian territory. The most important phase of the 'Great Game' ended with an Anglo-Russian frontier agreement in 1895.

Following the conclusion of the Hunza War in 1891 the Gilgit area was rendered comparatively safe, and in 1892 arrived the first private enterprise with scientific aims. Sir Martin Conway's sizeable and well organized expedition received grants from The Royal Society and from the Royal Geographical Society, and was royally led by one of the foremost mountaineers of the time. Conway, a talented and versatile man - among other things an art collector and critic - had much experience of climbing in the Alps. He divided mountaineers into two types: the 'centrists' settled into some big local 'centre' and climbed everything within reach. The 'eccentrists' (of whom he was undoubtedly one) kept on the move, preferring passes to peaks and never willingly staying more than one night in one place. Conway brought a Swiss guide with him, and was accompanied by Charles Bruce, later famous on Everest, and a party of Bruce's Gurkhas. Working from Gilgit, Conway explored the glaciers south-east and north of Rakaposhi. From there he made his way via the Hispar onto the Biafo Glacier, notching up a 'first' when he crossed the Hispar Pass. He followed the Kero Lungma down to Arandu, observing and mapping, leaving a trail for others to follow. Among the problems left for his successors to solve was what he described as the 'Snow Lake' an apparently outsize expanse near the head of the Biafo Glacier. It was even thought it might be an ice cap, a feature unknown outside the Polar regions. Bill Tilman was later to investigate 'Snow Lake' and was disappointed to find that it was no more than 20 square miles in extent instead of Conway's estimated 300, and could be identified as the eastern basin of the Biafo. Conway started something with his Swiss guide and Alpine experience, and also with his independent position. He was out to study the lie of the land, not the possible invasion routes of imaginary Russians, far less the fabulous markets of Cathay. He initiated much that is common form in today's expeditions, noting the effects of altitude and diet, the rates of pulse beats, the wear and tear on equipment. He brought with him tents proved by Whymper and Mummery in the Alps, the Andes and elsewhere. In short, he opened the Karakoram to private enterprise.

In Conway's tracks came something quite new in the Himalaya - an intrepid woman mountaineer, in the person of Fanny Bullock Workman accompanied by her husband Dr William Hunter Workman. The Workmans introduced a slight note of comedy into the awe-inspiring world of the high peaks. They were middle-aged Americans, rich and energetic, who visited the Karakoram eight times between 1898 and 1912, climbing, discovering and surveying, and scandalizing the 'old guard' of British India by the vigour of their researches and the confidence of their claims. They arrived on the scene early in the hot weather of 1898 having bicycled from Cape Comorin to Peshawar - Fanny wearing a topi adorned with the badge of the Touring Club de France, and with a tea-kettle perched on her handlebars. Their expeditions were well organized and staffed by Alpine guides and on several occasions by professional map makers; they were courageous and persistent; they made some impressive ascents; they took excellent photographs - one of Fanny displaying a 'Votes for Women' placard on the summit of a Karakoram peak. On the other hand they lacked sympathy with local people and paid scant attention to the work of their predecessors. In 1902 they explored the Chogo Lungma Glacier from Arandu, and in 1908, in the footsteps of Martin Conway, they worked their way over the Hispar Pass and down the Biafo Glacier to Askole. Fanny fell into a crevasse in Conway's Snow Lake, and later on the same trip lost what she called 'the Memsahib's treasured topi, though fastened with elastic...' (quoted in Middleton, pp. 85-6) blown off on one of their numerous ascents. The Workmans' exploration of the Siachen Glacier in 1911 and 1912, with a professional lent by the Survey of India, was their most useful contribution to the geography of the area.

The Siachen was really discovered by Tom Longstaff three years before the advent of the Workmans. Longstaff had long and impressive experience of the Himalaya when he came to the Karakoram in 1908, drawn by what appeared to him not so much a blank as a distortion in the official map of the south-eastern reaches of the range. Younghusband's line of march over the Mustagh Pass in 1887 was made to appear as having been over the Saltoro peaks, thus placing the Saltoro Pass on the watershed – and therefore on the Kashmir/Turkestan frontier; the whole area was thus fore-shortened. In a series of sallies north from the Shyok river to midway on the Siachen, and then up the Nubra river and on to the Siachen at a lower point, Longstaff put the record straight. He also identified the Teram Kangri group of peaks. His exploration, for which he received an RGS Gold Medal, proved beyond doubt that Younghusband's Mustagh Pass lay north of the Siachen on the true watershed, while the Saltoro Pass lay south at a lower height. The Siachen was found to be 45 miles long, not 12 as shown on the official map. As he wrote triumphantly in his autobiography: 'We had stolen some 500 square miles from the Yarkand river system of Chinese Turkestan and joined it to the waters of the Indus and the Kingdom of Kashmir' (Longstaff, p. 192).

Conway and Longstaff, and to some extent the Workmans, did much to throw the Karakoram open to the scientific world of Europe. The region now ceased to be a jealously guarded British preserve and became what it has been ever since – a field for investigation and for mountaineering feats for all nations. I should be going beyond my time and my knowledge were I to give a detailed account of the expeditions which tackled the Karakoram in the years before and after the First World War, but no historical survey of Karakoram exploration would be complete without at least mention of the leading names. The year 1909 saw the first of a series of important Italian expeditions, the one led by the Duke of the Abruzzi who reconnoitred the approaches to K2 and worked on the upper Baltoro. With him was Filippo De Filippi who, in 1913, led what was then the most highly organized scientific venture to date into the Karakoram. Interests represented in his team included geodesy and geophysics, geology, meteorology and climatology, and photography; a Survey of India team also joined the party, and 'multi-disciplinary' seems to be the word to describe the whole. The chosen route was east of the Siachen, to the main source of the Shyok and to the headwaters of the Yarkand river where Hayward had been in 1869. The Duke of Spoleto led an expedition sponsored by the Royal Italian Geographical Club in 1928 up the Baltoro with a view to exploration on and beyond the water-shed. Ardito Desio, geologist and mountaineer, was with them and made the first crossing of the Mustagh Pass since Younghusband. It will be remembered, no doubt, that in 1954 Desio led the first successful ascent of K2. Ashraf Aman, the first Pakistani mountaineer to reach the summit of K2, was a member of the Project's survey team, and it is pleasant to record his presence in the hall this afternoon.

ERIC SHIPTON (1907 – 1977) ON CONWAY'S
"SNOW LAKE" IN 1939

(Reproduced by kind permission of the RGS)

In 1930 Giotto Danielli, one of De Filippi's team in 1913, was again on the Siachen Glacier making a geological and botanical survey of its whole length. He was accompanied by Miss Ellen Kalau, an expert skier. A husband and wife team from the Netherlands, Dr and Mrs Visser, also did interesting work in the Karakoram between the wars. In 1922, with two Swiss guides they visited the south-east of the range, travelling by the Nubra. Three years later they were at the other end, investigating the tributaries of the Hunza river in the western Karakoram. On this expedition they had with them Afraz Ghul Khan of the Survey of India, described by Kenneth Mason as 'probably responsible for a larger area of detailed survey in high mountains than any other surveyor in the world' (Mason, p. 89). The two head valleys of the Khunjerab river were explored and mapped and the Khunjerab Pass, which today carries the Karakoram Highway, was reached. In 1929 and 1935 the Vissers were again on the Nubra.

Longstaff had disentangled the wilderness of glaciers, peaks and passes which make up the south-eastern reaches of the Karakoram. Beyond the range was another important 'blank on the map' to be filled in by Eric Shipton just before the outbreak of the Second World War. The Shaksgam Expedition of 1937 was a fitting climax to the period of primary exploration which has been the subject of my talk this afternoon. The political scene was to change after the war with the coming of Independence to Pakistan and to India; the 'exploration' scene changed also with the development of new techniques in scientific research and in mountaineering. The pioneering days were over.

Shipton's blank was 'an unknown region, several thousand square miles, surrounding the basin of the Shaksgam river, immediately beyond the main continental divide.' (Shipton, 1969, p. 100). As I have said Younghusband had been there before them but made no detailed survey, while Kenneth Mason's Survey of India expedition in 1926 had run into difficulties. Mason had to leave unsolved a geographical puzzle concerning the Shaksgam tributary of the Yarkand and the so-called 'Zug' (or 'False') Shaksgam joining the Yarkand further down stream. Shipton with H. W. Tilman, John Auden and Michael Spender, approached the region not, as had De Filippi in 1913 and Mason in 1926, from the south, but straight over the range at the head of the Baltoro. They had in mind Younghusband's Mustagh Pass, but Ardito Desio, when he had been that way in 1928, had spotted another possible and easier saddle. This they found and descended into their 'blank' where they succeeded in clearing up a number of problems. They came down the Sarpo Laggo Glacier and reached Younghusband's Aghil Pass three weeks later. In the course of a remarkably short time, splitting the party for different tasks, much was accomplished. Mason's 'Zug' Shaksgam was traced to its junction with the Yarkand, the glaciers descending from the north face of K2 were explored and the Skamri Glacier (which Conway had called the 'Crevasse Glacier') was surveyed. In his Abode of Snow (the definitive guide to date on Himalayan exploration) Kenneth Mason has no hesitation in stating that 'as a result of Shipton's travels a large amount of new survey was added to the map of the Karakoram' (Mason, p. 246).

Incidentally, some lesser puzzles were cleared up or, as Bill Tilman called them, 'legends'. Conway's 'Snow Lake' as I have already said, was found to be no more nor less than the headwater of the Biafo Glacier and the Workmans' 'Cornice Glacier' (which they thought to be a glacier without an outlet) was found, to quote Tilman, 'to behave as others do' (Tilman in Shipton, 1938, pp. 194-226), with a normal flow. Tilman regretfully marked his map with the words 'Martin's Moonshine' and 'Fanny's Fancy' (Shipton, 1969, p. 113). Less easily disposed of was the venerable legend of the yeti whose footprints Tilman's Sherpa companions identified on the surface of the Biafo Glacier. Is it not rather a happy note on which to end this brief talk, that although the march of science has penetrated to every corner of the Karakoram, still in its furthest reaches the Abominable Snowman may yet hold sway?

<u>REFERENCES AND WORKS CONSULTED</u>

<u>Dictionary of National Biography:</u> on Lord Conway of Allington (1856–1937) and on Sir Francis Younghusband (1863–1942).

<u>Geographical Journal</u> 118, 3 (1952) 293–5.

<u>Journal of the RGS</u> 19 (1873) ccv.

Keay, John (1977) <u>When Men and Mountains Meet</u>. London: John Murray.

Keay, John (1979) <u>The Gilgit Game.</u> London: John Murray.

Longstaff, T. G. (1955) <u>This my Voyage.</u> London: John Murray.

Mason, Kenneth (1955) <u>Abode of Snow</u>. London: Rupert Hart-Davis.

Middleton, Dorothy (1965) <u>Victorian Lady Travellers</u>. London: Routledge and Kegan.

Murphy, Dervia (1977) <u>Where the Indus is Young</u>. London: John Murray.

Shipton, Eric (1938) <u>Blank on the Map</u> (Tilman's section). London: Hodder and Stoughton.

Shipton, Eric (1969) <u>That Untravelled World.</u> London: Hodder and Stoughton.

Vigne, G. T. (1844) <u>Travels in Kashmir, Ladak, Iskardo: the countries adjoining the mountain-course of the Indus...</u> (2nd ed.) 1844. London: Colburn.

<u>ACKNOWLEDGEMENTS</u>

Permission has kindly been given by the publishers for quotations from: <u>When Men and Mountains Meet</u> by John Keay and <u>This my Voyage</u> by T. G. Longstaff (John Murray); and from <u>Blank on the Map</u> and <u>That Untravelled World</u> by Eric Shipton (Hodder and Stoughton).

Mountaineering in the Karakoram in the late 1970's

P. Nunn – Sheffield Polytechnic

ABSTRACT

The late 1970's were a time of intensive mountaineering activity in the Karakoram as in the Himalayan main chain. Successful ascents and significant developments are surveyed and placed within the context of changes in world mountaineering practice. Ascents and some possibilities for the future are critically interpreted. In particular the diminishing returns from large expeditions and their likely detrimental effects upon the climbing experience and the environment are emphasized.

INTRODUCTION

By 1976 Messner and Haberler had completed their two man alpine style ascent of Hidden Peak (26470 feet: 8068 m) and the Trango Tower was at last climbed despite formidable rock difficulties. Gasherbrum III had had its first ascent by a party including Polish women and an all female team had made the first such ascent of Gasherbrum II, the first 8000 m peak to be climbed by a women's party.

Nevertheless Galen Rowell found almost fifty unclimbed Karakoram peaks between 22740 and 25610 feet, "Below that height possibilities were legion on many technically difficult mountains as well as easier summits, and there remain a host of subsidiary peaks unlisted which have not been explored thoroughly. Also a tendency was emerging of attacking the more difficult faces of peaks previously ascended, following the usual evolution of alpinism." (1)

As early as 1892 Martin Conway encountered the West Biafo Wall "... as wonderful an avenue of peaks as exists among the mountains of the world. On both sides of the glacier for some fifteen miles they rise one beyond another, a series of spires, needle sharp, walled about with precipices on which no snow can rest... The aiguilles of Chamonix are wonderful... but these needles outjut them in steepness, outnumber them in multitude, and out-

FIG. 1.   PEAK 6960 (BAINTHA BRAKK OR OGRE 2) AND
BAINTHA BRAKK (OGRE 1).
The Ogre is the rock tower peeping over the left hand rock ridge.
The impressive triangular peak is the West Summit of 6960 m.

reach them in size." (2)

By 1981 only two peaks in this group had been climbed (Conway's Ogre almost to the summit 1980, Sosbun Brakk 21100 feet 1981). (See Fig. 1)

The West Biafo Wall is not untypical of the challenges remaining throughout the Karakoram region. Since 1976 a great deal has been done (see Appendix), but the ratio of success for expeditions is not high. The mountains are becoming far better known, which makes the possibilities of future ascent greater once an attempt has been made. (3)

Among major summits reached for the first time have been Pumarikish. The Ogre and all four of the Latoks; see Fig. 2. Many of the bigger mountains have been climbed several times, sometimes by new routes. In 1977 a large Japanese expedition climbed the Abbruzzi Ridge of K2 with a Pakistani, Ashraf Aman, in the summit party, and in 1979 that route was repeated by an Austro-Italian party. The Americans broke the baleful spell of forty years and climbed a new route on K2 in 1978 which utilized earlier pioneering by the Poles, and in 1981 a Japanese party succeeded on the West Ridge which has defeated several talented and well resourced attempts, again with a Pakistani, Nazir Sabir, in the summit party. (4)

Increasingly though, style is valued as well as success even on the highest mountains. The ultimate of this was the solo ascent of the Diamir Face of Nanga Parbat by Messner in 1978. (5) Others have solved smaller things but none compare with this tour de force, and Messner's success is part of a wider trend indicated by his later solo of Everest from the north and Roger Baxter Jones' near solo of Makalu in 1980. A change in values is permeating even to Japan. "Gloomy human relations" are said to have troubled the mammoth 1977 K2 Expedition, and it is being realized that large parties with little limit on resources can climb most mountains, but at a high cost in terms of the decay of human spirit and destruction of the environment within which climbing is a pleasure. The recent solo of Dhaulaghiri by Kanaki illustrates the strength of this trend. (6)

On smaller peaks the story has been chequered and the rapid developments of the Ogre Latok group are illustrative. In 1974 and 1975 Japanese parties tried the Ogre, Latok II West Ridge and Latok I, following in the footsteps of Don Morrison's 1971 attempt to reach the Ogre from the Simgang Glacier in the North, and a Japanese exploration of the same year. The Shizuoka Tohan party of 1974 ascended the col between the Ogre and Peak 6960 m and the col west of the Ogre was reached. In 1975 Makato Hara's party attempted Latok I but were driven back by storms and avalanches and Naoki Takada's fifteen-man party tried the West Ridge of Latok II but were defeated by steep rock ridges. Two British parties contented themselves with small peaks west of the Biafo and an ascent of a small summit above the Uzzum Blakk Glacier (Rowlands-Richardson 1975). (7)

Two years later the British attacked the Ogre and Latok II. (8) Morrison's West Ridge party retreated after his death from Latok II, but Bonington and Estcourt reached the West summit of the Ogre in a first push and subsequently Scott and Bonington reached the top (9)(10). Serious injury in an abseil accident and bad weather tested their resources to the limit and they escaped only with tremendous effort and help from their companions Anthoine and Rowlands. (11) Late in the summer Bergamaschi's party climbed Latok II from the Baintha Lukpar Glacier. Twelve climbers and 12000 feet of fixed rope were employed and the Italian expedition also climbed a number of lesser peaks (12). The stage was then set in this group for attempts on the more difficult faces of peaks already ascended and on the harder remaining

FIG. 2. TED HOWARD ON THE SUMMIT DURING THE FIRST ASCENT OF THE FOREPEAK OF THE GHUR RANGE, BIAFO GLACIER 1975.
Behind from left to right are Conway's Ogre, the Uzzum Blakk Spires and the Ogre (Baintha Brakk) behind.

mountains.

The Shizuoka Tohan Club climbed the Ogre from its South Face (7) in 1978, Makato Hara climbed half of the South Ridge of Latok III on very steep rock, and Nunn, Thexton and Wilkinson went higher on the West Ridge of Latok II, but failed in bad weather. Most remarkable was the American attempt on Latok I North Ridge, the Walker Spur of the Karakoram. The Lowe Brothers, Mike Kennedy and Jim Donini came within a hair's breadth of success but were also defeated by the weather. After twenty-six bivouacs they retreated from a point close to 23000 feet, after almost making the first ascent of the peak by its most trying and aesthetically attractive route. (13) See Fig. 3.

Elsewhere Bettembourg and Seigneur did Broad Peak alpine style, and a British party went very high on the formidable West Face of Gasherbrum IV. (Boysen, Anthoine, Minx, Barker). A subsequent attempt by a Japanese party failed in 1981 with three deaths.

At last in 1979 the Latoks yielded more plums to the Japanese with swift ascents of Latoks III and I, from the Baintha Lukpar Glacier. Elsewhere thirty or so expeditions took the field. Many tried to repeat existing routes on known mountains. Success though came on Gasherbrum V, Spantik (repeat) Pasu Peak by the Pakistan Army, Haramosh (repeat) and the technical Uli Biaho and Trango Great Tower (American). One recent development involved the descents of Baltoro Kangri and Broad Peak on ski in 1980 (British and French) and a hangglider stunt from the slopes of K2 by a large French expedition. (14)

Summary of divergent trends is difficult. Dinosaur expeditions rumble on, huge, unwieldy and fuelled by national pride and/or commercial money. Small expeditions suffer some of the same problems and sometimes merely concentrate commercial gain in fewer hands. The large expeditions have had great successes, with the Japanese on the North/West face of Rakaposhi or the West Ridge of K2, but smaller groups succeed too, on Dastaghil Sar in 1980 (Poles), on Gasherbrum II, Conway's Ogre and Sosbun Brakk. (15)

Possibilities remain legion still and fortunately smaller objectives provide scope as adventurous as the 8000 m peaks while being less expensive. There are dozens of unclimbed faces and summits in the 20000 - 22500 feet range, most of them of a high order of difficulty even when approached by their easiest routes. Nor should it be forgotten that the Karakoram is a wilderness more fierce than most mountain ranges, a high altitude desert empty of inhabitants. As the redoubtable General Bruce said years ago of the Karakoram, it is "a country to see and travel in, but essentially not a country to live in." (16) See Fig. 4.

FIG. 3.   THE UNFINISHED AMERICAN NORTH RIDGE OF LATOK 1.

38

REFERENCES

1) ROWELL, G., (1976). "Baltoro Odyssey", Mountain 49, Mountain Magazine Ltd., Sheffield, pp 18 - 34.

2) CONWAY, M., (1920). Mountain Memories, Cassell and Co. Ltd., p 148.

3) NIXON, J., (1981). British West Karakoram Expedition (attempt on Thaime Chish). Preliminary Report to the Mount Everest Foundation.

4) REICHARDT, L., (1979). "K2 The End of a Forty Year American Quest", Amer. Al. Jour., Vol. 22, No. 1:53, pp 1 - 18.

5) MESSNER, R., (1980). Solo Nanga Parbat. Kaye and Ward.

6) The Iwa to Yuki, No. 76, (August 1980), Yama-Kei Publishers Co. Ltd., Tokyo, p 108.

7) The Iwa to Yuki, No. 51 (1976), p 136; No. 52 (1976), pp 8 - 15.

8) MORRISON, D., (1971; 1975). Yorkshire Himalayan Expedition Reports, to the Mount Everest Foundation.

9) ANTHOINE, J. V., (1978). "The British Ogre Expedition 1977", Al. Jour., Vol 83, No. 327, pp 3 - 7, and

10) BONINGTON, C., (1978). "The Ogre", Amer. Al. Jour., Vol. 21, No. 2:52, pp 412 - 435.

11) ROWLANDS, C., (1979). "The Ogre", Yorkshire Ramblers' Club Jour., Vol. XI, No. 38 (1979), pp 216 - 226.

12) BERGAMASCHI, A., (1978). City of Bologna KK Expedition Report. Private Expedition Publications, Bologna.

13) KENNEDY, M., (1979). "Latok I", Amer. Al. Jour., Vol. 22, No. 1:53, pp 24 - 28.

14) The Iwa to Yuki Annual (1979), pp 149 - 152.

15) Mountain (1980). Information No. 74, p 14; No. 76, p 13; No. 77, p 11.

16) BRUCE, G. C., (1934). Himalayan Wanderer, A. Maclhose & Co., p 102.

FIG. 4. THE VIEW NORTH EAST FROM THE WEST RIDGE OF LATOK 2. THE CHOKTOI GLACIER
AND THE SHAKSGAM WATERSHED OF SINKIANG.

APPENDIX

KARAKORAM AND PAKISTAN HINDU KUSH SUCCESSFUL EXPEDITIONS 1974 - 81

OFFICIAL EXPEDITION NUMBER	NAME OF THE EXPEDITION	PEAK	NUMBER OF PERSONS
1974			
5	FRENCH MOUNTAINEERING KK EXPEDITION LED BY MR WALTER MICHLE	5877 M	10
11	JAPAN KYOTO UNIVERSITY EXP LED BY MR GORO IWATSUBO	K-12 (7468 M)	6
15	JAPAN UNPYO ALPINE CLUB EXP LED BY MR TOSHIYUKA MIYOSHI	KOROL ZOM (THUI I) (6660 M)	11
17	POLISH GERMAN ACADEMIC KK EXP LED BY MR KUROZAB	SHISPAR PASSU PEAK (7619 M)	16
19	SWISS EISELIN HINDU KUSH LED BY MR MAX EISELIN	TIRICH MIR (7766 M)	12 (WEST SIDE)
20	UK IMPERIAL COLLEGE EXP LED BY MR COLIN ROBERT	KOYOZOM (22545 FT) (6872 M)	6
1975			
3	AUSTRIAN KARAKORAM EXPEDITION LED BY DR KOBLMALLER	CHOGOLISA (7561 M)	8
5	AUSTRIAN KARAKORAM EXPEDITION LED BY MR HANNS SCHELL	HIDDEN PEAK (8068 M)	7
6	AUSTRIAN KARAKORAM EXPEDITION LED BY MR PETER BAUNGARTNER	GARMUSH (6534 M)	6
7	FRENCH LOYAN PREMIER LED BY MR JEAN PIERRE FRESAFOND	GASHERBRUM NO. II (8035 M)	18
9	ITALIAN "CITTA DI LECCO" EXPEDITION LED BY MR GUILIO FIOCHI JR	BALTORO CATHEDRAL (5010 M)	8
13	JAPAN HIROSHIMA KARAKORAM EXPEDITION LED BY MR KEIJI ENDA	KAMPIRE DIOR (7145 M)	
15	JAPAN SHIZUOKA UNIVERSITY KARAKORAM EXP LED BY MR KINYA OHTA	TERAM KANGRI NO. 1 (7465 M)	11
18	JAPAN ICHIKAWA KARAKORAM EXP LED BY MR YOSHIHIKO YAMAMOTO	K-12 (7468 M)	9

20	JAPAN "IWATE" KARAKORAM EXP LED BY MR KASH HARA	MALUBITING (7260 M)	11
21	JAPAN HEKIRYOU KARAKORAM LED BY MR TOMIAMU ISHIKAWA	LAILA PEAK (6986 M)	10
23	KYOTO KARAKORAM EXPEDITION LED BY MR SHINICHI HOTTA	PURIANSAR (6239 M)	10
24	JAPAN YOKOHAMA BERNIA ALPINE CLUB LED BY MR MASAO OKEBA	BUNI ZON (6551 M)	8
26	POLISH WOMEN KARAKORAM EXP LED BY MISS WANDA RUTKIEWIEZ	GASHERBRUM NO. III AND II (7925 M), (8035 M)	20
27	POLISH MOUNTAINEERING EXP LED BY MR JANASZ FERNSKI	BROAD PEAK (8040 M)	16
31	SWISS EISELIN SPORTS HINDUKUSH EXP LED BY MR REISER BURGEARNSERSTR	TIRICH MIR (7700 M)	41
36	UK EDINBURGH UNIVERSITY EXPEDITION LED BY MR COHEN	THUI NO. I (6662 M)	6

1976

1	AUSTRIAN RUPAL NANGA PARBAT EXPEDITION LED BY MR HANNS SCHELL	NANGA PARBAT (8125 M)	5
5	GERMAN GOPPINGER HIMALAYA KK EXPEDITION LED BY DR ALEXANDER SCHLEE	BATURA I (7785 M)	6
14	GAKUSHUIR UNIV KK EXP LED BY MR AKIKATA FUNAHASHI	SKYANG KANGRI (7544 M)	10
15	JAPAN OSAKA UNIVERSITY MOUNTAINEERING CLUB KK EXPEDITION	APSARASAS (7100 M)	8
17	JAPAN TOHOKU UNIV KK EXP LED BY MR HARUOSATO	SINGHI KANGRI (7202 M)	10
21	THE KOBE UNIV MOUNTAINEERING EXPEDITION LED BY MR KAZUMAS HIRAI	SHERPI KANGRI (7380 M)	10
22	THE IIDA CLUB ALPINE KK EXPEDITION LED BY MR HARUKI SUGIYAMA	GHURKUN (7720 M)	5

27	JAPAN SHIBAURA INSTITUTE OF TECHNOLOGY KK EXPEDITION LED BY MR TOMOYA AKIYAMA	BALTORO KANGRI NO. I (7312 M)	9
30	EISELIN SPORTS HINDUKUSH EXP LED BY MR EISELIN MAX	TIRICH MIR (7706 M)	24
32	BRITISH TRANGO TOWER EXPEDITION LED BY MR J V ANTHOINE	TRANGO TOWER (6257 M)	6
35	PAKISTAN SERVICES EXP LED BY MAJ GEN GHULAM SAFDAR BUTT	PAIJU (6666 M)	8

1977

2	ALL JAPAN K-2 EXPEDITION LED BY MR I YOSHIZAWA	K-2 (8611 M)	52
3	BRITISH OGRE EXPEDITION LED BY MR D SCOTT	OGRE PEAK (BAINTHA BRAKK 7285 M)	6
5	ITALIAN OPERAZIONE "BIAFO 77" HIMALAYA EXPEDITION LED BY MR A BERGAMASCHI	LATOK II (7120 M)	7
9	YUGOSLAV ALPINE EXPEDITION LED BY MR L JANEZ	GASHERBRUM I (HIDDEN PEAK 8086 M)	12
11	JAPAN DOSHISHA UNIVERSITY HINDUKUSH KARAKORAM EXP LED BY MR TAKAHIRO MATSUMURA	GARMUSH (6244 M)	5
15	ITALIAN SPEDIZIONE CAI BARALLO SESIA HINDUKUSH EXP LED BY MR TULLIOVIDONI	TIRICHMIR WEST IV (7738 M)	7
16	JAPAN IBARAKI UNIVERSITY EAST HINDUKUSH EXPEDITION LED BY MR HIROJI HIRAI	Mt UDREN (7080 M)	9
17	JAPANESE RAILWAY WORKERS KK EXPEDITION LED BY MR EIROAKI AKIYAMA	CHOGOLISA W PEAK (7000 M)	13
18	JAPAN OSAKA CLIMBERS CLUB KARAKORAM EXPEDITION LED BY MR NABO KUWAHARA	TAHU RUTUM (6651 M)	9
22	THE JAPANESE (AICHIGAKUIN UNIVERSITY) SCIENTIFIC AND MOUNTAINEERING EXPEDITION LED BY MR MICHIO YUASA	BROAD PEAK (8047 M)	19

23	UK CHARLOTTE MASON COLLEGE HIMALAYAN EXPEDITION LED BY MR R RUTLAND	PHUPARASH (8047 M)	6

1978

1	AUSTRIAN ALPINE CLUB EXP LED BY MR BROSING GEORG	GASHERBRUM NO. II (7952 M)	4
2	1978 AMERICAN K-2 EXP LED BY MR J W WHITTAKER	K-2 (NORTH EAST RIDGE 8611 M)	14
4	UPPER AUSTRIAN HIMALAYA EXP 1978 LED BY MR R WURZER	NANGA PARBAT PEAK (8125 M)	6
10	NORTH LONDON HINDUKUSH EXP LED BY MR NICHOLAS TRITTON	THUI II (6523 M)	4
14	CZECHOSLOVAKIA HIMALAYA EXP TO NANGA PARBAT LED BY MARIAN SAJNOHA	NORTHERN PEAK OF NANGA PARBAT (7816 M)	11
15	FRENCH BROAD PEAK EXP LED YANNICK SEIGNEUR	BROAD PEAK (8047 M)	4
17	ITALIAN NANGA PARBAT EXP LED BY MR R MESSNER	NANGA PARBAT (WEST SIDE 8125 M)	2
19	JAPAN ACADEMIC ALPINE CLUB OF HOKKAIDO LED BY MR AKINARI ISHIMURA	UN-NAMED PEAK (6447 M)	8
21	JAPAN REIHO ALPINE CLUB KK EXPEDITION LED BY MR YOSHIHARU MURATA	GENISH CHISH (SPANTIK 7027 M)	6
24	JAPAN 'HOSHI TO ARASHI' CLUB EXP LED BY MR KOUMEI NAKAMURA	SPANTIK PEAK (7027 M)	7
25	JAPAN FUKUOKA KARAKORAM EXP LED BY MR KATSUTOSHI IKEBE	TIRICHMIR PEAK (7708 M)	10
27	JAPAN HINDUKUSH EXPEDITION OF ISHIKAWA PREFECTURE WORKER'S ASCENT UNION LED BY MR KOICHIRO NISHI	MT ISTOR-O-NAL (7365 M)	7
30	JAPAN THE SECOND KK EXP OF OF THE SHIZUOKA TOHAN CLUB LED BY MR YOKIO KATSUMI	MT BAINTHA BRAKK (7285 M) SOUTH FACE	7
31	JAPAN-PAKISTAN DEFENCE ACADEMY ALPINE CLUB EXPEDITION LED BY LT ISAO FUKURA	PASSU PEAK (P55) (7284 M)	6

32	JAPAN SENOAI ICHIKO ALPINE CLUB KK LED BY MR TOHRU SHIBASAKI	GANCHEN PEAK (6462 M)	8
33	SANGAKU JUNREI CLUB KK EXP LED MR SADAMASA TAKAHASHI	TRINITY PEAK (6700 M)	6
34	JAPAN SHOWA ALPINE CLUB KARAKORAM EXP LED BY MR KENJI SHIMAKATA	HARAMOSH PEAK (7409 M)	9
35	JAPAN KANSAI ACADEMIC ALPINE CLUB KARAKORAM EXPEDITION LED BY MR HARUTOSHI KOBAYASHI	MT GHENT NO. II (7343 M)	10
37	JAPAN HAJ KARAKORAM EXP LED BY MR MITSUAKI NISHIBORI	BATURA II (7730 M)	10
38	POLISH-YUGOSLAVIAN EXP LED BY MR STANISLAW ANTONI RUDZINSKI	TIRICHMIR PEAK (7708 M)	16

1979

4	AUSTRIAN STYRIAN KK EXP LED BY MR HANNS SCHELL	GASHERBRUM II (8035 M)	7
6	FIRST CHILEAN KK EXP LED BY MR GASTON OYARZUN	GASHERBRUM II (8035 M)	9
12	ITALIAN SPEDIZIONO CAI-SEZIONE EXP LED BY MR CALCAGNO GIOVANNI	TIRICHMIR EAST (7692 M)	6
13	ITALIAN K-2 SOUTH FACE EXP LED BY MR REINHOLD MESSNER	K-2 (SOUTH) (8611 M)	7
14	JAPAN KWANSEI GAKUIN UNIVERSITY KK EXP LED BY MR MICHIO KAWAZOE	SIA GROUP PEAK (7024 M)	7
15	JAPAN KYOTO KK EXP LED BY MR RYUJU HAYASHIBARA	SIA KANGRI (7422 M)	10
16	JAPAN HIROSAKI UNIVERSITY KK EXPEDITION LED BY SUMITO HANADA	MT TERAM KANGRI III (7382 M)	12
17	JAPANESE BIAFO KK CLIMBING EXP LED BY MR NAOKI TAKADA	LATOK NO. I (7145 M)	10
18	JAPAN HOKKAIDO ALPINE ASSOCIATION KK EXP LED BY MR TAKAP SASAKI	PUMARI CHHISH (7492 M)	12

20	JAPAN ACADEMIC ALPINE CLUB OF HOKKAIDO KK EXP LED BY MR KOHEY EXHIZENYA	KUNYANG CHHISH NORTH PEAK (7108 M)	8
21	JAPAN YOKOHAMA ALPINE CLUB KK EXP LED BY MR TADASHI KAMEI	SKAMRI PEAK (6736 M)	7
22	JAPAN AKITA MOUNTAINEERING FEDERATION KK EXP LED BY MR BUNJI FUKUDA	CHOGOLISA (7640 M)	10
23	JAPAN HOSEI UNIVERSITY KK EXP LED BY MR MASAKATSU COI	LUPGHAR SAR NO. II (7100 M)	8
24	JAPAN WASEDA UNIVERSITY GRADUATE ALPINE CLUB KK EXP LED BY MR EIHO OTANI	RAKAPOSHI (7788 M)	7
25	JAPAN HIROSHIMA MOUNTAINEERING CLUB 7 JAPAN KK HIMALAYA EXP LED BY MR YOJI TERANISHI	LATOK III (6949 M)	
28	JAPAN HINDUKUSH EXP KAWASAKI TEACHER'S PARTY LED BY MR TADAKIYO SAKAHARA	BINDU-GOL-ZOM (6216 M)	
30	PAKISTAN-POLISH EXP LED BY CAPTAIN MUHAMMAD SHER KHAN	RAKAPOSHI (7788 M)	7 + 6 (13)
31	SPANISH KK EXP LED BY MR ANDRES FERNARDEZ	DIRAN PEAK (7266 M)	8
33	THE GREAT BRITAIN BORDERS HIMALAYA EXP LED BY DR IAN TATTERSALL	THUI NO. II PEAK (6523 M)	2
35	AMERICAN KK EXP LED BY MR JOHN FENTON CHARLES ROSKELLEY	ULI BIAHO TOWER (6083 M)	4
36	ANGLO-AMERICAN PAIJU EXP LED BY MR I R WADE	PAIJU PEAK (6757 M)	4
37	WEST GERMAN TEGERNSEER HIMALAYA KK EXP LED BY MR GLOGGNER JOHANN	LUPGHAR SAR I (7200 M)	8

1980

1	FRENCH EXPEDITION LED BY MR IVAN GHIRARDINI	MITRE PEAK (6010 M)	2
2	GREAT BRITAIN EXPEDITION LED BY MR JEREMY RICHARD STOCK	BALTORO KANGRI (7312 M)	7

6	GERMAN KK EXPEDITION LED BY DR BERNHARD SCHERZER	MT GHENT (7400 M)	5
8	FRENCH KK EXPEDITION LED BY MR BARRARD MAURICE	GASHERBRUM I (8086 M)	2
9	JAPAN YOKOHANA BERNINA ALPINE CLUB EXP LED BY MR HIDEO SATO	GASHERBRUM II (8035 M)	7
11	JAPAN KYOTO GAKUJIN EXP LED BY MR TATESHI SUDO	MANGO GUSOR 29 (6288 M)	3
12	AMERICAN KK EXPEDITION LED BY MR STEVE SWENSON	GASHERBRUM IV (7925 M)	7
19	BRITISH CONWAYS EXP LED BY MR ALEXANDER C DICKSON	UZUN BRAKK/ CONWAYS (6422 M)	3
21	JAPAN SHIGAKU KK EXP LED BY MR TADAO SUGIMOTO	YUIMARU SAR (7330 M)	5
23	POLISH KK EXPEDITION LED BY MR RYSZARD KOWALEWSKI	DISTAGHEL SAR (7885 M)	5

1981

1	AUSTRIAN GASHERBRUM II EXP LED BY MISS BINDER GABRIELE	GASHERBRUM II (8035 M)	6
6	JAPAN NAGANO KK EXP LED BY MR MESAHIRO MAEZAWA	GASHERBRUM I (8068 M)	12
7	JAPAN SHIGAKUKAI SOSBUM BRAKK EXP LED BY MR SEIICHI KAWAUCHI	SOSBUM BRAKK (6413 M)	4
12	ITALIAN SPEDIZIONE KK EXP LED BY MR GIOVANNI CALCAGNO	PAIJU PEAK (6600 M)	4
14	SWITZERLAND DANY EXP LED MR ROMOLO NOTTARIS	GASHERBRUM II (8035 M)	5
15	SPANISH CATALANA KK EXP LED BY MR ANTONI SORS FERRER	BROAD PEAK (8047 M)	5
16	JAPAN WASEDA UNIVERSITY ALPINE CLUB K-2 EXP LED BY MR TERUO MATSUURA	K-2 (8611 M)	14
17	FRENCH GASHERBRUM II EXP LED BY MR BEAUD ERICH	GASHERBRUM II (8035 M)	3

18	POLISH KK EXPEDITION LED BY MR PIOTR MLOTECHI	MASHERBRUM (S.W.) (7806 M)	9
22	JAPAN C.I.T. KK EXP LED BY MR MASAYOSHI FUIJII	KANJUT SAR (7760 M)	10
25	DUTCH MOUNTAINEERING EXP LED BY MR RONALD EDWIN NAAR	NANGA PARBAT (S.W. RIDGE) (8125 M)	8
32	ITALIAN CITTA DI BERGAMO EXP LED BY MR AUGUSTO ZANOTTI	NANGA PARBAT (DIAMIR FACE) (8125 M)	16

A note on Rakaposhi 7788 m (25,550 ft)

M. Vyvyan – Trinity College, Cambridge

Rakaposhi's North and West sides are contained by the Hunza River before and after this makes its left-angled turn towards the Gilgit valley; the major glaciation is on the East and South. The name is that used by Shina speaking people to the South whereas the Burushaski speakers who see its abrupt North face call it, or used to call it, Dumani. Martin Conway visited its Southern approaches as far as the Bagrot Glacier in his splendid and superbly recorded reconnaissance of the Karakoram in 1892 when he thought the "SSW arete seemed to offer an easy route to the summit". (1) Allowing for uncertainties in the orientation of the ridge, he got the first climbing route right in spite of the alpine scale of his judgement of difficulty. Rakaposhi does not seem to have been reconnoitred again from a climbing point of view until Secord and Vyvyan went to the Dara Khush (Happy Valley) below the Biro Glacier in 1938 and via its Western spur climbed the North-West peak which should most properly be named 'Jaglot Peak' from the village below it from which they started. (2)(3) Its height of little more than 6000 m was greatly overestimated by the climbers in their accounts and presumably following these by no less an authority than the late Kenneth Mason (4).

In 1947 Secord went back to Rakaposhi with Tilman, Gyr and Kappeler, the last two of whom had explored other approaches to the peak in 1946. (5) They began with an attempt on the South-West ridge via the Monk's Head, a major feature so named by the 1938 pair on account of its cowl-like appearance when they concluded that it lay on the most hopeful route. Unpromising weather deterred the party at about 20,000 feet so Tilman and Gyr returned to the Biro Glacier and climbed up to a point on the North-West ridge which connects the Jaglot Peak to the summit and to which Secord and Vyvyan had not descended. The route was not pursued owing to the standing danger of avalanches on the approaches to the ridge. (6) Two more unsuccessful attempts were then made in 1954 and 1956 on the SW branch of the South ridge – or 'spur' of this as it is less appropriately called, the considerable Kunti Glacier being contained between the two branches which diverge above the Monk's Head.

In 1954 Tissières, Band and Fisher, after reclimbing Jaglot Peak rejoined the rest of their party on the South-West ridge. Following it they reached

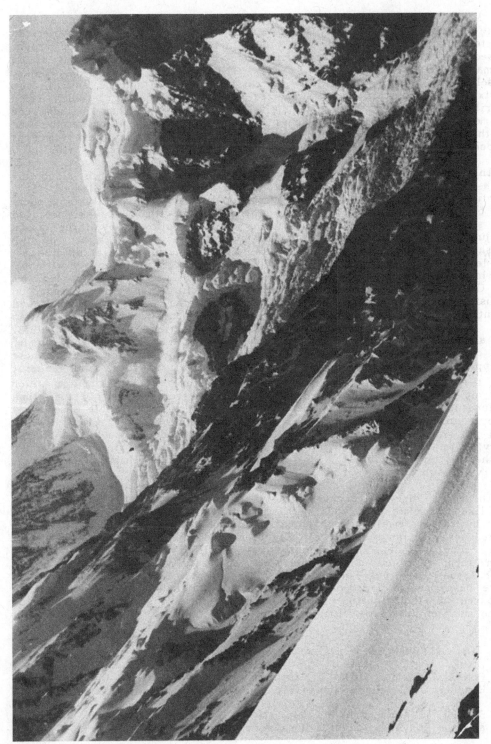

FIG. 1. PART OF THE SOUTH-WEST RIDGE OF RAKAPOSHI, SHOWING THE MONK'S HEAD WITH AN AVALANCHE FALLING FROM THE UPPER BIRO GLACIER.

the Monk's Head but the weather then forbade further reconnaissance, even visual. (7) (8)   Two years later Banks's powerful British-American quartet (Banks, MacInnes, Swift and Irvin) using minimum porterage, passed this landmark but were frustrated by avalanche conditions at an estimated 7160 m (23,500 ft.).   (9)   Then on June 25th, 1958, Banks, himself, with Patey was finally successful in reaching the summit from a sixth camp on this route supported by a party of eight which faced 'terrible conditions of snow and drift'.   Both Banks's teams included one or more active Pakistani climbers and official co-operation emphasised by the patronage of the C in C of the Pakistani Army may have somewhat compensated for the severe weather by easing logistical problems.   (10)

In 1964 Lynam's Irish party made a very slight advance along the North-West ridge reaching it, like Tilman, from the Biro Glacier, (11). Eventually in 1979 this route was completed by a Polish-Pakistani party led by Kowalewski and Sher Khan.   These two with Pietrowski achieved the second ascent of the mountain on July 1st followed by three others on July 2nd.   Then on July 5th the summit was reached again by two Polish girls, Czerwinska and Palmowska climbing individually at an interval and leaving the men, it seems, at Camp 3 (12)(13). Almost simultaneously with the Polish success a Japanese expedition from Waseda University, led by A. Otani put a party (Otani and Yamashita) on the summit on August 2nd, thus making the third ascent of the peak.   They followed a buttress on the daunting North side of the mountain where a route had been opened up by Herrligkoffer's Bavarian parties in 1971 and 1973, but abandoned it on each occasion some thousand metres below the summit, time or fine weather having run out.   (14)   Like the Germans, the Japanese equipped the severe route very thoroughly with fixed ropes and ladders, and the technical difficulties of the main feature, the rock ridge which held up the Germans can hardly have been eased by starting the climb in June instead of August as Herrligkoffer had done.   Valuable time seems however to have been gained on the glacier approaches.   Otani's party used six camps with two of their four high altitude Hunza-Nagar porters carrying to Camp 4; the summit pair used in addition one ice cave bivouac on the way up and down.   (15)   It was undoubtedly a climbing achievement of a very high order.

The North ridge was a challengingly elegant route rather than an obvious one like the North-West and South-West ridges.   The South ridge and the East buttresses remain topographically, if not technically, possible further approaches.   So far the mountain has been kind; has there been any equally high, equally independent, equally well fortified mountain which has so often been attempted unsuccessfully or successfully without exacting any fatal casualties?   But Rakaposhi has been treated with exemplary respect. There have been a disproportionate number of judicious strategic withdrawals compared to the fighting retreats through encircling storm which make up so much of Himalayan climbing history.

REFERENCES

1)   CONWAY, W. M., (1894).   Climbing and Exploration in the Karakoram-Himalayas.   T. Fisher Unwin, London.   p 136.

2)   SECORD, C. and VYVYAN, M., (1939).   'Reconnaissances of Rakaposhi and the Kunyang Glacier', Himalayan Journal XI, p 156.

FIG. 2. THE SUMMIT OF RAKAPOSHI AND THE NORTH-WEST RIDGE, FROM JAGLOT (NW) PEAK.

3)   VYVYAN, M., (1939).   'A Journey in the Western Karakorum', Alpine Journal LI, p 236.

4)   MASON, K., (1955).   The Abode of Snow: Rupert Hart-Davis, London. p 256.

5)   Information from Toni Hiebeler, Internationales Bergarchiv.

6)   TILMAN, H. W., (1947-1948).   'Rakaposhi', Alpine Journal LVI, p 329.

7)   CHORLEY, R., (1955-1956).   'To the Monk's Head on Rakaposhi', Himalayan Journal XIX, p 109.

8)   BAND, G., (1955).   Road to Rakaposhi; Hodder and Stoughton, London.

9)   MACINNES, H., (1957).   'Renewed Attack on Rakaposhi', Himalayan Journal XX, p 62.

10)   BANKS, M. P., (1958).   'Rakaposhi Climbed', Himalayan Journal XX, p 55.

11)   LYNAM, J. P. O'F., (1965).   'Rakaposhi, the North-West Ridge', Himalayan Journal XXVI, p 70.

12)   Alpine Journal LXXXV, (1980).   Notes, Asia, p 222.

13)   American Alpine Journal 1980, pp 651 - 652.   (Reference by courtesy of Internationales Bergarchiv).

14)   HERRLIGKOFFER, K. M., (1978).   Bergsteiger Erzählen; Bayerische Verlagsanstalt, Bamberg.   p 185.

15)   KODAMA, S., (1981).   'Rakaposhi from the North', Alpine Journal LXXXVI, p 185.

A science of mountains

R. Muir Wood – Trinity Hall, Cambridge

ABSTRACT

This article considers the beginnings of the science of mountains and the role of the IKP in the integration of several disciplines concerned with that science. The hope is expressed that the IKP may help stimulate more projects to further that integration towards a more unified approach to the study of earth sciences.

A SCIENCE OF MOUNTAINS

In order to see how the individual Karakoram Projects fit together to provide a scientific whole – or at least to find a matrix that can contain all the various lines of research one needs to be ambitious and perhaps optimistic. How can one relate glaciology to earthquakes or salt-weathering to the geology of Kohistan ? Yet to do research in mountains is to work towards some kind of science of mountains – but what does one mean by this term – is it more than just an umbrella grouping of disparate work? I believe not. There is already a century old precedent to the designation of a science based around a topographic feature and that is oceanography. Mountains and oceans present similar problems and for both their most obvious features are not their most significant. If a science of mountains is to be of any significance it should be able to illuminate the interconnections and perhaps provide new conceptual ways of seeing which are the most fundamental problems.

What is a science of mountains based upon? On first consideration it might appear to be geology – but the rocks themselves do not make mountains unique or distinguish mountains from other regions of the Earth's crust. On first impression the shapes of mountains appear to possess within their disorder considerable regularity but this is only a product of their having been raised at all. The key underlying concept behind the science is that of dynamics – of mountain building – that is better known as literally mountain carpentry – tectonics. The difference between the forms of mountain scenery and tectonics is the difference between the architectural ornamentation of cathedrals and the religion that brought them into existence.

The highest mountains can only remain high because they continue to be built. The faster they go up the faster they will correspondingly come down. The maximum height of mountains tolerated on our planet is partly a result of erosion overtaking uplift and partly at what depth the over-thickened continental crust will suffer widespread melting. The higher the mountains grow the colder their microclimate becomes and the more one is dealing with the snow and ice, rather than the water-based erosional economy. As the landscape becomes more transient so it becomes more dangerous. As the erosional journey from the mountain to the river speeds up so human settlements must be prepared for rapid adaptation. The drinking water that is heavily laden with tiny leaves of mica is a product of the same dynamism as the landslide that blocks the road.

With regard to this water - that David Giles the IKP Doctor has I believe correctly identified as being the most important concern for immediate assistance - I have a minor success to report that has resulted from the articles written in New Scientist (1 - 6). A group of researchers at the Department of Applied Mathematics Research School of Physical Sciences at the A.N.U. at Canberra (7) have been investigating interactions between solid surfaces using surfactant solutions and have discovered that it is possible to coagulate mica sheets with concentrations of 10^{-4} Molar of cationic surfactants such as Cetyl trimethylammonium bromide. The normal electrolytes used for flocculating clay minerals have no effect on the mica. Later this year, as a result of the IKP work, they are testing the procedure for effectiveness in the Himalayas - and will if sent a sample find the most appropriate means for cleaning up Hunza water. Once chronic problems such as these are eliminated - and the only reason that they have not previously been eliminated is that they have not been identified as being of great importance - then the acute problems of the mountains - the landslides, floods and earthquakes become highlighted.

The concept of dynamism, that is at the core of the mountain science, has a central concern - where and how is the movement localised. At shallow crustal levels we know that movement takes place along segregated zones of weakness that pass through the rock-masses - the faults. The existence of such faults as revealed both in major earthquakes and in background microseismics is definitive proof that mountains articulate - that different sections of crust move in relation to one another.

Where active fault systems break through to the surface they may impose this articulation on the topography and on the pattern of recent sediments. While it may be possible to identify the existence of the active fault zones through the kind of seismological programme that was operated on the IKP, to gain more information about the history of large fault movements (and therefore the size and frequency of major earthquakes) can sometimes be achieved by studying the geomorphology. The two subjects of geomorphology and seismology become inter-related - each providing information that feeds back into the other. But while conventional geographical geomorphology looks at a landscape as being reasonably isotropic in formation if not in appearance - as a product of regional uplift and altitude-induced climate - the seismotectonic approach to geomorphology investigates only those locations where the landscape is anisotropic, where, as along the base of the hills that suddenly raised themselves by up to 3 metres during the El Asnam earthquake - the landscape articulates.

From the core of our science of mountains – tectonics – we then move out to mountain destruction and erosion, and into the hazards associated with the primal war that is being fought between the forces of <u>up</u> and the forces of <u>down</u> within the mountains. There then is our science – ordered in terms of causality – as with oceanography an understanding of the location and the nature of earthquakes on the ocean-floor precedes our protection against tidal waves – tsunamis.

The Royal Geographical Society in putting its faith in the International Karakoram Project for its 150th Anniversary Expedition has begun what I hope is an important new commitment. There must be some fellows of the RGS who wonder what a geographical society is doing supporting geological and geophysical research. I would therefore like to explain by using some history. The IKP represented a deliberate move to embrace the earth sciences – a move that would have been impossible up until the last few years because the Earth Sciences did not exist until around 1970. Before that time there was geology that was notoriously parochial and there was a separate geophysics that preferred an abstract Earth without rocks. To say that there was a revolution in the Earth Sciences in the 1960s is misleading in that Earth Sciences – that is the science of a whole planet – only came into existence as a result of the discoveries of sea-floor spreading and continental drift. As an indication of just how parochial geology was – when the hypothesis of continental drift first arrived in England from Germany in 1922 a whole meeting was given over to a discussion of the topic at the Royal Geographical Society while the subject was never raised at the Geological Society. The problem for geology had always been one of scale – a founding principal of the new science of geology in the early 19th century was that talking on a scale beyond what one could see from a mountain peak – or what one could map in a few years – was purely speculative and not geological. Those who wished to talk on a wider scale and whose vision was globalised had to become cosmologists, travellers and geographers.

Now that the older subdisciplines have found common ground in the Earth Sciences there are still the older conservative geological institutions in existence that are inappropriate for the directions of modern research. Research finance and opportunities for large scale interdisciplinary meetings have been found only at the Royal Society. The new global geology at first concentrated on the oceans. Now that phase of research seems almost to be over and more and more attention is being given to the continents. The living parts of continents are the mountains and associated zones of high seismicity.

The overlap between seismotectonics, geomorphology and hazard studies has I believe become the most fruitful area of study in the modern Earth Sciences. It is also one that is ideally suited for small expeditions. Even the impossibility of detailed regional mapping of a mountain range as inaccessible as the Karakoram can be put to good advantage as it forces geologists to conceive on a grand scale – on a scale approaching that of global geology. I hope and believe that the Royal Geographical Society will build on the success of the IKP to encourage more defections over from the Earth Sciences and to emphasise the GEO in its own name that is all our common concern – knowledge of the outer shell of the planet on which we live and in particular translating that knowledge into human needs. It took geologists more than a century to approach the global world-view already possessed by some of the 19th century geographers. The enthusiasm and the potential for scientific creativity offered by the RGS should, and I hope will, draw the Earth Sciences more and more into its fold.

56

REFERENCES

1) MUIR WOOD, R., (1980). 'Science Goes to the Karakorum', New Scientist, 88 (1266), pp 374 - 377.

2) MUIR WOOD, R., (1981). 'Islands at the top of the mountains', New Scientist, 89 (1238), pp 274 - 277.

3) MUIR WOOD, R., (1981). 'The Moving Mountains', New Scientist, 89 (1242), pp 540 - 543.

4) MUIR WOOD, R., (1981). 'Decay in the Karakorum', New Scientist, 89 (1246), pp 820 - 823.

5) MUIR WOOD, R., (1981). 'Hard times in the mountains', New Scientist, 90 (1253), pp 414 - 417.

6) MUIR WOOD, R., (1981). 'Highway to the top of the world', New Scientist, 91 (1270), pp 656 - 658.

7) PASHLEY, R. M. and ISRAELACHVILI, J. N., (1981). A comparison of surface forces and interfacial properties of mica in purified surfactant solutions, Colloids and Surfaces, 2, pp 169 - 187.

Geology

The geographical and geological domains of the Karakoram

R.A.K. Tahirkheli and Q.M. Jan – Geology Dept., Peshawar University, Pakistan

ABSTRACT

This paper deals with some of the existing geographical and geological problems of the Karakoram and the authors present their views for their solution and better understanding.

The western and eastern boundaries of the Karakoram with the Hindukush and the Himalaya, respectively, still remain an unsettled issue. The authors on the basis of their observations have extended the western boundary to the high ranges forming the main divide between the Gilgit and the Yarkun rivers, lying southwest of Mastuj between Long. 72° and 73°E. In the north this boundary passes slightly east of Broghil between Long. 73° and 74°E.

In the east, the limit of the Karakoram has been stretched deep into Ladakh; and the eastern extent of the Deosai – Ladakh range with its granodiorite batholith, between Long. 78° and 79°E, forms its eastern boundary with the Himalaya.

Under the context of plate tectonics, the rock sequence established by the authors in the western section of the Karakoram (Hunza valley), from south to north is as follows:

Kohistan island arc Mass	Ladakh Granodiorite Chalt Ophiolite Yasin Group
Main Karakoram Thrust	
Asian Mass	Darkut Group Karakoram Granodiorite
F A U L T	
	The rock sequence of the Tethyan folded belt lying towards north of the Main Karakoram Range

58

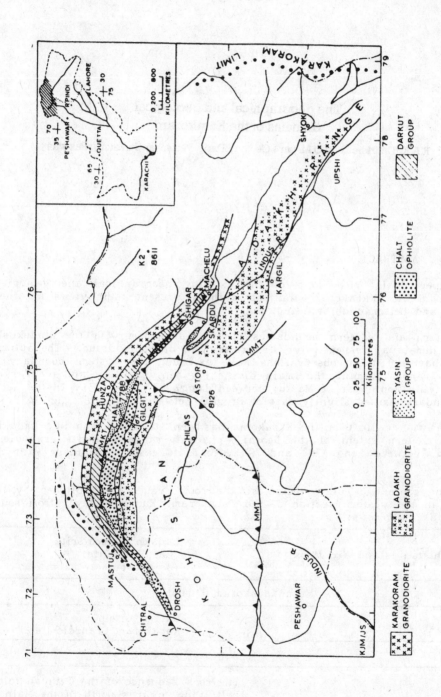

Fig. 1 A Preliminary Geological Map of the Karakoram

The Darkut Group, located on the south of the Main Karakoram Range, has so far remained undifferentiated. The authors propose a three-fold division (i) the Lower Meta-pelites (ii) the Middle Meta-Carbonate and (iii) the Upper Meta-Carbonate. The top formation has yielded fossils of Permo-Carboniferous age.

The flyschoid rocks representing the last remnants of the shrinking Tethys occur in several isolated patches in the Karakoram and Hindukush. These have been assigned various names in Baltistan, Gilgit and Chitral. It has been proposed to assemble them under a single stratigraphic nomenclature namely the Yasin Group, which will incorporate all these scattered flysch occurrences of Cretaceous-Lower Eocene age in the Karakoram and Hindukush.

Another change proposed by the authors for terminology in the Karakoram is to replace the old name "greenstone complex" with "Chalt ophiolite". The Chalt Schists/Formation was previously designated to a flysch occurrence in the Hunza valley. This has now been placed within the Yasin Group. The Chalt ophiolite on the basis of globutrunca fossils discovered in the Tissar section in the Shigar valley in Baltistan, has been placed in the Cretaceous.

INTRODUCTION

The Karakoram has been included as a part of the Trans-Himalayan system and constitutes the northwestern termination of this mountain chain. It extends for over 650 km with a northwest - southeast trend and lies between Long. 72° - 79°E and Lat. 35° - 36°N. Its mighty glaciers, widespread magmatism and intricate tectonic settings give it a unique status, which in the past could not be described adequately due to the limited and sporadic nature of the work in an area of difficult access.

The Karakoram inherits a fabulously rich terrain for scientists. Its selection by the Royal Geographical Society for conducting a multi-disciplinary scientific expedition to celebrate its 150th anniversary will help augment our scientific knowledge of the Karakoram.

The main theme of this paper will revolve around geography and geology of the Karakoram, and is based on the authors' past researches carried out during the last two decades. This topic is specially selected to acquaint the participating scientists with the already known rudiments of the Karakoram, thus providing them with a base on which to raise their edifice.

PAST RESEARCHES

When one raises his pen to write on the Karakoram, many prestigious names of those who have attempted to tame this mighty and majestic but unsympathetic terrain come to mind. While attempting to bring these stalwarts to the limelight, our pen must, out of necessity, be restrained to selecting only a few.

Father Ippolito Desideri da Pistoria was probably the first Italian explorer who penetrated the Karakoram on the eastern part during the early 18th century and published the first known travelogue of his trip. Francis Young-husband an English traveller had crossed the Karakoram in 1887 along the eastern Mustagh Pass (5422m) to reach India from China. H. H. Godwin-Austen is another revered British name who was the first to have a closer glimpse of K2, the second highest peak of the world, which is named after him. Marquis Osvaldo Roero di Cartanza, Duke Lante Grazioli della Rovere (Italy), Schlagintweit brothers (Germany), Martin Conway (England) are some other explorers who either made collective or solo attempts to explore the Karakoram between 1870 and 1910.

Before the First World War, a scientific expedition in the real sense, came to explore the Karakoram and was led by Dr. Fillippo De - Fillippi in the summer of 1913. It was a multi-disciplinary scientific expedition which included eight renowned scientists. This expedition mostly concentrated on the Baltistani part of the Karakoram and fifteen volumes covering the results of this work were published in Italian. This included three volumes on geodesy and geophysics, three on geology, three on petrology and paleont-ology, one on the history of the expedition, two on physical geography, two on anthropology and enthnology and one on the plants gathered above 4500m elevation.

Although very productive the results of this expedition were not well publi-cised and very little is known about its achievements outside Italy.

Kyoto University of Japan, in collaboration with the Punjab University and the Geological Survey of Pakistan, has also conducted a multi-disciplinary scientific expedition in the Karakoram during 1956-58. In this expedition, the senior author also got a chance to participate. Some of its published work has been translated in English which is of immense help to lay a base for new-comers to the Karakoram.

GEOGRAPHICAL DOMAIN OF THE KARAKORAM

The position of the Karakoram in relation to the other high mountain ranges of Central Asia is very vivid and well-carved. The Pamir complex forms the connecting link between the Hindukush, the Karakoram, the Kun Lun, the Tian Shan and the Trans-Alai ranges. These ranges fan out East and West into two vast virgations from the Pamir; the Tian Shan - Kun Lun - Karakoram and the Alai - Hindukush respectively.

The Karakoram occupies a pivotal position in the Himalayan arc system bord-ering the Indo-Pakistan subcontinent in the north. It is located on the northwestern margin of the Himalaya and forms a connecting link with the Hindukush and Pamir.

The extent of the Karakoram, i.e. its eastern and the western boundaries with the Himalaya and the Hindukush respectively, still remains an unsettled issue. The earliest workers had held the Shyok valley and the great Indus bend in the Gilgit district to form, respectively, the eastern and the western terminations of the Karakoram. Wadia who had worked in the Karakoram in the mid-thirties had also opted for this division.

Later, the western boundary of the Karakoram was shifted further west to the Yasin valley and the line of its western termination was drawn between Yasin and Mastuj in Chitral. Desio (1) supported this view and had upheld this division between the Karakoram and Hindukush in most of his subsequent publications.

The authors' observations in different sections of the Karakoram during recent years concerning the tectonics, magmatism and stratigraphy, along with detailed examination of the structural trends of the main range and its off-shoots, have brought some new revelations in support of its geographical domain. On the basis of this evidence the eastern boundary of the Karakoram, which was considered to be the higher ranges of the Shyok valley, is now to be shifted further east. The eastern limit of the Karakoram range is to be extended deep into the Ladakh area, as far east as the eastern limit of the Ladakh Granodiorite, called Ladakh intrusives by Frank et al (2). The Deosai-Ladakh Range running parallel to the Karakoram is considered by us as one of its sub-ranges on the south, sharing a border with the Kashmir Himalaya. These observations put the boundary between the Karakoram and the Himalaya between Long. 78°and 79°E.

The division between the Himalaya and the Karakoram may also be decided on the basis of the regional tectonics. The single Psang Po tectonic line bifurcates into two in the Ladakh area which are named the Indus Suture and the Shyok Suture (1). This bifurcation point of the suture coincides with the above mentioned division of the Karakoram from the Himalaya based on the orographic trend and the associated magmatism.

The western extent of the Karakoram may be decided by two factors, (a) the westward extension of the Main Karakoram granodiorite batholith which terminates southwest of Mastuj and (b) the structural trends of the Karakoram which take a sharp southwestern swing between Yasin and Mastuj, and with this trend they merge with the Hindukush. Thus the line of demarcation between the Karakoram and the Hindukush could be the high ranges on the west which form the drainage divide between the Yarkun river and the Ghizar river, and may be located between 72° and 73°E. With this trend in the north, this line of demarcation passes slightly west of Broghil, located between 73° and 74°E.

GLACIERS AND GLACIATION

It is an accepted fact that the Karakoram constitutes one of the most heavily glaciated regions outside the polar area. Its seven mighty glaciers, Biafo, Hispar, Baltoro, Gasherbrum, Choga Lungma, Siachen and Batura span over 350 square km. Very little research work has been done on these glaciers. These glaciers show great variation in their mechanism and no single glacier among the seven cited above can be taken as a base for interpreting their ice conditions. According to Danielli (3), the Karakoram glaciers show little evidence of any regular periodic variations corresponding with any supposed weather cycle. The velocity of the glaciers and their to and fro movements are largely based on the topographical factors and have little dependence on climatic conditions.

The lower limit of the movement of the snouts of the Karakoram glaciers as indicated by the distribution of ancient and fresh moraines, ranges from 1500 to 2000m, but the clastic deposits of fluvio-glacial origin could be transported to the lower levels in the major valleys.

Four glacial terraces are differentiated in the Karakoram. Their elevation ranges from 1000-1200m, 1800-2000m, 2500-3000m and 3500-4000m. The first two terraces are zones in which most of the population of the region is concentrated. The third one usually forms the grazing grounds during the summer months and provides temporary shelter to the people and their cattle. The last one, if well developed could also form grazing grounds during the summer months, otherwise it usually lies barren.

Karakoram contains nineteen mountain peaks over 7500m elevation, six of them are over 7900m which include K2 (8611m) with four Gasherbrum peaks and Broad peak. All of them rise within 25 km of each other above the Baltoro glacier.

The range south of this crest is known to geographers as the Lesser Karakoram and contains Rakaposhi, Haramosh and Saltoro groups, among them Rakaposhi (locally known as Dumani) occupies a dominant position at the western end of this range. Minapin (7272m), locally known as Diran, constitutes a long ridge extending for over 24 km.

GEOLOGICAL SETTING

The stratigraphic sequences encountered in the Karakoram in the western (Yasin valley) and the eastern (Ladakh area) sections will be discussed here to provide an insight into the paleogeologic conditions and associated magmatism which played their roles in raising this mighty range. In this attempt the Karakoram will be taken as an independent geographical entity. Its tectonic aspects give it a different stature which will be described separately.

In the western sections, the north – south trending Yasin and Hunza valleys which are about 70 km apart, expose more or less the same sequences of the rocks and may be considered as type sections of the western part of the Karakoram.

Among the earliest workers who conducted detailed and systematic study in these valleys the authors would like to produce the stratigraphic sequences established by Ivanac et al (4) and Stauffer (5)

Ivanac et al (1956) Hunza and Yasin valley	Age	Stauffer (1968) Hunza valley
Ladakh Granodiorite Karakoram Granodiorite Darkut Pass Granodiorite	Tertiary	Kailash Granodiorite Karakoram Granodiorite
Yasin Group Greenstone Complex Darkut Group	Cretaceous Triassic Permo–Carboniferous Precambrian	Chalt Schists Greenstone Complex Baltite Group Salkhala Series

These rock sequences were established before the application of plate tectonic concept in the northwestern sector of the Indo–Pakistan subcontinent underlying the Karakoram and the Hindukush ranges.

The rocks equivalent to the Salkhala Series have been reported by Wadia (6) Stauffer (5) and others in the vicinity of Jaglot in the Hunza valley. Recent investigations by the authors reveal them to belong to the flyschoid rocks, namely the Chalt Schists/Formation. The rocks have now been placed in the Cretaceous, and the higher grade of metamorphism is attributed to their involvement in the regional tectonics.

The Indus Suture was first described in the Psang Po area and was traced westward up to the eastern part of Ladakh. Beyond this area the extension of this tectonic line was tentatively marked and discussed by various workers and its trace in the Karakoram and Hindukush provided some confusion. Desio (1) had shown two faults, namely Shyok and Chalt (which more or less follow the surficial trace of the present Indus Suture) in the Karakoram, yet these faults were never interpreted as the extension of the Indus Suture. The first attempt to delineate the extension of the Indus suture zone in the Karakoram and Hindukush, based on systematic field work during several field visits, was made by Tahirkheli and Jan (7), in a coloured geological map of northern Pakistan. In this map the extension of the Indus suture in the Karakoram and Hindukush is named as the Northern Megashear, which subsequently was called the Main Karakoram Thrust by Mattauer et al. (8).

Under the context of plate tectonics, the rock types established by the authors in the western sector of the Karakoram, south to north in the Hunza valley, is as follows:-

Kohistan Island arc Mass:	Ladakh Granodiorite Chalt ophiolite Yasin Group
Northern Megashear	
Asian Mass	Darkut Group Karakoram Granodiorite
F A U L T	
	The rock sequences of the Tethyan Folded Belt lying towards north of the Main Karakoram Range (not discussed)

THE DARKUT GROUP

The Darkut Group is the oldest name coined in the geological literature for an assemblage of meta-pelites, psammite and carbonate rocks forming a thick sequence on the southern flank of the Main Karakoram Axial Belt. Other names assigned to this sequence in the vicinity of Hunza are "Baltit Group" (5) and "Dumordu Schists" (9).

This group so far has not been differentiated, but the authors consider it to have at least three mapable lithologies which, from the older to the younger, are: Meta-pelites; Meta-carbonate No. 1 and Meta-carbonate No. 2; the latter having suffered a relatively lower grade of metamorphism than the other two.

The Meta-pelites consist of phyllitic slate, biotite schist, garnet-staurolite schist, amphibolite and para-gneisses with alternations of subordinate quartzitic bands. This sequence is considered to be the oldest in the Darkut Group and on the south is involved in the fault tectonics of the Northern Megashear, along which it obducts the southern Kohistan Mass. Because of the higher grade of metamorphism, no fossils have been found in these rocks. This unit may have an approximate thickness of 5000 - 6000 m and its type section lies in the upper reaches of the Hunza valley, and the section starts on the south of the river near Minapin.

Overlying this unit with a gradational contact is a carbonate bed over 300 m thick, which is again well developed in the upper reaches of the Hunza valley southwest of Karimabad, located slightly away from the road. The lithology of this carbonate bed ranges from medium crystalline limestone to marble. On the weathered surfaces it is light grey, yellowish brown and pale white, whereas the fresh faces are white to light grey. It is medium to very coarse grained and thick bedded to massive. Near its contact with the Karakoram Granodiorite, the apophyses of pegmatite, aplite and minor granite bodies have intruded this sequence and imparted spinel/ruby mineralizations of quite wide extent. So far, no fossils have been reported from this carbonate sequence.

The youngest carbonate unit in the Darkut Group is semi to medium crystalline limestone, its section is exposed in the upper reaches of the Yasin Valley between Darkut village and the Darband Pass. The limestone is fine to medium grained, thin to thick bedded; and contains alternations of slates and phyllites. This limestone is fossiliferous and of the two localities from where fossils were collected, one lies in the vicinity of Darkut. On the basis of fossils, a Permo-Carboniferous age has been assigned to this unit.

The authors are not yet certain about the contact relationship of this limestone with the other two lower lithologies, because the three units may not necessarily be present in all the sections. However, since the youngest unit of the Darkut Group is dated to be of Permo-Carboniferous age on the basis of fossils, then the other two which underlie it should be older and could be Early Paleozoic or Precambrian.

The Darkut Group is developed on the southern margin of the Eurasian Plate. The authors have traced its extension right from Shyok valley (east), across the Shigar, Hunza and Yasin valleys to the western termination of the Karakoram. Among the above mentioned three lithologies, the metapelites are the thickest and show consistent development. Basic and acid igneous intrusions in the form of sills and dykes are quite common.

THE GREENSTONE COMPLEX

Greenstone complex is a field name assigned by earlier workers to a heterogeneous assemblage of rocks of dominantly green colour. According to these workers, this formation is largely composed of volcanics which were interbedded with the sedimentary rocks and have been metamorphosed to rocks rich in hornblende, epidote and serpentine minerals.

The rocks differentiated and identified are: basalt, andesite, bedded tuffs; greenschists with abundant hornblende, epidote and talc; serpentinite; epidiorite; dolerite; quartzite and marble. Stauffer (5) reports a typical greenschist to contain about 50% hornblende, 15% epidote and 30% quartz with garnet being present in very minor amounts.

A type section, which is easily accessible along the Karakoram Highway displaying typical morphological characteristics of the greenstone complex, is located south of Chalt. Its southern and northern contacts with the Ladakh Granodiorite and Darkut Group are located near Manim Das and north of Jaglot villages respectively. Its contact in the south with the granodiorite is quite sharp and is easily discernible but the northern contact with the Yasin Group is unconformable, and will be discussed later.

The authors have made observations on the greenstone complex in the Karakoram in the following sections:

(i)   South of Machelu village in the Hushe nala in the Shyok valley

(ii) At Hushupa and Tissar villages located on the eastern and western bank of the Shigar valley

(iii) Downstream of Chalt in the Hunza valley

(iv) South of Yasin village in the Yasin valley

(v)   South of Mastuj in Chitral; the western geographical limit of the Karakoram.

All these sections showed marked variation in the lithology of the greenstone complex. The greenstone complex was held by previous workers to belong to the Triassic. This age was based on the fossiliferous Darkut Group which was the next older sequence encountered in the immediate vicinity of these rock outcrops. In 1978, during a joint field programme with geologists of Montpellier University, France, a thin bedded semi-crystalline grey limestone associated with the greenstone complex near the top of the sequence was found to contain globutrancana near Tissar village in the upper reaches of the Shigar valley in Baltistan. This fossil find restricts the age range of the greenstone complex to the Middle Cretaceous. This age fits well with the concept of the shrinking Tethys, before the Indo-Pakistan-Eurasian continental collision during which flysch sedimentation was in progress, and was interrupted by the volcanic flows which interbed the sequence at the base. Pillowed lava associations with the greenstone complex further authenticate the volcanic outflow under suboceanic conditions.

The emplacement and lithological characteristics of the greenstone complex in the context of plate tectonics is more clear than was postulated by the earlier workers. It forms the base of the Yasin Group and incorporates the extension of the ophiolitic melange in the Karakoram and Hindukush part of the Indus Suture zone (called Northern Megashear or Main Karakoram Thrust), which has already been reported by earlier workers (10) at Psang Po and was traced westward up to Ladakh.

The authors assign the greenstone complex the name of Chalt Ophiolite. The name Chalt is already coined as Chalt Schists and Chalt Formation in the published geological literature yet the ophiolite sequence exposed near Chalt is easily accessible and is well-developed with the characteristic lithology of the other sequences exposed in the vicinity. Thus this section demands a special treatment to typify the ophiolitic mélange of the Karakoram belt.

All the flysch sequences including those of Chalt, having association with the Main Karakoram Thrust, have been variously named by different workers, such as Yasin Group, Chalt schists, Katzarah Formation and Burji La Formation. Similarly in the western extension of this megashear, the Reshun Formation and Drosh Orbitolina Limestone are names which have been assigned to their counterparts in the Hindukush. These are some of the known flysch occurrences but there are some more which have yet to be named. This multivarious terminology for the flysch sequences having the same depositional history and similar lithology has created problems in deciphering the regional structure and stratigraphy in the northwestern sector of the Indo-Pakistan plate, in the context of plate tectonics.

Among all the flysch deposits mentioned above, Yasin has a central location in the Karakoram and Hindukush, and constitutes one of the earliest established sequences in this part of the Indo-Pakistan subcontinent. It contains fossils on the basis of which the age assigned to it is also authentic. Its development compared to the sequences elsewhere is rather thick and lithologically well-preserved. Thus the authors are of the view that all the flysch deposits should be grouped under one terminology and the name Yasin Group is considered to be more appropriate.

Under this new proposal the name Chalt will be discontinued in favour of the new name Yasin Group proposed for all the isolated flysch occurrences in the Karakoram. Therefore in future the Chalt ophiolite will signify the type section of the ophiolitic melange associated with the Main Karakoram Thrust which is considered to be the western extension of the Indus suture zone in northern Pakistan.

THE YASIN GROUP

The Yasin Group is a name assigned (4) to a group of sedimentary and volcanic rocks which are 800-1000 m thick forming an isolated bed at Yasin village. It is developed on both the flanks of the valley and may have a strike length of 6-8 km. The rock assemblage constitutes slaty shales, slates, phyllites, phyllitic schists, limestones, sandstones and conglomerate, interbedded with the volcanics which are comprised of basalt, andesite, rhyodacite, tuffs and agglomerate. Acid and basic igneous bodies forming sills and dykes are also present. The sedimentary rocks are fossiliferous and the ones diagnosed are Hippurites, Corals and Orbitolina of Lower to Middle Cretaceous age.

Desio (11) had proposed a three-fold division of the Yasin Group, namely the lower Gojalty Formation, the Middle Manich Sandstone and the Upper Taus Shales. This division is based mainly on the frequency of lithology encountered in each formation. This division may hold good as long as observations in the field are made, but for the purpose of mapping, the three formations individually do not make a mapable unit. The authors instead suggest a two-fold division. The lower unit is comprised of thin-bedded light grey slates and phyllites, flaggy sandstones with conglomerate lenses which interbed with basalt, andesite, rhyodacite, tuffs and agglomerate; the upper constitutes semi to medium crystalline limestone and fossiliferous grey and maroon slaty shales and slates incorporating conglomerate lenses. The frequency of volcanics substantially decreases in the upper horizon.

The degree of metamorphism in the rocks of the Yasin Group and its equivalents elsewhere in the Karakoram and the Hindukush is variable. It is generally noticed that the rocks in this group at Yasin and in the western sections in the Hindukush are less metamorphosed and most of them retain sedimentary textures, but their equivalents exposed in the eastern sections of the Karakoram in the Hunza Valley and the Baltistan are relatively more metamorphosed. This pertinent metamorphic variation could be attributed to the regional metamorphism induced by the stresses generated during the wedging of the Indo-Pakistan Plate into Eurasia which created the Nanga Parbat - Haramosh loop. Most of the sections where the rocks show a higher grade of metamorphism are located in the vicinity of this mega-antiform.

The Yasin Group, for correlation purposes in the Karakoram and Himalaya will incorporate all the known sequences such as Orbitolina Limestone (Drosh) Reshun Formation, Chalt Schist/Formation, Katzarah Formation, Burji La Formation, Thalichi Schists and several others which have been discussed in the geological literature but have not yet been assigned any name. Their age is Cretaceous in the Karakoram and Cretaceous to Lower Tertiary in the Hindukush. These flyschoid rocks are the remnants of the shrinking Tethys before the continent - continent collision. These deposits are involved in the regional tectonics which have disrupted their continuity. As a result the Yasin Group is not consistently developed along the Main Karakoram Thrust.

GRANODIORITE BATHOLITHS

Two parallel east-west trending batholiths, the Karakoram Granodiorite and the Ladakh Granodiorite form two intimate magmatic associations with the Karakoram, which are encountered all along the geographical limit of the range. The Karakoram batholith has developed along the Main Axial Belt of the range, whereas the Ladakh Granodiorite marks its southern fringe. East of the Nanga Parbat-Haramosh loop, this batholith constitutes the main rock type of the Deosai-Ladakh range. The Karakoram Granodiorite starts from the high mountains of the Shyok Valley in the east and terminates southwest of Mastuj in Chitral in the west. It separates the northern Tethyan Folded Belt from the southern part of the Karakoram. The Ladakh batholith commences from the upper reaches of the Gilgit river and extends to Ladakh through the Deosai plateau in Baltistan, where it attains a great thickness. A third batholith, namely the Darkut Pass Granodiorite, has been described by Ivanac et al (4) in the upper reaches of the Yasin Valley in the Karakoram Range, who considered it a separate entity, but the others considered it a part of the Karakoram Granodiorite batholith.

The Karakoram Granodiorite intrudes the rocks of Darkut Group, dated as Permo-Carboniferous, but the authors consider part of the group to be older. This group occupies the southern margin of the Eurasian plate. The Ladakh Granodiorite, on the other hand, intrudes the greenstone complex (Chalt ophiolite) and the flyschoid rocks belonging to the Cretaceous. Thus the location of the Ladakh Granodiorite in the context of Eurasia-Indo-Pakistan Plate boundary is in the Kohistan island arc.

In general, the granodiorites are pale grey, medium to coarse, porphyritic and biotite-rich. They are massive but usually show foliations on the margins. Karakoram Granodiorite contains a lot of acid intrusives consisting of granite, pegmatite, aplite and quartz on the southern margin, which form apophyses in the Darkut Group along the contact. The Ladakh Granodiorite is rather rich in mafic minerals and displays lit-par-lit injections of pegmatite aplite and quartz near Parri, upstream of the confluence of the Gilgit-Indus rivers. Beside diorites, hornblendite and syenite also form irregular masses.

The Ladakh Granodiorite has been dated by the Rb/Sr method (13) and has yielded an age of 48 m.y. The Karakoram Granodiorite samples collected from Biafo, Hispar and Hunza areas (13) and analysed by the Rb/Sr method has given an age of 13, 24 and 8.6 m.y. respectively.

Many papers have been contributed on the geology of Ladakh, covering the eastern sector of the Karakoram, but the most recent one is that by Frank et. al. (2). This paper pertains to the geological mapping of about 25,000 sq. km. area in Ladakh and has come up with some new data on its geology and two suture zones.

In this paper, the rock types are divided into the following six zones, which from north to south are:-

(i)   Ladakh intrusives (Ladakh Granodiorite) of Upper Cretaceous to Eocene and which are predominantly tonalitic in character.

(ii) The Indus Molasse; Hemis conglomerate of Upper Tertiary age.

(iii) The Indus Flysch, Namica La Flysch and the Lamayuru Flysch of Upper Cretaceous to Paleogene.

(iv) Dras Volcanics.

(v)   Ophiolitic Melange.

(vi) The Zanskar platform overlain by the Paleozoic to Mesozoic sediments.

REGIONAL TECTONIC SETTING

The existence of a calc-alkaline magmatic island arc on the southern margin of the Karakoram has given a new dimension to this sector of the Indo-Pakistan Plate under the purview of the global plate tectonics. Earlier, Kohistan had been differentiated as a tectonic zone of the Karakoram by Desio (1).

This arc separates the northern marginal mass of the Indo-Pakistan Plate from the Karakoram constituting the southern margin of the Eurasian continent. Only regional tectonics are intended to be discussed here, therefore, for a comprehensive overview of the geology of the Kohistan arc, the authors recommend readers to obtain guidance from the following publications: Tahirkheli et al. (14), Tahirkheli and Jan (15) and Tahirkheli and Jan (7).

It was mentioned earlier that the single Indus suture located at Psang Po, in its western extension at Ladakh, bifurcates into two sutures, named by Frank et. al. (2), as the Indus suture and the Shyok suture. The southern of these two sutures is named the Main Mantle Thrust by Tahirkheli et. al. (14) and the northern suture is called the Northern Megashear by Tahirkheli and Jan (7) which was subsequently named as the Main Karakoram Thrust by Mattauer et al. (8).

The building of the Kohistan island arc had started in Mesozoic, most likely during late Jurassic by northward intraoceanic subduction of the oceanic part of the Indo-Pakistan Plate. This process continued till the end of the Paleocene.

The Tethyan fossiliferous sedimentary remnants, emplaced along the Main Karakoram Thrust indicate the closure of the sea during the Cretaceous and the arc-continent collision in the late Cretaceous period. Typical ophiolitic sequences are recorded at the Chalt and Yasin sections, elsewhere they are present in isolated patches, and in several sections have been disrupted by shearing.

The southern suture, called the Main Mantle Thrust, welds the Indian Plate with the island arc. The northwestern Indo-Pakistan continental margin is subducted under the arc. The closure of the sea along this suture zone was probably post-Paleocene and the island arc-continent collision occurred some time between Lower and Middle Eocene. An east-west trending fossiliferous sedimentary wedge containing Paleocene faunas in the trench - arc gap is one of the major pieces of evidence in support of this view.

The sequential order of the deformational history in the Karakoram with respect to its southern neighbours, the Kohistan arc and the Indo-Pakistan Plate may be summarised as below:-

a. Building of the Kohistan island arc by the intraoceanic subduction in the Mesozoic, most likely starting in the late Jurassic and continuing until the end of the Paleocene.

b. The first contact of the arc with the Eurasian Plate by under-thrusting of the Kohistan island arc along the Main Karakoram Thrust during the Post-Cretaceous period.

c. The Indo-Pakistan Plate and the arc collision between the Lower and the Middle Eocene along the southern suture, the Main Mantle Thrust.

d. Formation of the Nanga Parbat-Haramosh loop during post-Mid-Eocene, due to subsequent stresses generated by wedging of the Indo-Pakistan Plate into the Kohistan island arc. These stresses were responsible for deforming the Main Mantle Thrust and generating regional meta-morphism in the surrounding rocks.

e. Creation of the Main Boundary Thrust during Middle to Upper Miocene.

f. Formation of the Hazara-Kashmir syntaxis during Miocene-Pliocene orogenies.

g. The last major tectonic scar created in the frontal part of the Karakoram-Himalaya is called the Marghala Thrust by Tahirkheli (16) disrupting the Murrees (Mid. Oligocene - Mid. Miocene) and the Siwaliks (Pliocene - Pleistocene) which on the basis of Palaeomagnetic studies (Keller et. al. (17)) has been dated to be 1.8 m.y. old.

REFERENCES

1) DESIO, A., (1964). Tectonic Relationship between the Karakoram, Pamir and Hindukush (Central Asia) XXII IGC, Part XI, 1964, pp 192 - 213.

2) FRANK, W., GANSSER, A. and TROMMSDORF, V., (1977). Geological Observations in the Ladakh Area (Himalayas); a preliminary report. SMPM, 57, pp 89 - 113.

3) DANIELLI, G., (1914). Italian Expedition to the Himalaya and Karakoram Vols. I-XIII, Bologna.

4)   IVANAC, J. F., TRAVES, D. M. and KING, D., (1956).   The Geology of the N. W. Portion of the Gilgit Agency.   GSP Vol. VIII, part 2, pp 1 - 27.

5)   STAUFFER, K. W., (1968).   Geology of the Gilgit-Hispar Area, Gilgit Agency, West Pakistan.   USGS/GSP.   PK-19, pp 1 - 38.

6)   WADIA, D. N., (1932).   Note on the geology of Nanga Parbat and adjoining portions of Chilas Gilgit district, Kashmir.   Rec. Geol. Surv., India, Vol. 66, part 2.

7)   TAHIRKHELI, R. A. K. and JAN, M. Q., (1979).   A preliminary geological map of Kohistan and the adjoining area, Northern Pakistan. Peshawar University Report.

8)   MATTAUER, M. et. al., (1979).   Main Karakoram Thrust.   Personal communication.

9)   DESIO, A. and ZANETTIN, B., (1956).   Sulla Costituzione Geologica del Karakoram Occidentale (Himalaya) Cong. Geol. Intn. XX Session, Section XV.

10)   GANSSER, A., (1974).   The ophiolitic melange a worldwide problem and Tethyan Example.   Eclogea. Geol. Helv. 67/3, pp 479 - 507.

11)   DESIO, A., (1963).   Review of the geology formations of the western Karakoram (Central Asia) Riv. Ital. Paleont. and Stratigr. 69, Milano, pp 475 - 501.

12)   MISCH, P., (1949).   Metasomatic granitization of batholithic dimension. Amer. Jour. Sci. Vol. 247.   No. 10, pp 673 - 705.

13)   DESIO, A., TONGIORGI, E. and FERRARA, G., (1964).   On the geological age of some granites of the Karakoram, Hindukush and Badakshan (Central Asia) XXII Sess. IGC. India, 11, pp 479 - 496.

14)   TAHIRKHELI, R. A. K., MATTAUER, M., PROUST, F. and TAPPONNIER, P., (1979).   The India-Eurasia suture zone in Northern Pakistan: synthesis and interpretation of recent data at plate scale.   In Geodynamics of Pakistan GSP, pp 125 - 130.

15)   TAHIRKHELI, R. A. K. and JAN, Q. M., (1978).   Geology of Kohistan, Karakoram Himalaya, Northern Pakistan.   Cent. Excell. Geol. Pesh. Univ., pp 1 - 187.

16)   TAHIRKHELI, R. A. K. (in press).   Major tectonic scars of Peshawar Vale and adjoining areas, and associated magmatism.   Proc. Int. Comm. Geod. Cent. Excell. Geol. Pesh. Univ. Pakistan.

17)   KELLER, H. M., TAHIRKHELI, R. A. K., MIRZA, M. A., JOHNSON, G. D., JOHNSON, N. M. and OPDYKE, D. N., (1977).   Magnetic polarity stratigraphy of the Upper Siwalik Deposits, Pabbi Hills, Pakistan, Earth Planet. Sci. Letters, 36, pp 187 - 201.

Geology of the South Central Karakoram and Kohistan

M.P. Coward,* Q.M. Jan,*** D. Rex,* J. Tarney,**
M. Thirwall,* and B.F. Windley.**

* Leeds University, U.K., ** Leicester University, U.K.,
*** Peshawar University, Pakistan

ABSTRACT

The Indus suture in northern Pakistan splits into two branches that bring rocks of diverse nature in contact. The northern suture consists of a mega-mélange passing northwards into pelitic and calcareous rocks which show a rapid increase in grade of metamorphism away from the suture. These rocks are intruded by the calc-alkaline Karakoram batholith. Sedimentary rocks to the north of the batholith are mostly Palaeozoic in age and are dominated by slates and shales with some calcareous and arenaceous rocks, and extend up to the Chinese border. The southern suture, with associated ultramafics and a wedge of blueschists, is bordered on the south by the rocks of the Indian plate consisting of a basement of psammites and schists intruded by Cambrian granites and overlain by isoclinically folded and low grade metamorphosed cover of carbonates and shales.

Between the two sutures occur rocks of the Kohistan sequence represented by, from north to south, tightly folded pillow-bearing volcanics and sediments; early foliated and late post-tectonic tonalites and diorites with small but abundant intrusions of aplites and pegmatites and basic to inter-mediate dykes; the Chilas complex, a stratiform cumulate body over 300 km long and 8 km thick; an amphibolite belt with a complex mixture of other rocks; and the Jijal complex, a 200 km^2 tectonic wedge of high-pressure mafic granulites and chromite-layered ultramafics. The Chilas complex is folded by an F_1 isoclinal anticline and the upper part of the Kohistan plutonic belt by a major F_2 syncline. The whole of the Kohistan sequence with its two phases of isoclinal folds was tilted during the Himalayan collision so that the structures are now subvertical.

INTRODUCTION

Northern Pakistan is remarkable in that it hosts three of the world's most prominent but geologically little known mountain chains – the Karakoram, Himalaya, and Hindukush. The newly constructed Karakoram Highway provides a complete section across the NW Himalaya and Karakoram whilst the Hindukush cross section can be studied along the Dir-Chitral road. This Karakoram-Kohistan region was investigated locally in the north by the British Geological Survey of India (1) (2), followed by the Geological Survey of Pakistan since 1947 (3). A number of Italian expeditions under the leadership of A. Desio have also provided significant information on the Karakoram Mountains over the past 50 years. The southern part of the region – Kohistan proper – has been locally investigated by geologists from Punjab University and from the Geological Survey of Pakistan, and more extensively and intensively by geologists from Peshawar University led by R. A. K. Tahirkheli.

The Karakoram-Kohistan terrain, having the highest relief in the world, is difficult, unsympathetic, and geologically complex. A number of foreign earth scientists along with their Pakistani colleagues are actively involved in unravelling the complex history of continental collision, rock accretion, metamorphism, uplift and cooling which has occurred during the last 100 million years or so. It is only recently that a more coherent picture has begun to emerge as new data on various aspects of the geology of this area have become available.

The Indus suture, which separates the Indian and the Eurasian plates (4), divides westwards into two, the northern suture (Main Karakoram Thrust, MKT) and the southern suture (Main Mantle Thrust, MMT). These enclose the Kohistan sequence which is bounded on the west by the Chaman transform fault. The precise timing of continental collision and of the closure of the suture in the Karakoram-Himalaya is uncertain and radiometric dates range from 85 to 5 m.y. Although continental collision culminated in the Miocene-Pliocene, tectonic activity is still in progress. The Himalayas are considered to have reached great heights in the late Pleistocene. The uplift rate over the last 25 m.y. has been estimated to be 0.7 - 0.8 mm per year (5), whilst for the Mt. Everest area Chinese scientists have estimated a rise of 10 mm per year for the past 100,000 years; see (6).

In this paper we describe the principal structural pattern of the Kohistan sequence between the two sutures and the principal units of rocks from north to south along the Karakoram Highway (KKH) between Khunjerab and Abbottabad (Fig. 1).

THE GEOLOGY OF THE REGION

The geology of the Karakoram-Himalayan region in northern Pakistan can be grouped under three headings:

1. The Eurasian Plate

2. The Kohistan Sequence

3. The Indian Plate

These three associations of rocks have distinctly different lithologies and tectonic settings and are separated by the two branches of the Indus suture. Both the sutures are marked by the occurrence of a mélange including ultramafic rocks, the southern one also having a wedge of garnet granulites considered to have recrystallized at more than 40 km depth (7). The rocks making up the Kohistan sequence, between the two sutures, are predominantly calc-alkaline plutonics and volcanics with subsidiary volcano-sedimentary and sedimentary rocks. Tahirkheli et al. (8) and Bard et al. (9) have suggested that the Kohistan sequence represents the crust and uppermost mantle of an extended island arc turned on end during the collision of the Indian-Asian landmasses. Our most recent studies

(10) have shown that the structure of the area is too complex for such a simple interpretation and requires a detailed analysis before a final conclusion can be reached about its nature.

The Eurasian Plate

In the Karakoram area, Gansser (11) distinguished three tectonic zones: i) a northern Karakoram Tethyan zone, ii) a central metamorphic zone with plutonic rocks, and iii) a southern volcanic schist zone. Among these, the last one is now considered to be a part of the Kohistan sequence occurring to the south of the MKT, whilst the first two occur to the north of the MKT.

i) Rocks of the northern sedimentary zone occupy the northern part of the Hunza Valley between Pasu and Khunjerab and have been divided by Tahirkheli (12) into six principal formations:

The Misgar Slates occur in the northernmost part of the Hunza Valley. They constitute a 3500 m thick monotonous sequence of slates and phyllitic slates with thin carbonaceous and quartzitic bands. They are intruded by an early phase of gabbros and diorites and a later phase of more abundant quartz porphyrys, granites, aplites and pegmatites. The siliceous igneous rocks are especially common along the Pakistan–China border in the Khunjerab area where they form a linear belt, not yet fully investigated. The Misgar Slates have been assigned a Palaeozoic age by Tahirkheli (12).

The Kilik Formation (13), 2000 m thick, consists of limestone and dolomite with thin partings of arenaceous slates and red quartzite. On the basis of crinoidal stems, they have been assigned to a Devonian–Lower Carboniferous age and correlated with the Shogram Limestone of the Chitral (14).

The Gircha Formation, exposed in the vicinity of Khaiber, is 4000 m thick (14). The upper part is argillaceous with thin calcareous intercalations, whilst the lower part is sandstone with intercalation of calcareous-arenaceous shales. Brachiopods, corals, and bryozoans of Lower Permian age have been found in this formation.

The Pasu Slates (15) reach a maximum thickness of over 1500 m and occur around Pasu, Gulkin and Batura. They consist of slates and phyllitic slates, with dark limestone and subordinate quartzite and have been placed in the Upper Palaeozoic. They have a tectonic contact with the Karakoram batholith and the Gujhal formation.

The Gujhal Formation is a 500 m thick mass of dolomite, dolomitic limestone and limestone with thin bands of siliceous schists and two conglomerate horizons, one at the base and the other in the middle part. The formation is exposed in the Batura and Pasu areas and has been assigned a Permo-Triassic age (14).

The Shanoz Conglomerate, occurring between Pasu and Khaiber, consists of calcareous fragments in a siliceous matrix. It has been considered to be the lower Cretaceous equivalent of the Reshun conglomerates in the Chitral area.

ii) The Karakoram Batholith. The batholith constitutes an arcuate belt of plutonic rocks up to 10 km broad and over 350 km long. A whole range of textural and lithological variations occur. Gneissose biotite granodiorite is the principal rock-type of the batholith but adamellite, quartz diorite and hornblende granodiorite also occur. These have been intruded by granitic, aplitic and pegmatitic dykes and rare dolerites. Along the KKH, the following sequence has been recorded: Gneissose biotite granodiorite with an early foliation is cut by coarse-grained adamellite/granite and pegmatite-aplite. The pegmatite-aplite may be banded, with some thin bands rich in either garnet or tourmaline and at places tightly folded. These are intruded by grey, medium-grained adamellite and by a younger phase of medium-grained leucogranite. The final phase of intrusions are aplite-pegmatites with rare beryl.

Radiometric ages on the rocks range from about 50 to 7 m.y. (16). Four K/Ar ages on biotite determined by us range from 47 to 5 m.y. At least some of the young ages might reflect metamorphic/uplift events. The data of Blasi et al. (16) show that the granodiorite-adamellite rocks have low ratios of $(Na_2O+K_2O)/Al_2O_3$ and $(Na_2O+K_2O)/CaO$, typical of calc-alkaline compositions found in orogenic zones.

iii) <u>Rocks to the south of Karakoram Batholith.</u> These are mainly a series of pelitic sediments starting in the northern suture zone and showing a rapid northwards increase in grade of metamorphism through chlorite-biotite slates, garnet-chloritoid phyllites with syntectonic garnets and garnet-staurolite schists with prominently rotated garnets (17), to silli-manite schists. Other rocks found in this zone are biotite-garnet gneisses (having migmatitic aspect in the north), calc-silicate rocks, amphibolites, thinly bedded grits and graphitic schists. An important unit of this zone is marked by marbles containing sapphire, ruby corundum, and a variety of other minerals including an exceptionally aluminous pargasitic amphibole. The corundum-bearing rocks occur in a 50 km long by 4 km broad belt, but the corundum is of low quality (18).

Okrusch et al. (19) have suggested that the corundum-bearing marbles crystallized at a total fluid pressure of about 7 Kbar at around 600° C, the temperature increasing up to 700° C towards the northwest. This is consistent with our finding the staurolite-out metamorphic zone to the south of the marble occurrence. Deformed two-mica granites form augen gneisses interlayered with metapelites, although there are also late tectonic granites which cross-cut the structure and are gently folded, and others which cross-cut but are undeformed.

In the neighbouring Ishkoman valley to the west, garnet-staurolite schists have not been found. Here the principal rocks are conglomerates, rhythmically banded black graphitic shales and slates, with minor quartzites, calcareous rocks, gritstones and phyllites.

The Northern Suture

In the section from Hunza to Chalt there is an almost chaotic arrange-ment of large lenses, each several kilometres long and several tens of metres wide, of limestone, sandstone, conglomerate and mafic and ultra-mafic rocks in a matrix of chloritoid slates. The basic rocks with prominent volcanic breccias and greenschists are rich in epidote, chlorite and actinolite. The ultrabasics consist of serpentine, talc-chlorite schists, talc-carbonate schist, calcite-chlorite schists, chromite-chlorite schist, and minor relict harzburgite. Ultrabasic masses are apparently more abundant to the west of Chalt. There are large lenses of quartzite which may have formed <u>in situ</u> or which may be tectonic blocks and lime-stone intermixed with other sediments. The whole assemblage has the appearance of a major melange with no simple repetitions, as expected in an imbricate zone. The structures in the high grade metamorphic rocks contrast with those in the main melange to the south in which the transport direction was mainly updip as seen from mineral lineations and folds with curvilinear hinges.

This tectonic zone is considered to mark the suture between the Kohistan sequence and the Eurasian plate to the north. There is no evidence of blueschists, of obducted high pressure granulites or of an ophiolite, but instead large tectonic lenses of a melange.

The Kohistan Arc

There is a tectonic break between rocks of the northern suture and the volcanics and sediments belonging to the Chalt Group to the south which make up the northern crust of the arc. This group contains meta-greywackes and slates, epidotic grits and tuffs, hornblende-bearing tuffs, chlorite schists, schistose amphibolites, amygdaloidal pillow-bearing basalts (Fig. 3A) and fragmental basic volcanics. The total thickness of deformed and weakly metamorphosed sediments and volcanics reaches several kilometres but this may involve repetitions by folding and thrust-ing. The rocks are folded by large upright, tight to isoclinal anticlines and synclines which plunge east or west and they are cut by thin granitic dykes and by muscovite pegmatites which are discordant to both cleavage and bedding.

Further south near Gilgit and Raikot, there are graded psammites and pelites and locally thick piles of deformed pillow lavas, but these occur as screens between large plutons of diorite and tonalite. South of the Chalt Group there is a plutonic belt making up the middle to lower crust of the arc which contains in a southward sequence:

a) Tonalites and Diorites with quartz, plagioclase, biotite and hornblende. They vary from mafic to felsic and there is more than one generation; early ones may be foliated and gneissic and latter ones are discordant and homogeneous with undeformed xenoliths of igneous and sedimentary material. In any one section there may be up to five generations of intrusion (Fig. 3B). The rocks change rapidly from outcrop to outcrop and there are large discrete plutons. The diorites tend to be weakly to non-foliated, occasionally foliated in narrow shear zones, and a few have intrusion breccias with angular fragments of gneiss, tonalite, pegmatite and calc-silicate pods.

b) Aplites and Pegmatites which often form impressive swarms of sheets up to 50m thick, locally making up over 30% of the rock volume, especially near the confluence of the Gilgit and Indus rivers. They are post-tectonic, most dip gently and have a N-S strike. Pegmatites cutting the metasediments have muscovite and garnet, but those cutting the tonalites have biotite with rarer garnet. There are also interlayered aplite-pegmatite dykes.

c) Basic dykes which are most common in the Gilgit Valley. They range from a few metres to over 10m thick, mostly trend NE and dip steeply south, range from ultrabasic to intermediate in composition and their metamorphism varies with the deformation. Almost all dykes post-date the foliation of the tonalites, but many are parallel to the foliation. Some are undeformed, some are deformed throughout, and a few are folded with an axial planar cleavage parallel to the regional foliation of the tonalite, but generally the foliation is parallel to the dyke margins and oblique to the regional fabric. The foliation of the dykes is shown by chlorite, biotite and hornblende, but often there is post-tectonic growth of hornblende.

d) <u>The Chilas Complex</u> is a vast stratiform cumulate body over 300 km long and 8+ km thick. It contains an upward sequence of hypersthene gabbro, major chromite-layered dunite, norite, gabbro, minor troctolite, harzburgite and dunite, and at the top, norite. Particularly impressive are rhythmically-alternating phase-graded cumulate layers up to 10 cm thick, slump folds, syn-sedimentation faults, and sedimentary breccias. Some layers up to about 30 cm thick are of almost pure anorthosite. Both homogeneous and layered rocks are cut by dykes of pyroxene-hornblende anorthosite. The lower dunites are up to 1 km thick and contain 3 m thick compact chromitite seams. All these rocks show evidence of several phases of deformation. Isoclinal folds in norites have hypersthenes orientated in axial planar fabrics and the penetrative mineral fabric in the norites is parallel to the axial planes of folded pyroxene amphibolite dykes. These relationships suggest a tectonic origin for the main mineral fabrics in the complex.

e) <u>An Amphibolite Belt</u> which contains rather a wide range of lithologies. There are banded amphibolites with or without garnet, hornblendites, hornblende schists, garnet gabbros, anorthositic gabbros and anorthosites, diorites, tonalites and granites and thin garnet quartzites and calc-silicate lenses. The proportion of amphibolite is commonly low. The belt is distinctive in that most intrusive rocks are concordant and parallel to the regional trend and have been intensely deformed, many of the coarser leucocratic types becoming augen gneisses.

f) <u>The Jijal Complex</u> between Patan and Jijal which occupies about 200 km^2. In the north are garnet-clinopyroxene-plagioclase rocks containing relics of norite, and so it is likely that these are high pressure metamorphic equivalents of the Chilas complex. The grain size is similar to that of the norite but garnets continued to grow after the deformation and locally grew to over 8 cm especially in leucocratic veins (Fig. 3C). Hornblendites may contain hornblende-garnet, garnetite and garnet plagioclase. Towards the southern boundary of the complex there is an increase in proportion of clinopyroxenes and hornblendites, until the main ultramafic body is reached, which consists of clinopyroxenites, and dunites which have lenses of layered chromitite up to 5 m thick. Jan and Howie (7) concluded that both the granulites and the dunites suffered granulite grade metamorphism at $690 - 770^{\circ}$ C and $12 - 14$ kb and at $800 - 850^{\circ}$ C and $8 - 12$ kb respectively. The Jijal Complex is probably a tectonic fragment of the Chilas Complex that was subducted or down-thrusted to a substantial depth against the MMT.

Structure of the Kohistan Arc

The structure of the Kohistan Arc is dominated by a series of EW-trending, tight to isoclinal, upright folds with hinges plunging to the NE (Fig. 2). There is a westward-facing anticline near Gilgit and a major westward-facing syncline south of Jaglot. This latter structure has a half wavelength of well over 50 km; with its numerous parasitic folds it affects the rocks from Gilgit to Jijal. This is one of the largest tight synclines yet known deforming the whole crust. It is a second phase structure which deforms an early foliation in some of the tonalites and metasediments. In the Chilas Complex the mineral fabric is folded

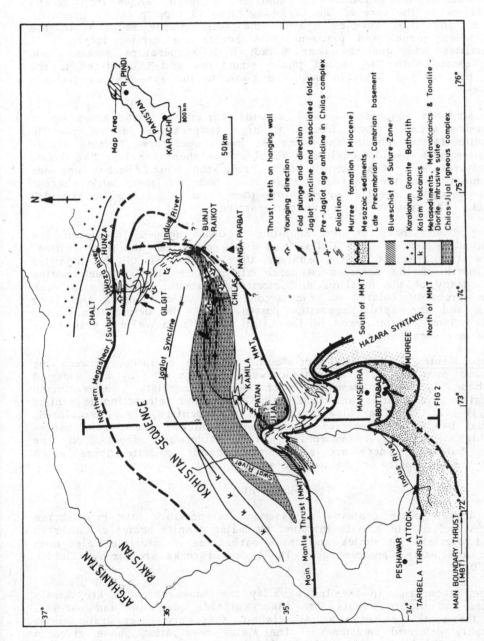

Fig. 1. Regional map showing the main lithologies and structures
in the Kohistan région.

by an earlier isoclinal fold phase which duplicates the stratigraphy in an F_1 anticline. There should be a complementary isoclinal syncline passing through the amphibolite belt between the Chilas and Jijal complexes.

The metamorphic grade in the Kohistan sequence ranges from green-schist facies in the core of the Jaglot syncline to high pressure granulite facies in the Jijal Complex against the MMT in the south. In the Jijal Complex garnet and pyroxene grow across the earlier fabric. In the tonalites and gneisses near Raikot high temperature metamorphism is synchronous with the second phase structures and is confined to the central part of the Kohistan sequence close to the syn- to post-tectonic intrusives.

The Kohistan sequence contains a wide variety of shear zones up to a kilometre wide. Some formed at high temperatures giving rise to amphibolite facies schists and some at low temperatures forming ultra-cataclasites with pseudotachylite. NE of Chilas there are late NNE-trend-ing fractures and shear zones which bring the Nanga Parbat gneisses into contact with the Kohistan sequence. Some of these affect recent alluvial deposits, locally overturning them and producing slickensides on faults. The area is still tectonically active.

The igneous rocks of the Kohistan sequence formed throughout the deformation sequence. The Chilas and Jijal complexes, the amphibolites, tonalites and granites which separate them, and some of the tonalites in the north of the Kohistan sequence all predate the first deformation phase. Many of the tonalites and diorites, however, show only a weak foliation probably related to the second deformation phase, and some tonalites and the aplite-pegmatites post-date all the deformation. The basic and intermediate dykes in the Gilgit and Hunza valleys even post-date the late shear zones.

The Main Mantle Thrust (MMT) or Southern Suture contains, between the Indus and Swat Valleys, a tectonic wedge up to 10 -15 km wide of glaucophane schists, piemontite schists, greenschists, serpentinised harzburgites and dunites and local talc-schists and talc-actinolite schists (20) (21) (22). The blueschists and greenschists are intercalated, individual bands being several tens of metres thick. These rocks under-went high pressure metamorphism at considerable depths on the Southern Suture. There are a few post-tectonic dolerite dykes which crosscut the schistosity in the tectonic wedge.

The Indian Plate

The Indian plate contains a basement of possibly late Precambrian psammites and biotite schists intruded by major granite bodies of Cambrian age and overlain by shales and thick carbonates of mid-late Palaeozoic, Mesozoic and early Cenozoic age. The basement rocks are strongly folded (Fig. 3D).

a) The Basement. In the Indus valley the basement rocks are mainly psammites and biotite schists with minor sulphide quartzites and marbles, but to the south, as near as Abbottabad, there are lower grade slates and weakly deformed mudstones. The slates near Attock have given a K/Ar age of 554 ± 14 my (23). North of Abbottabad the slates are intruded by large bodies of coarse porphyritic homogeneous cordierite granite, the most important of which is the Mansehra granite, which has a Rb/Sr whole rock isochron age of 516 ± 16 my (24). This forms part of a belt

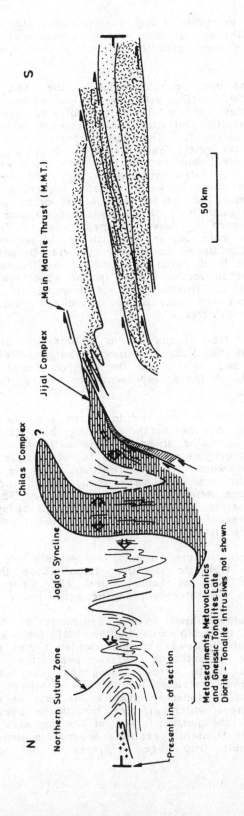

Fig. 2. Simplified section through the Kohistan sequence. Ornament as Figure 1.

of Cambrian granites extending along the southern side of the Himalaya. NW of Mansehra a cleavage in the slates is cross-cut by granite veins in the aureole of the main granite indicating that some deformation is of Precambrian age.

b) The Cover. The basal conglomerates of the cover sequence can be seen 8 km SW of Besham resting with angular unconformity on the basement gneisses. The pebbles contain quartzite, foliated granite and diorite, granitic pegmatite, phyllite and mica schist, most of which can be matched with the underlying basement. The conglomerates are a few metres thick and pass upwards into grits, and then into graphitic sulphide-bearing schists which are several hundred metres thick and which also contain some conglomerates. The latter are matrix-supported and contain pebbles of both basement and cover rocks. Similar sequences of sediment, but with variable thicknesses of marble and schist occur as synclinal keels in the basement throughout most of the Swat-Indus-Kunar region. Further south the cover sequence is less prominent, has suffered lower grade metamorphism, and is more calcareous with thick sequences of limestone near Abbottabad. Limestones containing Silurian-Devonian fossils occur near Nowshera, 35 km N of Peshawar, and Carboniferous nautiloids occur in limestones in Buner in southern Swat in the Abbottabad area. Calkins et al. (25) group all the limestones as of Mesozoic age. The correlation of these different types and ages of cover rocks on the northern margin of the Indian plate is uncertain.

South of a line from Peshawar to Abbottabad and north-east of Abbottabad adjacent to the Hazara syntaxis along the Kunar valley there are red sandstones and shales of the Murree Formation which is of Oligocene-Miocene age. These represent the early Himalayan molasse deposits.

The overall structure of the Indian plate consists of a wide zone of intense deformation in the north affecting both cover and basement rocks in the form of a major shear zone beneath the overriding Kohistan thrust slab. This deformation, which is of Himalayan-collision age decreases southwards where it is confined to narrow zones; near Abbottabad it is in the form of a major thrust which carries the gneisses over Mesozoic calcareous metasediments to the south. This Tarbela thrust is probably a continuation of the Main Central Thrust system of northern India (4) (24). These Mesozoic rocks are carried across the Oligocene-Eocene molasse-type deposits by the MBT.

The sediments of the Murree Formation show only one phase of intense deformation producing tight to isoclinal folds above the Kunar Valley. These structures also fold the Tarbela and MBT thrusts into the major Hazara Syntaxis, which is a steeply plunging anticline.

The grade of metamorphism in the sediments of the Indian plate increases from south to north towards the MMT from almost unmetamorphosed rocks south of Peshawar to greenschist facies near Abbottabad, to sillimanite-kyanite amphibolite facies near the MMT. Calkins et al. (25) defined a single series of isograds, but our studies suggest that there was regional metamorphism of the basement schists prior to intrusion of the Cambrian Mansehra granite, as well as regional metamorphism of Himalayan age, which affected the Mesozoic cover, as well as the basement rocks in the northern part of the region. An E-W trending shear zone crosses the Mansehra granite, separating undeformed porphyritic granite to the south from augen gneisses to the north, as well as

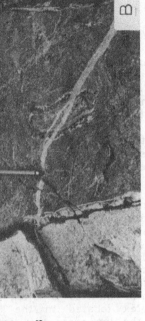

FIG. 3

A and B

C and D

Metavolcanic rocks with epidote lenses, south of Chalt. Rakaposhi is dominantly made up of the Chalt volcanics.

Zoned calc-silicate lens in biotite schist, cut by aplitic intrusions. 17 km SE of Gilgit.

Anorthosite vein cutting a folded vein with garnet porphyroblasts in host garnet granulite. South of Patan.

Z-folds in the gneisses of the Indian plate in the vicinity of the Main Mantle Thrust near Jijal.

separating strongly metamorphosed cover to the north from weakly metamorphosed cover to the south.

DISCUSSION

A cross-section through the Karakoram-Kohistan-Indian plate sequence
is shown in Fig. 2. The MMT is clearly the most important structural
boundary separating rocks of different origin and type and high pressure
blueschists and granulites are localised along it; we consider this to
be the main Tethyan suture. In contrast, the chloritoid slates of the
Hunza valley pass right through and across the northern suture, from
which we conclude that this suture represents a closed small marginal
basin north of the Kohistan arc. This is supported by the structural
evidence which indicates no major shear zone or uplifted lower crustal
rocks along this northern suture.

According to our interpretation of the structure the volcanic rocks
at Chalt and the sediments at Gilgit are folded around the Jaglot syncline
by a second phase structure, and similar amphibolites of probable volcanic
origin are repeated to the south of the Chilas complex by a first phase
structure. This would mean that the granulites of the Chilas complex
represent the lowermost unit of the Kohistan sequence, whereas the
granulites of the Jijal complex represent a tectonic slice of the bottom
of the Chilas complex and the chromite-layered dunites a slice of the
upper mantle. The base of the Chilas complex located in the centre
of the first phase anticline is not seen, but the top contact is intrusive
into sediments and amphibolites. The fact that a granulite-grade layered
complex forms the base of the crustal sequence makes it comparable to
the Ivrea Zone of the Italian Alps (27).

The earliest rocks in the Kohistan sequence seem to be the gabbros
and basic volcanics and associated greywackes of the Gilgit-Chalt area
and their tectonically repeated equivalents to the south. These were
intruded by voluminous early tonalites and diorites, generally foliated
and gneissic. These rocks have not been observed in the Chilas complex,
and its intact cumulate stratigraphy would prevent the passage of granitic
liquids of such a scale. We therefore conclude that the Chilas complex
is one of the latest igneous rocks that pre-dates the first phase major
structures and that it was emplaced into the lavas and the sediments
when they were still in a sub-horizontal position.

The whole Kohistan sequence and its two phases of tight isoclinal
folds was tilted, so that the structures now appear vertical. For a
discussion on the origin of this tilting see Coward et al. (10). Many
of the late tonalites, diorites, granites and pegmatites were intruded
after the intense earlier F_1 and F_2 deformation. Many of the intermediate
to basic dykes are synchronous or post-date shear zones which are later
than the tilting. Therefore much of the northern part of the Kohistan
sequence is late post-tectonic and it is not easy to relate it to the
primary development of the arc sequence. This conclusion is supported
by the wide range in age of the granitic bodies in the region near and
west of Gilgit from 54 my to 19 my (28). Geochemistry will no doubt
throw light on the mode of origin and source materials of this wide
spectrum of intrusive rocks.

Acknowledgements

We would like to thank Professor R. A. K. Tahirkheli for his considerable help with logistic support and the RGS for inviting M. Q. Jan to participate in the Karakoram Project. Financial support from the Natural Environment Research Council (Grant GR3/4242) is gratefully acknowledged.

References

(1)   HAYDEN, H. H. 1915. Notes on the geology of Chitral, Gilgit and Pamirs. Rec. Geol. Surv. India 45, 271 - 335.

(2)   WADIA, D. N. 1933. Notes on the geology of Nanga Parbat (Mnt. Diamir) and adjoining portions of Chilas, Gilgit district, Kashmir. Rec. Geol. Surv. India 66, 212 - 233.

(3)   IVANAC, J. F., TRAVES, D. M. and KING, D. 1954. On geology of the northwest region of Gilgit Agency. Rec. Geol. Surv. Paki. III, 3 - 27.

(4)   GANSSER, A. 1980. The significance of the Himalaya suture zone. Tectonophysics, 62, 37 - 52.

(5)   MEHTA, P. I. C. 1980. Tectonic significance of the young mineral dates and the rates of cooling and uplift in the Himalaya. Tectonophysics, 62, 205 - 217.

(6)   JACKSON, P. 1980. On the Tibetan Plateau. Nature 287, 486 - 487.

(7)   JAN, M. Q. and HOWIE, R. A. 1981. The mineralogy and geochemistry of the metamorphosed basic and ultrabasic rocks of the Jijal complex, Kohistan, NW Pakistan. J. Petrol., 22, 85 - 126.

(8)   TAHIRKHELI, R. A. K., MATTAUER, M., PROUST, F. and TAPPONIER, P. 1979. The India-Eurasia suture zone in Northern Pakistan: synthesis and interpretation of recent data at plate scale. In Geodynamics of Pakistan. Farah, A., & DeJong, K. A., (Eds) Geol. Surv. Pakistan, Quetta, 125 - 130.

(9)   BARD, J. P., MALUSKI, H., MATTE, P. and PROUST, F. 1980. The Kohistan sequence: crust and mantle of an obducted island arc. Geol. Bull. Univ. Peshawar. Sp. Issue 13, 87 - 94.

(10)   COWARD, M. P., JAN, M. Q., REX, D., TARNEY, J., THIRWALL, M. and WINDLEY, B. F. (in press). Structural evolution of a crustal section in the Western Himalaya. Nature, London.

(11)   GANSSER, A. 1964. Geology of the Himalaya. Wiley Inter-science. London.

(12)   TAHIRKHELI, R. A. K. 1979. Geology of Kohistan and adjoining Eurasian and Indo-Pakistan continents, Pakistan. In Geol. Bull. Univ. Peshawar, Sp. Issue 11, 1 - 30.

(13)   DESIO, A. and MARTINA, E. 1971. Geology of Upper Hunza valley, Karakoram, W. Pakistan. Bull. Soc. Geol. Ital. 91, 284 - 314.

(14)   DESIO, A. 1963. Review of geological formations of the West Karakoram (Central Asia). Riv. Ital. Pal. Estr., 69, Milano, 475 - 501.

(15)   SCHNEIDER, H. J. 1959.   Tektonic und magmatisms in NW-Karakorum. Geol. Rundschau 46, 426 - 476.

(16)   BLASI, A., BRAJKOVIC, A., BLASI, C. De POL, FORCELLA, F. and MARTIN, R. F. 1980.   Contrasting feldspar mineralogy, textures and composition of Tertiary anorogenic and orogenic granites, Gilgit area, northwestern Pakistan.   N. Jb. Miner. Abh 140, 1 - 16.

(17)   POWELL, C. McA. and VERNON, R. H. 1979.   Growth and rotation history of garnet porphyroblasts with inclusion spirals in a Karakoram schist.   Tectonophysics, 54, 25 - 43.

(18)   FARUQI, S. H. 1978.   Hunza ruby deposits and their genesis. Paper presented in Int. seminar on Min. Expln., Technol., Peshawar, Dec. 1978.

(19)   OKRUSCH, M., BUNCH, T. E. and BANK, H. 1976.   Paragenesis and petrogenesis of a corundum-bearing marble at Hunza (Kashmir). Mineral. Deposita, 11, 278 - 297.

(20)   DESIO, A. 1977.   The occurrence of blueschists between the middle Indus and the Swat Valleys as an evidence of subduction (North Pakistan). Renc. Accad. Naz. Lincei, 62, 1 - 9.

(21)   SHAMS, F. A., JONES, G. C. and KEMPE, D. R. C. 1980.   Blue-schists from Topsin, Swat district, NW Pakistan.   Mineralog. Mag. London, 43, 941 - 942.

(22)   JAN, M. Q. and SYMES, R. F. 1977.   Piemontite schists from Upper Swat, northwest Pakistan.   Mineralog. Mag. 41, 537 - 540.

(23)   KEMPE, D. R. C., 1978.   Acicular hornblende schists and associated metabasic rocks from northwest Pakistan.   Mineralog. Mag. 42, 405 - 406 and M33 - M36.

(24)   LE FORT, P., DEBON, F., and SONET, J. 1980.   The "Lesser Himalayan" cordierite granite belt, typology and age of the pluton of Mansehra, Pakistan.   Geol. Bull. Univ. Peshawar. Sp. Issue 13, 51 - 62.

(25)   CALKINS, J. A., OFFIELD, T. W., ABDULLAH, S. K. M. and TAYYAB ALI, S. 1975.   Geology of the southern Himalaya in Hazara, Pakistan and adjacent areas, US geol. Surv. Prof. Pap. 716C, 29p.

(26)   VALDIYA, K. A. 1980.   The two intracrustal boundary thrusts of the Himalaya. Tectonophysics, 66, 323 - 348.

(27)   RIVALENTI, G., GARUTI, G., ROSSIE, A., SIENA, F. and SINIGOI, S. 1981.   Existence of different peridotite types and of a layered igneous complex in the Ivrea zone of the western Alps. J. Petrol., 22, 127 - 153.

(28)   CASNEDI, R., DESIO, A., FORCELLA, F., NICOLETTI, M. and PETRUCCIANI, C. 1978.   Absolute age of some granitoid rocks between Hindu Raj and Gilgit Rivers (Western Karakoram). Rend. Accad. Naz. Lincei, Ser. 8, 53, 204 - 210.

Glaciology

Ice-depth radio echo-sounding techniques employed on the Hispar and Ghulkin glaciers

G. K. A. Oswald

ABSTRACT

The impulse radar technique used originally to sound the temperate Vatnajökull, Iceland ice cap and successfully employed on the Hispar and Ghulkin glaciers during the Karakoram Project is reviewed. Possible improvements to the system are discussed, with particular reference to the transmitting and receiving systems and to their implementation in the 'Subscan' impulse radar system. Attention is given to techniques for integrating successive output signals in order to reduce the effective noise level, and the statistical theory of a fast and simple digital integration technique is given. A brief description of the Subscan digital signal discrimination and recording systems concludes the paper.

INTRODUCTION

In the early stages of the Karakoram Project, the intention was to operate from a helicopter over the glacier, in order to achieve very wide coverage with a minimum of logistical effort. This would require a compact radar system, capable of penetrating the ice without extensive antennas. However, it was realized that the availability of a helicopter should not be absolutely relied upon, and that it would be wise to have in addition a system for use on the surface of the glacier. This was the equipment developed by Cumming, Ferrari and Owen[1] at the Cambridge University Engineering Laboratories and used successfully on the Vatnajökull ice cap in Iceland[2]. At this time, work was beginning at Sensonics Ltd. on the Subscan impulse radar system, which would be capable of operation from a helicopter, and included features designed to improve the perform- ance over that of existing equipment. In the event, the decision to take the "Vatnajökull" equipment was shown to be more than justified. All the real ice depth measurements made during the Project were made using this equipment, and though the experiments performed with the Subscan equipment were useful and will contribute to its modification for future work, we were not successful in using it to penetrate the temperate glacier ice.

THE VATNAJÖKULL EQUIPMENT

The Vatnajökull equipment has been described in detail elsewhere[1], but a brief account is in order here. The system consists of transmitting and receiving electronic subsystems with their respective antennas, and recording and monitoring equipment in the form of an oscilloscope and Polaroid camera.

The transmitter is triggered internally and generates a unipolar current pulse using an avalanche transistor, with a charged delay line as the element which defines the duration of the impulse. The impulse is approximately $0.5\mu S$ long and a peak output power of several hundred watts is delivered.

The transmitting and receiving antennas were developed from designs originated by Wu and King[3] and consist of wire dipoles, loaded with resistors along their length. The design has undergone various changes since the first experiments in Iceland, resulting in a physical simplification which has been shown empirically to have little effect on performance. Their main characteristic is that of very broad frequency response, arising from the damping achieved by the resistive loads. The result is a very clean and 'ring-free' pulse.

The receiver is triggered by detecting the direct signal from the transmitter, which travels along the ice surface. In order to avoid saturation of the receiver amplifier, the received signal is blanked for a time after the initial reception of the direct pulse. The horizontal sweep of an oscilloscope is also initiated by the direct pulse, and the received signal amplitude is displayed as the 'Y'-deflection of the CRT trace. The equipment therefore displays the difference in delay between the direct pulse and the echo from the base of the glacier. Simple geometry can then be used to derive the actual range of the echo, assuming a value for the velocity of electromagnetic waves in ice.

The equipment is simple to operate and maintain, and has shown excellent reliability over several field expeditions. Its limitations are those of a cumbersome physical construction for the operation on a moraine covered glacier and a receiving system which could be improved in sensitivity. Other improvements could be made resulting in increases in complexity which might or might not be justifiable in view of the particular application. The antenna could be improved, from the standpoint of the operator, by allowing easier deployment. The present antennas are constructed from heavy wire using resin-encapsulated loading resistors at intervals along their length. Under the conditions of a rough-surfaced glacier, these have to be unrolled and then rewound for each sounding station, which operation occupies the greater part of the time taken for each sounding. A spring-loaded canister containing a lighter, more flexible wire would greatly improve the speed with which soundings could be taken.

It has been found in different circumstances that the Vatnajökull system is capable of sounding temperate ice to depths up to 900 m. It would be of value to increase the range of the system. This must largely be done by attention to the receiving and recording systems.

Some improvement could be achieved by using an amplifier with time-varying gain. In this way, instead of switching on the amplifier after gating out the high-amplitude direct pulse, the amplifier could be gradually brought up to its maximum gain, at which level the system noise would occupy a significant proportion of the available amplitude and weak signals returning from reflectors at long range could be detected more easily. Such an amplifier would be well within the capability of existing technology, and would represent a useful enhancement of the system.

THE SUBSCAN SYSTEM

The transmitter

We will here make a brief reference to the Subscan transmitting antenna, which is described more fully in Appendix A. The technique of broadening the frequency response of a dipole antenna by means of resistive loading has been mentioned above. It operates essentially by controlling the current distribution in the conductor in order to eliminate reflections from the ends. However it has the disadvantage of dissipating much of the input power in the load resistors. The Subscan transmitter was designed to take advantage of the principle, while avoiding the inefficiency, by using active devices placed along the conductors in place of the loading resistors. As a simple first design, it was decided to use a loop antenna, where the current could be uniform around the loop, controlled in elements very short compared with the wavelengths involved, but with a total dimension of the same order as those wavelengths.

The antenna could be shown to generate the correct form of current pulse, but facilities were not available to test its performance fully over the range of pulse widths required. It is now felt, in addition, that the complexity implied by such design is not justified in view of the relatively small gain in efficiency of about 6 - 8 dB which would, at best, be available.

The signal enhancement system

The Subscan receiver achieves a more ambitious level of signal enhancement and though adding greatly to the complexity of the system, would be capable of greatly improving its performance. Though designed to operate with the transmitter mentioned above and with signals of much higher frequency than those used in the Vatnajökull equipment, it could be used in the same way with the low-frequency antenna and amplifier. The enhancement is achieved by a technique of integrating the signals received from successive pulses, thus enhancing coherent echo signals over the incoherent system noise. The Subscan receiver system is based on a statistical approach to electronic signal analysis, on the principle that no signal is undetectable except in the sense that it is obscured by unknown signals. These may be caused by thermal noise in conductors, by thermally generated radiation, or by broadcast sources. At present we will combine all these under the term 'noise'. In the case of radar, this can be stated as follows:

For a one-shot system, a signal is said to be present if the instantaneous detected amplitude at any given delay exceeds, by a specified factor, the average expected noise, that is

$$V_{ti} > \kappa V_{rms}$$

where V_{ti} is the instantaneous amplitude at time t and V_{rms} is the root mean square noise amplitude. In this case κ will be of the order 2.

In most systems, some degree of averaging of successive receiver output traces is achieved, allowing improved reliability of signal detection, effectively improving the signal-to-noise ratio. In many cases the averaging process is done by superimposition of successive traces on an oscilloscope, using the persistence of vision of the eye to derive a reference noise distribution and detecting small signals as shifts in that distribution. This process may be approximated by the condition for signal detection:

$$V_{ti} > \kappa \varepsilon V_{rms}/\sqrt{n}$$

where n is the number of traces produced within one 'response time' of the eye, and ε is an unknown, < 1, representing the subjective efficiency of the eye in discriminating shifts in the observed intensity distribution.

This averaging process can, of course, in principle be carried out quantitatively using various techniques of integration, both analogue and digital.

(i) Analogue integration

(a) Delay-line integration.

This is a technique where the analogue signal is stored as a travelling wave in a looped transmission line. The signal is synchronised with the travel-time round the loop, and a new return is added to the stored signal for each traverse. This system has many disadvantages, in particular arising from drift, reflections and attenuation inherent in the repeated use of the delay line. It has the advantage of efficiency in terms of signal energy.

(b) Sampling integration

A less efficient, but simpler system is that of sampling and holding the instantaneous signal received in a 'window' of the trace, and integrating over successive returns. The window is scanned slowly from one extreme of delay to the other, and a low-frequency integrated analogue of the return is built up. Though simple, this method is slow, and wasteful of signal energy, since only a small portion of each return contributes to the final output. This approach has been used by Theodorou[4].

(ii) Digital integration

Digital integration itself is a simple matter of arithmetic adders and data storage devices, with appropriate digital control circuitry. The fundamental technique central to digital integration is the conversion itself, from analogue to digital, of information in the sampled signal.

The main variables in this technique refer to the precision and rate of recurrence of the conversion process, both of which contribute to the data rate capacity required of the converter. The precision required defines the number of data bits contained in the binary number into which the received signal voltage is converted. This may vary from one to, normally, 16, and is in practice generally greater than or equal to 4. Inaccuracies in the conversion can generally be said to add to the effective noise level, since information is lost concerning the instantaneous value of the signal voltage. However, the more precise the conversion, the longer it takes, and the longer must be the interval between samples.

The advantages of digital integration are those of stability (implying high limits on the useful period of integration) and accessibility to sophisticated control, analysis and display.

(iii) Binary integration

The Subscan receiver employs a method of digital integration which has received relatively little attention, and is known as <u>binary</u> or <u>single-bit</u> integration. In this technique, the multiple-bit <u>A/D</u> converter of the conventional system is replaced by a simple voltage comparator giving a single binary bit as its output, whose state depends on the polarity of the analogue input signal. This apparent lack of precision in the instantaneous conversion process can be justified as follows:

We have already seen that the process of integration is only of importance when we are dealing with a signal which is not large compared with the ambient noise. For most of such signals, the instantaneous signal amplitude arising from the coherent radar signal is small compared with that due to the random fluctuations; we can therefore say that accurate knowledge of the amplitude is irrelevant since the majority of that instantaneous signal carries no useful information. In this case, the 'signal' has no measurable existence as an instantaneous voltage, but only as a tendency in the statistical distribution of amplitude for that value of delay.

Thus what is required of the integrating system is not accurate instantaneous A/D conversion, but a sensitive detector of coherent variations in the distribution of amplitude. The first statistical parameter to be considered is the central tendency of the noise distribution. This may most simply be characterised by a voltage level such that the instantaneous signal level exceeds it for 50% of the time. Thus for a large number of samples of the signal, 50% would be expected to exceed that value.

It can quickly be seen that, if the sampler output is fed to a voltage comparator whose reference input is held at this 'null' level, the output will be expected to be '1' for 50% of samples. If the comparator output is summed digitally over a large number of N samples, the result will be expected to be $S = N/2$, with a standard error of $(N/2)^{1/2}$. The number S is a measure of the central tendency of the distribution around the null level (V_o).

We can see that, if a constant shift is applied to the distribution at a particular delay, while maintaining the reference input at V_o , the number S will change according to the magnitude of the shift and the exact form of the noise distribution. Let us now investigate the 'efficiency' of the number S as an indicator of the probable underlying signal strength. 'Efficiency', εs is defined here as the ratio of the precision of estimate of signal strength obtained by this indicator after N samples, compared with the analogue integration technique. For that technique, the signal-to-noise ratio is

$$\sum v_s / \sum v_n = N v_s / \sum v_n$$

For a normal distribution, $\sum v_n \to N^{1/2} v_{rms}$ as $N \to \infty$

so $\sum v_s / \sum v_n \to N^{1/2} v_s / v_{rms}$

Efficiency of the Single-bit Integrator

We will assume a general distribution of amplitude due to noise, $p_n(V)$ such that the probability of V lying between V_1 and $V_1 + \delta V$ is $p_n(V_1)\delta V$, and where $\int_{-\infty}^{\infty} Vp_n(V)dV = 0$ and $\int_{-\infty}^{\infty} p_n(V)dV = 1$. We now superimpose a small signal, which is coherent with the radar repetition interval, $V_s(t_k)$ and therefore a function of the delay time t_k. The amplitude distribution is now

$$p(V,t_k) = p_n(V-V_s(t_k))$$

Let us now assume a particular form for the noise distribution p_n such that

$$p_n(V) = A.\exp(-V^2/(2V_o^2))$$

where V_o is an rms value and $A = 1/(V_o\sqrt{2})$

Superimposing $V_s(t_k)$,

$$p(V) = A.\exp(-(V-V_s(t_k))^2/2V_o^2)$$

For a large number of samples, N, the number S may now be deduced from

$$\frac{S}{N} \rightarrow \frac{\int_0^{\infty} p(V)dV}{\int_{-\infty}^{\infty} p(V)dV} \qquad \text{as } N \rightarrow \infty$$

or, for finite samples,

$$\frac{S}{N} = \frac{\sum_{n=0}^{\infty} Np(V_n)\delta V}{\sum_{n=-\infty}^{\infty} Np(V_n)\delta V}$$

where $V_n = n\delta V$.

The distribution of S is defined by the binomial distribution

$$p(S) = \frac{N!\left[\int_0^{\infty} p(V)dV\right]^S}{S!(N-S)!\left[\int_{-\infty}^{\infty} p(V)dV\right]^{N-S}}$$

We now have a new 'signal', S, which is derived from the original radar returns via the integrating process, and has a mean value

$$\bar{S} = N\int_0^{\infty} p(V)dV$$

and standard deviation

$$\sigma = \left[N \int_0^\infty p(V)\,dV \int_{-\infty}^0 p(V)\,dV \right]^{1/2}$$

$$= \left[\frac{\bar{S}(N-\bar{S})}{N} \right]^{1/2} = N^{\frac{1}{2}} \left[\frac{\bar{S}}{N} \frac{(N-\bar{S})}{N} \right]^{1/2}$$

The 'signal-to-noise ratio' achieved by this process is determined as follows:

Consider a superimposed coherent signal $V_s(t_k)$, giving an initial signal-to-noise ratio

$$\rho n = \frac{V_s(t_k)}{V_{rms}}$$

This produces a slight shift in the mean value of S such that

$$\frac{S}{N} = \int_0^\infty A\exp\left[-\frac{(V-V_s(t_k))^2}{2V_{rms}^2} \right] dV$$

We define the new signal-to-noise ratio as

$$\rho s = \frac{|\bar{S}-\bar{S}_0|}{\sigma} = |\bar{S}-\bar{S}_0| \left[\frac{N}{\bar{S}(N-\bar{S})} \right]^{1/2}$$

where $\bar{S}_0$ is the mean value of S for $V_s = 0$, normally 0.5. It can be seen that for a fixed ratio $\bar{S}/N$, the signal-to-noise ratio

$$\rho_s \propto N^{1/2}$$

as would also be expected for the case of analogue, or multi-bit digital integration. Let us consider the case where $V_s = +V_{rms}$. Here S/N is 0.841 (from standard tables of the cumulative error function) whilst $S_0 = 0.5$. Thus

$$\rho s = \frac{0.34 N^{1/2}}{(0.84 \times 0.16)^{1/2}} = 0.93 N^{1/2}$$

For $V_s = V_{rms}/10(-20dB)$, S/N = 0.54

Hence

$$\rho s = \frac{0.04 N^{1/2}}{(0.54 \times 0.46)^{1/2}} = 0.08 N^{1/2}$$

For $V = V_{rms}/2$, $S/N = 0.69$

Hence $\rho s = \dfrac{0.19N^{1/2}}{(0.69 \times 0.31)^{1/2}} = 0.41N^{1/2}$

For $V_s = 2V_{rms}$ $S/N = 0.97$

Hence $\rho s = \dfrac{0.47N^{1/2}}{(0.97 \times 0.03)^{1/2}} = 2.75N^{1/2}$

When we re-convert from the digital sum $\bar{S}$ to real signal voltages, we arrive at an efficiency figure for the process in the following way:

For $V_s = +V_{rms}$, $S/N = 0.84$, the probable value of V is V_{rms} and the standard deviation $\pm \sigma/N = 0.37/N$ corresponds to an uncertainty given by

$$\delta V = \sigma/(N^{1/2}P)$$

where P is the probability density of V_s. Thus the S/N ratio is

$$\frac{V_s}{\delta V} = \frac{PV_s}{\sigma/N^{1/2}}$$

The efficiency of the conversion can be said to be

$$E = \frac{PV_{rms}}{\sigma}$$

This tends to $P/0.5$ as $V_s \to 0$, or 0.798 (≈ 0.8).

As a function of V_s/V_{rms}

$$E = \frac{\dfrac{1}{V_o\sqrt{2\pi}} \exp(-V_s^2/2V_o^2)}{\dfrac{1}{V_o\sqrt{2\pi}}\left[\displaystyle\int_{V_s}^{\infty} \exp(V^2/2V_o^2)dV \int_{-\infty}^{V_s} \exp(-V^2/2V_o^2)dV\right]^{1/2}}$$

Using the four values previously chosen we have:

for $V_s = 2V_{rms}$, $P = 0.055/V_{rms}$, $\sigma = 0.156$, $E = 0.35$

for $V_s = V_{rms}$, $P = 0.24/V_{rms}$, $\sigma = 0.367$, $E = 0.65$

for $V_s = V_{rms}/2$, $P = 0.35/V_{rms}$, $\sigma = 0.46$, $E = 0.76$

for $V_s = V_{rms}/10$, $P = 0.39/V_{rms}$, $\sigma = 0.50$, $E = 0.78$

The efficiency of conversion for this noise distribution tends to be approximately 0.8. This limit is a function of the noise distribution employed. A rectangular distribution such that $P = 0$ for $V < -V_o$ and $V > V_o$ and $P = 1/2V_{rms}$ for $-V_{rms} < V < V_o$ would give rise to efficiency of 1 for $-V_{rms} < V_s < V_{rms}$, but a restricted upper limit to the signals which can usefully be sampled in this way. (This would be realised by a triangular wave of fixed amplitude.)

It can be seen that for signals comparable with or less than the noise amplitude, a single-bit sampling system can provide an efficient method of conversion and integration. There are various problems which are more serious for this method of sampling than for analogue, or multi-digit systems. The first is that a signal which is coherent with the transmitted pulse, but arises not from reflectors which are of interest, but from internal reflections or signals within the system, may still be large compared with the noise distribution, and prevent a fixed-reference, single-bit comparator from usefully integrating or averaging the random noise. Secondly, the 'transfer function' of the system, relating the signal voltage V to the output S, depends entirely on the total random noise distribution observed in the received signal. This includes a stationary thermal noise distribution, and a component arising from broadcast radio interference which varies over periods comparable with the period of variation of real echo signals in the received signal train. Thus it may be necessary to introduce a known noise source into the received signal to reduce the effect of such variations. Radio interference can also be reduced by improving the directionality of the receiving antenna.

Signal discrimination

The receiving system is, however, a cheap method of both enhancing signals and converting them to digital form. This in turn lends itself to automatic recording, including the possibility of a degree of intelligence in the recording system in terms of discriminating real echoes and processing them in order to deduce further information (for instance in a moving system, the processor could perform direct deconvolution of the received echoes.

A first step towards such an intelligent system was implemented in the Subscan receiving system. A digital signal related to the quantity S mentioned above was passed to a microprocessor-controlled digital recorder developed at the British Antarctic Survey by G. Musil. The processor was programmed to monitor the incoming measures of S, and to compare them with tables to deduce the likelihood of the presence of a signal, and its amplitude. The value passed to the processor represented the difference between the observed value of S for each window of the received trace, and a 'null' value which was stored in the receiver unit and could be updated. In this way it was possible to cancel the effect of stationary system-generated signals of low amplitude.

The processor then took each value and read from a table the probability of its occurring by chance, without the presence of a coherent signal. The system could be set to accept different levels of probability as a threshold below which there was said to be a signal present within the time window, and would write on to tape the range corresponding to such windows, and the probable amplitude of the signals as compared with the rms noise level.

The processor also could be programmed to demand navigational or identification data to record with the range and amplitude information, thus allowing all necessary information to be stored in digital form on a single tape cartridge. This system can also be shown to operate satisfactorily, and represents a powerful addition to the existing tools of radio-echo sounding.

REFERENCES

1. CUMMING, A. D. G., FERRARI, R. L. and OWEN, G. (1980). <u>Electronic design and performance of an impulse radar ice-depth sounding system on the Vatnajökull ice-cap, Iceland,</u> Proc. KK Conference Vol. 1.

2. BISHOP, J. F., CUMMING, A. D. G., FERRARI, R. L. and MILLER, K. J. (1980). <u>Results of impulse radar ice-depth sounding on the Vatnajökull ice-cap, Iceland,</u> Proc. KK Conference Vol. 1.

3. WU, T. T. and KING, R. W. P. (1965). <u>The cylindrical antenna with non-reflecting resistive loading,</u> IEEE Trans. on Antennas and Propagation, AP-13, p. 369.

4. THEODOROU, E. A. (1979). <u>The design of a radar sounding system for rocks and soils,</u> Ph. D. Dissertation, University of Cambridge.

ACKNOWLEDGEMENT

The author gratefully acknowledges financial support from the Natural Environment Research Council.

APPENDIX A

The Subscan Transmitter

The unit was designed to exploit the principle of controlling, at points close together compared with the waveband employed, the current density in a radiating conductor, without the low efficiency inherent in a resistively-loaded antenna. In the Subscan transmitter the loading resistors were replaced by active devices which could achieve the same degree of control of currents with reduced power dissipation. A second advantage, in principle, of this approach is that all signal feeding can be done at low level between well-matched sources and loads, and the actual power to be transmitted can be delivered directly to the radiating conductors by mounting the active devices on the conductors themselves. This form of distributed transmitter is particularly attractive in view of recent developments in rugged solid-state high frequency devices which can be linearly controlled. In view of the limited funds and time available, the simplest geometry was used to realise the principle; that is, a circular hoop of sufficient diameter to allow a good radiation impedance, divided into arcs, each of which would have a resonant frequency above the band in use. The loop had a diameter of 1 m, and was divided into 16 sectors. Four VMOS FET's were mounted on each sector, allowing a maximum pulse current around the loop of 12A and a voltage swing of up to 80V on each.

Fig. A1. illustrates the connections at each sector, showing how the pulse current I_p travels around the loop, but is effectively isolated from the radial conductors which carry the drive signal and d.c. power from the control circuits at the centre of the loop, to the peripheral radiators. In Fig. A1.; T_n represents a VMOS device allowing an 80V peak drain-source voltage and 3A peak pulse current. A number of devices can be connected in parallel in order to increase the effective current. The capacitor C_n maintains the d.c. rail potential at each device, and also forms a low impedance path for high-frequency currents passing around the loop. A small damping resistor R_n is needed, since the device output capacitance was sufficient to bring the resonant frequency of each element rather close to the signal frequency band.

The waveform produced at the drains of the FET's had the form shown in Fig. A2. The waveform first dips as the current through the inductor formed by the radiator increases, and then peaks as this is rapidly switched off. The linear nature of the active elements means that the circuit is capable of handling different drive signals as produced by the central drive unit. In this case, the pulse is shaped by allowing the loop current to increase slowly and then switching off rapidly, producing a high-voltage pulse of about 80V in each sector, equivalent to 1280V when summed around the loop. This is possible using only a 12V supply rail so that no dangerous power supplies were necessary. The pulse has the form shown in Fig. A2. By slowing the cut-off of the current, after allowing it to grow for a longer period, a slower impulse can be generated, while maintaining the amplitude of the voltage generated around the loop. This however has the disadvantage of reducing the radiation impedance and therefore the efficiency of the transmitter.

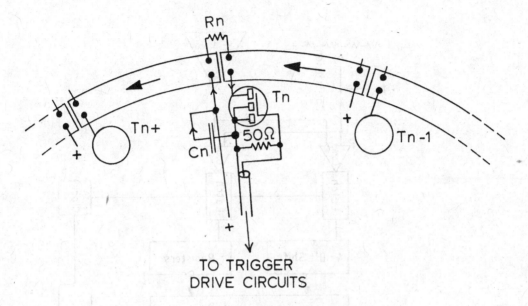

FIG. A1. Part of the Subscan transmitter antenna loop

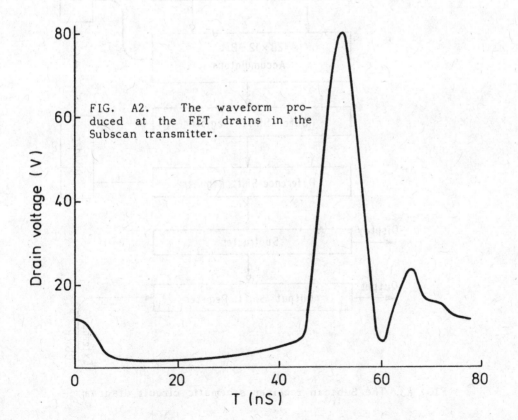

FIG. A2. The waveform produced at the FET drains in the Subscan transmitter.

98

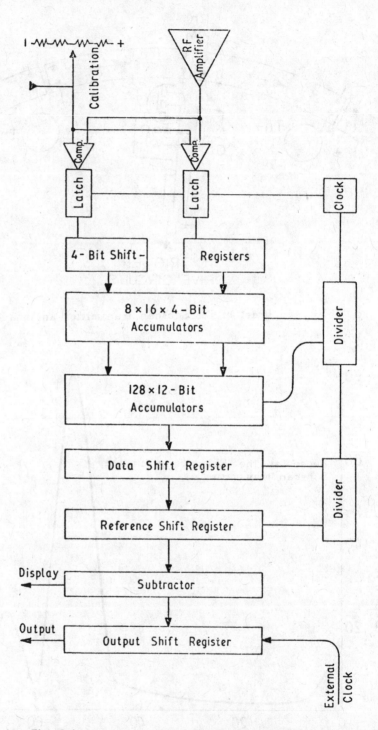

FIG. A3. The Subscan receiver schematic circuit diagram.

APPENDIX B

Details of the Subscan integrating receiver

The unit is capable of sampling at rates up to 100 MHz, and of recognising signals with duration longer than 2.5 nS. A total number of up to 65536 radar returns can be integrated, giving a theoretical reduction of 48 dB in the level of the minimum detectable signal. The unit feeds the most significant 12 bits of the integrator sum for each window to the microprocessor, and also outputs a D/A conversion of the top 8 bits. X-sweep and blanking signals are also provided for a monitor oscilloscope. The unit uses ECL, TTL and MOS circuitry. The clock rate is derived from a voltage-controlled multi-vibrator, and can be varied at will.

The data output is independent of the integrating process (that is it operated from a strobe pulse provided externally), subject to the condition that the correct number of words of data are requested by the processor for each cycle.

APPENDIX C

The Processor

The microprocessor-controlled recording system used an INTEL 8086 microprocessor, and was developed using an SDK-86 system design kit. The processor is a 16-bit device, capable of handling the 12-bit words from the integrator and performing rapid 16-bit arithmetic operations. The program written allows options giving either total recording of the output data, or selection of time windows containing significant signal levels. Navigation or station identification data can also be recorded. The system used a PEREX 8041 data cartridge recorder.

Impulse radar ice-depth sounding on the Hispar glacier

Dong Zhi-Bin – University of Lanzhou P.R.C.; **R. L. Ferrari** – Trinity College, Cambridge
M.R. Francis – BHRA, Cranfield; **G. Musil** – British Antarctic Survey;
G. K. A. Oswald – Cambridge Consultants Ltd ;
Zhang Xiangsong – Institute of Glaciology and Cryopedology, Lanzhou, P.R.C.

ABSTRACT

The work carried out by a radio-echo sounding team on the Hispar glacier is described. Results of impulse radar depth-soundings at eleven different sites are given. The measurements are discussed in relation to the effects of surface moraine and melt-water upon the echo signal.

INTRODUCTION

As part of the Karakoram Project a glaciology programme was carried out in two locations, the Hispar and the Ghulkin glaciers, and with two essential objectives. These were, to perform depth surveys of the two glaciers over as wide an area as possible, and to pioneer the use of new and sophisticated equipment to improve the efficiency of the radio echo-sounding technique. The Hispar work is described here whilst the Ghulkin results are given in an accompanying paper (1).

The Hispar expedition was based at a camp on the north side of the glacier between the mouths of the Khiang and Phumarikish glaciers, somewhat west of the Dachigan grazing ground as shown in Fig. 1. The camp was established on July 11 after leaving the Base Camp at Aliabad on July 5. Supplies and equipment were carried by 40 porters without serious mishap. An ablation valley some 70 – 100 metres above the surface of the glacier provided the site for the camp. A laboratory tent was set up, and a radio link to Aliabad established before testing the equipment. No obvious malfunctions were found to have developed during the walk in to the camp, and on July 17, an equipment (see Cumming, Ferrari and Owen (2)), previously used on the Vatnajökull ice cap, Iceland, by Bishop et alia (3) was taken on to the glacier for a live test. The results of the first two soundings at Sites 1 and 2 (see Fig. 1) approximately 400 and 700 metres from the side of the glacier are shown in Figs. 2(a) and 2(c). It appears that an echo is present in both cases at approximately 5 μs delay. If we assume that these are not due to an invariant internal signal, it would appear that they represent reflections from some point other than the side-walls of the glacier valley, since the delay has not changed greatly with a significant change in sounding position across the valley. As an illustration, if we hypothesise a parabolic shape for the valley cross-section, we could expect three normal reflections, of which that from the 'bottom' of the U would be least affected in range by changes in sounding position across the glacier.

An alternative explanation for these signals would be a strong reflection from points about 500 - 600 metres along the glacier surface, either up or down the glacier from the sounding point. Though this possibility was not disproved in the first day's test, it was considered unlikely, as was that of a fixed internal signal at 5 µs delay. These were in fact both discounted in the subsequent series of soundings, when the 'echo' could be seen to change in delay with our position, but not so as to indicate a centre of reflections near the surface of the glacier. It was significant that a technique which had worked on the relatively open and even surface of Vatnajökull in Iceland was also proved to be effective in the conditions of a moraine-covered glacier.

Tests were then made using a new 'Subscan' equipment (4) in an attempt to duplicate the results of the Vatnajökull set. This equipment was intended for use on cold ice, with a short, high resolution impulse. We were not successful in distinguishing an echo from the base of the glacier, and though this would have been unlikely in temperate ice, it was apparent that the characteristics of the receiving antenna and amplifier were not compatible with the high dynamic range provided by the rest of the system. In attempts to improve the performance, the impulse length was increased, and adjustments were made to the receiver, but without success.

A series of soundings with the Vatnajökull system was then undertaken, forming one transverse cross-section, and beginning a longitudinal profile up the glacier. These were successful, and indicated a depth of approximately 600 metres in places. When the return walk to the sounding point became too long for worthwhile work to be performed in a day, a subsidiary camp was established on the glacier itself some 6 km from the main camp. Further soundings were taken until it became apparent, after echoes had been detected from a depth of over 600 metres, that no echo was recognisable. The reason for this is probably that the depth of the ice had simply grown too great for successful penetration. In Iceland the equipment penetrated successfully over 900 metres. If this is the correct cause for the loss of echo, it must be inferred that the transmission properties of the Hispar ice are worse, because of a combination of content and salinity, by a factor of approximately 3:2 in terms of dB per metre, or a considerable margin in the coupling of the impulse into the ice. This is not unreasonable since the dielectric properties of ice depend radically upon the water content and salinity. (It should be remembered that the surface of the glacier, at which much ablation takes place in the summer, is covered with moraine which would provide a source of such salinity.)

However there is a small possibility that the loss of echo was caused by a malfunction of the equipment, but further investigation of this loss of penetration on the Hispar glacier was curtailed by injury and logistic problems.

An attempt was made to use the signal enhancing unit from the Subscan equipment to improve the Vatnajökull system's dynamic range. However this represented a fairly major modification for which the facilities available were minimal, and successful operation in this configuration was not achieved. It is to be expected, however, that with sufficient preparation the system could be used in this way with greatly improved performance, and automated recording.

The first objective of the Hispar expedition was achieved, though with much smaller extent of coverage than had been hoped. This was because of the logistic problems of movement, with equipment, over the difficult surface of this glacier.

The objective of performing experiments with new equipment in the real sounding environment was achieved, but with largely negative results. However, since the Vatnajökull equipment works well, and the signal enhancement and recording subsystem of the Subscan set was seen to operate correctly, we may conclude that a high-performance system could be achieved by a combination of the two.

THE HISPAR RADIO-ECHO DATA REDUCTION

General

During the course of the time spent on the Hispar Glacier, eleven independent ice-thickness measurements were taken, using the 'Vatnajökull' radio-echo sounder (2) at the sites shown in Fig. 1. A microprocessor

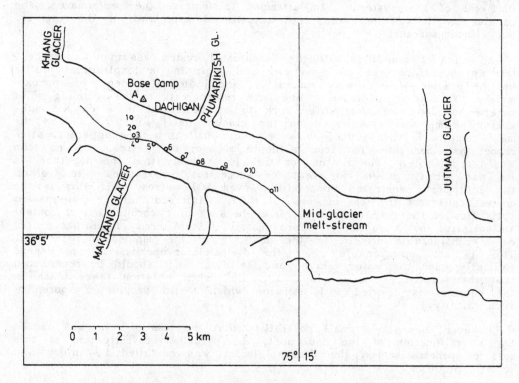

FIG. 1. Map of the Hispar glacier showing the positions of sounding sites 1 - 11 (based on E. E. Shipton's map (7)).

computing facility enabled ice thickness data to be evaluated at the Dachigan base camp. The broken-up and moraine-covered glacier surface limited the choice of possible measurement sites. Neither the presence of surface water, nor the existence of surface moraine up to a metre in depth seriously impaired the operation of the impulse sounder, and several of the ice thickness measurements were made under such conditions.

Data Reduction

The radio echo returns were recorded in the 'A-mode' on Polaroid film. A proprietary, lightweight, low-power consumption 'Scopex' oscilloscope was modified and employed for this purpose. A 'Tektronix' oscilloscope had been used to calibrate the Scopex timebase and to measure its time delay in X-deflection triggering. The A-mode photographs (Figs. 2 and 3) were used to evaluate the difference between the arrival times of the direct surface wave which triggers the Scopex timebase, and that reflected from beneath the glacier.

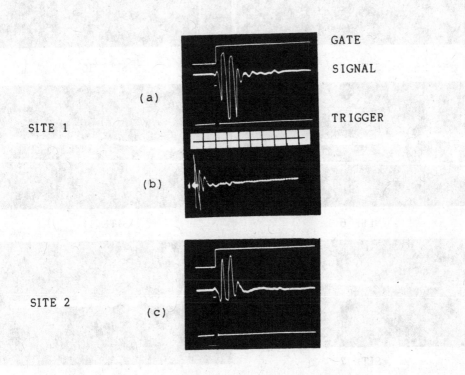

FIG. 2.   Radio-echo results for Sites 1 and 2:

      (a) Site 1 using a Tektronix oscilloscope.

      (b) Site 1 using the Scopex oscilloscope.

      (c) Site 2 using the Tektronix oscilloscope as in (a).

Horizontal scale on the Tekronix records: 1μs per division.

104

SITE 3

SITE 8

SITE 4

SITE 9

SITE 5

SITE 10

SITE 6

SITE 11

SITE 7

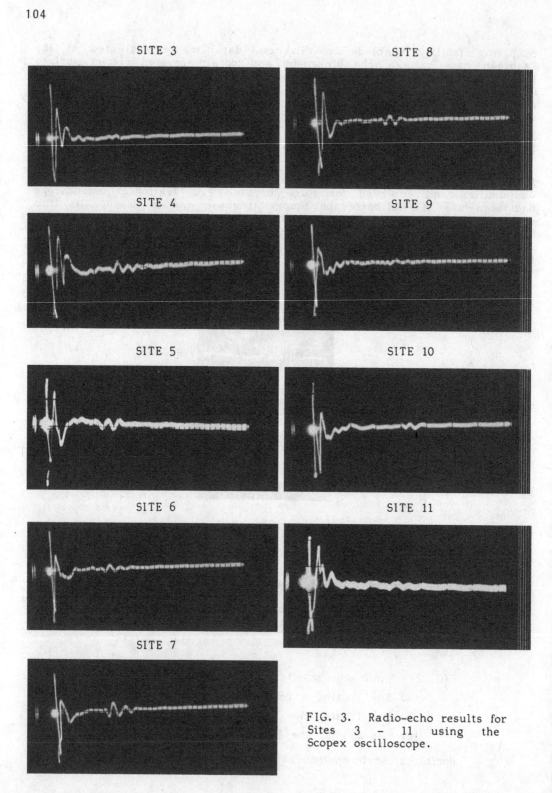

FIG. 3. Radio-echo results for Sites 3 - 11 using the Scopex oscilloscope.

The ice thickness separating the echo-sounder and the under-ice reflecting target can be evaluated using the equation

$$d = \frac{1}{2}\left[v^2(t + s/c)^2 - s^2\right]^{\frac{1}{2}}$$ (1)

where s = transmitter/receiver separation (m),
d = ice thickness (m),
t = difference between the surface and air wave arrival times (s),
c = speed of surface mode of propagation (m/s),
v = speed of electromagnetic wave propagation through bulk ice (m/s).
This may not necessarily be an accurate measure of the depth of ice at the sounding point, which is poorly defined if the bed is rough on a large scale. Here the 'first return' is always used to evaluate ice thickness. Table 1 shows the ice-depths reduced from the oscilloscope data of Figs. 2 and 3 using Eq. (1) with c = 208 m/μs, v = 169 m/μs (see Cumming et alia (2)) and s = 60 m.

Site of measurement	Time delay between direct transmitter pulse and reflected pulse (μs)	Distance to nearest bedrock reflector (m)
1	5.1	454
2	5.4	480
3	5.1	454
4	5.2	463
5	5.1	454
6	4.8	429
7	4.8	429
8	5.5	488
9	5.7	505
10	6.7	590
11	6.5	573
Estimated error	+6%	maximum +18 m

Table 1. Ice depth-sounding results from the Hispar Glacier. The last column derives from Eq. (1) with a 60 m transmitter/ receiver separation.

The locations of the sounding sites 1 - 11 shown in Fig. 1 were determined by routine survey methods using a theodolite and a base line of length 513 m set up near Dachigan camp. The base line itself was defined by triangulation on fourteen of the peaks marked on E. E. Shipton's map of the Hispar glacier (7). Figure 5 shows the resulting plots of ice-thickness, surface profile and bottom profile along the sounding path; see also Table 2.

106

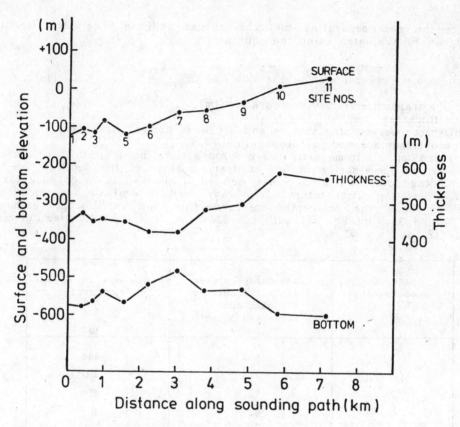

Fig. 5. Plot of ice-thickness, surface profile and bottom profile – datum through point A. (Base Camp)

Site no.	Eastings [m]	Northings [m]	Height above A [m]	Height OD [m]
1	-467	-768	-125	3801
2	-354	-1133	-106	3821
3	-358	-1445	-117	3809
4	-215	-1663	-84	3843
5	401	-1752	-120	3807
6	994	-1973	-98	3828
7	1727	-2263	-60	3867
8	2372	-2447	-53	3874
9	3368	-2680	-30	3897
10	4360	-2743	12	3939
11	5407	-3578	34	3961

TABLE 2. Ice sounding site co-ordinates. Origin at point A (by base camp, see Fig. 1), position 75° 10' 40" E, 36° 7' 54" N, height 3927 m above mean sea level. Orientation magnetic north. The location of A was derived by triangulation onto peaks marked on Shipton's map (7).

Estimation of the errors involved

The calibration of the Scopex was done at Site 1 (see Fig. 2). The first two photographs shown display the same echo signal on (a) the Tektronix oscilloscope, (b) the Scopex oscilloscope. The delay in the triggering is shown in Fig. 4, and is estimated to be $1.7 \pm 0.05\mu s$.

A

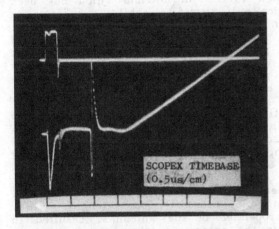

FIG. 4. Tektronix oscilloscope records:

(A) Measurements of the Scopex timebase delay ($0.5\mu s$/division).

B

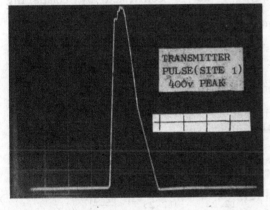

(B) Transmitter pulse recorded across the antenna at Site 1 ($0.2\mu s$/division).

The echo delay as observed in Fig. 2(a) is estimated to be 4.9 ± 0.15 μs. Looking at the Scopex trace, and taking the centre of the white dot as its origin, the trace length from the origin to the beginning of the echo return represents $3.2 \pm 0.2\mu s$. Using this result, estimation of the ice distances at various sites can be made from the remaining Scopex photographs shown in Fig. 3. Figure 2(c) shows the record at Site 2, where observations were made with the Tektronix instrument only.

The effect of moraine and melt-water on the echo signal

Measurements taken at Sites 1, 2 and 3 involved sounding through various depths of moraine. The worst case was at Site 1, where it is estimated that the moraine was up to one metre deep. The results indicate an increase in the initial receiver ringing and some degree of signal attenuation. The possible causes are:

(a) Change in radiation resistance.

Moraine, depending upon its composition, can have quite a high dielectric constant. The most common types of rock present were thought to be gneiss and argillite, both of which have relative permittivities of about 8 (see Clark (5)). An antenna which has been matched to the properties of ice will therefore exhibit a change in radiation resistance if placed in close proximity to a layer of moraine. Depending upon the degree of the antenna-receiver mismatch, caused by the inappropriate dielectric properties of the antenna surroudings, the system performance will be degraded and 'ringing' in the receiver may occur. The receiver ring will limit the shallowest ice thickness obtainable, if not obscuring the echo signal altogether.

The antennas used were resistively loaded dipoles as described by Cumming et alia (2) and having an efficiency of about 9%. These are broadband antennas, matched to the dielectric properties of clean ice. The effect of introducing a different dielectric medium would be to alter the electrical length of an antenna.

The .maximum moraine thickness encountered was in fact small (electrically) when compared with the transmitted pulse length. This, together with the broadband properties of the antennas, resulted in a relatively small degradation in the antenna match to the transmitter and receiver respectively. The outcome therefore was merely a minor lengthening of the receiver 'ringing', as is confirmed by comparing the received signals at Sites 1 and 7, where respectively maximum and virtually zero moraine thicknesses were encountered.

(b) Increase in the transmission loss through reflections at the air/moraine and moraine/ice boundaries.

The two-way power transmission loss through the moraine surfaces due to reflection at dielectric interfaces, for normal incidence and assuming zero conductivity and vacuum permeability each for air, moraine and ice, is given by Linlor and Jiracek (6):

$$\text{Net power loss (dB)} = -20\log_{10}\left[1 - \frac{(z_{12}+z_{23})^2 - 4z_{12}z_{23}\sin^2\left(\frac{2\pi d}{\lambda 2}\right)}{(1+z_{12}z_{23})^2 - 4z_{12}z_{23}\sin^2\left(\frac{2\pi d}{\lambda 2}\right)}\right] \quad (2)$$

where

$$z_{12} = \frac{\sqrt{\varepsilon_m} - \sqrt{\varepsilon_a}}{\sqrt{\varepsilon_m} + \sqrt{\varepsilon_a}} \qquad z_{23} = \frac{\sqrt{\varepsilon_i} - \sqrt{\varepsilon_m}}{\sqrt{\varepsilon_i} + \sqrt{\varepsilon_m}}$$

and

ε_m = relative permittivity of moraine,

ε_a = relative permittivity of air,

ε_i = relative permittivity of ice,

d = moraine thickness,

λ_2 = equivalent pulse wavelength in the moraine layer.

By assuming the relative dielectric constants to be 8.0, 3.2 and 1.0 for the moraine, ice and air respectively, the moraine to be 1 m thick and λ_2 = 63 m corresponding to a centre frequency of the transmitted pulse being 1.69 MHz (see Cumming et alia (2)), then from equation (2) the net power transmission loss is found to be 0.77 dB. If moraine were not present, the net power transmission loss would be predicted as 0.72 dB.

The effect of moraine is this case appears to be negligible. Indeed, even if large quantities of surface water were assumed to be present and the effective relative dielectric constant of the moraine-water mixture was assumed to be as high as say 50, the net power loss, as predicted by Eq. (2) would come to no more than about 1.3 dB.

(c) Increase in the dielectric losses due to moraine.

The increase in the dielectric losses by the inclusion of a moraine layer is difficult to evaluate in this case. Although no data has been collected regarding the moraine electromagnetic characteristics, it is reasonable to expect that no great contribution would be made by the moraine to the dielectric losses of the ice by such an electrically thin layer. It is thought that the presence of water with its possible salinity may have a greater effect than that of dry moraine alone.

(d) Presence of water within the ice.

It is possible that the water content of a temperate glacier has some effect upon the absorption of signal energy as it passes through. Water may be distributed in large-scale pockets and intra-glacial streams, an in microscopic regions of high stress at grain boundaries in the ice. In either case, the rate of absorption in water will be a function of the conductivity of the water, which will depend on the concentration of dissolved impurities. In a moraine-covered glacier, this can be expected to be greater than in an ice-mass such as Vatnajökull, where the surface is free from debris. Though we have no quantitative measure of the water conductivity this may have contributed to the reduced penetration observed in the Hispar glacier, as compared with the Vatnajökull ice cap.

ACKNOWLEDGEMENTS

The authors are indebted to R. Charlesworth, K. J. Miller, P. Nunn and T. Riley who provided logistics support on the glacier. They also gratefully acknowledge financial support from the Natural Environment Research Council.

REFERENCES

1) FRANCIS, M. R., MILLER, K. J. and DONG, Z. B., (1981). Impulse radar ice-depth sounding of the Ghulkin Glacier, this volume, pp 111 - 123.

2) CUMMING, A. D. G., FERRARI, R. L. and OWEN, G., (1980). Electronic design and performance of an impulse radar ice-depth sounding system used on the Vatnajökull ice-cap, Iceland. These proceedings, Vol. 1, pp 111 - 125.

3) BISHOP, J. F., CUMMING, A. D. G., FERRARI, R. L. and MILLER, K. J., (1980). Results of impulse radar ice-depth sounding on the Vatnajökull ice-cap, Iceland. These proceedings, Vol. 1, pp 126 - 134.

4) OSWALD, G. K. A., (1981). Ice-depth radio echo-sounding techniques employed on the Hispar and Ghulkin glaciers, this volume, page 86.

5) CLARK, S. P. (editor), (1966). Handbook of physical constants, Geological Society of America Memoir 97, pp 565 - 571.

6) LINLOR, W. I. and JIRACEK, G. R., (1975). Electromagnetic reflection from multi-layered snow models, J. Glaciology, Vol. 14, pp 501 - 515.

7) SHIPTON, E. E., (1939). Map of Hispar-Biafo glacier regions (Karakoram Himalaya), Royal Geographical Society, London.

Impulse radar ice-depth sounding of the Ghulkin glacier

M. R. Francis – B.H.R.A. Cranfield
K. J. Miller – University of Sheffield
Dong Zhi-Bin – University of Lanzhou, P.R.C.

ABSTRACT

The second stage of the glaciology programme is described. The results of ice-depth sounding of the Ghulkin glacier at thirty-three locations is presented comprising one longitudinal central profile and four transverse sections.

INTRODUCTION

Damming of the Hunza river by glacial action, rock fall, landslip or mud flash is now well documented, e.g., (1). A major lake formed just downstream of the Pasu and Ghulkin villages in 1972 after a rockfall. Between 1973 and 1974 the Batura glacier meltwater stream suddenly shifted its course a few hundred metres, the consequence of which was the removal of a section of the Karakoram Highway embankment and the destruction of a major bridge. In 1977 the Balt Bare glacier above the village of Shishkat released a surge of melt water that on its descent collected boulders, stones, grit and soil. The resulting mud flash took away several houses from the village before spreading across the valley floor to block off the river. The 20 m high dam created a lake that stretched back almost to Pasu and in the process drowned several kilometres of newly constructed road. When the dam finally yielded to the water pressure (bombing by the Pakistan air force being ineffective) one important bridge and a long stretch of roadway was buried in accumulated lake bed silts but by this time a completely new road section and bridge was under construction.

The treacherous behaviour of the Hunza glaciers was again dramatically illustrated in April 1980 when the main meltwater river of the Ghulkin glacier, much smaller than that of the Batura, shifted without warning and cut across the Karakoram Highway narrowly squeezed between the edge of the Hunza river and the steepening moraines of the glacier. Soon the road and its foundations were eroded away by a new moraine fan of boulders and cobbles containing, but not constraining, a new frequently changing meltwater channel. The old bridge stood forlorn a few hundred metres to the south spanning a now empty river bed.

Of concern to the Frontier Works Organisation (FWO) of the Pakistan Army was the future behaviour of the Ghulkin glacier. Was its present behaviour a warning of a surge that could take the glacier across the valley; the most dangerous of all the natural hazards to be reckoned with by the local population. There is much evidence that the Ghulkin glacier undergoes periodically high flow velocities due largely to a gradient of 1 in 3.7 from 7000 m at its head down to only 2500 m at river level. It is deeply entrenched in a great trough cut through the granite gneiss bedrock with hanging glaciers perched 600 m above the glacier level to complement the huge scree and snow cones that have been truncated by the rapid movement of the ice. Further evidence is provided by the huge series of crevasses, thirty metres wide and probably more than 100 m deep (2). Geomorphological studies relating to this glacier have recently been reported (3).

It was therefore decided that an examination be made of the fickle behaviour of the glacier. The FWO needed to find both a short and a long term solution to the problem of maintaining communications across the breach which at the height of the summer melt could only be crossed during the early hours before dawn. The first part of the study was to determine the thickness of the glacier and its profile using our impulse radar equipment. From such base-line data estimates could be made on future trends and comparisons made with data collected by future workers.

Our equipment had been previously tested on the Hispar glacier the detailed results of which are published elsewhere in these proceedings (4). Fortunately the equipment which had been successfully employed (5) on the Vatnajökull ice-cap of Iceland was successful in penetrating the warm moraine-covered Karakoram ice of the Hispar glacier thereby resolving several important academic points but now it was important to use this equipment in a more practical situation namely to obtain data from which the FWO could make sensible judgements on future actions to safeguard the Karakoram Highway.

The Ghulkin glacier is a relatively small glacier, see Figs. 1 and 2 by Karakoram standards being approximately 12 km long and nowhere more than 1 km wide. On 23rd August a camp was established on the south side of the glacier with the aid of men from the FWO and porters from Ghulkin village. The camp was located some two hours walk above the village. Membership of the party was now reduced from eight to three members by a combination of accident, illness and design but two villagers were elected to stay to help carry the radio-echo sounding equipment across the glacier. Their assistance was invaluable when it came to relocating the positions of the ice-sounding stations by the surveyors in the chaotic moraine mounds of broken rock strewn all over the glacier surface.

The equipment was first assembled and tested at a point on the glacier immediately opposite the camp site (sounding 1); see Fig. 3. Meanwhile a network of markers were set out on the north and south flanks of the glacier to provide survey control. Originally it was intended to use standard survey techniques (6) to locate all ice-depth sounding stations.

The ice-sounding programme started with minimum delay moving first up stream to the centre of what would become the upper transverse section, (sounding 2). Over the next ten days the radar equipment was progressively moved towards the snout down the centre line of the glacier and a total of some thirty soundings recorded including depths taken across the transverse sections. The stations were later located with respect to the control network by two members of the main survey party, the detailed

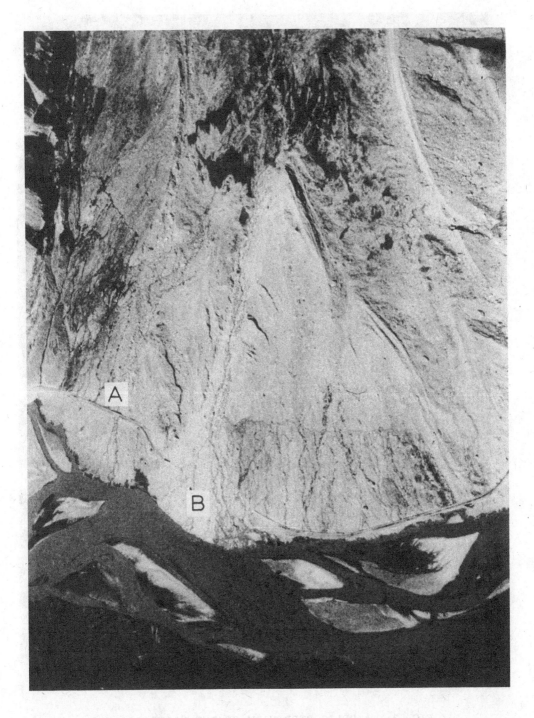

FIG. 1. A HELICOPTER VIEW OF THE SNOUT OF THE GHULKIN GLACIER;
(A) OLD BRIDGE; (B) NEW MORAINE FAN AND MELT CHANNEL.

114

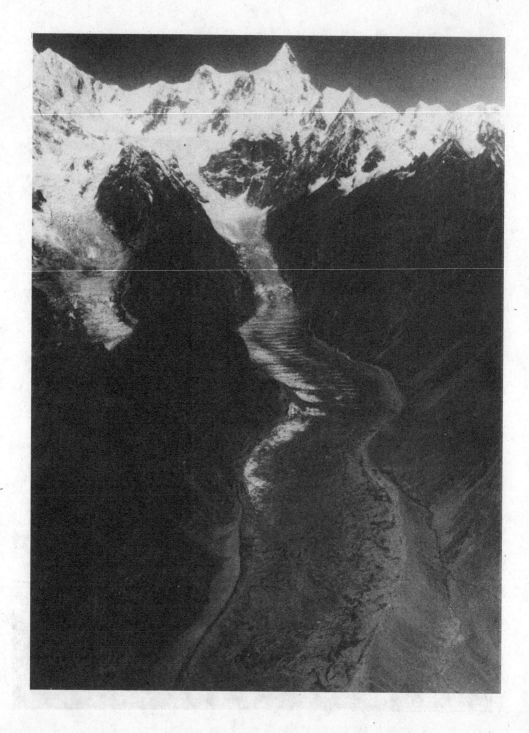

FIG. 2. A HELICOPTER VIEW OF THE UPPER REACHES
OF THE GHULKIN GLACIER.

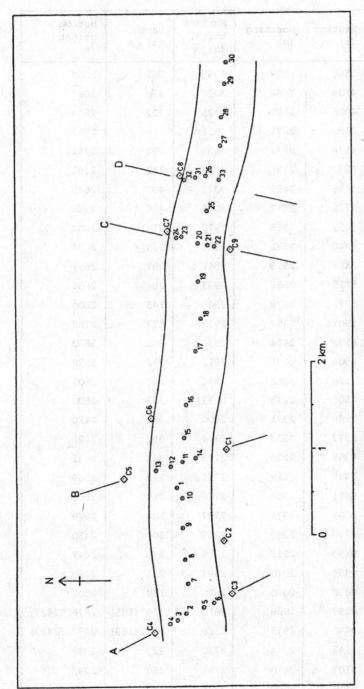

FIG. 3. MAP OF THE GHULKIN GLACIER SHOWING CONTROL POINTS (◊) AND SOUNDING (○).
The position of the edge of the glacier is approximate.

results being shown in Fig. 3 and the Table below. This final work is being completed with the aid of tellurometers and theodolites.

Station number	Easting (m)	Northing (m)	Surface height (m)	Depth (m)	Bottom height (m)
1	5569	2589	2945	382	2563
2	4105	2484	3037	370	2667
3	4026	2582	3025	222	2803
4	3980	2675	3011	127	2884
5	4176	2272	3033	292	2741
6	4230	2155	3015	222	2793
7	4456	2459	3027	402	2625
8	4736	2492	3006	428	2578
9	5099	2516	2978	374	2604
10	5446	2521	2965	338	2627
11	5866	2513	2942	289	2653
12	5818	2656	2935	284	2651
13	5773	2828	2945	145	2800
14	5907	2357	2939	177	2762
15	6238	2484	2931	301	2630
16	6506	2471	2911	333	2578
17	7128	2362	2854	253	2601
18	7506	2295	2833	333	2500
19	7944	2321	2800	381	2419
20	8377	2323	2774	352	2422
21	8355	2208	2773	341	2432
22	8339	2119	2791	222	2569
23	8411	2508	2777	214	2563
24	8389	2571	2787	108	2679
25	8749	2209	2753	303	2450
26	9150	2217	2719	270	2449
27	9494	2049	2693	233	2460
28	9820	2030	2670	169	2501
29	10199	1986	2612	136 (185)	2476 (2427)
30	10440	1975	2588	111 (165)	2477 (2423)
31	9142	2344	2730	229	2359
32	9107	2470	2754	159	2593
33	9114	2060	2725	140	2585

DATA REDUCTION

The operational arrangement of transmitting and receiving antennas was the same as that used during the Hispar study, with a transmitter-receiver separation of 60 m and the antennas in line. The equation relating ice depth and signal travel time is as follows:

$$d = 1/2 \sqrt{\{c^2(t_d + t_m)^2 - s^2\}}$$

where d = depth

t_m = time delay measured from the start of the recording in µsec

s = transmitter-receiver separation, 60 m

c = velocity of propagation in ice, 169 m/µsec

and $t_d = \dfrac{s}{v} + t_o$

where v = velocity of propagation along ice/air interface, 208 m/µsec

t_o = instrument delay, 1.7 µsec

The orientation of the antennas was generally along the axis of the valley. On the upper part of the glacier the reason for this was the hope that several reflections could be separately detected. For instance, if a parabolic shape to the valley is assumed three reflections should be obtained. The antennas generate a field approximately radially symmetric but in the plane of the antenna the radiation pattern follows a cosine law, thus with an orientation across the valley reflections from immediately below would be favoured over those from the side.

Further down the valley the ice surface was broken by numerous ice cliffs oriented along the glacier which effectively ruled out a transverse orientation of the antennas.

ERROR ANALYSIS

The error in the instrument delay of 1.7 µsec may be taken as ± 0.05 µsec. Provided that the echo is correctly identified the error in measurement from the photographic records is perhaps ± 0.5 mm or ± 0.07 µsec. This gives a combined error of ±0.12 µsec representing ±10 m.

There is a further source of error in misinterpretation. Figure 4a shows the clearest recording taken on the Ghulkin. The various positive peaks in the echo are labelled P and the negative peaks N. In some of the more noisy recordings it is possible that P2 and P3 have been confused and similarly N1 and N2. In particular this is the case with reflections from shallow depths. At the beginning of the trace can be seen a signal resulting from the strong direct wave through the air/ice interface covering the receiver. With shallow depths the reflected signal begins to merge with the receiver ring with consequent interpretation problems. At a depth of less than 165 m, the whole echo can no longer be seen as it has moved off to the left. It is at this point that the

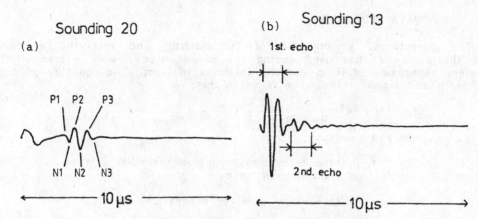

FIG. 4. EXAMPLES OF RECORDINGS
(a) SOUNDING 20 (CLEAR RECORDING)
(b) SOUNDING 13 (DOUBLE ECHO)

problem of misinterpretation is most serious, mistaking P2 for P3 or N1 for N2 leads to a further error of 0.55 μsec or 46 m depth.

RESULTS

Examples of recordings from several ice-sounding stations can be seen in Fig. 5 and the interpreted depths for all soundings in Table 1. Fig. 6 gives the longitudinal profile of the glacier and the four cross-sections are shown in Fig. 7. The deepest sounding was at station 8 where the depth was 428 m. The depth can be seen to be generally decreasing towards the snout. The surface slopes fairly evenly in the same direction but the glacier bottom appears rather more rugged. This is partly due to error as previously discussed. The error involved in measuring the height of the sounding points above the arbitrary datum is negligible in comparison with the error in measurement of the depth. Most of the soundings taken along the longitudinal are fairly clear however. So it is felt that this profile is a reasonable representation of the actual glacier profile. The most doubtful region must be the lower region and in particular the last two soundings 29 and 30. It is possible that the depth has been underestimated at these two points and for this reason the alternative depth is shown as a dotted line.

The four glacier sections, A, B, C and D shown in Fig. 7 were constructed by simply taking a smooth envelope around circles constructed on the sounding points with radii representing the measured depths. These indicate a general shape but with only four or five measurements per section this cannot be regarded as being highly accurate. For instance it is possible that the valley bottom is far more U-shaped than indicated here. This would not show up in this survey and would have required far more measurements to clarify.

A problem of interpretation arose with the sections because records taken towards the sides of the glacier tended to be rather noisier. This was assumed to be because of several reflectors of varying strengths at different depths across the section. On the Hispar Glacier this had not been observed because of its far larger size. In that case the reflections tended to come from immediately below the sounding point, other reflections within the 10 μsec time window being precluded by the broader valley shape.

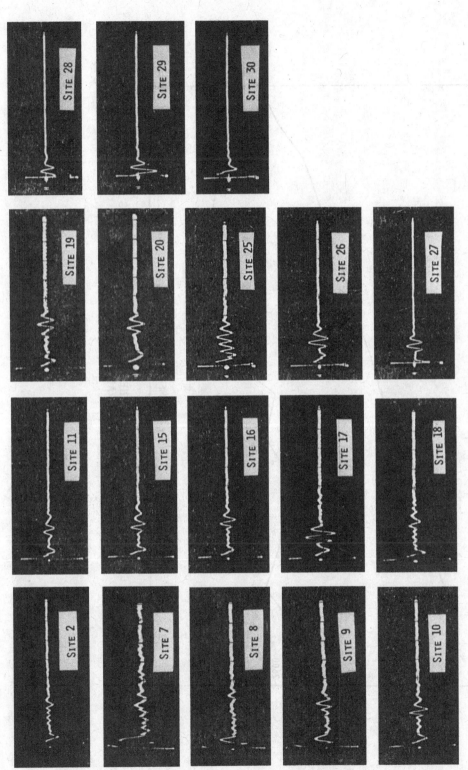

FIG. 5. PHOTOGRAPHIC RECORDS FOR THE LONGITUDINAL PROFILE (TRACES FOR SITES 7, 8 AND 9 ARE INVERTED).

120

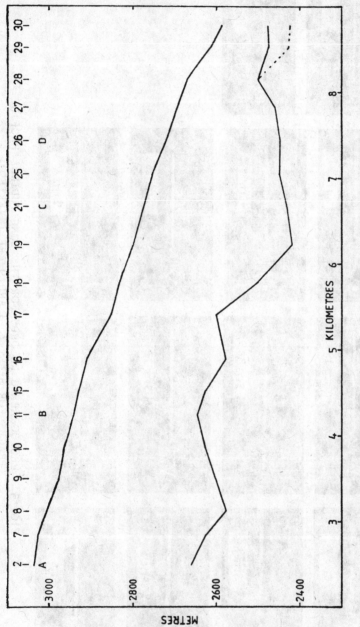

FIG. 6.   THE LONGITUDINAL PROFILE OF THE GHULKIN GLACIER SHOWING HEIGHTS ABOVE
AN ARBITRARY DATUM.

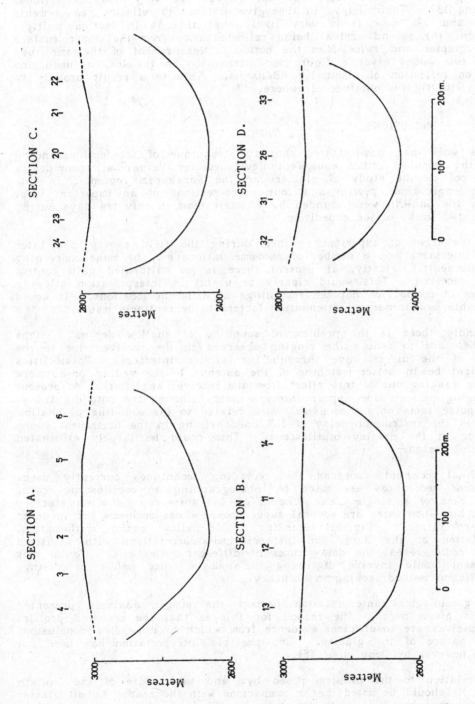

FIG. 7. TRANSVERSE PROFILES OF THE GHULKIN GLACIER WITH HEIGHTS ABOVE AN ARBITRARY DATUM.

One particular trace (see Fig. 4b) showed two very clear reflections, (sounding 13). Their delays in time give distance to reflector measurements of 145 and 294 m. It is very likely that this is in fact a multiple reflection, the second arrival being reflected once from the upper surface of the glacier and twice from the bottom. Measurement of the amplitude of the two waves gives a figure for attenuation in the ice, assuming no losses on reflection of about 1.6 dB/100 m. This is a result broadly in keeping with figures obtained elsewhere, (7).

CONCLUSIONS

This work has demonstrated that the technique of ice-depth sounding using the impulse radar equipment developed for Vatnajökull provides a useful tool in the study of glaciers in the Karakoram region. In this case a longitudinal profile and four cross-sections of an important local glacier, the Ghulkin, were sounded by a small team in only ten days during the last two weeks of the expedition.

In the light of experience gained during the field survey and later during interpretation, a number of recommendations may be made concerning the equipment. Firstly, at present there is no calibrated gain control on the receiver. This would clearly be useful in interpretation allowing a degree of comparison between recordings at different locations. It would also enable measurement of attenuation factors to be routinely made.

Secondly there is the problem of sounding of shallow depths. Steps could be made to reduce the ringing observed in the receiver due to the arrival of the direct wave through the air/ice interface. Possibilities here might be in better matching of the antenna to the medium or in more effective masking out of this effect from the receiver amplifiers. At present the ringing persists for approximately 2 μsec although the outgoing trans-mitter pulse lasts only 0.6 μsec. Also related to the sounding of shallow depths is the instrument delay of 1.7 μsec arising in the horizontal sweep generator in the display oscilloscope. This could be largely eliminated by careful design.

A final comment concerns the recording techniques currently used. At present recordings are made by photographing an oscilloscope trace. In the case of a single clear echo (Fig. 4a) this may be a satisfactory method, but when there are several superimposed echoes problems are incurred in interpretation. Digital recording would allow easier mathematical manipulation of the data, for instance cross-correlations with idealized signals could reveal the delay times of different reflections. To do this at present would involve digitizing the analogue trace before processing, an inefficient method lacking in accuracy.

No glaciological interpretation beyond the simple analysis presented above is given here. The reason for this is that the revealed profiles by themselves are insufficient evidence from which to draw firm conclusions on the nature of the glacier. One possible interpretation has been put forward however by Dong et al (8).

In relation to the problem posed by, and the nature of the Ghulkin glacier, it should be noted that a comparison with the nearby Gulmit glacier by our geomorphologists indicates that the Ghulkin glacier is not unique in its behaviour. Here too there is evidence of several meltwater channels, very steep overall gradients and a history of surges. However the

Gulmit glacier terminus lies far above the Karakoram Highway and its fickle behaviour poses no threat since minor alterations in snout position and meltwater streams can be quickly accommodated.

The simple but difficult problem of the Ghulkin is that there is no room for manoeuvre by the road builders between the steep slopes of the terminal moraine and the Hunza river.

The records presented in this paper will be of value to future glaciologists monitoring changes in the Ghulkin glacier.

ACKNOWLEDGEMENTS

The radio-echo party would like to thank Steve Redhead and Robert Muir Wood for their help in carrying out the echo-sounding programme and Jon Walton and Tom Crompton for their work in surveying the positions of the sounding stations. Finally thanks are due to the assistance given to this programme by the FWO in terms of a helicopter survey and manpower to move the radar equipment onto the glacier.

REFERENCES

1) These proceedings, Volumes 1 and 2 (1983).

2) ANDREWS, R. M., McGREGOR, A. R. and MILLER, K. J., (1983). 'Fracture Toughness of Glacier Ice', Vol. 1, these proceedings, p.147.

3) DERBYSHIRE, E. and MILLER, K. J., (1981). 'Highway beneath the Ghulkin', Geographical Magazine 53 (10), pp 626 - 635.

4) DONG, Z. B., FERRARI, R. L., FRANCIS, M. R., MUSIL, G., OSWALD, G. K. A. and ZHANG, X. S., (1982). 'Impulse radar ice-depth sounding on the Hispar glacier', Vol. 2, these proceedings, p.100.

5) CUMMING, A. D. G., FERRARI, R. L. and OWEN, G., (1980). 'Electronic design and performance of an impulse radar ice-depth sounding system used on the Vatnajökull ice cap, Iceland', Vol. 1 of these proceedings.

6) MILLER, K. J., (1978). 'Simple surveying techniques for small expeditions', Royal Geographical Society Special Publication.

7) WESTPHAL, W., (1963). Private communication to S. Evans, see Robin, G. de Q., Evans, S. and Bailey, J. T., (1969). 'Interpretation of radio echo-sounding in polar ice sheets', Phil. Trans. Roy. Soc. A 265 (1166).

8) DONG, Z. B., ZHANG, X., LI, J. and XU, S., (1981). 'Radio-echo sounding on the Ghulkin glacier in the Karakoram Mountains', Lanzhou University.

Survey

The survey work of the
International Karakoram Project

Chen Jianming – Institute of Glaciology and Cryopedology, Lanzhou, P.R.C.
T.D. Crompton – University College, London
J.L.W. Walton – University College, London
R. Bilham – Lamont-Doherty Geological Observatory, U.S.A.

ABSTRACT

In 1913 a triangulation network was set up to connect the survey systems of Russia and India. This chain crosses the Karakoram, a region of intense tectonic deformation. The southern section of the 1913 link was remeasured by the survey party of the International Karakoram Project with a view to detecting continental movement over the intervening 67 years. The field work was completed successfully despite a number of setbacks including the death of one of the team. The network has been analysed in a number of ways in order to find the best method for the analysis of possible deformation. Figural distortions have been detected and a geophysical interpretation of these has been essayed.

INTRODUCTION

Land surveying techniques are used in many aspects of scientific field work. The survey team of the International Karakoram Project included 6 trained and experienced land surveyors and it was expected that a proportion of their time would be spent carrying out 'service work' for other scientific disciplines. This proved to be the case and by the end of the summer every member of the team had spent a period working for one of the other programmes. That work will be discussed later.

The survey team did, however, have a separate programme of their own. It was extensive enough to be beyond the scope of the other scientific groups yet its subject is intrinsically linked with the studies of the seismologists and geomorphologists. The various stages of this part of the programme will be described here.

HISTORICAL APPRAISAL

The main task of the survey programme was the re-observation of as much as possible of a triangulation connection, originally observed in 1913, between what was then India and Russia. This survey must surely rank as one of the epic pieces of 20th century surveying. Carried out across largely unknown territory, linking at the northern end with a similar triangulation party who had started from Russia, the report by Kenneth Mason[1] makes compulsive reading and gives a valuable insight into the techniques and equipment used and the generally arduous conditions under which the work was carried out. The aim of our re-observation was to obtain estimates of the tectonic displacements that have taken place in this active plate boundary zone during the intervening 67 years.

PROPOSALS FOR 1980 RE-OBSERVATION

The original proposal for this re-observation was made in 1979; much detailed research and analysis was then carried out to determine the feasibility of carrying out an ambitious programme in a single short summer season and also whether such a re-observation could possibly produce any significant results.

It was calculated[2] that the deformation of Shunuk station relative to Gashushish station (Fig. 1.) could be determined to an accuracy of 1.64 m (error ellipse semi-major axis), by a re-observation. This error ellipse was produced by combining the ellipses for the 1913 and proposed 1980 surveys. The estimated error of the 1913 survey alone was 1.63 m; this is largely scale error caused by only observing (necessarily) a pure triangulation chain of single short base. It was proposed to re-observe the network using combined trilateration/triangulation, measuring as many distances as seemed necessary. From the network analysis it was shown that if 10 or so selected distances were measured to an accuracy of 7 ppm the error ellipse dimension could be reduced to 0.52 m while measurement of every distance would reduce it to 0.43 m. Both of these error ellipses, when combined with the 1.63 of the 1913 survey, produced an estimated error for the deformation of 1.64 m. If no distances were measured at all, the estimated error would be 2.30 m.

As an estimated deformation between the Indian and Russian plates was 2.5 m [3], it was clear that Electromagnetic Distance Measuring instruments would have to be used in the re-triangulation in order to produce the best possible results. However, it was also clear that distances measured to 1 ppm would not produce significantly better results for _deformation_ since 1913 than if they were measured to 7 ppm.

Further research and discussions with Peter Mott, who had surveyed in this area in 1939[4] indicated that there was every chance of Mason's 1913 stations remaining intact. Located as they are in lofty and inaccessible positions vandalism was unlikely - the only destructive element would be the harsh and variable climate.

It was estimated that a party of 6, plus porters, could re-observe the network from Gashushish up to Sirsar during the time available in the field. To progress further would entail departing from the route of the Karakoram Highway and logistic requirements would quickly snowball.

To keep the number of trained personnel to a minimum and to reduce the number of ascents to stations it was decided to observe to opaque beacons wherever possible, backing these up with heliographs. The beacons were constructed in London before departure using a design given in [5].

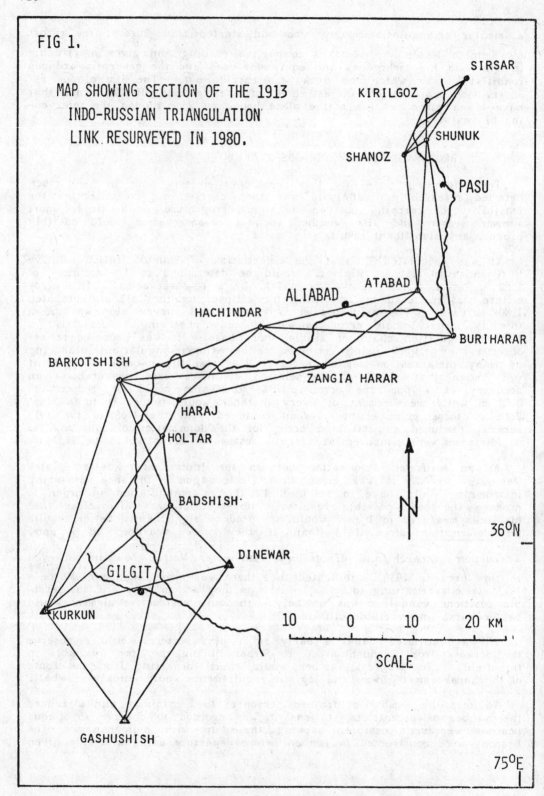

FIG 1.

MAP SHOWING SECTION OF THE 1913
INDO-RUSSIAN TRIANGULATION
LINK RESURVEYED IN 1980.

SIRSAR

KIRILGOZ

SHUNUK

SHANOZ

PASU

ATABAD

ALIABAD

HACHINDAR

BURIHARAR

BARKOTSHISH

ZANGIA HARAR

HARAJ

HOLTAR

N

BADSHISH

DINEWAR

36°N

GILGIT

KURKUN

10   0   10   20 KM

SCALE

GASHUSHISH

75°E

It was also proposed to use a satellite doppler receiver to fix independently the positions of some of the triangulation stations and to compute some of the geodetic parameters in the area.

FIELD WORK

In 1913 Mason's party left Rawalpindi on 14 April and arrived in Aliabad (Fig. 1) on 20 May. They had used numerous means of transport and large numbers of porters. Our party left Rawalpindi at 0730 on 26 June 1980 and after a flight to Gilgit and a short drive along the Karakoram Highway reached base camp in Aliabad by 1300. This two lane surfaced highway is a remarkable feat of construction and provided us with a communications lifeline through the region. When the road was free from landslips and washouts our Land Rovers could move easily from one end of the network to the other. For this reason alone we were able to attempt and complete the ambitious programme of observations.

Most of the triangulation work was carried out by 2 Kern DKM3 theodolites although the smaller Wild T2 theodolites were used for some of the more inaccessible stations. The high resolution reflecting telescope of the DKM3 proved ideal for observing to opaque beacons, which provided a perfect target at distances of up to 40 km. Original calculations of logistics assumed that 25 km would be the maximum range over which these beacons could be observed; in the event 30 - 35 km range was often observed and on one occasion 6 rounds of angles were observed to a beacon 52 km distant. The heliographs did not have to be used at all. Vertical angles were observed from every station. Distances of up to 43 km were measured with 3 pairs of lightweight Tellurometer CA1000s measuring to an accuracy of about 7 ppm. These proved satisfactory both to use and to carry and a total of 26 distances were observed in the network (Fig. 2.).

Equipment that proved invaluable to the programme were our Storno VHF radios. Weighing only ½ kg they provided our scattered parties with an excellent communications network at distances of up to 50 km. The use of these greatly speeded up our mountain-top activities.

The season started slowly, all party members having to get used to altitude and other, less predictable, problems associated with local living. The northern quadrilateral was measured first. This is thought to lie across a geological fault line and is of interest to the geophysicists even if we were not able to link up with it from the south. Its small size also provided a good testing ground for our observing procedures. Much comfort was gained from finding all 4 stations intact!

Attention was then turned to the southern end of the network. Beacons had to be erected on a number of stations so they could be observed from others. It was at this stage, on 14 July 1980, that Jim Bishop, the most experienced mountaineer of the group, fell to his death while climbing to erect a beacon on Kur Kun. Although this accident delayed progress and deeply shocked everyone, the work continued. A number of problems connected with combinations of bad weather, incorrect navigation, fasting porters (it was the Moslem month of Ramadan) and flat instrument batteries created further delays at the southern end and by 30 July only 2 distances and a single set of horizontal angles had been observed. However, between 30 July and 28 August a total of 21 ascents to stations were made and the network was satisfactorily completed as far

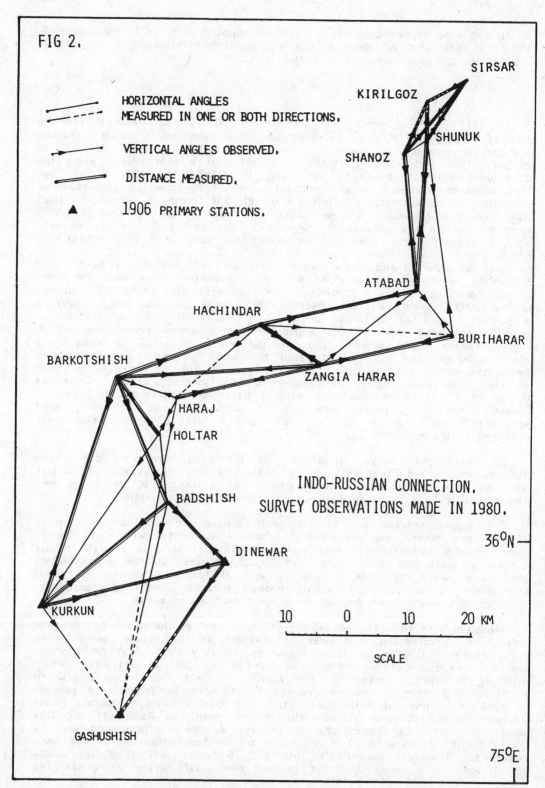

FIG 2.

HORIZONTAL ANGLES
MEASURED IN ONE OR BOTH DIRECTIONS.

VERTICAL ANGLES OBSERVED.

DISTANCE MEASURED.

▲ 1906 PRIMARY STATIONS.

SIRSAR

KIRILGOZ

SHUNUK

SHANOZ

ATABAD

HACHINDAR

BURIHARAR

BARKOTSHISH

ZANGIA HARAR

HARAJ

HOLTAR

INDO-RUSSIAN CONNECTION.
SURVEY OBSERVATIONS MADE IN 1980.

BADSHISH

36°N—

DINEWAR

10   0   10   20 KM

KURKUN

SCALE

GASHUSHISH

75°E

as Sirsar. Even if time had permitted, we were not allowed to move any further north than this station so there a halt was called and the necessary field checks made.

The stations of Hachindar and Buriharar, both well marked by large cairns and visited recently by the Survey of Pakistan (SOP), were noted to be well out of position. This became obvious when on neither of the lines Hachindar - Barkotshish and Buriharar - Atabad were the stations intervisible. Yet the lines had been observed in 1913. In both cases the correct stations were located - up to 2.3 km away in the case of Hachindar. The relevant information was passed on to the SOP.

Throughout the season the JMRI satellite doppler receiver had been in operation and good fixes were obtained at valley sites linked into the stations of Barkotshish, Zangiaharar and Shanoz. Severe problems were encountered at times with the receiver in such a steep sided valley but it was not really practicable to carry it up to the triangulation stations themselves.

The survey team left base camp on 15 September, fit and content after a successful survey season yet sadder and wiser at the terrible price that had been paid for the acquisition of the data.

COMPUTATIONS

All observations and the resulting triangular misclosures were checked in the field. The misclosures are shown in Fig. 3. after triangles had been corrected for spherical excess. Only one triangle was unacceptable - this included observations to a beacon only 4 km away which was a poor target.

Vertical angles were measured wherever possible. The final total included 23 reciprocal pairs and single ended observations. Height differences were computed and all station heights were within 1 m of those listed in 1913. No further details of the vertical heighting network are included in this paper.

Measured distances were corrected for meteorological conditions. Although wet and dry bulb temperatures were taken at both ends of each line measured, the small Thommen altimeters we had were not really accurate enough. However, the geoidal height of each station was known and pressures were computed assuming standard conditions at sea level. Field observations showed that day to day variations of atmospheric pressure were small and errors incurred because of this approximation are unlikely to exceed 2 ppm.

Comparison tests were carried out for our 3 pairs of Tellurometer CA1000s on a 700 m base line established by Aga 6BL near base camp before and after the survey. This line was measured using all nine Master/Remote combinations in turn. Results of these tests showed good consistency, all measured distances lying within the ± 20 mm quoted for the instrument. All distances were corrected for slope and curvature and reduced to the geoid. The JMRI satellite doppler receiver was used at stations near Dinewar, Barkotshish, Zangia Harar and Shanoz and linked into these stations. Unfortunately an instrument error on the site near Dinewar meant no fix could be computed but Geoid/Spheroid separations have been calculated for the other three stations and are

FIG 3.   INDO-RUSSIAN CONNECTION. 1980 SURVEY
TRIANGULAR MISCLOSURES.

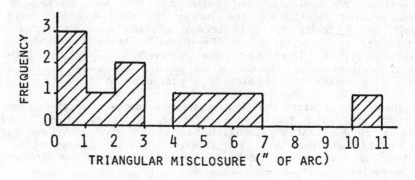

FIG 4.   INDO-RUSSIAN CONNECTION.
1980 SURVEY.
OBSERVATION EQUATION RESIDUALS.

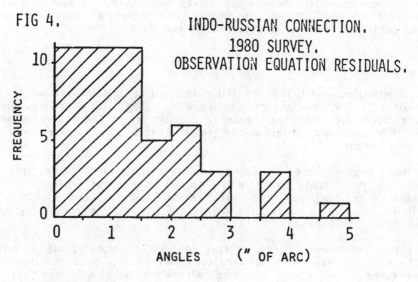

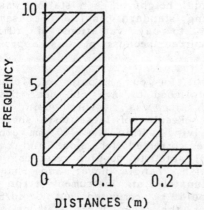

42 m, 31 m and 28 m respectively.   A mean separation of 34 m produces a further 5 ppm correction to all distances.

Observations were then ready for the computer adjustment procedure but before this could be carried out two problems had to be clarified.

1)   Control for the network.   If the retriangulation was being carried out purely to produce a set of geographical co-ordinates the Doppler stations, if properly weighted, could have been considered as suitable control.   However, it is network deformation that is being studied with the absolute co-ordinates not important.   As the Doppler fixes have standard errors of ± 3 m and the expected deformations are not greater than 2.5 m this control system cannot be used.   The three southernmost stations, Gashushish, Kurkun and Dinewar are primary control stations of the 1906/1907 Gilgit series and surveyed more accurately than the secondary 1913 Indo-Russian Link.   It is probable that these three stations lie outside the main deformation zone so the (changing) geometry of this triangle was looked at in detail.   This analysis is included in Appendix A.   There is no significant change of shape of the triangle, in fact, the consistency is remarkable, so for preliminary control the stations Gashushish and Kurkun were assumed to have co-ordinates unchanged since 1913.   This analysis does not take into account the possibility of the rotation or translation of the entire triangle.

2)   Selection of reference spheroid.   The Everest spheroid was originally defined in terms of the Indian foot.   However, the quoted length of the standard bar representing the Indian foot has changed significantly over the years[6].

In 1913 the Indian foot was taken to be 0.30479841 m which fixed the axes for the spheroid at 6377276 m and 6356075 m.   All modern work in Pakistan is based on the revised Indian foot with the resulting spheroid having axes of 6377301 m and 6356100 m.   An added complication for us was that all Doppler fixes were computed in terms of WGS72.

Although the newer dimensions for the Everest Spheroid are in current use and are probably more accurate, it was decided for this investigation to use the 1913 axes.   Correspondingly, various spheroid to spheroid transformations were carried out and all subsequent computation based on the spheroid with axes;

Semi-major   6377276 m

Semi-minor   6356075 m

All observations and their estimated standard errors ($\pm 2.0''$ for angles and ± 0.03 m $\pm$ 7 ppm for distances) were set and a computer adjustment carried out with stations Gashushish and Kurkun fixed.   The output from the computer included adjusted co-ordinates, observation equation residuals and error ellipse axes.   The mean (mod) residuals are shown in Fig. 4. σ_0 was computed to be 1.01 so the estimate of standard errors was reasonable.

ANALYSIS OF DEFORMATION

One way of examining deformation is to compare the 1913 and 1980 sets of co-ordinates.   When comparing co-ordinates it is much simpler to work on a plane.   A special transverse Mercator Projection –

"The Karakoram Grid" - was defined and co-ordinates computed for both epochs. Details of this projection are given in Appendix B. By placing the central meridian close to the middle of the study area, scale factor and t-T corrections were kept to a minimum. The full list of co-ordinates for 1913 and 1980 is in Table 1, together with apparent movements.

It is quite clear that most of this apparent movement is due to a 'swing' of the long survey chain about a short baseline; the various vectors are in more or less the same direction and of magnitude proportional to their distance from the fixed end of the network. However, the vectors for 3 of the stations do not fit into this pattern.

Buriharar

All survey observations were made to and from an incorrect station. Only after the last observations were finished was the 1913 station located about 400 m further up the ridge. Lack of time coupled with poor observing conditions meant that repeat observations could not be made so the position of the old station was measured using a crude vertical subtense method, with a base of 2 m. The subtended angle was approximately 18' and was measured on two settings of the Kern DKM3. The standard error of this angle should be about 8", thus the distance should be correct to 1 part in 140 or about 2.5 m. The discrepancy between sets of co-ordinates is 25 m. Presumably a mistake was made somewhere in the subtense observations which is regrettable.

Shunuk

The cause of the large apparent movement here is a mystery. The station was first visited by Walton and was found intact and in good condition on top of a rock pinnacle. It is not possible that this station has been misidentified - it is equally impossible for it to have moved by natural causes some 20 m along and up the ridge! Possibly the 1913 co-ordinates have been incorrectly transcribed at some stage but the station descriptions published by the Survey of India include computed angles and distances which correspond to the listed co-ordinates. It must be remembered that the northern end of the 1913 survey agreed remarkably well with the Russian triangulation system.

Zangia Harar

This discrepancy is neither so large nor so obvious as the other two but it does indicate that there is some deformation occurring in the vicinity of this station.

ANALYSIS OF CHANGES OF LENGTH

Another possible way of approaching the problem is to study each triangle of the network, comparing side lengths in 1913 and 1980. This was done and the results shown in Fig. 5.

If the sides relating to the stations Buriharar and Shunuk are discounted, a pattern emerges which indicates considerable scale error between the two surveys, increasing as one goes north. Because distances have been measured in 1980 it is likely that scale error (amounting to 200+ parts per million at the northern end of the network) has been propagated as a result of the (necessarily) poor geometry of the 1913 triangulation. It is clear that a sensible analysis of deformation cannot

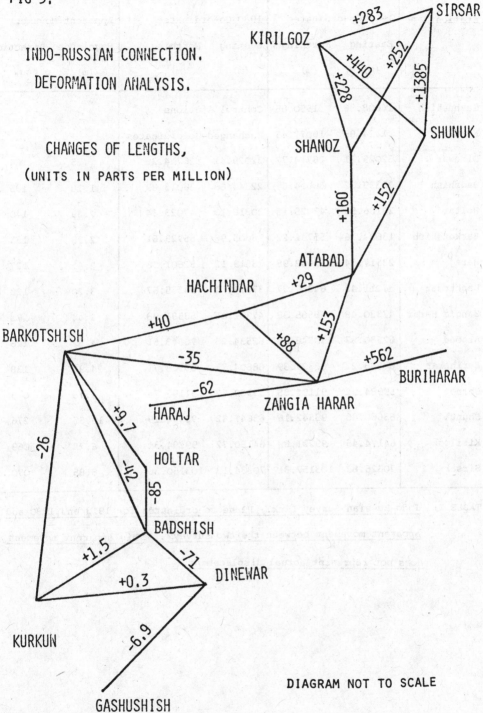

FIG 5.

INDO-RUSSIAN CONNECTION.
DEFORMATION ANALYSIS.

CHANGES OF LENGTHS,
(UNITS IN PARTS PER MILLION)

SIRSAR

KIRILGOZ   +283

+440   +252   +1385

+228

SHUNUK

SHANOZ

+190   +152

ATABAD
+29

HACHINDAR

+40   +153

BARKOTSHISH   +88

-35   +562

-62   BURIHARAR

HARAJ   ZANGIA HARAR

+9.7

-26   -42   HOLTAR

-85

BADSHISH

+1.5   -71

+0.3   DINEWAR

KURKUN

-6.9

GASHUSHISH

DIAGRAM NOT TO SCALE

| Station Name | 1913 Co-ordinates | | 1980 Co-ordinates | | Apparent Movement | |
| | Easting | Northing | Easting | Northing | Magnitude | Direction |
	(m)	(m)	(m)	(m)	(m)	deg. °
Gashushish	14608.86	1550.85	Control stations			
Kurkun	1424.85	19077.45	Unchanged Co-ordinates			
Dinewar	32029.31	26714.72	32029.41	26714.38	.35	164
Badshish	22339.82	36014.85	22340.66	36013.89	1.18	135
Holtar	20186.55	47025.16	20818.10	47023.34	2.39	140
Barkotshish	13604.36	55731.22	13605.98	55729.61	2.28	135
Haraj	23517.34	52602.98	23519.12	52601.06	2.61	137
Hachindar	37011.43	64337.37	37014.14	64335.67	3.20	122
Zangia Harar	47330.48	58585.38	47331.05	58581.98	3.44	171
Atabad	62530.57	70546.08	62534.28	70543.51	4.52	126
Buriharar	68588.22	63327.39	68604.77	63308.71	24.96	138
Shanoz	59984.92	91192.85	59989.62	91193.74	4.78	079
Shunuk	63666.86	93342.16	63647.42	93344.19	19.55	276
Kirilgoz	64174.45	99291.89	64180.72	99294.31	6.72	069
Sirsar	70325.82	103157.31	70334.13	103160.36	8.85	070

TABLE I: Indo-Russian Survey Link. Plane co-ordinates for 1913 and 1980 and apparent movement between the two surveys. This apparent movement does not represent actual displacement.

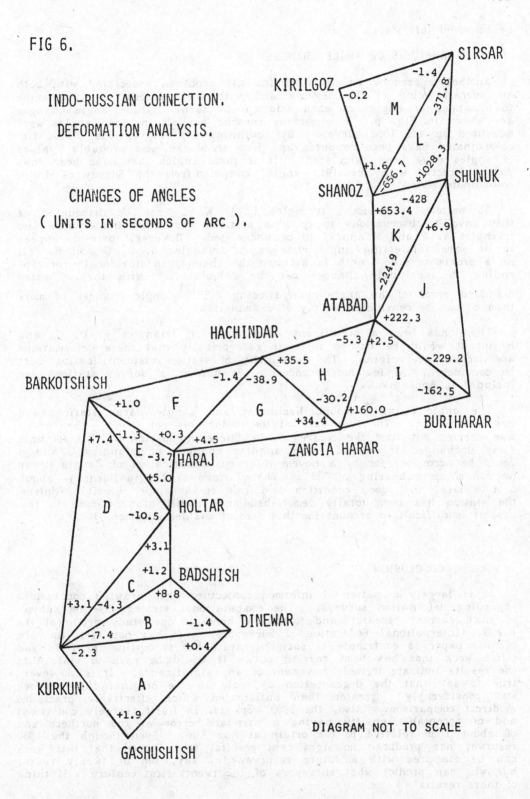

FIG 6.

INDO-RUSSIAN CONNECTION.

DEFORMATION ANALYSIS.

CHANGES OF ANGLES

(UNITS IN SECONDS OF ARC).

DIAGRAM NOT TO SCALE

be achieved this way.

ANALYSIS OF ANGLE CHANGES

Another approach, one that avoids all problems associated with both an overall swing of the network and with scale errors, is to examine the changes of angles of each triangle in turn. These angle changes are shown in Fig. 6. For various reasons not all included angles were measured in the 1980 survey. By combining all observations, 'best fit' co-ordinates have been computed and from these the most probable values of angles have been calculated. It is these angles that have been compared directly with the 1913 angles computed from the Survey of India co-ordinate lists.

No pattern is obvious, triangles I, J, K, L, can be discounted as they involve observations to or from stations Buriharar and Shunuk and triangles A, F and M appear to be unchanged. However, there do appear to be some interesting angle changes in triangles B, D, G and H. If an approximate side length is assumed the changes in side length corresponding to the angle changes can be calculated. With the estimated standard error of an angle change being 2.5"[2], angle changes of more than 8" can be considered worthy of examination.

There has been a significant deformation in triangles B, D, G, and H but it would be unwise to state categorically that these deformations are tectonic in origin. The possibility of station misidentification must be considered. A few notes concerning a number of survey stations are included in Appendix C.

The angle changes around Hachindar and Zangia Harar stations are most interesting. The fine mark at Hachindar was not found – the survey was carried out from the centre of the Surveying platform. Yet triangle F is unchanged in shape. The angular changes in triangles G and H could be accounted for by a movement or misidentification of Zangia Harar by 2.8 m on a bearing of 068°. The platform at Zangia Harar is about 4 m square, in good condition and lies on a minor summit. Unless the station has been totally demolished and rebuilt at some time in the past it is difficult to account for this size of misidentification.

CONCLUSION

It is largely a matter of informed conjecture when drawing conclusions regarding deformation surveys. The computational strategy is the subject of much current research and forms the brief of one study group of the F.I.G. (International Federation of Surveyors). It has not been the aim of this paper to contribute to such literature but to outline the field and office work that has been carried out. It would be rash to state that the results indicate tectonic movement of any significance. It is, however, true to say that the degeneration of scale in the original 1913 network was considerably greater than anticipated which effectively precludes a direct comparison. Also, the 1980 work is, in itself, largely consistent and of acceptable quality having a standard error at the northern end of about 1 m relative to the origin at Kur Kun. Even though the 1980 resurvey has produced no significant results, it is hoped that this work can be compared with a future resurvey in, say, ten or twenty years, but who can predict what surveyors of the twenty-first century will think of these results?

APPENDIX A

Check on stability of Primary Triangle

Stations Gashushish, Kurkun and Dinewar

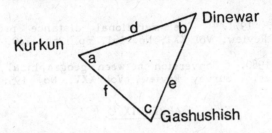

Compare sides and angles, 1913 – 1980.

Angles	(a)	67 03 39.0	67 03 38.6	-0.4"
	(b)	41 17 36.3	41 17 36.8	+0.5"
	(c)	71 38 44.9	Not measured	
	(d)	31542.60	31542.53	-0.07 m
	(e)	30605.13	30605.00	-0.13 m
	(f)	21931.10	Not measured	

Compute (f) from observed quantities to check on consistency of triangle.

$$(f) \ 1980 \ = \ 21931.11 \ m$$

$$(f) \ 1913 \ = \ 21931.10 \ m$$

For preliminary control take stations Gashushish and Kurkun as fixed.

APPENDIX B

The Karakoram Grid

The Karakoram Grid is based on a transverse Mercator Projection of the Everest Spheroid.

The central meridian is 74° 40'00" E and the origin of northings is 35° 00'00"N.

False origin of co-ordinates is situated at;

$$74°40'00" \ E \ = \ 50,000 \ m \ E.$$

$$35° \ 00'00"N \ = \ -80,000 \ m \ N.$$

The actual semi-axes used are; major, 6377,276 m

minor, 6356,103 m

Karakoram Grid co-ordinates have been produced from Mason's published co-ordinates.

Meridional distances were computed using the technique and constants given in (a) below and co-ordinates computed using an extension of the method given in (b).

a) T. Vincenty, 1971. The meridional distance problem for desk computers. Survey Review, Vol XXI, No. 161, pp 136 - 140.

b) N. J. Field, 1980. Conversion between geographical and Transverse Mercator Co-ordinates. Survey Review, Vol XXV, No. 195, pp 228 - 230.

APPENDIX C

Kurkun. 1906 Primary station. Recessed concrete pillar. On steep summit and in excellent condition.

Dinewar. 1906 Primary station. Recessed concrete pillar - partly crumbled and fine mark missing. Relocated by estimation of centre of outer annulus of pillar.

Badshish. Revisited by SOP recently. Fine mark re-cut by them. Large platform and mark in excellent condition.

Holtar. Platform in good condition on flattish summit. Fine mark re-cut. Station visited by SOP in 1980.

Barkotshish. On small rock platform on crenellated ridge. Fine mark intact and solid, re-cut by SOP. Visited in 1980 by SOP.

Haraj. Revisited by SOP 1 day prior to visit by IKP party. Fine mark re-cut by them. Platform in excellent condition. No beacon erected - observations made to aluminium pole set into cairn above fine mark.

Hachindar. Platform in excellent condition but fine mark not found. Re-located by estimation of centre of platform.

Zangia Harar. Platform on sharp summit and in excellent condition. Visited by SOP in 1980, fine mark re-cut by them.

Atabad. Small platform in reasonable condition on top of tall steep rock pillar. Not found since 1913. Fine mark found. Beacon left in place.

ACKNOWLEDGEMENTS

These are many and space precludes the authors from listing them all by name. This paper is a summary of 18 months of preparation, field work, computation and interpretation. Thanks must go to those who provided advice and encouragement when the programme was first proposed, to those who provided financial assistance to make a field season possible and to those who lent the necessary (and very costly) survey equipment free of charge.

Special thanks must go to Dr. A. L. Allan and all the staff of the Department of Photogrammetry and Surveying, University College London;

whose collective skills and facilities have been at our disposal throughout the project. Also special thanks to Professor K. J. Miller, leader of the International Karakoram Project 1980 and his administrative team who stage managed the whole venture so well. Thanks must also go to Major Rana, A. Razzaq Awan, Ashraf Aman, Ted Smith, Nigel Atkinson, Alan Colvill and John Allen, who with the authors formed the survey team of the Project and who climbed some 200,000 m during the summer of 1980 in order to retrieve the data presented here.

Final thanks must go to the late Jim Bishop, who provided advice and encouragement from the start and gave all that he had that the programme might be successful.

REFERENCES

1) K. Mason, (1914). "Completion of the link connecting the triangulations of India and Russia." Survey of India, Vol. VI.

2) T. O. Crompton. "The Pakistan to Russia Triangulation Connexion: past and projected error analyses." Vol. 1 these proceedings, pp 171-185.

3) J. Armbruster, L. Seeber and K. H. Jacob, (1978). "The North-Western termination of the Himalayan mountain front; active tectonics from micro-earthquakes". Journal of Geophysical Research. No. 83, pp 269 - 282.

4) P. G. Mott, (1950). "Karakoram Survey 1939 - A new map." Geographical Journal 116, pp 89 - 95.

5) C. A. Biddle, (1956). "A new Topo Survey Beacon." Empire Survey Review, Vol. XIII, No. 101, pp 335 - 338.

6) G. Bomford, (1939). "The readjustment of the Indian Triangulation." Survey of India professional paper No. 28, pp 100 - 105.

The calculation of crustal displacement and strain from classical survey measurements

T. O. Crompton – University College, London

ABSTRACT

This paper identifies some of the problems inherent in the determination of planimetric displacement and strain in the earth's crust by repeated survey measurements. An approach which avoids the use of co-ordinates is developed. The technique should be familiar to the surveyor and, at the same time, useful to the geophysicist. After some discussion of the relative accuracies of 1913 and 1980 measurements, the paper concludes with a simple application using data acquired during the International Karakoram Project.

INTRODUCTION

The basic philosophy behind the detection of crustal displacement and strain by survey measurements is straightforward. Observations are carried out at two different epochs and their calculation results in two distinct sets of co-ordinates. Comparison of these co-ordinates then gives a measure of strain and displacement that has taken place between epochs. Indeed, if the observations at both epochs were errorless, this approach would probably be the best and is certainly an easily visualised concept. However, even with idealised observations, the pattern of displacement vectors can be difficult to interpret in all but the simplest situations. Figure 1 depicts such vector patterns for a simple survey network of six stations at three levels of complexity. (Although fictitious, the geometry of this network has been chosen to be similar to that of IKP 80). Figure 1(a) shows the vector field associated with simple displacement across a fault FF' and is an easily interpreted pattern. Figure 1(b) shows the displacement vectors associated with a uniform strain field operating over the area of the survey network without any fault displacement and is less obviously interpreted. Figure 1(c) gives the combined effect of the displacement of Figure 1(a) with the strain of Figure 1(b) resulting in an apparently more random pattern of movement. In all three cases, point 1 has been held fixed as origin.

Once the survey measurements at each epoch have been contaminated by random observing errors then the picture becomes even more confused. Moreover, it is likely that systematic errors will then be present in co-

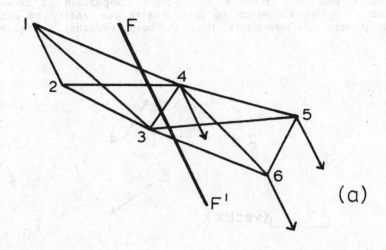

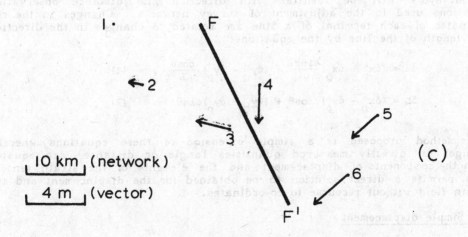

FIGS. 1(a), (b), (c). VARIOUS PATTERNS OF DISPLACEMENT VECTORS.

ordinates and this renders the simple comparison of co-ordinates almost useless. Systematic error in co-ordinates can easily be confused with the very pattern of movements that is being sought. Figure 2 shows the

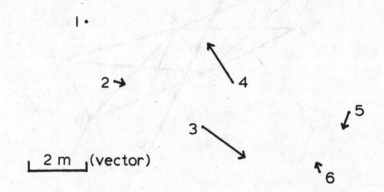

FIG. 2.   VECTORS OF FIGURE 1(c) AFTER TRANSFORMATION.

residual vector movements of Figure 1(c) after a transformation has been applied in order to remove "systematic" differences. The transformation consists of a rotation of 11".7 and a scale change of 22 ppm. Both the rotation and scaling are small yet their effect is to greatly decrease the co-ordinate differences. This demonstrates the general inadequacy of co-ordinate comparison and the possibly misleading effects of transformation. The purpose of this paper is to present a method which enables displacement and strain to be estimated without the use of co-ordinates as an intermediate step.

THE METHOD OF OBSERVED DIFFERENCES

Surveyors will be familiar with direction and distance observation equations used in the adjustment of survey networks. Changes in the co-ordinates of each terminal of a line are related to changes in the direction and length of the line by the equations:

$$\delta\theta = (\delta x_p - \delta x_Q)\frac{\sin\theta}{L} + (\delta y_Q - \delta y_p)\frac{\cos\theta}{L} \quad , \quad (1)$$

$$\delta L = (\delta x_Q - \delta x_p)\cos\theta + (\delta y_Q - \delta y_p)\sin\theta \quad . \quad (2)$$

The method proposed is a simple extension of these equations whereby changes in directly measured quantities (angle or distance) are equated with the components of displacement and the elements of the strain tensor. This permits a direct solution to be obtained for the displacement and the strain field without recourse to co-ordinates.

(a) Simple displacement

Consider a survey line PQ of length L and direction θ which is crossed by a fault at which displacement has occurred between epochs. Let the components of the displacement be f_x and f_y assumed small compared to L. The changes in direction and length of the line are given directly from (1) and (2) as:

$$\delta\theta = -f_x \frac{\sin\theta}{L} + f_y \frac{\cos\theta}{L}, \quad (3)$$

$$\delta L = f_x \cos\theta + f_y \sin\theta. \quad (4)$$

Each measured direction and distance which crosses the fault will give an equation of type (3) or (4). It is more usual to measure angle than direction and, assuming both limbs of the angle cross the fault, the change is given by:

$$\delta\alpha = \left(\frac{-\sin\theta_2}{L_2} + \frac{\sin\theta_1}{L_1}\right) f_x + \left(\frac{\cos\theta_2}{L_2} - \frac{\cos\theta_1}{L_1}\right) f_y. \quad (3a)$$

Evidently the location of the fault must be known in order that the unknowns f_x and f_y may be attached to certain points and not to others. Also, the measurement of survey lines which do not cross the fault will contribute nothing to the solution. A unique solution is obtained by two measurements crossing the fault (either two angles, one angle and one distance, or two distances). It will however be more usual for redundancies to exist which can be dealt with by a standard least squares solution.

(b) Homogeneous strain

Suppose now that crustal deformation (without displacement) is taking place as a result of finite homogeneous strain acting over the area of a survey figure. Assume that the co-ordinates at epoch one (x, y) and epoch two (x', y') are related by the non-orthogonal transformation:

$$\begin{pmatrix} x' \\ y' \end{pmatrix} = \begin{pmatrix} a & b \\ c & d \end{pmatrix} \begin{pmatrix} x \\ y \end{pmatrix}$$

A survey line PQ has co-ordinates (x_p, y_p); (x_Q, y_Q); (x'_p, y'_p); (x'_Q, y'_Q) at the old and new epochs.

Define $\delta x_p = (x'_p - x_p)$, $\delta y_p = (y'_p - y_p)$ with similar expressions for δx_Q and δy_Q. Then

$$(\delta x_p - \delta x_Q) = (x'_p - x_p - x'_Q - x_Q)$$

$$= (ax_p + by_p - x_p - ax_Q - by_Q + x_Q)$$

$$= a(x_p - x_Q) + b(y_p - y_Q) + (x_Q - x_p).$$

Putting $(x_Q - x_p) = \Delta x$ and $(y_Q - y_p) = \Delta y$ gives:

$$(\delta x_p - \delta x_Q) = -a\Delta x - b\Delta y + \Delta x.$$

Similarly:

$$(\delta y_Q - \delta y_p) = (cx_Q + dy_Q - y_Q - cx_p - dy_p + y_p)$$

$$= c\Delta x + d\Delta y - \Delta y$$

Substitution in (1) gives:

$$\delta\theta = -\Delta x(a - 1)\frac{\sin\theta}{L} - b\Delta y\frac{\sin\theta}{L} + c\Delta x\frac{\cos\theta}{L} + \Delta y(d - 1)\frac{\cos\theta}{L}.$$

But $\quad \Delta x/L = \cos\theta$ and $\Delta y/L = \sin\theta$ $\quad$ Hence:

$$\delta\theta = -(a - 1)\sin\theta\cos\theta - b\sin^2\theta + c\cos^2\theta + (d - 1)\sin\theta\cos\theta$$

$$\delta\theta = (d - a)\sin\theta\cos\theta - b\sin^2\theta + c\cos^2\theta.$$

Similarly, substitution in (2) and division throughout by L gives:

$$\frac{\delta L}{L} = \Delta x(a - 1)\frac{\cos\theta}{L} + b\Delta y\frac{\cos\theta}{L} + c\Delta x\frac{\sin\theta}{L} + \Delta y(d - 1)\frac{\cos\theta}{L}$$

$$= a\cos^2\theta + (b + c)\sin\theta\cos\theta + d\sin^2\theta - 1.$$

Using $\quad \cos^2\theta = 1/2(1 + \cos2\theta)$ and $\sin^2\theta = 1/2(1 - \cos2\theta)$ gives:

$$\delta\theta = (d - a)\sin\theta\cos\theta - 1/2b(1 - \cos2\theta) + 1/2c(1 + \cos2\theta)$$

$$\delta\theta = 1/2(d - a)\sin2\theta + 1/2(b + c)\cos2\theta + 1/2(c - b)$$

and:

$$\frac{\delta L}{L} = 1/2a(1 + \cos2\theta) + 1/2(b + c)\sin2\theta + 1/2d(1 - \cos2\theta) - 1$$

$$\frac{\delta L}{L} = -1/2(d - a)\cos2\theta + 1/2(b + c)\sin2\theta + 1/2(d + a) - 1.$$

It is now possible to convert to strain notation by putting $(a - 1) = \varepsilon_{11}$; $(d - 1) = \varepsilon_{22}$; $b = \varepsilon_{12}$; and $c = \varepsilon_{21}$. Then $(b + c) = (\varepsilon_{12} + \varepsilon_{21}) = \gamma_2$; $(d - a) = (\varepsilon_{22} - \varepsilon_{11}) = \gamma_1$; $1/2(c - b) = 1/2(\varepsilon_{21} - \varepsilon_{12}) = \omega$; and $1/2(d + a) - 1 = 1/2(\varepsilon_{11} + \varepsilon_{22}) = \Delta$.

This is Frank's notation (1) but note the change of sign for γ_1 and ω which results from a different choice of axes. The direction and distance equations may now be written in terms of strain parameters as:

$$\delta\theta = 1/2\sin2\theta.\gamma_1 + 1/2\cos2\theta.\gamma_2 + \omega, \qquad (5)$$

$$\frac{\delta L}{L} = -1/2\cos2\theta.\gamma_1 + 1/2\sin2\theta.\gamma_2 + \Delta. \qquad (6)$$

γ_1 and γ_2 are measures of shear, ω is a rotation and Δ the dilation. As stated earlier, it is more usual to measure angles than directions and for for the change in angle, $\delta\alpha$, equation (5) is rewritten as:

$$\delta\alpha = \gamma_1(1/2\sin2\theta_2 - 1/2\sin2\theta_1) + \gamma_2(1/2\cos2\theta_2 - 1/2\cos2\theta_1). \quad (5a)$$

Several points arise from consideration of these equations. Four unknowns are present and, in theory, a unique solution exists if four measurements are made. There are, however, certain restrictions. Rotation, ω, can only be obtained if directions are independently measured at both epochs. This is unusual survey practice and the re-measurement of angles gives no estimate of ω. Similarly, dilation, Δ, can only be obtained if repeat distance measurements are made. Put another way, the measurement of a pure triangulation network enables estimates of γ_1 and γ_2 to be made. The re-measurement of a pure trilateration network enables estimates of γ_1, γ_2 and Δ to be made. The re-measurement of a mixed triangulation/trilateration network adds nothing, but may be considered advantageous for reasons of accuracy and the introduction of redundancies. It is this argument that exposes the fallacy of co-ordinate comparison. Equations (5a) and (6) have been derived from the transformation

$$\begin{pmatrix} x' \\ y' \end{pmatrix} = \begin{pmatrix} a & b \\ c & d \end{pmatrix} \begin{pmatrix} x \\ y \end{pmatrix}$$

and, assuming that the two co-ordinate sets (x,y) and (x',y') are available, then the unknown transformation parameters a, b, c, d can be found and hence $\gamma_1, \gamma_2, \omega$ and Δ .

(c) Displacement combined with strain

This more general situation can be dealt with by combining equations (3a) with (5a) and (4) with (6) for any measurements which cross the fault. This results in a system of equations in up to five unknowns (f_x, f_y, γ_1, γ_2 and Δ) but measurements which do not cross the fault will not contain f_x and f_y .

GENERAL DISCUSSION

Although the method described is explicitly designed to avoid the use of co-ordinates, it must be remembered that approximate co-ordinates are required in order to calculate coefficients. The location and type of fault along which displacement is assumed to occur must be known in relation to the survey stations. This should present no great difficulty in networks that are specifically designed to study such a feature. If, however, the fault lies in an unknown position, then it seems possible that a trial and error appproach may locate the survey figure through which displacement has occurred but a positive identification by a geologist or geophysicist is to be preferred.

An interesting question is whether or not the observations at each epoch should be subjected to an adjustment in order to achieve consistency of results. Consider for example the three angles of a single triangle measured at each epoch, thus allowing estimates of γ_1 and γ_2 to be made. If each set of observations is made consistent by adjustment and the values of $\delta\alpha$ are obtained from these, then any two angles may be used to solve for

γ_1 and γ_2. The third equation is a linear combination of the other two and contributes nothing to the solution. Moreover, no estimate of the accuracy of γ_1 and γ_2 can be made as the solution is essentially unique. Once the technique is extended to a more complicated survey figure (such as a braced quadrilateral where eight angles are measured), then redundancy exists whether or not there has been a "pre-adjustment" and the usual least squares method can be used to obtain the best estimates of γ_1 and γ_2. It is the author's opinion that "pre-adjustment" should not be used and that the differences of directly measured quantities be used which should result in more realistic error estimates for the unknowns. It is, of course, prudent to perform a "pre-adjustment" at each epoch in order that unacceptable observations may be rejected, mistakes detected and sensible a priori estimates made of the standard errors of angle and distance differences.

The method described appears to possess several advantages over co-ordinate methods. The unknowns in the solution are the parameters which describe the strain field and displacement and a simple extension of the least squares solution gives their error estimates. The method also permits a flexible approach in modelling the strain field. By a suitable choice of unknowns, the strain field may be assumed uniform over large areas covering several survey figures or may be modelled figure by figure. Useful results can also be obtained either from disconnected observations or from a network with very weak geometry, both these situations being unsuitable for the co-ordinate approach. Perhaps the greatest advantage of the method is its computational efficiency. Co-ordinate based methods of evaluating the strain tensor require the computation of the "inner co-ordinates" of the network at each epoch followed by the computation of the transformation between the two sets of inner co-ordinates. The determination of inner co-ordinates requires a free net adjustment involving the use of the generalised matrix inverse to overcome the resulting rank deficiency. Such calculations are lengthy, even for modern computers, when dealing with a large network. In contrast, using the method given here, the complete analysis of displacement and strain (if assumed uniform over the whole region) is achieved by the inversion of a non-singular, symmetric matrix of order five.

APPLICATION TO THE I.K.P. 1980

Before applying the method to the I.K.P. data, a few general comments must be made about their accuracy. These comments will be brief. A more detailed analysis can be found elsewhere in these Conference Proceedings. The most important point is that the epoch one measurements are only available as results. This means that neither the observations nor the computations for Mason's 1913 (2) work can be consulted. Consequently, a comparison of measured quantities can only be made after the angles and distances have been recomputed from 1913 co-ordinates. Moreover, these angles and distances will then possess errors due to rounding of the geodetic co-ordinates to 0".01. Indeed, small random differences exist between azimuths computed from 1913 co-ordinates and the azimuths published in the triangulation lists. This phenomenon may result from the published azimuths having been computed from unrounded co-ordinates and, having no further evidence, these values are adopted. Similar small random differences also exist between published and computed distances, possibly for the same reason.

Fortunately it is possible to give a more detailed appraisal of the second epoch of observations undertaken on the International Karakoram Project 1980. A simultaneous adjustment using all observed angles and distances was carried out with a priori standard errors of $\pm$ 2" for angles and

(± 0.03 ± 7 ppm) m for distances. This resulted in a value of 1.0187 for σ_o which indicates that realistic a priori selections had been made. The largest angle residual was + 4".68 and the largest distance residual was ± 0.220 m on a 38 km line. The largest semi-major axis of the error ellipse of position was 0.994 m. This was at the northernmost end of the scheme (Sirsar) relative to an origin at the southern end (Kur Kun). It was realised during the fieldwork that observations involving two stations (Buri Harar and Shunuk) differed greatly from 1913 values. Buri Harar had been misidentified and the 1913 location of this point was found too late to include in the scheme. Shunuk was positively identified and the nature of the terrain precluded any other nearby location, yet the plan position and height of this point differ by 20 m and 10 m respectively. No satisfactory explanation has been found and neither Shunuk nor Buri Harar can be used in a comparison.

Inspection of the 1913 and 1980 co-ordinates reveals a variable trend in latitude differences and a systematic trend in longitude differences. When 1913 azimuths and distances are compared with 1980 values the following general conclusions may be drawn. There is:

(i)   a general deterioration of orientation between the two networks amounting to 16" at the northern end;

(ii)   a good overall agreement of angles in spite of the orientation problems; and

(iii) an extremely large variation in scale.

Despite the encouraging agreement of scale on the opening line (Kur Kun to Dinewar), large positive and negative scale errors rapidly accumulate. Differences of + 4.00 m (253 ppm) and – 0.94 m (85 ppm) occur in the lines Shanoz – Sirsar and Badshish – Holtar. These variations are so large that no useful comparison of distances can be made. This restricts the method to angles only. However, some interesting variations in angles occur in the vicinity of Zangia Harar and the technique for determination of displacement is now applied to a small portion of the network covering this region.

Figure 3 shows the relevant part of the network with angles numbered

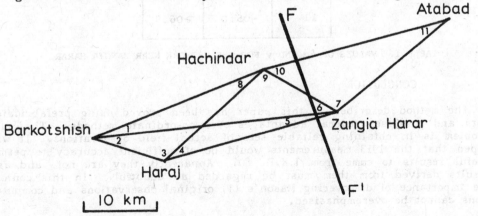

FIG. 3.   THE NETWORK NEAR ZANGIA HARAR.

1 – 11. The line FF' is the arbitrarily selected location of a discontinuity along which displacement is assumed to occur. Table 1 gives the difference in angles, $\delta\alpha''$, in the sense 1980 minus 1913. The 1913 angles are obtained from the triangulation lists and the 1980 angles are computed from adjusted co-ordinates. Table 1 also lists residuals, v'', in each of the angles after a least squares solution of nine equations of type (3a). Note that $\delta\alpha = -v$ for angles 3 and 8. Neither limb of these angles crosses FF' and, as no observation equation is generated, no change occurs. The least squares solution of the nine equations with equal unit weight gives values $f_x = -0.95 \pm 0.21$ m and $f_y = -2.85 \pm 0.32$ m with $\sigma_0 = \pm 4''.2$. This may be interpreted as a displacement of 3.01 m either at an azimuth of $71^\circ.6$ on the S.W. side of FF' or at an azimuth $251^\circ.6$ on the N.E. side of FF'. Examination of the residuals shows them to be typical of the angular differences in other parts of the scheme and consequently no attempt is now made to apply the strain field model to this example.

Angle	$\delta\alpha''$	v''
1	+07.4	-03.1
2	-08.1	+03.8
3	+03.1	-03.1
4	+05.8	-03.9
5	-00.6	+03.0
6	+34.6	-00.1
7	-34.8	-04.0
8	-02.4	+02.4
9	-39.4	+00.6
10	+39.7	-02.9
11	-05.0	+06.9

TABLE 1.   VALUES OF $\delta\alpha$ AND v FOR THE NETWORK NEAR ZANGIA HARAR.

CONCLUSION

The method described in this paper has been proved using prefabricated data and possesses several advantages over co-ordinate methods. The main problem is in obtaining reliable sets of actual field observations. It was hoped that the 1913 measurements would be of sufficient accuracy to permit useful results to come from I.K.P. 80. Apparently they are not, and any results derived from them must be regarded as doubtful. In this context the importance of discovering Mason's (1) original observations and computations cannot be overemphasised.

ACKNOWLEDGEMENTS

The author is indebted to all the members of the I.K.P. survey team, the administrative staff of the Royal Geographical Society and his colleagues in the 'Department of Photogrammetry and Surveying at University College London. Finally, a special acknowledgement is reserved for Kenneth Mason and his contemporaries in the Survey of India, without whose original survey there would have been no justification for the I.K.P. 80 survey programme.

REFERENCES

1) FRANK, F. C., (1966). 'Deduction of earth strains from survey data', Bulletin of the Seismological Society of America 56(1), pp 35 - 42.

2) MASON, K., (1914). Records of the Survey of India, Volume VI. Completion of the link connecting the triangulations of India and Russia 1913. Survey of India, Dehra Dun. 121 pp.

BIBLIOGRAPHY

1) ASHKENAZI, V., (1980). 'Least squares adjustment: signal or just noise?', Chartered Land Surveyor/Chartered Minerals Surveyor 3(1), pp 42 - 49.

2) BRUNNER, F. K., (1979). 'On the analysis of geodetic networks for the determination of the incremental strain tensor', Survey Review XXV (192), pp 56 - 67.

3) CROMPTON, T. O., (1980). 'The Pakistan to Russia triangulation connexion: past and projected error analyses'. Paper presented at the Islamabad Conference; Recent Advances in Earth Sciences. June 1980 , Vol. 1, these proceedings, pp 171-185.

4) JAEGER, J. C., (1971). Elasticity, fracture and flow. Chapman and Hall, London. 268 pp.

Earthquakes

A microearthquake survey in the Karakoram

G. Yielding*, S. Ahmad,** I. Davison, J.A. Jackson*, R. Khattak***,
A. Khurshid†, G.C.P. King* and Lin Ban Zuo‡

* Department of Geophysics, Cambridge
** Quaid-i-Azam University, Islamabad, Pakistan
*** Atomic Energy Commission, Pakistan
† Geological Survey of Pakistan
‡ Institute of Geophysics Beijing, P.R.C.

ABSTRACT

A portable seismic network was operated in the Karakoram area of northern Pakistan for over two months during the summer of 1980. The principal objectives were to record possible subcrustal earthquakes beneath the Karakoram Range, and to monitor crustal activity throughout northern-most Pakistan. Of the 371 earthquake locations presented here, about two-thirds lie in the Hindu Kush intermediate-depth seismic zone. One intermediate-depth earthquake (at a depth of about 140 km) was recorded from beneath the Karakoram Mountains. Shallow crustal activity also occurs in the Karakoram Range, with events recorded from near the north-west end of the Karakoram Fault and near a number of smaller lineaments in Baltistan. Seismicity in Kohistan appears to be largely confined to the upper crust, though occasional earthquakes at 65 km depth (approximately the base of the crust) occur beneath areas of higher topography (Nanga Parbat, Rakaposhi). An area close to the Hamran (1972) and Darel (1981) earthquakes showed intense activity in the depth range 0 - 20 km, similar to that involved in the mainshocks themselves.

TECTONIC SETTING

The northward movement of the Indian sub-continent during the Cenozoic is now well-documented (1 - 3). Analysis of Indian Ocean spreading rates as defined by magnetic anomaly patterns suggests that the convergence rate between India and Asia slowed from 100mm/yr to approximately 50mm/yr perhaps forty million years ago; about the same time as the two continental masses (or their marginal island arcs) first came into direct contact. The earlier convergence was accommodated by subduction of the Tethys oceanic lithosphere beneath the southern margin of Asia, but the arrival of the Indian sub-continent at the subduction zone resulted in a profound change in the mode of deformation. Average continental crust, 35 kilometres thick and almost 20 per cent léss dense than the mantle, is both too thick and too light to be carried down into the asthenosphere. Thus the buoyancy of continental crust probably prevents it from being subducted into the mantle, and India and Eurasia sutured themselves together to form a single larger continent. The Indus-Tsangpo ophiolite belt in southern Tibet, north of the Himalayas, appears to mark the approximate boundary between

151

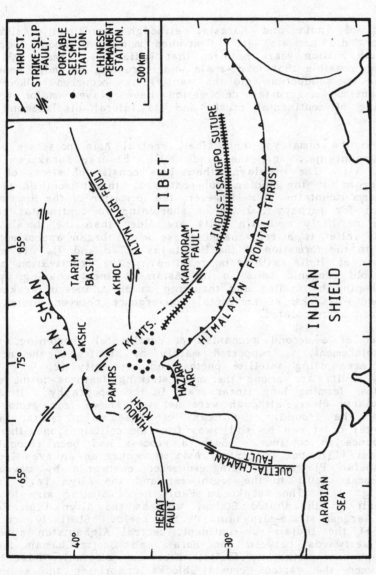

FIGURE 1. SIMPLIFIED MAP OF THE TECTONICALLY ACTIVE FEATURES OF CENTRAL ASIA. KSHC and KHOC are the Chinese seismic stations at Kashgar and Khotan respectively.

the two continents (4), (Fig. 1). Further west, in Pakistan, the suture zone splits into two, sandwiching between them the Kohistan Complex: it has been suggested that this is an old island arc, caught between the colliding continents (5, 6). Note that in Figure 1 the major strike-slip systems slide continental material out of the collision zone. The seismic stations used in the present study are also shown as black dots whilst KSHC and KHOC are the Chinese seismic stations at Kashgar and Khotan respectively.

The suturing of India and Eurasia, although significantly slowing their convergence, did not stop it. Continued motion at about 50mm/yr for the past forty million years requires that India has rammed 2000 km into Asia. It is possible that large-scale underthrusting of India beneath Asia (as opposed to subduction into the mantle) has occurred. However, two other mechanisms of crustal deformation have played major roles: vertical thickening of continental crust, and its lateral displacement out of the collision zone.

In addition to the Himalayas and Tibet, central Asia possesses other chains of high mountains, e.g., the Hindu Kush, Pamirs, Karakoram and Tian Shan (Fig. 1). The Himalayas themselves consist of slices of the old northern margin of the Indian sub-continent, thrust southwards on one another to form mountains (7). However, the formation of the Himalayas can only account for perhaps 300 km of shortening in continental crust (2). The more northerly mountain belts are older than the Himalayas, but their current relief is a rejuvenated phase which began approximately simultaneously with the formation of the Himalayas (2, 8 and 9). It seems that the collision of India with Asia resulted in the reactivation of a whole series of old orogenic belts in the Asian continent causing uplift and crustal thickening by folding and thrusting on old planes of weakness. In this way about 1000 km of horizontal convergence between India and Asia may have been accommodated.

The importance of a second mechanism of continental shortening, that of sideways displacement, is supported mainly by studies of the active tectonics of the area using satellite photos and seismicity (2, 9 - 12). Major strike-slip faults are among the most striking features to be seen on satellite photos, forming long linear scars in the topography. In Asia they are particularly clear, although were not well-known from geological reconnaissance on the ground. By means of these major faults, large areas of continental crust can be slid away from the collision zone, thereby allowing convergence to continue. Such a process had been recognized in the Middle East (13), but in central Asia it occurs on an even larger scale. The Tibetan Plateau is being squeezed eastwards by movement along the Karakoram Fault to the south-west and the Altyn Tagh Fault to the north (Fig. 1). The Karakoram Fault is of similar size to the San Andreas Fault in the United States, whereas the Altyn Tagh Fault is probably the largest strike-slip fault in the world. Similarly, on the western margin of the Indian sub-continent, central Afghanistan appears to be sliding westwards between the Herat and Quetta-Chaman faults (2, 14). Although now strike-slip faults, these may originally have been the sutures between the various crustal blocks comprising the southern margin of Asia (15).

At the junction of the Karakoram and Altyn Tagh Faults are the Karakoram Mountains, and similarly the Hindu Kush Mountains lie adjacent to the junction of the Herat and Quetta-Chaman Faults. The Hindu Kush are unusual amongst intra-continental mountain belts in possessing a highly active belt of intermediate-depth seismicity (70 - 300 km), located in a sub-vertical slab which may possibly represent a recently-subducted intra-continental basin (15 - 17). Occasional intermediate-depth earthquakes have also been reported teleseismically beneath the Karakoram Mountains (USGS, ISC; see also Lin and Shu, this volume), though unlike the Hindu Kush this region has not previously been studied by a local (and hence more accurate) seismic network. Also, the nature of the north-west end of the Karakoram Fault is uncertain. The clear lineament seen on satellite imagery does not extend much beyond 35°N, although other major strike-slip faults occur north-west of the Karakoram Mountains (25) (see Fig. 1). It is possible that within the Karakoram Mountains themselves the motion is taken up on numerous smaller faults. The Karakoram Fault itself has shown relatively little teleseismically recorded activity in recent years.

South of the large strike-slip systems, active thrusting continues along the Himalayan Frontal Thrust (18). At its north-west termination, the surface faulting swings round to the Hazara Arc to the north of Islamabad (Fig. 1), but local recording of microearthquakes (by an array near Tarbela Reservoir) has revealed that the seismic zone continues north-west into Kohistan (the Indus-Kohistan Seismic Zone) (19, 20). In this area the large thrust faults seen at the surface appear to be inactive features, whereas active seismic deformation occurs in the basement, mostly at 10 - 30 km depth but possibly extending down to 70 km.

OBJECTIVES OF THE MICROEARTHQUAKE SURVEY

A major initial objective of the project was to assess activity on the north-western end of the Karakoram strike-slip fault, and to examine the intermediate-depth earthquakes thought to occur under the Karakoram Range itself (along the Pakistan-China border). Unfortunately, we were unable to install stations on the northern flanks of the Karakoram Range, and hence coverage of this area is not good. However, we have kindly been provided with arrival-time data from the Chinese permanent seismic stations at Khotan and Kashgar at the edge of the Tarim Basin, northeast of the Karakoram Range (Fig. 1). This data, in conjunction with the stations we operated near Pasu and Skardu, permits at least some examination of possible deep activity.

Two earthquakes of significant size have occurred in northern Pakistan in recent years, near Hamran (1972) and Patan (1974) (Jackson and Yielding, these proceedings). Stations were positioned in order to examine any continuing activity in these regions. Subsequent to our fieldwork, on 12 September 1981, a magnitude 6.7 earthquake occurred in the Darel valley, only a few tens of kilometres south-east of the 1972 Hamran event).

Landsat imagery of northern Pakistan reveals numerous minor lineaments, such as straight escarpments bordering alluvium-filled valleys (e.g., the Shigar valley, north of Skardu, see Fig. 2, which is particularly prominent). These lineaments are probably active faults but teleseismic data does not show any activity in this area.

In addition to the specific topics listed above, our array has provided reconnaissance information of a large part of northern Pakistan that was previously unexplored seismically. Although some historical earthquakes are known (e.g., Gilgit 1943 (26)), teleseismic observations over the last twenty years generally indicate an apparent lack of activity.

154

FIGURE 2. SATELLITE (LANDSAT) PHOTOGRAPH SHOWING THE LINEAMENTS IN THE BALTISTAN AREA.

OPERATION OF ARRAY, AND DATA REDUCTION

During the period 6 July to 14 September, 1980, a variable number of portable seismic stations were operated throughout northern Pakistan. A total of twenty-four station positions were occupied, with a maximum of twelve stations operating simultaneously. The data presented in this paper covers the period 17 August to 15 September, 1980. This is the second half of the recording period, and the time during which the station distribution was at its greatest extent (with stations in Baltistan and at Pasu, as well as throughout much of Kohistan). Principal station positions are shown in Figure 3, with those used here shown as solid triangles.

Station siting caused some problems. The optimum station distribution, as defined by the project objectives, had to be modified according to the existence (or otherwise) of jeepable tracks. Wherever possible, seismometers were located on bedrock to improve sensitivity. The rugged topography, and the need to carry replacement batteries, forced a compromise with the desire to site stations away from roads to avoid traffic noise. The violent mountain rivers were also a major noise source. Thunderstorms, part of an unusually north-reaching monsoon incursion, produced a lot of noise on some records (one seismometer was swept away by a large mudslide).

The instruments used were all smoked drum recorders (four Lamont-type and eight Sprengnether MEQ-800) and were coupled with Willmore Mk II and Mark Product L4-C vertical component short period seismometers. The filter settings were adjusted to give a flat frequency response between 1 and 10 Hz. The recorders were set to rotate at a speed of 60 mm/min, which gave two-day records for the MEQ-800 instruments and five-day records for the Lamont instruments. Clock-drifts were checked regularly with radio time-signals (usually Delhi or Moscow) so that linearly interpolated corrections could be applied to arrival-time readings. The positions of the stations were determined using 1:250000 Topographic Maps (Gilgit, Churrai, Baltit and Mundik sheets) and are believed to be accurate to within 500 m.

The acquired data consist of over 260 seismograms, containing about 7,500 P-arrivals in total. In order to obtain the maximum amount of information from the data, all arrivals are being read, to produce as many locatable events as possible (usually in such studies a low magnitude cut-off is set). To facilitate this process, a semi-automated process developed by Soufleris et al (21) is being used. This involves the picking of arrivals by means of a high-resolution (0.025 mm) digitizing table, and subsequent conversion into arrival-times by computer. A further computer program is used to match arrival-times from different stations, and to format these in preparation for the location program.

Local events (within the array) generally had impulsive P-arrivals and could be read to within 0.1s. Their S-arrivals were also impulsive, though being in the P-code are less easy to read. However, more distant events (mostly from the Hindu Kush) tended to have emergent P-arrivals, with impulsive S-arrivals; in these cases the S-arrival was actually easier to pick than the P, as the S-P interval was about 30 - 40s. For the location program, P- and S-arrivals were initially assigned equal weight. All the locations presented here have been obtained using the HYPO71 location program (23).

156

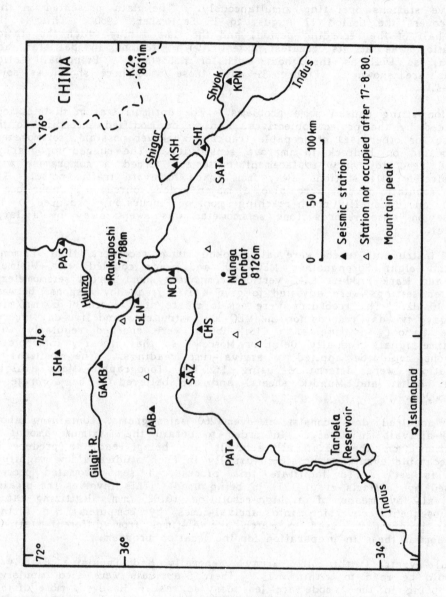

FIGURE 3. MAP OF THE KARAKORAM REGION SHOWING THE POSITIONS OF THE SEISMIC STATIONS OF THE 1980 MICROEARTHQUAKE PROJECT.

INITIAL LOCATIONS

Information for a crustal model of the area is available from Beloussov et al (22), who performed a refraction profile stretching about 1000 km from Srinagar (India) in the south to Dzhambul (USSR) in the north, passing directly through the Nanga Parbat massif. They find a thick layered crust with a P-wave velocity averaging 6.3 - 6.7 km/s and thickness about 65 - 70 km, and an upper mantle velocity of 8.05 km/s. Chatelain et al (16), investigating local seismicity in the Hindu Kush, used a velocity model having a crust 45 km thick, with 6.0 km/s P-wave velocity, overlying a mantle of P-wave velocity 8.0 km/s. A group of seventy earthquakes (representing five days recording) was located using these various models. It was found that the model giving the smallest RMS of travel-time residuals for the locations was that having the Moho at 45 km, with P-wave velocities 6.0 km/s above the Moho and 8.0 km/s below it. This model was then used for the location of all recorded events in the period 17 August - 14 September, 1980 (371 events). These locations are shown in Figure 4.

The most striking feature of Figure 4 is the intense belt of activity to the north-west of the array, containing about two-thirds of all the locatable events. This is the Hindu Kush zone of intermediate depth seismicity, known to be highly active from teleseismic data and previous local studies (16, 17 and 24). Although the data here are only a one-month sample, the general planform of activity in the belt is very similar to that seen in the teleseismic locations for the past two decades: the zone is widest and most active in eastern Afghanistan, tapering north-east into the Russian Pamirs and then taking up a more easterly trend. This similarity between a short sample recorded locally and the many years of teleseismic locations was also noted by Chatelain et al (16) after their field study in eastern Afghanistan.

The depths given by the location program for the Hindu Kush events are mostly in the range 50 - 200 km, with the deepest at 280 km. Because all of these events are well outside our array, control on their depths is not good. However, the observed range is in good agreement with that found by Chatelain et al (16).

The second principal group of locations shown in Figure 4 is that in northern Pakistan: a scattering of seismicity occurred throughout the area of the array and also to the south of it. In contrast to the Hindu Kush, these events are predominantly shallow (crustal). A number of locations deeper than 45 km are shown, but most of these are not reliable. (Fig. 4 shows all initial locations, regardless of quality. The location program, which is iterative, used a starting depth of 50 km, and if an event is poorly recorded the depth is held fixed while the epicentre is calculated). Relocations for these 'central' events are presented and discussed in a subsequent section.

A small number of events were located along the Karakoram Range (the Pakistan-China border region). These are discussed in the next section.

THE KARAKORAM RANGE AND CHINA

The combination of our portable array with the Chinese permanent seismic stations at Kashgar (KSHC) and Khotan (KHOC) in the Tarim Basin effectively gives an extended array covering both flanks of the Karakoram Mountains (see Fig. 1). However, the Chinese stations are considerably further away

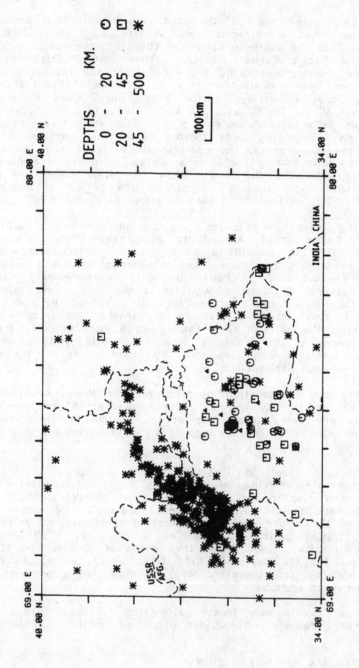

FIGURE 4. MAP SHOWING ALL LOCATABLE EVENTS RECORDED IN THE PERIOD 17 AUGUST - 14 SEPTEMBER, 1980 (NOTE THAT ALL LOCATIONS ARE SHOWN, REGARDLESS OF QUALITY). The velocity model used was an upper layer (45 km thick) of P-wave velocity 6 km/s, overlying a half-space of 8 km/s. The starting depth for location procedure was 50 km. Earthquakes of different depths are plotted as different symbols. Frontiers are shown as dashed lines, seismic stations as filled triangles.(▲).

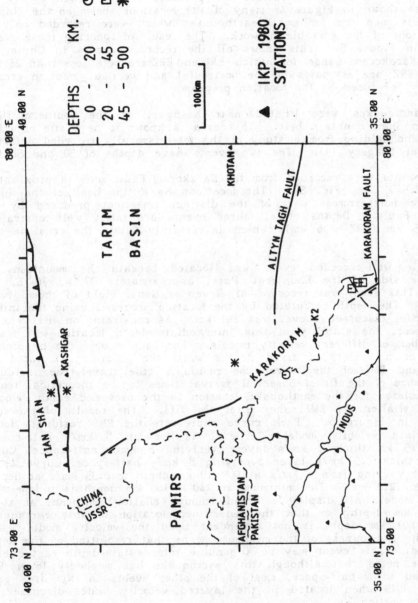

FIGURE 5. MAP SHOWING THOSE LOCATIONS NORTH-EAST OF THE PORTABLE ARRAY
FOR WHICH ERH AND ERZ ARE LESS THAN 25 KM.
Major strike-slip faults are shown, and also the Tian Shan frontal thrust belt.
Other notation as Figure 4.

from the Karakoram than the portable stations, and hence were less well placed for recording Karakoram earthquakes. Events which occurred in this area, but were not recorded on the Chinese stations, cannot be located accurately as they are well outside the portable array. Such events are among those shown in Figure 4: many of the locations shown on the Chinese area of this map are for small earthquakes which were recorded only on a few stations of the portable network. The result of ignoring these events is shown in Figure 5. This shows all the recorded events in China and the high Karakoram Range for which ERH and ERZ are both less than 25 km. (ERH and ERZ are estimates of the horizontal and vertical location errors, respectively, produced by the location program.)

Two earthquakes were located near Kashgar, on the southern flank of the Tian Shan mountain belt. This area is known to be a site of active thrusting and folding from a study of the teleseismically recorded activity, and Landsat imagery (9). The two events have depths of 50 and 54 km.

Three events were recorded from the Karakoram Fault zone (approximately 35.3°N, 77.5°E, see Fig. 5). The locations lie at the head of the Nubra valley, the northernmost part of the distinct lineament produced by the Karakoram Fault. Depths of all three events are fairly well-constrained at 34 - 35 km (ERZ = 6 km), which is certainly within the crust in this area.

A single well-recorded event was located beneath the mountains on the Chinese side of the Khunjerab Pass (approximately 37°N, 76°E, see Fig. 5). This event was recorded at eleven stations, eight of them giving S-arrivals. The depth calculated by the location program, using the initial velocity model discussed above, was 141 km. As constraint on the velocity model is poor, the stability of this intermediate depth location was tested for a number of different velocity models. For each model, the hypocentre was held at a variety of fixed depths while the program calculated the epicentre and RMS of the travel-time residuals (the travel-time residuals are a measure of the fit of observed arrival times to the theoretical travel times, calculated for the earthquake location in the presumed earth velocity model; the smaller the RMS, the better the fit). The results of this test are shown in Figure 6. Each curve represents the RMS residuals found for a certain velocity model. Curve 1 is for an 8 km/s half-space. Curve 2, 45 km thick 6 km/s layer overlying 8 km/s half-space. Curve 3, 65 km thick 6.5 km/s layer overlying 8 km/s half-space. Curve 4, 6-layer model varying from 6 km/s at 0 - 10 km depth to 8.35 km/s at depths greater than 210 km. The minimum at 140 - 160 km depth is independent of velocity model. All display a broad (though shallow) minimum at about 140 - 160 km depth, so that the determined location of this earthquake at about 140 km depth is not dependent upon the velocity model used. However, the shallowness of the minimum means that resolution of the depth is not good. This event may be a genuine intermediate-depth earthquake. (It may be noted that although this earthquake has markedly lower RMS when located in a half-space, most of the other events in this area gave a minimum RMS when located in the layered velocity model discussed in the previous section.)

About eighty kilometres to the south of this earthquake, (at approximately 36.3°N, 76°E, see Fig. 5), is an event which by contrast is quite shallow, located at 17 km depth. This was well-recorded (P+S at eight stations, though not in China), and the location is quite well constrained (ERZ = 2.5 km, ERH = 4.0 km, RMS = 0.75 secs). Such upper crustal activity may be associated with the presumed smaller-scale faulting

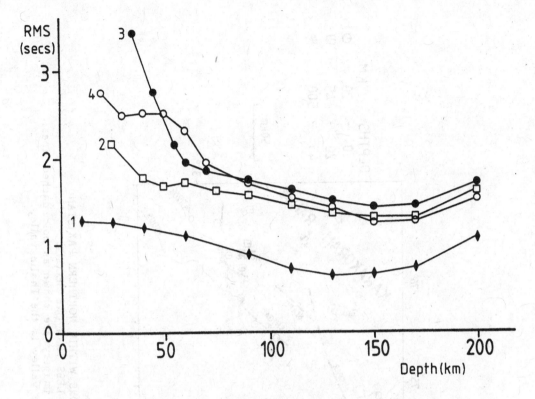

FIGURE 6.   PLOT OF RMS OF TRAVEL-TIME RESIDUALS AGAINST DEPTH
FOR AN INTERMEDIATE-DEPTH EARTHQUAKE EAST OF THE KHUNJERAB PASS
(APPROXIMATELY 37°N, 76°E, SEE FIG. 5)
Different curves show results using different velocity models; see text.

joining the Karakoram Fault to those strike-slip faults further north-west
(see Fig. 5).   The position of this event is in fact close to a straight,
15 km long, NW-trending valley, adjacent to the Braldu Glacier (east of
the Shimshal valley).   The trend of this valley may be controlled by active
faulting.   (See also Fig. 7 for position of earthquake and valley).

KOHISTAN AND BALTISTAN

The initial locations shown in Figure 4 include seventy-two events
within northern Pakistan, either within the array or close to it.   The
velocity model used for these locations was arrived at using a representative
sample, of all recorded events, as described in a previous section.   As
two-thirds of the events were located to the north-west in the Hindu Kush,
this velocity model is biased towards the earth structure between the
Hindu Kush and the seismic stations.   Because of the heterogeneous nature
of the geology within the area of the array (Kohistan Complex and
Nanga Parbat massif), the seventy-two events within the array were used
to gain more control on the local velocity model.   They were located in
six different velocity models (with crustal layers of different thicknesses
and velocities), and the mean RMS of all the locations was used as a
guide to the "best" velocity model.   As many of these earthquakes were
relatively shallow, they were located using a starting depth of 20 km

162

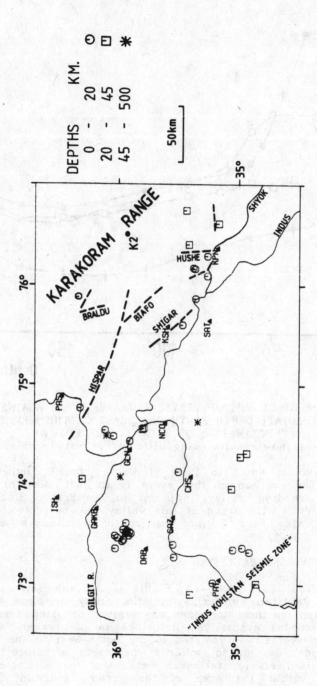

FIGURE 7. MAP OF LOCATIONS WITHIN NORTHERN PAKISTAN
FOR WHICH RMS WAS LESS THAN ONE SECOND.
Selected lineaments visible on Landsat imagery are shown as bold dashed lines;
the minor lineament west of the Hushe valley is the Thalle valley.

(as opposed to 50 km for the initial locations). Also, as they occurred relatively close to the recording stations, the S-arrival is within the coda of P-waves and therefore harder to read: for this reason the S-arrivals were now given half the weight of P in the location procedure.

Of the models tried, that used for the initial locations still gave the lowest mean RMS (6 km/s crust, Moho at 45 km, 8 km/s mantle). The residuals were not very sensitive to the thickness of the crust: lowering the model Moho from 45 km to 60 km only increased the mean RMS from 1.00s to 1.03s (location depths increased slightly also). However, increasing the crustal velocity from 6 km/s to 6.5 km/s increased the mean RMS from 1.00s to 1.07s or 1.09s (for 45 km and 60 km thicknesses respectively). These results are summarized in Table 1. Considering the probable lateral

Model	Upper layer		Half-space	Mean RMS (72 events)
	Thickness	Vp	Vp	
1	-	-	8 km/s	1.75
2	-	-	6	1.12
3	45 km	6 km/s	8	1.00
4	60	6	8	1.03
5	45	6.5	8	1.07
6	60	6.5	8	1.09

TABLE 1. RESULTS OF TEST FOR VELOCITY MODEL WITHIN THE AREA OF THE ARRAY.

All 72 locatable events within Northern Pakistan were relocated using six different velocity models, and the mean RMS for all events was taken as a measure of the fit of the model to the arrival-time data. Model three was chosen as the "best" velocity model.

variability in this large (500 km across) and tectonically complex area, further attempts at refining the one-dimensional velocity model were considered unjustified. The model adopted for the location of events within the array was therefore the same as that used initially; a 45 km thick upper layer, of P-wave velocity 6 km/s, overlying a half-space of 8 km/s. Events which gave an RMS of residuals of over one second were rejected as being unreliable. The remaining forty-five events are shown in Figure 7.

With three exceptions, all the events are shallower than 45 km. Two of the exceptions occurred near Gilgit (station GLN on Fig. 6), one to the north-west at 48 km depth, and the other to the north-east (beneath Rakaposhi) at 64 km depth. The third event, also at 64 km depth, was beneath Nanga Parbat (south of station NCO on Fig. 7). These depths are quite well-controlled, as shown by a plot of RMS against depth (Fig. 8). The depths obtained with other velocity models are similar, though have greater errors. The Rakaposhi and Nanga Parbat events lie very close to the refraction line of Beloussov et al (22), and clearly

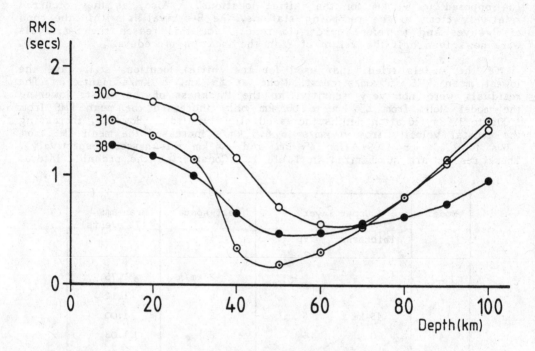

FIGURE 8.   PLOT OF RMS AGAINST DEPTH FOR THE THREE DEEPEST
EARTHQUAKES WITHIN THE ARRAY.
No. 30 was south of station NCO, no. 31 was west of station GLN,
and no. 38 north-east of GLN (see Fig. 7).   The RMS minima are in
the range 45 – 65 km, i.e. near the base of the crust in this area.

lie near the base of the crust in this area (the Moho may be shallower
west of the Nanga Parbat massif).   Five other locations lie near GLN,
but these are all in the range 17 – 26 km depth, and may not be related
to the deeper events.

A dense cluster of events was located at the head of the Hamran valley
(between stations GAKB and DAB in Fig. 7).   This is close to the Hamran
(1972) and Darel (1981) earthquakes, and is shown in more detail in
Figure 9.   (Many small local earthquakes were recorded at stations GAKB
and DAB, but could not be located as they were not recorded at other
stations).   The cluster occupies an area about 30 km long, trending WNW-
ESE; horizontal errors for individual events are 1 – 3 km.   Depths are
in the range 0 – 18 km, with the deeper ones more accurately located
(including one at 15.3±2.3 km depth, recorded as 11 P- and 10 S-arrivals).
It is noteworthy that the Hamran and Darel earthquakes nucleated at about
10 km depth, and that the Hamran aftershocks had a depth range of about
7 – 21 km (Jackson and Yielding, these proceedings).   Thus in this area
seismic deformation appears to be restricted to the top 20 km of the crust
both during important earthquakes and during "background" activity.

The relocations for the 1972 and 1981 events (shown in Fig. 9), may
be in error by 10 – 15 km, as the relocation procedure does not give an
absolute geographical position.   It is possible that the area of the Hamran
aftershocks swarm, in fact coincides   with the cluster of locally recorded

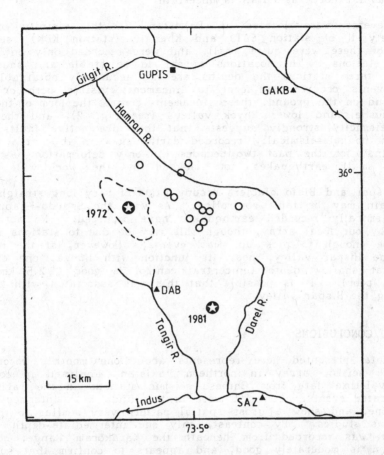

FIGURE 9.   MAP OF EARTHQUAKE LOCATIONS IN THE HAMRAN-DAREL AREA.
LOCATIONS FROM THE 1980 PROJECT ARE SHOWN AS OPEN CIRCLES.
The Hamran (1972) and Darel (1981) earthquakes are shown as stars,
and the extent of the Hamran aftershock cloud as a dashed line
(from Jackson and Yielding, these proceedings).

events (i.e. the 1972 and 1981 epicentres shown in Fig. 9 should be
displaced 10 - 15 km ENE).   Alternatively, the locally recorded events
(1980) may indicate a propagation of activity between the 1972 and 1981
fault zones, in which case the 1972 and 1981 epicentres should be shifted
about 15 km to the north.

Figure 7 shows a number of events in the vicinity of Patan (station
PAT, in south-west of map).   These correspond to the "Indus Kohistan
Seismic Zone" (IKSZ) of Armbruster and Seeber (19 and 20), who used
a seismic array around Tarbela Reservoir to study local seismicity.   The
IKSZ is believed to be a basement continuation of the Main Boundary Thrust
of the Himalayas.   The events located in the zone by this study have
a depth range of 0 - 31 km (individual events +7 km or less).   Armbruster
and Seeber (19 and 20) show most of the IKSZ activity to be in the depth
range 5 - 25 km, but with some extending down to 70 km.   However, they
do not discuss the accuracy of their locations, and the reliability of their

locations as far north as Patan is uncertain.

Nine events were located in Baltistan, in the vicinity of Skardu (immediately N of station SAT) and Khapalu (station KPN), see Figure 7. All of these were quite small, and were recorded only on the three Baltistan stations. The locations appear to be stable and shallow, but with only three stations the depths are not accurately obtainable. Most of the events occurred adjacent to lineaments visible both on Landsat imagery and on the ground; these lineaments control the form of the Shigar, Thalle, Hushe, and lower Shyok valleys (see Fig. 2), and the location of the seismicity strongly suggests that they are active faults. There have been no teleseismically recorded earthquakes in this area, so it is possible that for the past two decades the only deformation has occurred by very small earthquakes not detected by the world-wide network.

The Hispar and Biafo glaciers occupy parts of very long straight valleys which again may be fault controlled. As with the Skardu-Khapula area, no teleseismically recorded earthquakes have occurred. No activity was recorded by our local array, though this may be due to stations not being sited close enough to pick up small events. However, at the north-west end of the Hispar valley, near its junction with Hunza, one event was recorded at shallow depth (epicentral control is good ($\pm$ 2.5 km), depth control is poor). It is possible that this was associated with an active fault along the Hispar valley.

CONCLUSIONS

The data presented here represents about one month's recording by a portable seismic array in northern Pakistan, analysed in conjunction with arrival-time data from Chinese seismic stations in the Tarim Basin. Of 371 located events, about 250 were in the Hindu Kush intermediate-depth seismic zone, and show a gross spatial pattern very similar to that found in previous studies. By contrast, only one intermediate-depth (140 km) earthquake was recorded from beneath the Karakoram Range; control on its location is moderately good, and appears to confirm that sub-crustal seismic activity occurs here, as suggested by teleseismic data.

The prominent lineament of the Karakoram Fault becomes indistinct as it approaches the Karakoram Mountains, though other prominent faults occur further north-west near the China-USSR border (see Fig. 5). Although numerous minor lineaments occur through the Karakoram Range and Baltistan, little crustal activity has been recorded here teleseismically. Our array recorded three well-located events at the north-western end of the Karakoram Fault, one near a lineament in the high Karakoram Range, and nine in Baltistan, also close to lineaments. This small-scale activity suggests that these lineaments are indeed active faults, and that the lack of teleseismically recorded earthquakes is due to a short sampling time rather than a lack of seismogenic faulting.

In contrast to the intermediate-depth activity beneath the Hindu Kush and Karakoram Ranges, seismicity within Kohistan appears to be confined to the crust. Of thirty-six well-located events, only three lie in the depth range 45 - 65 km (the rest are shallower), and it may be significant that two of these were beneath some of the highest topography in the area (Nanga Parbat and Rakaposhi). Some shallow activity was recorded from the Patan area of the Indus Kohistan Seismic Zone, but the principal site of shallow seismicity was the Hamran valley, close to the epicentre of the Hamran (1972) earthquake. This cloud of events occupied an area 30 km long, with a depth range of 0 - 20 km; these depths are similar to those of the aftershocks of the Hamran earthquake, and consistent with the figure of 10 km found for the nucleation depths of the Hamran mainshock and the Darel (1981) earthquake. Owing to the lack of accuracy in the teleseismic locations of the 1972 and 1981 events, it is not clear exactly how they relate spatially to the locally recorded events. However, constraint on the range of depths is good, and it seems that in the Hamran-Darel area seismicity is restricted to the top 20 km of the crust.

ACKNOWLEDGEMENTS

This work was funded by the Royal Society, the Wood Fund, and the Natural Environment Research Council. Shane Wesley-Smith, John Allen, and A. Awan helped in the running of the seismic network. This is Cambridge University Department of Earth Sciences contribution no. 216.

REFERENCES

(1)   MCKENZIE, D. P. and SCLATER, J. G., (1971). 'The Evolution of the Indian Ocean since the late Cretaceous', Geophys. J. Roy. Astr. Soc., Vol. 25, pp 437 - 528.

(2)   MOLNAR, P. and TAPPONNIER, P., (1975). 'Cenozoic tectonics of Asia: Effects of a continental collision', Science, Vol. 189, pp 419 - 426.

(3)   POWELL, C. McA., (1979). 'A speculative Tectonic History of Pakistan and Surroundings: Some constraints from the Indian Ocean', in "Geodynamics of Pakistan" ed. by Farah and De Jong, Geological Survey of Pakistan, pp 5 - 24.

(4)   GANSSER, A., (1974). 'The ophiolitic melange, a world-wide problem on Tethyan examples', Eclogae Geol. Helvetiae, Vol. 67, pp 497 - 507.

(5)   TAHIRKHELI, R. A. K., MATTAUER, M., PROUST, F. and TAPPONNIER, P., (1979). 'The India Eurasia Suture Zone in Northern Pakistan: Synthesis and Interpretation of Recent Data at Plate Scale', in "Geodynamics of Pakistan", ed. by Farah and De Jong, Geological Survey of Pakistan, pp 125 - 130.

(6)   COWARD, M. P., JAN, D., REX, D., TARNEY, J., THIRLWALL, M., and WINDLEY, B. F., (1982). 'Structural evolution of a crustal section in the western Himalaya', Nature, 295, pp 22 - 24.

(7)   STOCKLIN, J., (1980). 'Geology of Nepal and its regional frame', J. Geol. Soc., Vol. 137, pp 1 - 34.

(8)   KRAVCHENKO, K. N., (1979). 'Tectonic Evolution of the Tian Shan, Pamir and Karakoram', in "Geodynamics of Pakistan" ed. by Farah and De Jong, Geological Survey of Pakistan, pp 25 - 40.

(9)   TAPPONNIER, P. and MOLNAR, P., (1979). 'Active faulting and Cenozoic tectonics of the Tian Shan, Mongolia, and Baykal regions', J. G. R., 84, pp 3425 - 3456.

(10)   TAPPONNIER, P. and MOLNAR, P., (1977).   'Active faulting and
        tectonics in China'.   J. Geophys. REs., 82, pp 2905 - 2930.

(11)   MOLNAR, P. and TAPPONNIER, P., (1978).   'Active Tectonics of
        Tibet', J. Geophys. Res., 83, pp 5361 - 5375.

(12)   NI, J. and YORK, J. E., (1978).   'Late Cenozoic Tectonics of the
        Tibetan Plateau', J. Geophys. Res., 83, pp 5377 - 5384.

(13)   McKENZIE, D. P., (1972).   'Active tectonics of the Mediterranean
        region', Geophys. J. Roy. Astr. Soc., 83, pp 109 - 185.

(14)   PREVOT, R., HATZFELD, D., ROECKER, S. W., MOLNAR, P., (1980).
        'Shallow earthquakes and active tectonics in Eastern Afghanistan',
        J. Geophys. Res., 85, pp 1347 - 1357.

(15)   TAPPONNIER, P., MATTAUER, M., PROUOST, F., CASSAIGNEAU, C.,
        (1981).   'Mesozoic ophiolites, sutures, and large-scale tectonic move-
        ments in Afghanistan', Earth and Planet. Sci. Letters, 52,
        pp 355 - 371.

(16)   CHATELAIN, J. L., ROECKER, S., HATZFELD, D., MOLNAR, P., (1980).
        'Microearthquake seismicity and fault plane solutions in the Hindu
        Kush region and their tectonic implications', J. Geophys. Res.,
        85, pp 1365 - 1387.

(17)   ROECKER, S. W., (1981).   'Seismicity and tectonics of the Pamir-
        Hindu Kush region of Central Asia'. Ph. D. Thesis, M. I. T.,
        Cambridge, Massachusetts, U. S. A.

(18)   MOLNAR, P., CHEN, W. P., FITCH, T. J., TAPPONNIER, P.,
        WARSI, W. E. K. and WU, T. F., (1977).   'Structure and tectonics
        of the Himalaya: A brief summary of relevant physical observations',
        Proc. CNRS Symp. Geol. Ecology Himalayas, Paris, pp 295 - 300.

(19)   ARMBRUSTER, J., SEEBER, L. and JACOB, K. H., (1978).   'The north-
        western termination of the Himalayan Mountain Front: Active tectonics
        from microearthquakes', J. Geophys. Res., 83, pp 269 - 282.

(20)   SEEBER, L. and ARMBRUSTER, J., (1979).   Seismicity of the Hazara
        Arc in Northern Pakistan: Decollement versus basement faulting',
        in "Geodynamics of Pakistan", ed. by Farah and De Jong, Geological
        Survey of Pakistan, pp 131 - 142.

(21)   SOUFLERIS, C., JACKSON, J. A., KING, G. C. P., SPENCER, C. P.,
        PAPAZACHOS, B. C. and SCHOLZ, C. H., (1982).   'The 1978 earthquake
        sequence near Thessaloniki (northern Greece)', Geophys. J. Roy.
        Astr. Soc. 68, pp 429 - 458.

(22)   BELOUSSOV, V. V., BELYAEVSKY, N. A., BORISOV, A. A.,
        VOLVOVSKY, B. S., VOLVOVSKY, I. S., RESVOY, D. P.,
        TAL-VIRSKY, B. B., KHAMRABAEV, I. Kh., KAILA, K. L.,
        NARAIN, H., MARUSSI, A. and FINETTI, J., (1980).   'Structure of
        the lithosphere along the deep seismic sounding profile: Tian Shan-
        Pamirs-Karakoram-Himalayas', Tectonophysics, 70, pp 193 - 221.

(23)   LEE, W. H. K. and LAHR, J. C., (1975).   'HYP071 (Revised): a
        computer program for determining hypocentre, magnitude and first-

motion pattern of local earthquakes', U. S. G. S. Open File Report pp 75 - 311.

(24)   BILLINGTON, S., ISACKS, B. L. and BARAZANGI, M., (1977). 'Spatial distribution and focal mechanisms of mantle earthquakes in the Hindu Kush–Pamir region: a contorted Benioff zone', Geology, 5, pp 699 - 704.

(25)   BURTMAN, V. S., PEIVE, A. V. and RUSHENTSEV, S. V., (1963). 'The main strike–slip faults of the Tian Shan and Pamir', Faults and Horizontal Movements of the Earth's Crust, (in Russian), Trans. Geol. Inst. Acad. Sci. USSR, 80, pp 152 - 172

(26)   QUITTMEYER, R. C. and JACOB, K. H., (1979).   'Historical and modern seismicity of Pakistan, Afghanistan, Northwestern India and Southeastern Iran', Bull. Seis. Soc. Am., 69, pp 773 - 824.

Source studies of the Hamran (1972.9.3), Darel (1981.9.12) and Patan (1974.12.28) earthquakes in Kohistan, Pakistan

J. Jackson and G. Yielding – Department of Geophysics, University of Cambridge

ABSTRACT

Long period body waves are examined to show that the Hamran (1972.9.3), Darel (1981.9.12) and Patan (1974.12.28) earthquakes in Kohistan had focal depths of about 8 to 10 km. All involved high angle reverse faulting (thrusting) and had seismic moments of about 2.2 to 2.7 x 10^{25} dyne-cm. These shallow depths contrast with the deeper hypocentres found in the Hindu Kush and northeast Karakoram to the north and in Hazara to the south. The Hamran and Patan shocks were assigned depths of 45km by the ISC, indicating that even well-recorded events in this region may have focal depths in error by 30 km. The accurate focal depth control in Kohistan provided by the microseismic network operated during the International Karakoram Project will be important in understanding how the deeper seismicity further north is related to that in the south.

INTRODUCTION

Within the area of the seismic network operated during the International Karakoram Project only three earthquakes of significant size have occurred since the installation of the World Wide Standard Seismograph Network (WWSSN) in the early 1960's and the advent of routine and reasonably accurate epicentre and magnitude determinations. These are the Hamran (1972.9.3, m_b 6.3), Patan (1974.12.28, m_b 6.0) and Darel (1981.9.12, m_b 6.1) earthquakes, whose epicentres are shown in Figure 1. Indeed, teleseismically recorded activity has been scarce throughout the whole of Kohistan, Baltistan and the Northern Areas for the last 20 years, and epicentres reported by the U. S. Geological Survey (USGS) and International Seismological Centre (ISC) during this period are mainly aftershocks of the Hamran and Patan earthquakes. Historically, however, the region is known to have been active, and in 1871 and 1943 damaging shocks occurred near Gilgit (1, 2).

Of particular interest are the focal depths of earthquakes in this region. To the northwest, in the Hindu Kush, earthquakes are common to depths of about 300 km (3, 4), though, rather surprisingly, very low levels of seismicity are observed shallower than 70 km. Only southwest of Chitral, towards the Kunar – Chaman fault system of eastern Afghanistan, does shallow activity increase (5). To the northeast, beyond the highest peaks of the Karakoram and in the Aksai Chin, intermediate depth earthquakes are found down to about 100 km (6, 7), though these are much less frequent than in the Hindu Kush. Shallow earthquakes also occur in the same area. No reliable teleseismic locations are known from the region in which the highest peaks of the Karakoram lie; immediately east of the Hunza river and north of the Skardu – Gilgit section of the Indus. To the south of the Patan region, in the Hazara arc, local seismic networks have reported focal depths down to 50 km (8, 9).

A cursory glance at the hypocentres reported in the Kohistan region by the USGS indicates apparently intermediate depth seismicity. In particular, depths of greater than 200 km have been reported for shocks near Patan (1965.1.23), Gilgit (1964.1.24) and Khaplu (1977.9.13), which is deeper than the crustal thickness of 60 – 65 km determined by Beloussov et al (10). Closer examination shows these events to be very poorly recorded, by few stations (21, 5 and 7 respectively) with a bad azimuth and distance distribution from the epicentre. Without good station coverage, preferably including local stations, focal depths determined from arrival times alone may be very inaccurate (11, 12). These three deep events are certainly untrustworthy. An alternative method of estimating focal depth involves modelling the long period P waveforms observed on WWSSN instruments. This technique is discussed in detail by Langston and Helmberger (13) and Jackson and Fitch (14). It can only be used for shocks of magnitude greater than about 5.6 and in this study is applied to the Hamran, Patan and Darel earthquakes. The first two of these were located at 45 km by the ISC, who have not yet located the Darel shock (for which the USGS reported an unresolved depth of 33 km). Other shocks in the Kohistan area with reported depths down to 100 km are too small to be subjected to this type of analysis.

Fault plane solutions, indicating the type and orientation of faulting during earthquakes, are also rare in this area. Armbruster et al. (8) and Seeber and Armbruster (9) show strike slip and thrust faulting solutions in the Hazara arc to the south of Kohistan. Pennington (15) and Ambraseys et al. (16) present a thrust solution for the Patan shock, and Chandra (17) shows thrust faulting in the Gupis-Hamran area. Chandra (17) and Shirokova (18) both show a normal faulting solution for a shock 50 km south of Gupis (1965.1.29, USGS location: 35.6° N, 73.6° E, m_b 5.7), which was certainly too small to be read on long period WWSSN instruments, and a search by the first author could not find unambiguous short period onsets either. As both of them publish only the finished solutions and not the individual polarity readings on which they are based, the quality of their results is not clear. Shirokova (18) presents (without polarities) a normal faulting solution for another shock on 1968.3.3, 34.7°N, 72.3°, m_b 5.2, and does show individual polarities for the 1972 Hamran mainshock; which contain so many inconsistencies that a fault plane solution is not even attempted. In view of this the reliability of these normal faulting mechanisms (surprising in an area dominated by thrust and strike slip tectonics) must be open to doubt.

THE HAMRAN EARTHQUAKE OF 1972.9.3

This shock, of m_b 6.3, killed at least 100 people and destroyed more than 1000 houses (1). The most intense destruction was in the Hamran valley (see Figures 2 and 3) in the villages of Hamran, Ushkur and Hangrus, with less damage east of Hangrus and in Darel (19). It was felt strongly in Kabul, Rawalpindi, Peshawar, Jhelum and Lahore. It was followed by many aftershocks of magnitude greater than 5.0, the largest

172

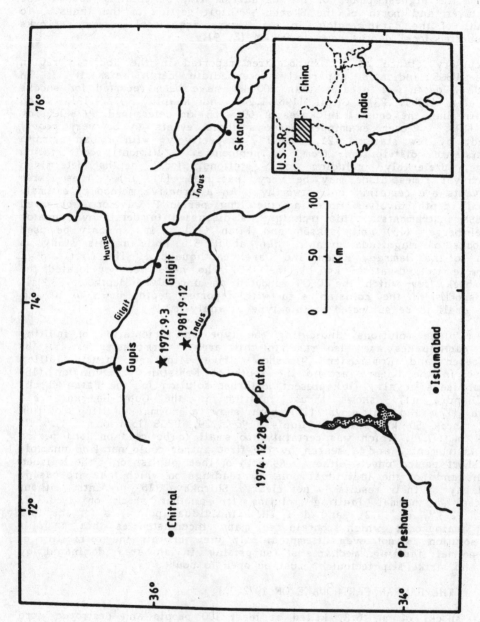

FIGURE 1. LOCATION MAP OF KOHISTAN AND SURROUNDING AREAS.
EPICENTRES ARE MARKED BY STARS.

of which were relocated using the technique of Jackson and Fitch (20) and are shown in Figure 2 and Table 1. This technique relocates after-shocks relative to a reference shock (the mainshock in this case) and produces a more reliable pattern of epicentres as the bias from an assumed earth velocity model is reduced. Figure 2 omits error bars for the sake

Fig. 2.	Date	OT	m_b	Lat.	Long.	Depth Km	N
M	1972 9 3	1648	6.3	35.940	73.330	10	
2	1972 9 3	2303	5.6	35.952	73.281	7	175
3	1972 9 4	0050	5.3	35.982	73.278	14	123
4	1972 9 4	0123	5.4	35.926	73.376	20	131
5	1972 9 4	0236	5.2	35.984	73.368	9	121
6	1972 9 4	0351	5.3	35.965	73.317	11	122
7	1972 9 4	1337	5.2	35.922	73.327	7	116
8	1972 9 4	1342	5.8	35.920	73.376	21	176
9	1972 9 5	0914	5.1	35.919	73.392	11	104
10	1972 9 17	1737	5.4	35.969	73.273	13	147
11	1972 10 12	0021	5.2	35.971	73.294	15	106
12	1972 10 13	0504	5.2	35.924	73.344	20	91
13	1973 12 9	0236	5.0	35.969	73.356	10	103
Darel	1981 9 12	0715	6.1	35.726	73.551	8	101
Patan	1974 12 28	1211	6.0	35.060	72.910	10	

Table 1 Hypocentres of the Hamran sequence relocated relative to the mainshock (M), for which the ISC location was assumed. OT is the origin time (GMT), m_b the USGS body wave magnitude, and N the number of stations used in the relocation. Depths are in km and are converted from relative to absolute values using the mainshock focal depth of 10 km found in this paper. Numbers in the left hand column identify the shocks in Figure 2. The Darel earthquake is shown relocated relative to the Hamran mainshock, with its depth of 8 km from waveform modelling (Figure 5). Also shown is the ISC location for the Patan earthquake, with the 10 km depth found in this paper.

of clarity, but the dashed line encloses the whole aftershock area and includes the error bars for each location, which are less than 5 km. Also shown is the fault plane solution for the main shock (Figure 4), with the compressional quadrant shaded. Figure 2 shows that the after-shocks extend over about 15 km, with the mainshock in the middle of the pattern. The northwesterly trend of the aftershocks is in agreement with the fault plane solution in Figure 4, which shows a high angle reverse fault (thrust) mechanism for this earthquake. Molnar et al. (21) give a similar mechanism with a more easterly strike. In Figure 4 compressional first motions are filled, dilatational open, and nodal readings marked with crosses. P and T axes are marked by open circles with crosses. Slip vectors are marked by open circles with dots on the nodal planes

174

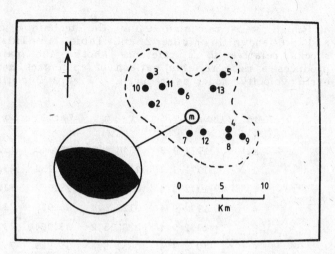

FIGURE 2. AFTERSHOCKS OF
THE HAMRAN EARTHQUAKE
NUMBERS IDENTIFY EACH
SHOCK IN TABLE 1 AND ARE
SEQUENTIAL.

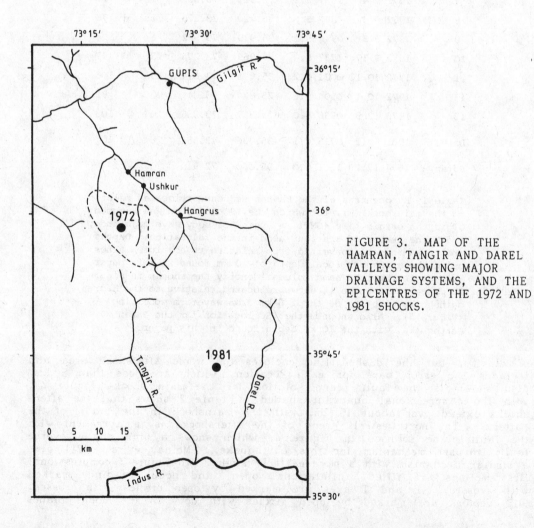

FIGURE 3. MAP OF THE
HAMRAN, TANGIR AND DAREL
VALLEYS SHOWING MAJOR
DRAINAGE SYSTEMS, AND THE
EPICENTRES OF THE 1972 AND
1981 SHOCKS.

and their horizontal projections are shown as arrows. Faulting could have occurred on either plane 1 or 2, and this ambiguity can not be resolved with the data discussed in this paper. Also shown are the observed (top) and synthetic (bottom) long period vertical P waveforms at various WWSSN stations. Synthetics were calculated with the program of Langston and Helmberger (13) using a value of $t^* = 1.0s$. The value of the seismic moment (M_o) at each station is shown in units of 10^{25} dyne-cm. Also shown in Figure 4 are the observed and synthetic long period P waveforms at selected WWSSN stations in the distance range 26° to 80°. These constrain the focal depth to about 10 km. Uncertainties in the source time function and velocity structure above the source as well as the fit of observed to synthetic seismograms all contribute to error, which may be conservatively estimated as ±4 km (14). In this study a uniform P velocity of 6.1 kms^{-1} above the source was assumed. The best fit was found using a triangular time function of two seconds rise and fall time giving a maximum total duration of four seconds faulting, which, at a typical rupture velocity of 3 kms^{-1} implies a fault length of up to 12 km. This is consistent with the aftershock distribution in Figure 2.

By comparing observed with synthetic amplitudes the seismic moment, M_o, may be estimated. Values at each station are shown in Figure 4. The average value (excluding KOD which, at 26°, is too close to be reliable) is 2.2×10^{25} dyne-cm. If we assume a fault length of 12 km and faulting to a depth of 10 km at a 45° dip, this would imply a mean seismic displacement of 44 cm and a stress drop of 12 bars (Table 2). No surface

	Nodal strike	Planes dip	Depth (km)	M_s	m_b	M_o dyne-cm	A (km^2)	$\bar{u}$ (cm)	σ (bars)
Hamran	104	36S	10	6.2	6.3	2.2×10^{25}	12x14	44	12
	104	54N							
Darel	104	36S	8	5.9	6.1	2.4×10^{25}	(12x14)	(48)	(13)
	104	54N							
Patan	104	45N	10	6.2	6.0	2.7×10^{25}	(12x14)	(48)	(12)
	124	47S							

Table 2 Fault parameters for the Hamran, Darel and Patan main-shocks. Strikes and dips of the nodal planes in Figures 3, 5 and 6 are shown. M_s is the USGS surface wave magnitude, A is the assumed area of faulting, $\bar{u}$ the mean seismic slip, and σ the stress drop. These quantities are related by:

$$M_o = \mu \bar{u} A, \quad \sigma = \mu \bar{u}/L$$

where $\mu = 3 \times 10^{11}$ dyne-cm and L is the shortest fault dimension, in this case assumed to be 12 km. Values of A, $\bar{u}$ and σ for the Darel and Patan earthquakes are shown in brackets, because there is no reliable evidence on which to base an estimate of fault length, and hence A.

faulting has been reported from this earthquake. Estimates of seismic moment from body waves are often smaller than those from both longer period surface wave analyses and direct field measurements where surface faulting is visible (22 - 24). The mean displacement and stress drop

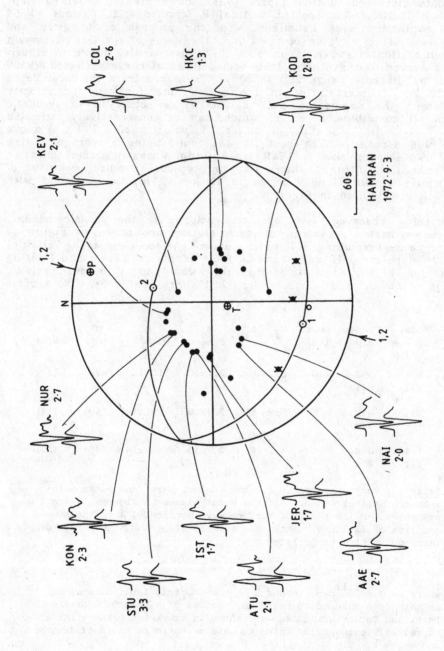

FIGURE 4. FAULT PLANE SOLUTION AND SYNTHETIC WAVEFORMS FOR THE HAMRAN EARTHQUAKE. POLARITIES WERE READ ON WWSSN LONG PERIOD INSTRUMENTS AND ARE PLOTTED ASSUMING A P VELOCITY IN THE CRUST BELOW THE SOURCE OF 6.8 KMS⁻¹.

reported here are therefore also likely to be underestimates.

Chandra (17) shows fault plane solutions for two aftershocks of this series (Numbers 2 and 8 in Figure 2 and Table 1) but shows no polarity readings for either. Both were too small for reliable long period solutions to be read, but clear short period onsets at a number of stations in Africa, Europe, North America and India as well as long period dilatations at Nilore (NIL) to the south require thrust solutions, and are consistent with the solution for the mainshock in Figure 4.

The ISC epicentre for the mainshock is about 10 km southwest of the Hamran valley (Figure 3). The Hamran valley (draining north) is separated from the Tangir and Darel valleys (draining south) by a watershed at 14 - 16,000 ft. The 1972 mainshock is shown in its ISC location, with the extent of its aftershock distribution indicated by the dashed line. The 1981 Darel shock is shown in its relocated position (5 km north of its USGS epicentre). The trend of the aftershocks and the fault plane solution of the mainshock suggest that the valley might be fault controlled. However, the ISC location could certainly be misplaced by 10 - 15 km, so it cannot be assumed that the valley marks the outcrop of a thrust dipping southwest (rather than northeast).

THE DAREL EARTHQUAKE OF 1981.9.12

The Darel earthquake (m_b 6.1) caused severe destruction in the Darel and Tangir valleys, killing an estimated 222 people (25). The waveform data currently available is shown in Figure 5. Symbols are the same as in Figure 4. The smaller filled circles are first motions reported to the USGS by WWSSN stations and were not read by the authors. First motions require a fault plane solution involving predominantly thrusting, and waveforms can be modelled using the same fault orientation as the Hamran mainshock (Figure 4). Again a triangular time function of four seconds total duration was used. Synthetic seismograms best fit the observed at a depth of 8 km with a seismic moment of 2.4×10^{25} dyne-cm. This shock therefore appears to be of similar character and size to the 1972 Hamran earthquake. Its relocated epicentre relative to the 1972 Hamran mainshock is about 31 km to the southeast and is shown in Figure 3 and Table 1. If the 1972 shock is fixed at its ISC epicentre this places the relocated Darel shock 5 km north of the epicentre located for it by the USGS. Although the Darel earthquake apparently occurred on a southeastern continuation of the Hamran valley, and 1972 aftershock trend, this may be fortuitous. The topography shows that the upper Hamran valley curves eastwards with no apparent continuation to the southeast. Also the strike of the nodal planes in the Darel fault plane solution are not sufficiently well controlled to ascertain whether it occurred on this northwest trend.

Only one sizeable aftershock was recorded teleseismically in the month following the Darel mainshock (on 1981.9.18, 17h 04m GMT, m_b 5.1). Its USGS location is the same as that of the mainshock and it was not recorded by enough stations to make relocation worthwhile.

THE PATAN EARTHQUAKE OF 1974.12.28

The Patan earthquake (m_b 6.0) was very destructive, killing 1,000 people and destroying 18,000 houses (1, 16). It was followed by numerous aftershocks, most of which were smaller than those at Hamran, and were not recorded by sufficient stations to make relative relocations worthwhile. Figure 6 shows a fault plane solution which is compatible with that of

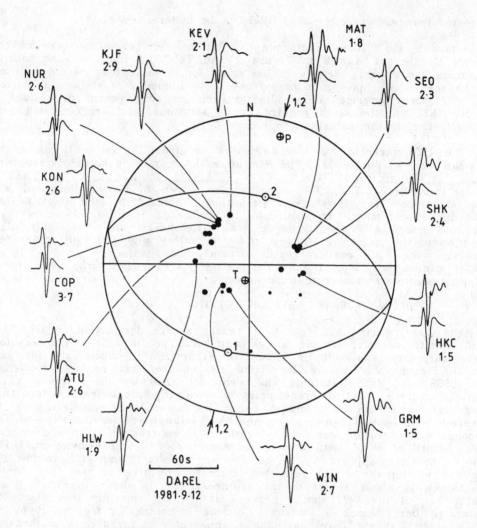

FIGURE 5.   FAULT PLANE SOLUTION AND SYNTHETIC WAVEFORMS
FOR THE DAREL EARTHQUAKE

Pennington (15) and Ambraseys et al. (16) who were able to use local
stations near Tarbela Dam as an additional constraint.   This again shows
high angle reverse faulting (thrusting) with a northwesterly strike and,
as at Hamran, synthetic waveforms best fit the observed at a focal depth
of 10 km with a triangular time function of four seconds total duration.
No surface faulting was observed following the earthquake (1, 16).   An
average seismic moment of 2.7 x 10^{25} dyne-cm is close to that at Hamran
and Darel and would imply a similar stress drop and seismic displacement
if we again assume a fault length of 12 km.   This is consistent with the
USGS aftershock epicentres, but they are all small (mostly less than m_b
5.0), and may not be reliable.   A clear downward pulse is visible following
the downward sP peak at KEV, COP, ATU and BUL in Figure 6 and may
be indicative of a second rupture.   If this is the case it is another reason
why the moment value given here is likely to be an underestimate.

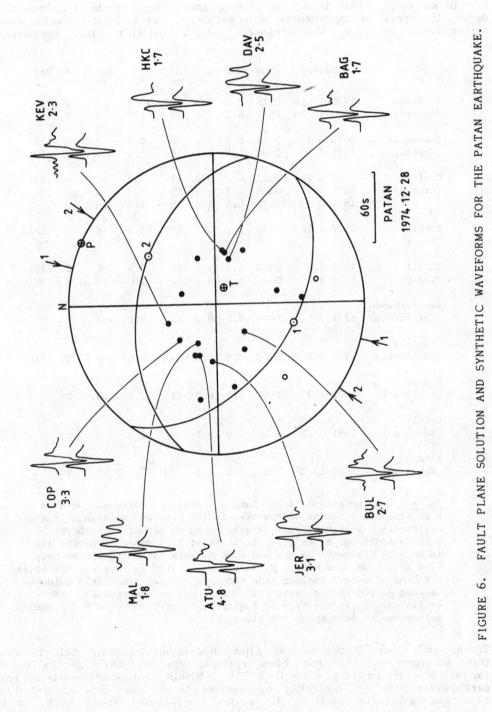

FIGURE 6. FAULT PLANE SOLUTION AND SYNTHETIC WAVEFORMS FOR THE PATAN EARTHQUAKE.

DISCUSSION

All three of the mainshocks discussed here had shallow foci about 8 - 10 km deep. This is not surprising and appears to be a common focal depth in areas of continental deformation. Table 3 lists some recent earthquakes in the Mediterranean and California for comparison.

	Date	time	M_s	m_b	L Km	M_o	depth Km	F	Ref
El Asnam	1980 10 10	1225	7.3	6.5	25	25.0	10	T	(22)
(Algeria)	1980 10 10	1539	6.0	6.1	12	2.4	8	T	(22)
Thessaloniki	1978 5 23	2334	5.6	5.6	8	0.6	6	N	(26)
(Greece)	1978 6 20	2003	6.5	6.1	16	5.2	6	N	(27)
Corinth	1981 2 24	2053	6.7	5.9	15	7.3	10	N	(23)
(Greece)	1981 2 25	0235	6.4	5.5	15	1.7	8	N	(23)
	1981 3 4	2158	6.4	5.9	15	1.0	8	N	(23)
Tabas (Iran)	1978 9 16	1535	7.7	6.5	80	44.0	8	T	(28)
Friuli	1976 5 6	2000	6.5	6.0	20	2.9	8	T	(29)
(Italy)	1976 9 15	0921	5.9	5.4	10	1.0	8	T	(29)
San Fernando (California)	1971 2 9	1400	6.5	6.2	10	8.6	13	T	(30)
Oroville (California)	1975 8 1	2020	5.8	5.6	10	0.6	6	T	(31)
Hamran (Kohistan)	1972 9 3	1648	6.3	6.3	12	2.2	10	T	
Darel (Kohistan)	1981 9 12	0715	5.9	6.1	?	2.4	8	T	
Patan (Kohistan)	1974 12 28	1211	6.2	6.0	?	2.7	10	T	

Table 3   A comparison of the Hamran, Darel and Patan earthquakes with some other recent earthquakes in the Alpine-Himalayan belt and California. L is the fault length, based on either aftershocks or surface faulting, M_o is given in units of 10^{25} dyne-cm, depths are in km. In all cases depths and moments were estimated by the body wave technique used here. M_o values calculated in this way are often a factor of 2 or 3 smaller than those calculated from field evidence or long period surface waves (22 - 24), and so are probably under-estimates. F is the type of faulting; T for thrust and N for normal. References to each shock are given.

These, and other shocks in the Alpine-Himalayan mountain belt to which this waveform analysis has been applied typically have depths in the range 6 - 15 km (e.g. 14, 32 - 35). Within continental collision zones earthquakes with well-controlled hypocentres deeper than this are relatively rare and generally confined to a few well-known areas such as the

Hindu Kush. Yet they do occur in southern Tibet (36) as well as the north-eastern Karakoram and Pamir (6, 7) and Armbruster et al. (8) report depths of 50 km in the Hazara arc south of Kohistan. The relation between the seismicity of Kohistan and the apparently deeper seismicity surrounding it is not clear. Although the Hamran, Darel and Patan shocks are shallower than 15 km this does not rule out the possibility that smaller shocks in the region may occur at greater depths. This has recently been demonstrated in the Oroville area of California (37), where seismicity occurs to depths of 40 km even though the Oroville earthquake itself had a focal depth of 6 km (31) and its aftershocks extended only to 12 km (38, 39). The Oroville region of the Sierra Nevada foothills has anomolously low heat flow for California (40) and there exists the intriguing possibility that the maximum depth of seismicity within the crust is closely linked to the regional heat flow, which in turn controls the variation of the rock strength with depth (37, 41). Although the Hamran aftershocks are all apparently shallow (Table 1), the relocated depths of these smaller shocks are probably not accurate to better than 10 km. Both the Hamran and Patan mainshocks were assigned depths of 45 km by the ISC, indicating that even for well recorded shocks, whose errors should be smaller (11), routinely determined focal depths in this region can be wrong by 30 km. The precisely determined maximum depth of seismicity in this part of Kohistan will be one of the most interesting results of the microearthquake survey carried out during the International Karakoram Project.

ACKNOWLEDGEMENTS

We would like to thank W-P. Chen and S. Roecker for kindly making their work available to us prior to publication, R. Sibson for drawing our attention to the Oroville earthquakes, Dieter Illi for telling us of his field investigation of the Hamran earthquake, and Russ Needham of NEIS for sending arrival times of the Darel earthquake. This work was supported by the Natural Environment Research Council and is Cambridge University Department of Earth Sciences contribution No. 187.

REFERENCES

(1)   AMBRASEYS, N., LENSEN, G. and MOINFAR, A., (1975). 'The Patan earthquake of 28 December 1974'. UNESCO Technical Report, Paris.

(2)   QUITTMEYER, R. C. and JACOB, K. H., (1979). 'Historical and modern seismicity of Pakistan, Afghanistan, Northwestern India and Southeastern Iran'. Bull. Seism. Soc. Am., 69, pp 773 - 824.

(3)   CHATELAIN, J. L., ROECKER, S. W., HATZFELD, D. and MOLNAR, P., (1980). 'Microearthquake seismicity and fault plane solutions in the Hindu Kush region and their tectonic implications'. J. Geophys. Res., 85, pp 1365 - 1387.

(4)   ROECKER, S. W., SOBOLEVA, O. V., NERSESOV, I. L., LUKK, A. A., HATZFELD, D., CHATELAIN, J. L. and MOLNAR, P., (1980). 'Seismicity and fault plane solutions of intermediate depth earth-quakes in the Pamir-Hindu Kush region'. J. Geophys. Res., 85, pp 1358 - 1364.

(5)   PREVOT, R., HATZFELD, D., ROECKER, S. W. and MOLNAR, P., (1980). 'Shallow earthquakes and active tectonics in Eastern Afghanistan'. J. Geophys. Res., 85, pp 1347 - 1357.

(6)   CHEN, W-P. and ROECKER, S. W., (1981), personal communication.

(7)   SHIROKOVA, Y. I., (1979). 'Mechanism of the foci of the Pamirs earthquakes with depths of 100 – 140 km'. Bull. (Izv.) Acad. Sci. USSR. Earth Physics, 15, pp 819 – 824.

(8)   ARMBRUSTER, J., SEEBER, L. and JACOB, K., (1978). 'The northern termination of the Himalayan mountain front: active tectonics from micro-earthquakes'. J. Geophys. Res., 83, pp 269 – 282.

(9)   SEEBER, L. and ARMBRUSTER, J., (1979). 'Seismicity of the Hazara arc in northern Pakistan: decollement vs. basement faulting'. in: A. Farah and K. A. DeJong (Editors), Geodynamics of Pakistan. Geol. Survey Pak., pp 131 – 142.

(10)   BELOUSSOV, V. V., BELYAEVSKY, N. A., BORISOV, A. A., VOLVOVSKY, B. S., VOLVOVSKY, I. S., RESVOY, D. P., TAL-VIRSKY, B. B., KHAMRABAEV, I. K., KAILA, K. L., NARAIN, H., MARUSSI, A. and FINETTI, J., (1980). 'Structure of the lithosphere along the deep seismic sounding profile: Tian Shan – Pamirs – Karakorum – Himalayas'. Tectonophysics, 70, pp 193 – 221.

(11)   JACKSON, J. A., (1980). 'Errors in focal depth determination and the depth of seismicity in Iran and Turkey'. Geophys. J. Roy. astr. Soc., 61, 285 – 301.

(12)   ELLSWORTH, W. L. and ROECKER, S. W., (1982). 'Sensitivity of the earthquake location problem to network geometry'. Geophys. J. Roy. astr. Soc., (in press).

(13)   LANGSTON, C. A. and HELMBERGER, D. V., (1975). 'A procedure for modelling shallow dislocation sources'. Geophys. J. Roy. astr. Soc., 42, pp 117 – 130.

(14)   JACKSON, J. A. and FITCH, T. J., (1981). 'Basement faulting and the focal depths of the larger earthquakes in the Zagros mountains (Iran)'. Geophys. J. Roy. astr. Soc., 64, pp 561 – 586.

(15)   PENNINGTON, W. D., (1979). 'A summary of field and seismic observations of the Patan earthquake, 28 December 1974'. in A.Farah and K. A. DeJong (editors), Geodynamics of Pakistan. Geol. Survey Pak., pp 143 – 147.

(16)   AMBRASEYS, N., LENSEN, G., MOINFAR, A. and PENNINGTON, W., (1981). 'The Patan (Pakistan) earthquake of 28 December 1974: field observations'. Q. J. eng. Geol., 14, pp 1 – 16.

(17)   CHANDRA, U., (1978). 'Seismicity, earthquake mechanisms and tectonics along the Himalayan mountain range and vicinity'. Phys. Earth and Planet. Int., 26, pp 109 – 131.

(18)   SHIROKOVA, Y. I., (1974). 'A detailed study of the stresses and fault planes at earthquake foci of Central Asia'. Bull. (Izv.) Acad. Sci. USSR, Earth Physics, 11, pp 22 – 36.

(19)   ILLI, D., (1981), personal communication.

(20)   JACKSON, J. A. and FITCH, T. J., (1979).   'Seismotectonic implica-
       tions of relocated aftershock sequences in Iran and Turkey'.
       Geophys. J. Roy. astr. Soc., 57, pp 209 - 229.

(21)   MOLNAR, P., CHEN, W-P., FITCH, T. J., TAPPONNIER, P.,
       WARSI, W. and WU, F., (1976).   'Structure and tectonics of the
       Himalaya: a brief summary of relevant geophysical observations'.
       Himalaya: Coll. int. du C. N. R. S., 268, pp 269 - 294.

(22)   YIELDING, G., JACKSON, J. A., KING, G. C. P., SINVHAL, H., VITA-
       FINZI, C. and WOOD, R. M., (1981).   Relations between surface
       deformation, fault geometry, seismicity, and rupture propagation
       characteristics during the El Asnam (Algeria) earthquake of
       10 October 1980'.   Earth Planet. Sci. Lett., 56, pp 287 - 304.

(23)   JACKSON, J., GAGNEPAIN, J., HOUSEMAN, G., KING, G., PAPADIMITRIOU,
       P., SOUFLERIS, C. and VIRIEUX, J., (1982).   'Seismicity, normal
       faulting, and the geomorphological development of the Gulf of Corinth
       (Greece): the Corinth earthquakes of February and March 1981'.
       Earth Planet. Sci. Lett., (in press).

(24)   NIAZI, M. and KANAMORI, H., (1981).   Source parameters of the
       1978 Tabas and 1979 Qainat, Iran, earthquakes from long period
       surface waves'.   Bull. Seism. Soc. Am., 71, pp 1201 - 1214.

(25)   'Pakistan Red Crescent (1981)', preliminary report, October 8th
       1981.

(26)   SOUFLERIS, C. and KING, G. C. P., (1981).   ' A source study of
       the largest foreshock (May 23) and the mainshock (on June 20) of
       the Thessaloniki 1978 sequence', in B. C. Papazachos and P. Carydis
       (Editors), The Thessaloniki 1978 earthquakes (in press).

(27)   SOUFLERIS, C. and STEWART, G. S., (1981).   'A source study of
       the Thessaloniki (northern Greece) 1978 earthquake sequence'.
       Geophys. J. Roy. astr. Soc., 67, pp 343 - 358.

(28)   BERBERIAN, M., (1981).   'Continental deformation in the Iranian
       plateau'.   Ph.D. thesis, University of Cambridge, 120 pp.

(29)   CIPAR, J., (1980).   'Teleseismic observations of the 1976 Friuli,
       Italy earthquake sequence'.   Bull. Seism. Soc. Am., 70, pp 963 -
       983.

(30)   LANGSTON, C. A., (1978).   'The February 9, 1971 San Fernando
       earthquake: a study of source finiteness in teleseismic body waves'.
       Bull. Seism. Soc. Am., 68, pp 1 - 29.

(31)   LANGSTON, C. A. and BUTLER, R., (1976).   'Focal mechanisms of
       the August 1, 1975 Oroville earthquake'.   Bull. Seism. Soc. Am.,
       66, pp 1111 - 1120.

(32)   BUTLER, R., STEWART, G. S. and KANAMORI, H., (1979).   'The
       July 27, 1976 Tangshan, China earthquake: a complex sequence
       of intraplate events'.   Bull. Seism. Soc. Am., 69, pp 207 - 220.

(33)   KRISTY, L., BURDICK, L. J. and SIMPSON, D. W., (1980).   'The
       focal mechanisms of the Gazli, USSR earthquakes'.   Bull. Seism.
       Soc. Am., 70, pp 1737 - 1750.

(34) HARTZELL, S., (1980). 'Faulting process of the May 17, 1976 Gazli, USSR earthquake'. Bull. Seism. Soc. Am., 70, pp 1715 - 1736.

(35) CIPAR, J., (1979). 'Source processes of the Haicheng, China, earthquake from observations of P and S waves'. Bull. Seism. Soc. Am., 69, pp 1903 - 1916.

(36) CHEN, W-P., FITCH, T. J., NABELEK, J. L. and MOLNAR, P., (1981). 'An intermediate depth earthquake beneath Tibet: source characteristics of the event of September 14, 1976'. J. Geophys. Res., 86, pp 2863 - 2876.

(37) MARKS, S. M. and LINDH, A. G., (1978). 'Regional seismicity of the Sierran foothills in the vicinity of Oroville, California'. Bull. Seism. Soc. Am., 68, pp 1103 - 1115.

(38) LESTER, F. W., BUFE, C. G., LAHR, K. M. and STEWART, S. W., (1975). 'Aftershocks of the Oroville earthquake of August 1, 1975'. Calif. Div. Mines and Geol., Spec. Report 124, pp 131 - 138.

(39) LAHR, K. M., LAHR, J. C., LINDH, A. G., BUFE, C. G. and LESTER, F. W., (1976). 'The August 1975 Oroville earthquakes'. Bull. Seism. Soc. Am., 66, pp 1085 - 1100.

(40) LACHENBRUCH, A. H., (1968). 'Preliminary geothermal model of the Sierra Nevada'. J. Geophys. Res., 73, pp 6977 - 6989.

(41) SIBSON, R. H., (1982). 'Fault zone models, heat flow and the depth distribution of earthquakes in the continental crust of the United States'. Bull. Seism. Soc. Am., (in press).

Stress field and plate motion in the Karakoram and surrounding regions

Shu Pei Yi and Lin Ban Zuo – Institute of Geophysics, Beijing, P.R.C.

ABSTRACT

The spatial distribution of earthquakes in the Karakoram and surrounding regions is studied, along with focal mechanisms, macroseismic evidence and faulting patterns. From these data the stress field in the region is inferred. Finally, relations with plate motions are discussed.

INTRODUCTION

The Karakoram region is at the apex of the India-Eurasian convergence, with complex geological structures and tectonics, and a high level of seismicity. During June-September 1980, scientists from the U.K., China and Pakistan explored the geomorphology, seismicity, glaciers, possible fault displacements, and housing methods in the area, collecting much new material. Using data from earthquakes, geology, and so on, this paper attempts to outline the stress fields of the Karakoram and surrounding regions so as to provide background evidence for the investigation of continental plate collision.

FEATURES OF THE STRESS FIELD

The Karakoram is the eastern part of the Hindu Kush-Pamir-Karakoram arc, where the Indian sub-continent wedges deeply into Eurasia. The whole arc is a seismically active region, with frequent large earthquakes. Using P-wave first motions to obtain a focal mechanism solution, one is able to obtain a stress direction for the focal region. Some earthquakes, especially those of larger magnitude, can initiate stress release over an area much larger than the focal region. Combining many large earthquake mechanism solutions, the basic outline of the modern stress field of the area can be understood. This paper collects 85 earthquake mechanism solutions, for major events ($M_S \geq 5.0$) which occurred in the area 25° – 40°N, 60° – 80°E after 1958. Aftershocks were not used as the manner of stress accommodation after large earthquakes may be complex. The data came from references 1 - 6. Events are numbered in chronological order in Table 1, and upper-hemisphere focal projections are shown in Figure 1.

186

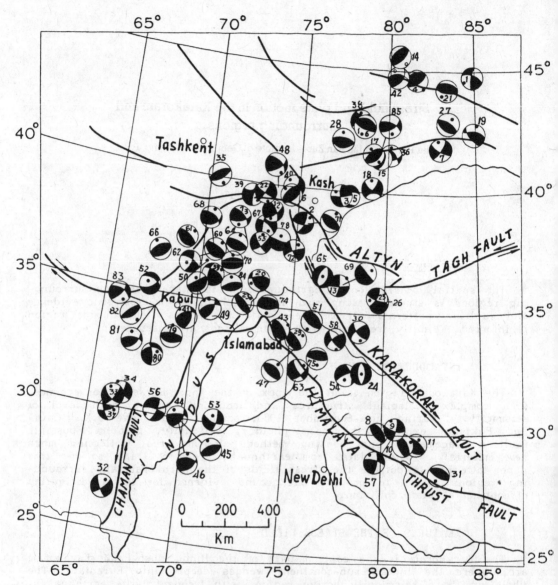

Fig. 1. Fault plane solution: distribution map of Pamir arc top
and surrounding regions.

From Figure 1, the dominant direction of the main compressive stress
axis is approximately north-south, with angles of plunge mostly $< 40°$.
The average trend of P-axes is $358°$, plunging $19°$ out of the ground,
as shown in Figure 2 (a Wulff net projection of all P-axes). Plotting
P-axes on a map (Fig. 4), we can see that the dominant directions change
across the region, from NNW in the western part to N – S in the centre
and NNE in the eastern area: these directions are generally orthogonal
to the arc trend. Collecting all the tension axes (Fig. 3, Wulff net),
most of them are seen to be nearly vertical, as is usual in thrusting.

No	Mo	Day	Yr	Lat. φ(N°)	Long. λ(E°)	Depth (km)	Region	M	Pole of Nodal Plane Az	Pl	A Pole of Nodal Plane Az	Pl	B Az	Pl	P Az	Pl	T Az	Pl	Reference (accuracy)
1	1	10	1958	44.8	84.9	10	Sinkiang	5.8	94 SW	60	180 E	85			229	15	134	23	B
2	11	15	1955	38.8	75.3	40	"	6.4	95 NE	60	170 SW	65			44	3	312	40	C
3	4	14	1961	39.3	77.8	20	"	6.8	64 SE	70	166 SW	60			297	6	203	36	B
4	8	20	1962	44.3	81.7	25	"	6.4	48 SE	30	48 NW	60			139	15	319	75	B
5	11	26	1962	39.9	77.4	49	"	5.7	3	90	93	90			318	0	228	0	A
6	8	29	1963	39.8	74.2	51	"	6.5	47 SE	43	47 NW	47			137	2	317	88	B
7	12	18	1963	41.9	82.6	30	"	5.1	25 NW	62	120 NE	80			346	27	250	12	B
8	9	26	1964	30.0	80.0	50	China–Nepal border	6.2	125 SW	50	125 NE	40			35	20	215	85	B
9	3	6	1966	31.4	80.5	50	Tibet	6.6	19 NW	45	151 NE	56			5	63	263	7	B
10	6	27	1966	29.7	80.9	33	"	6.3	130 NE	50	130 NE	40			40	5	220	85	B
11	6	27	1966	29.7	81.9	33	"	6.5	120 NE	85	300 NE				30	40	210	50	B
12	5	11	1967	39.3	73.8	10	Sinkiang	6¼	34 NW	75	120 SW	80			349	10	255	25	B
13	5	28	1967	36.1	77.7	28	"	5.9	78 NW	30	167 SW	85			33	9	301	16	A
14	8	20	1967	45.5	80.4	21	"	5.3	35 SE	30	35 NW	60			125	15	305	75	B
15	2	12	1969	41.5	79.3	16	"	6.5	54 SE	50	54 NW	40			324	5	144	85	B
16	6	5	1970	42.7	78.8	32	USSR–China border	6.8	105 NE	65	105 SW	25			195	20	15	70	B
17	3	23	1971	41.4	79.3	12	Sinkiang	6.0	48 SE	52	48 NW	38			318	7	138	83	B
18	1	16	1972	40.2	79.0	9	"	6.2	33 NW	75	124 NE	85			349	14	257	8	B
19	4	9	1972	42.3	84.8	21	Kashmir	5.6	87 SE	60	139 NE	44			20	17	127	61	B
20	9	3	1972	35.9	73.4	23	Sinkiang	6.5	80 SE	70	144 NE	40			15	30	127	50	B
21	6	3	1973	44.3	83.3	12	Pamir	6.0	107 NE	75	107 SW	15			197	30	72	60	A
22	8	11	1974	39.4	73.8	7	Pakistan	7.3	27	90	27	90			162	0	208	0	A
23	12	28	1974	35.0	72.9	45	Tibet	6.2	118 NE	37	118 SW	53			28	8	75	82	B
24	1	19	1975	32.4	78.6	36	Sinkiang	6.9	170 SW	53	170 NE	37			255	82	317	8	B
25	4	28	1975	36.0	79.9	0	"	6.1	1 NW	80	93 NE	84			227	2	136	10	B
26	6	6	1975	36.2	80.1	45	"	6.1	93 SW	80	6 NW	77			230	28	198	8	B
27	1	10	1976	42.5	83.2	33	USSR Issk Lake	5.8	108 SW	70	108 NW	20			18	25	220	65	B
28	10	18	1965	42.0	77.6	12	Kashmir	5	85 SE	60	76 NW	30			351	15	190	83	1*
29	2	21	1967	33.7	75.4	41	Tibet	5.1	110 NE	30	110 SW	60			10	15	81	75	2*
30	2	11	1968	34.2	78.7	24	Nepal	5.1	136 NE	76	43 SE	64			177	8	3	28	1*
31	10	16	1973	28.4	83.0	34		5.1	93 NE	55	93 SW	35			183	10		80	B

1* Л.А. Мишарина   2* B.K. Rastogi

Table 1 (cont'd next page)

No	Mo	Dy	Year	Lat	Long	h	Region	M	Nodal plane 1	Nodal plane 2	Axes	Ref
32	10	4	1974	26.38	66.65	32	Pakistan	6.3	61 NW 80	332 NE 85	17 11 286 4	3*
33	10	3	1975	30.27	66.33	14	"	6.6	21 NW 85	292 NE 80	336 11 66 3	
34	10	3	1975	30.44	66.41	24	"	6.5	90	111 90	336 0 66 0	
35	1	31	1977	40.11	70.68	20	Tajik	6.3	68 SE 40	248 NW 50	158 5 338 85	
36	3	12	1978	41.7	80.1	26	Sinkiang	5.5	75 NW 45	287 SW 50	1 3 266 73	
37	3	16	1978	29.93	66.3	33	Afghanistan–Pakistan border	6.3	51 NW 62	0 F 40	120 12 7 60	
*	*	*	*	*	*	*	*	*	*	*	*	
38	3	24	1978	42.84	78.61	33	Alma-Ata	7.3	51 NW 58	293 SW 53	170 4 266 54	
39	11	1	1978	39.35	72.62	33	Pamir	6.7	277 SW 80	5 SE 75	142 18 50 4	
*	*	*	*	*	*	*	*	*	*	*	*	
40	4	15	1955	39.9	74.6	3	Sinkiang	7	103 SW 53	234 NW 48	346 3 254 62	(5)
41	9	16	1956	34	69.5	3	Middle Asia	6	188 NW 37	34 SE 57	292 8 167 73	(5)
42	12	21	1958	44.8	80.6	3	"	6	83 SE 72	200 NW 30	333 23 205 54	(5)
43	9	2	1963	33.9	74.7	44	Kashmir	5.1	145 SW 70	145 NE 20	55 25 235 65	(7)
44	1	24	1966	29.9	69.7	26	Pakistan	5.6	210 NW 27	81 SE 72	335 24 199 58	(7)
45	2	7	1967	29.9	69.7	10	"	6.0	237 NW 45	57 SE 45	327 0 57 90	(7)
46	2	7	1967	30.3	69.9	11	"	5.8	92 S 55	231 NW 43	343 6 230 68	(7)
47	2	21	1967	33.7	75.4	41	Kashmir	5.1	110 NE 30	110 SW 60	20 15 190 75	(5)
48	5	11	1967	39.3	73.8	2	Pamir	6	235 NW 55	105 SW 50	168 4 265 60	(7)
49	5	15	1969	34.6	70.4	22	Hindukush	5.6	234 NW 16	75 SE 75	340 30 173 60	(7)
50	12	28	1972	34.7	70.4	63	"	5.6	238 NW 3	58 SE 87	328 42 148 48	(5)
51	1	16	1973	33.7	75.3	18	Pakistan	5.6	87 SE 50	321 NE 55	205 3 110 59	(5)
52	6	8	1956	35.2	67.5	3	Middle Asia	6	260 NW 68	120 SW 26	327 64 172 22	(7)
53	2	7	1966	30.2	69.7	10	Pakistan	6½	185 W 54	115 NE 66	336 46 239 8	(5)
54	6	27	1955	32.5	78.5	3	Tibet	6	63 SE 80	159 SW 60	198 28 296 12	(7)
55	2	2	1965	37.5	73.2	15	Pamir	6	231 NW 76	142 NE 87	10 19 273 15	(5)
56	8	1	1966	30.1	68.6	33	Pakistan	6	97 SW 80	188 NW 83	142 1 233 11	(7)
57	12	17	1966	29.9	80.8	19	Tibet	6¼	67 NW 67	89 S 75	316 2 226 23	(5)
58	2	2	1968	34.2	78.8	24	"	5.1	136 NE 76	43 SE 64	177 8 81 28	(11)
59	4	6	1956	36.4	70.7	220	Hindukush	6.8	10 SE 47	120 SW 70	60 14 165 47	(11)
60	2	17	1958	36.5	70.7	210	"	6.6	120 SW 65	120 NE 25	30 20 210 70	(11)
61	1	9	1960	36.5	70.7	200	"	6.8	125 NW 70	125 NE 20	35 25 215 65	(11)
62	6	20	1961	36.5	70.5	220	"	6.6	20 NW 68	100 SW 66	241 33 150 1	(11)
63	6	17	1962	33.7	75.8	88	Kashmir	6.9	64 NW 60	155 NE 89	23 21 285 20	(11)
64	7	6	1962	36.6	70.4	203	Hindukush	6.9	97 SW 75	97 NE 15	7 30 187 60	(11)
65	6	26	1963	36.4	76.7	100	Sinkiang	6	33 SE 38	33 NW 52	306 83 124 7	(11)
66	1	28	1964	36.5	70.0	207	Hindukush	6.8	50 SE 62	50 NW 28	320 17 140 73	(11)

Table 1 (cont'd)

No.	Mo	Day	Year	Lat	Long	Depth	Location	M											Ref
67	3	23	1964	38.1	73.5	127	Hindukush	5.5	18	SE	50	120	SW	76	168	38	62	16	(5)
68	3	14	1965	36.3	70.7	219	"	6.6	119	SW	64	262	NW	31	16	:7	242	66	(7)
69	6	22	1965	36.2	77.6	107	West Kunlun mountain	5.7	129	SW	3	309	NE	87	39	48	219	42	(7)
70	6	6	1966	36.3	71.1	225	Hindukush	6.3	270	N	38	81	S	52	355	7	142	82	(7)
71	1	22	1968	38.2	75.7	132	Pamir	6.2	108	NE	42	308	SW	50	28	5	274	78	(5)
72	1	13	1972	37.7	75.1	84	"	5.6	246		90	336	NE	75	20	11	112	11	(7)
73	1	23	1954	37.4	72.5	3	"	6	262	NW	66	121	SW	30	185	20	321	63	(5)
74	9	3	1972	35.9	73.3	36	North west of Kashmir	6.2	272	NE	64	92	SW	26	182	71			Rastogi (1976)
75	10	24	1973	33.1	75.9	52	Kashmir	5.4											"
78	11	12	1972	38.4	73.7	33	Sinkiang	6.0	98	SW	10	278	NE	80	8	55	188	35	"
79	8	8	1969	36.4	70.9	198	Hindukush	5.8	82	SE	49	262	NW	41	352	4	172	86	"
80	6	24	1972	36.3	69.7	47	"	5.9	341	SE	12	81	SW	89	63	47	279	49	"
81	6	10	1965	36.1	70.5	85	"	5.8	85	SE	59	223	NW	39	336	8	222	67	"
82	1	20	1972	36.4	70.7	214	"	5.8	48		39	228		51	138	6	318	84	"
83	8	16	1966	36.5	70.8	197	"	5.5	220		66	98		40	153	12	266	56	"
84	7	8	1972	36.5	71.4	105	USSR Afghanistan border	5.6	75		50	255		40	345	5	165	85	"
85	6	15	1971	41.4	79.4	17	Sinkiang	5.4	221		61	95		40	157	10	270	63	"

3* Quoted from Yan Tia-quan et al.

Table 1. Collected seismic data. References (1) (2) (3) (4) (5) (6).

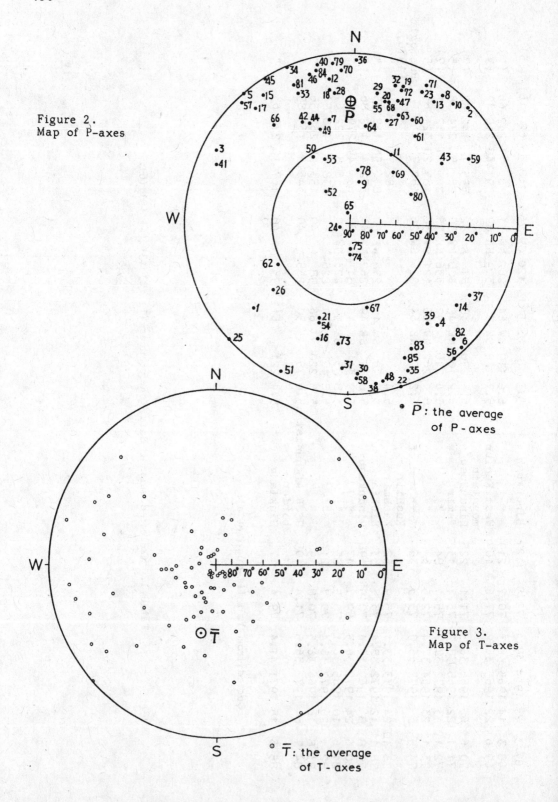

Figure 2.
Map of P-axes

$\overline{P}$: the average
of P-axes

Figure 3.
Map of T-axes

$\overline{T}$: the average
of T-axes

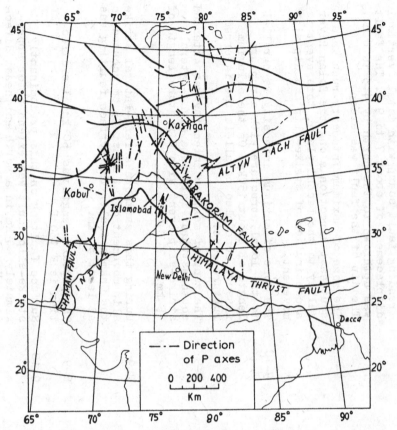

Figure 4.   Distribution of P-axes

Intermediate-depth events in the Pamir-Hindu Kush region have the same feature: when the focal depth is about 150 km, the tension axes in the fault-plane solutions are almost vertical, while the compressive axes are horizontal, trending north-south. This represents internal deformation of a subducted plate (8), and shows that different depths have the same stress regime. These facts indicate that the earthquakes in this region are mainly controlled by a north-south horizontal compression between the Indian and Eurasian plates.

From the geology and geomorphology, and patterns of faulting, we can see that in addition to the thrust zones at the apex of the Hindu Kush-Pamir-Karakoram arc, there are great strike-slip faults to the east and west. To the west there is the Chaman Fault, a left-lateral strike-slip feature. Wellman (9) considers that left-lateral displacement reaches 500 km, while Gawad (10) describes a translation distance of about 1000 km, left-lateral, along the Sulaiman Range – Kirthar Range line. To the east is the Karakoram Fault, a right-lateral strike-slip fault and a remarkably clear feature. This cuts the Indus and Yarlung Zangbo ophiolite belts, and if these homologous middle-late Cretaceous ophiolites were formerly continuous then about 250 – 300 km right-lateral motion

No	Date Mo Day Yr	Location	Lat.&Long.	M	Depth km.	Intensity	Description
1	6 16 1819	Rann of Kutch India	24.0°N 69.0°E	7¼-8¼		9-10+	This feature is an approximately east-west trending fault scarp with total vertical displacement of about 7 to 9 m. The fault was upthrown to the north by 4.5 to 6 m. The dip of the fault was steep.
2	8 27 1931	Mach of Pakistan	29.9°N 67.3°E	7.4		7-8	The zone of maximum intensity runs in a southeast direction along the Bolan River Valley to the Kachhi plain, this event is a combination of strike-slip and dip-slip movement.
3	5 30 1935	Quetta of Pakistan	30.2°N 67.0°E	7.5		9-10	The zone of maximum intensity is parallel to the local structure with a rupture length of about 150km. On the alluvium fissure zone also parallels the inferred rupture zone.
4	10 21 1909	Kachhi Plain of Pakistan	29.0°N 68.2°E	7.2		8-9	The zone of maximum intensity was extended southeastward. This is a right-lateral strike-slip event with a rupture length of 75 km.
5	12 20 1892	Chaman of Pakistan	31.1°N 66.4°E	6¾		8-9	During this earthquake left-lateral strike slip movement of at least 75 cm occured on the Chaman fault. The surface faulting was at least 30 km.
6	6 9 1956	Sayghan of Afghanistan	35.2°N 67.7°E	7.6		8-9	The rupture length was 50 km. Its direction is N 150 E.
7	7 6 1905	Paghmam of Afghanistan	34.6°N 68.9°E	7-7¾		9-10	Surface faulting on this predominantly strike-slip fault was extended for approximately 60 km in a north-northeast direction. Vertical displacement of several metres was recorded. It appears that this earthquake ruptured the northern portion

Table 2 (cont'd next page)

No.	Mo	Day	Year	Location	Coordinates	Magnitude		Intensity	Remarks
8	2	19	1842	Aligar Valley of Afghanistan	34.8°N 70.3°E			8-9	The rupture proceeded from the north-north-east to the south-southwest along a portion of the Gardz fault.
9	4	4	1905	Kangra of India	32.0°N 76.3°E	8.6		10+	The main faulting extended in a south-east direction from Kangra. The length of the rupture is about 100 to 150 km, the rupture extended from Kangra to the secondary centre of maximum intensity at Dehra Dun. Secondary faulting with a southeast strike was noted at Barwar Lake near Larji. It may have resulted from slip on a buried shallow dipping thrust fault.
10	7	5	1895	Tash Kurghan	37.7°N 75.1°E	7		8	From geological structure and seismicity of West Kulan.
11	4	15	1955	Sinkiang Wuqia	39°45'N 74°32'E	7	25	9	From figure set of Chinese isoseismal intensity line.
12	8	22	1902	Sinkiang Atusen	39°53'N 76°12'E	8¼	56	10	
13	7	26	1971	Sinkiang Shiker	39°52'N 77°16'E	5.6	39	7	
14	4	14	1961	Sinkiang Pachu	39°53'N 77°45'E	6.8	20	9	
15	1	16	1972	Sinkiang Kelpin	40.2°N 78.9°E	6.2	24	7	
16	2	12	1969	Sinkiang North east of Wushi	41°27'N 79°22'E	6½	16	8	
17	2	24	1949	Sinkiang Luntan	41.9°N 83.2°E	7¼		9	
18	4	9	1972	Sinkiang Luntan	42°13'N 84°34'E	5.6	26	7	
19	12	23	1906	Sinkiang Manass	43.9°N 85.6°E	8	24	10	

Table 2. Description and Analysis of Some Large Earthquakes.

has occurred (Fig. 5).

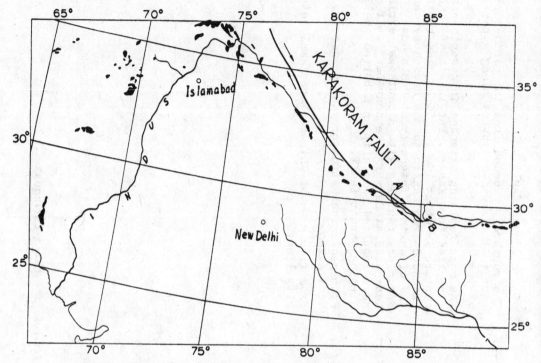

Figure 5. Distribution of ophiolite outcrops around the Hindu Kush –
Pamir – Karakoram arc.

The Middle–Late Cretaceous ophiolites of the upper Indus and
Yarlung Zangbo rivers are offset by the Karakoram Fault (A – B
on map). From Landsat imagery, the amount of relative movement on the fault is estimated to be 250 – 300 km.

Huan Wei Lin et al (11) think that at least 300 km right-lateral displacement has occurred. Hence the strike-slip faulting appears to be
symmetrical about the northern apex of the arc.

Macroseismic intensity data for large earthquakes also display such
symmetry. Chen Pei Shan (12) has shown that the direction of the long
axis of earthquake intensity isoseisms, especially the innermost isoseism,
is controlled by the local rupture trend. The earthquakes along both
sides of the arc show isoseismal trends sub-parallel to the local trend
of the arc (see Fig. 6). (A possible exception is e.g. no. 4 in Figure
6, where isoseisms may be controlled by the trend of a conjugate axis).
Trends of ground fissures and dislocations during large earthquakes show
the same pattern (13) (see Table 2).

Seismicity in this region is intense. A map of the epicentre distribution
since 1965 (11) is shown in Figure 7, and again displays symmetry about
the apex of the arc. A large number of moderate to strong earthquakes
are distributed along the top of the arc and along the east and west
sides (13). Since 1900, ten great earthquakes (Ms ≥ 8) have occurred along
the arc, a very large number for such a small region and brief time
interval. At the north-west part of the arc, beneath the Hindu Kush,

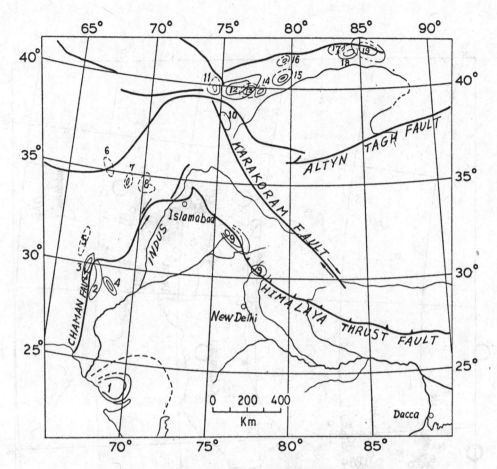

FIG. 6. Distribution of strike-slip fault and macroscope intensity isoseisms.

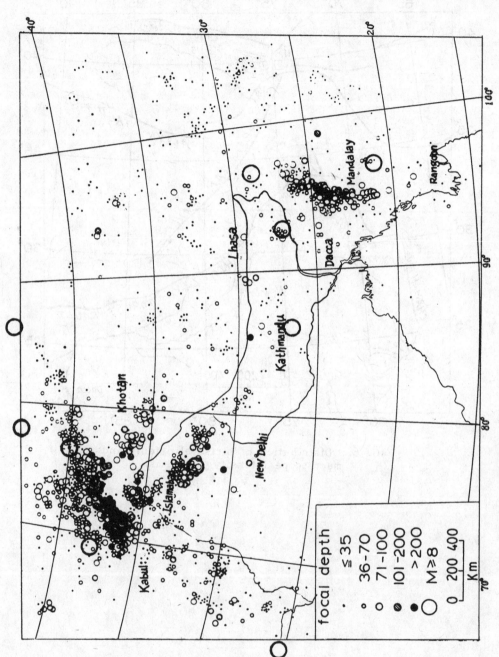

FIG. 7. Map of focal depths and great earthquakes.

is a concentration of intermediate-focus shocks, at depths up to 300 km (see Fig. 8). By contrast, there are far fewer events to the south in the Indian sub-continent.

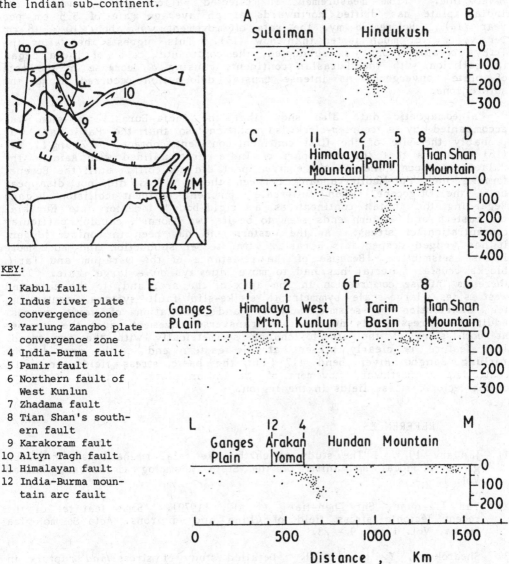

KEY:

1 Kabul fault
2 Indus river plate convergence zone
3 Yarlung Zangbo plate convergence zone
4 India-Burma fault
5 Pamir fault
6 Northern fault of West Kunlun
7 Zhadama fault
8 Tian Shan's southern fault
9 Karakoram fault
10 Altyn Tagh fault
11 Himalayan fault
12 India-Burma mountain arc fault

Fig. 8. Section of Multi-rock layer subduction of Qinghai-Xizang Plateau and surrounding regions.

DISCUSSION

The above four lines of evidence show that the collision of India with Eurasia has exerted a decisive control on the tectonic stress fields of the region.

According to the conclusions of the Chinese Qinghai-Xizang (Tibet)

Plateau expedition, the suture zone between the Indian and Eurasian plates lies along the Yarlung Zangbo River line, extending northwest to Indus River line. From measurement of collected palaeomagnetic samples, the Indian plate has drifted northwards at an average rate of 5.5 cm per year (14). Before 38 my, the rate of movement was about 10 - 18 cm per year, i.e. much faster than now (15). This suggests that after the Tethys oceanic crust was subducted, the continental crust of India began its collision with the Eurasian continent, causing a decrease in the rate of plate convergence as intense crustal deformation occurred along the suture zone.

Palaeomagnetic data also show that the India-Eurasia collision was accompanied by a counter-clockwise rotation, so that the Pamir arc was probably the site of the first continent-continent contact. Zhang Li Min (16) considers that the west part of India first collided with Asia in the Palaeocene epoch, whereas the eastern part did not collide until the Eocene, forming the "Pal-Mediterranean" between them. This did not disappear until the Oligocene, when full-scale continent-continent collision began. Regarding the Indian continent as a "rigid-body" indentor into Eurasia, the eastern and western ends seem to be its two "horns", causing particular concentration of stress. As the western end has been in contact longer, it has wedged deeper into Eurasia, with steeper subduction and much more intense seismicity. Because of the resistance of the Qaraqum and Tarim blocks crustal material has had to move sideways on a large scale. Thus there is intense compression in the apex of the arc, and, to the east and west sides, large-scale symmetrical strike-slip fault systems, controlling the distribution of seismic activity and orientation of fault planes. Additional stress fields due to this transverse movement cause the stresses at both sides of the arc to change symmetrically with the arc trend. This feature is clearly present at the eastern end of the arc, in the Yarlung Zangbo River bend (17) but the basic stress fields are still controlled by northward movement of the Indian plate. This is an outline of the tectonic stress fields in the region.

REFERENCES

1) January 1973. "The study of earthquake focal mechanisms", Research group of focal mechanism of the State Seismological Bureau, China. Vol. 1.

2) Yan Tia-quan, Shi Zhen-liang et al., (1979). Some features of the recent tectonic stress field of China and environs, Acta Seismologica Sinica, Vol. 1, pp 9 - 23.

3) Shearokova, Y. I., (1974). Detailed study of stress and rupture in seismic focus of mid-Asia, Geophysics, Vol. 11, pp 22 - 36.

4) Chandra, V., (1978). Seismicity, earthquake mechanisms, and tectonics along the Himalayan mountain range and vicinity, Phys. Earth and Planet. Int., Vol. 16(2).

5) Yeh Hung, Liang Yi-shan, et al., (1975). The analysis of the recent tectonic stress of the Himalaya mountain arc and its vicinities, Scientia Geologica Sinica, No. 1, pp 32 - 48.

6) Rastogi, B. K., (1976). Source mechanism studies of earthquakes and contemporary tectonics in Himalaya and nearby regions, Bull.

Inst. Sei. Eq. (Tokyo) 14, pp 99 - 134.

7) Yan Tia Quan. Features of recent tectonics of collision zone in the mid-Asia region (in the press).

8) Roecker, S. W., et al., (1980). Seismicity and fault plane solutions of intermediate depth earthquakes in the Pamir-Hindu Kush regions, J. G. R., Vol. 85, pp 1358 - 1364.

9) Nowroozi, A. A., (1971). Seismotectonics of the Persian plateau, Eastern Turkey, Caucasus and Hindu Kush regions, Bull. Seis. Soc. Am., 61, pp 317 - 341.

10) Gawad., M. A., (1971). Wrench movements in the Baluchistan arc and relation to Himalaya-Indian Ocean tectonics, B. G. S. A., 82, (5), pp 1235 - 1250.

11) Huan Wei-lin, Wang Su-yun et al., (1980). The distribution of earthquake foci and plate motion in the Qinghai-Xizang Plateau, Acta Geophysica Sinica, Vol. 23, No. 3.

12) Chen Pei-shan, Yeh Shou-min, (1975). Relation between the seismic source mechanism and the intensity distribution, Acta Geophysica Sinica, Vol. 18, No. 1.

13) Quittmeyer, R. C. and Jakob, K. H., (1979). Historical and modern seismicity of Pakistan, Afghanistan, northwestern India and southeastern Iran, B. G. S. A., 69, (3), pp 773 - 823.

14) Zhu Zhi-wen, Shu Xiang-yuan et al., (1981). Palaeomagnetic observations in Xizang and continental drift, Acta Geophysica Sinica, Vol. 24, 1.

15) Molnar, P. and Tapponnier, P., (1975). Cenozoic tectonics of Asia; effects of a continental collision, Science, Vol. 189.

16) Zhang Li-min and Fu Cheng-Yi, Collision of continental plates on Xizang (Tibet) plateau (in the press).

17) Zhang Li-min, Shu Pei-yi et al., (1980). The average stress fields of Zhano and Damxung regions in the Xizang Plateau, Acta Geophysica Sinica, Vol. 23, 3.

Housing and natural hazards

A critical review of the work method and findings of the Housing and Natural Hazards Group

I. Davis – Oxford Polytechnic, Oxford

ABSTRACT

A review of the objectives, work method and conclusions of this programme is presented. The paper, recognising the fact that this study was the first known integrated attempt to examine the topic in depth, makes a critical review of the approach adopted and analyses the limitations of the study as well as the progress that was made through a multi and interdisciplinary method. Certain recommendations emerge concerning the conduct of future projects relating to risk reduction which may be of value to governments and research personnel within disaster-prone areas.

CONTEXT OF PROJECT

At the 7th World Conference on Earthquake Engineering in Istanbul in 1980 a group of six leading engineering seismologists from India, Italy, Yugoslavia, Mexico, Peru and China reported on their studies on the seismic risk to rural structures. They comment in their 'State of the Art Report' that:

> 'it is unfortunate that this risk is increasing, rather
> than decreasing at most places on account of an increas-
> ing population, poverty of the people, scarcity of wood,
> cement and steel, lack of understanding of earthquake
> resistant features, etc.'

They then projected a future situation with the expectation:

> 'that the number of dwelling units may double in the next
> 20-25 years due to the explosion of population in developing
> countries.'

(Reference 1)

This increasingly vulnerable environment of poor families throughout disaster prone countries of the Third World has been recognised for many years. More specifically the relationship between casualties and unsafe houses in hazardous environments has become apparent, as well as the relationship of these natural hazards to the siting of houses. In addition the resistance of the basic form and mode of construction to hazards is only partially understood in general terms, and it is apparent that this has to be examined for each hazard-prone environment and the specific types of buildings within it. Therefore whilst knowledge in some aspects of this problem is well developed in other areas it remains rudimentary, and has yet to attract the concern of the relevant professions within disaster-prone countries, and the attention of the governments responsible for the safety of their citizens. Despite a growing awareness of the increasing scale of damage and casualties from disasters (from 1900 to 1976, 4.57 million people were killed and 232.55 million people injured or left homeless (2)), it would appear that prior to the Karakoram project no multi, and interdisciplinary study of housing risk reduction had been undertaken anywhere in the world. Field research to reduce housing vulnerability has, with certain isolated exceptions, occurred in the context of post-disaster reconstruction rather than pre-disaster mitigation and such studies have been confined to technical analysis of building performance or settlement location. A recent example of this restricted view has been produced by an august body, the International Association of Earthquake Engineering (3). They have set out the findings of some of the world's leading authorities in Engineering Seismology, and one part of the book is concerned with non-engineered construction. The standpoint of the study is that the vulnerability of such buildings is an engineering problem, no mention being made of social, cultural or economic criteria.

Expanding on such technical matters, the project team and their advisers formulated this investigation around a series of issues and questions that looked beyond this narrow technical frame of reference and enquiry to include such basic matters as the social, cultural and economic realities facing the occupants of houses in the Karakoram region as well as the expression of these values in house forms and settlement patterns. In effect, an investigation was proposed to study the composite risks facing vulnerable communities from natural hazards, as well as their own perception of such risks, and their concern to reduce or accept them, given their own needs, priorities and values. Thus the subject of vulnerability was examined within the context of normal living patterns, (see Fig. 1) and local attitudes to housing rather than from a remote standpoint coloured by a radically different set of values.

This was our objective, but we found in practice that it was extremely difficult, if not impossible, to climb into the shoes of the communities where we were working, and inevitably our findings have been shaped, despite our concern, by our own inherited value judgements.

OBJECTIVE OF PROJECT

The overall importance of this project lay in two broad areas, the general and the specific. The former was to examine the relationship between natural hazards and human response and adaptation to such hazards within marginal economies, and in the light of our study to suggest (probably for the first time) appropriate methods for the study of vulnerability relative to housing and settlements, leading to effective policies to reduce risks, (if this is in fact found to be possible given local perceptions and priorities.)

Secondly, in the specific context of this expedition, to understand

202

FIG. 1. TERRACED HOUSING IN BALTIT VILLAGE, HUNZA VALLEY.
Note various uses for roof surfaces, drying apricots, sleeping, etc.

the opportunities facing vulnerable communities to the natural hazards that effect the Karakoram region, and to suggest appropriate measures that can be adopted for the mitigation of future damage and casualties in the following five contexts:

1. Environmental mitigation measures to infrastructure that can be undertaken by the local community at minimal cost (i.e., walls, bunds, dykes, etc.)

2. Warning systems for floods/rockfalls/mudslides.

3. Improvements to make the existing building stock more resistant to hazards.

4. Improvements to the design and construction of new buildings in both post-disaster and normal conditions.

5. The final objective of the project came as a by-product from the first case study of the village of Barculti in the Yasin Valley. This is to advise on overall development within the valley in the light of information gathered by the research team.

In undertaking this project the group were fortunate in many respects. We were advised by several eminent men in the various related fields of this project who provided guidance on the overall strategy of the expedition, as well as assisting the team at an Interim Symposium held in Oxford in February 1981, and finally in reviewing the published papers that emerged from the Karakoram Conference (see Vol. 1 of these proceedings). Further assistance that made the project financially viable came from the Science Research Council, the Overseas Development Administration, the Office of the United Nations Disaster Relief Co-ordinator (UNDRO), and Ove Arup and Partners, as well as the sponsors of the overall expedition.

The formulation of this project team required skills and experience in various fields of knowledge which included: Anthropology, Architecture, Economics, Engineering Seismology, Geology, Planning, Social Geography and Sociology; a list of members is given in the Introduction of Volume I.

STUDY METHOD

Undertaking a project of this nature presented a challenge, since it was necessary for the team to evolve a working method that would synthesise the basic social, cultural, economic and technical dimensions of the problem of housing vulnerability. A basic approach was developed, which inevitably was dependent on ideal working conditions. The particular needs were to locate communities that would be hospitable and co-operative to the visiting team; secondly, to have effective translators to assist us, and thirdly, to have team members with appropriate skills and experience available to make the particular studies needed in a given location.

Just as it was essential to develop a study method that was based on optimum conditions, it was nevertheless inevitable that such conditions would be unlikely to be fully satisfied. Following a reconnaissance visit by Ian Davis to Pakistan to establish contact with local officials and Professor Israr ud Din, it was decided to attempt case studies in three diverse contexts. Firstly, an area that had suffered a major disaster in recent years – to determine whether the reconstructed houses and settlements had been modified in the light of previous failures (Pattan Earthquake of 1974). Secondly, an area that had not suffered any recent disaster, and thirdly an area that had an awareness of disaster but not received any major damage or casualties. However, due to circumstances and opportunities this pattern was only followed in a very general way. A total of four principal case studies was undertaken, i.e., (1) Barculti– Yasin Valley; (2) Patan – Indus Valley; (3) Ghorahabad – Indus Valley; and Ganesh – Hunza Valley. The location of the general areas of work can be seen on Fig. 2. A summary of what transpired in each case study is included in Appendix 2.

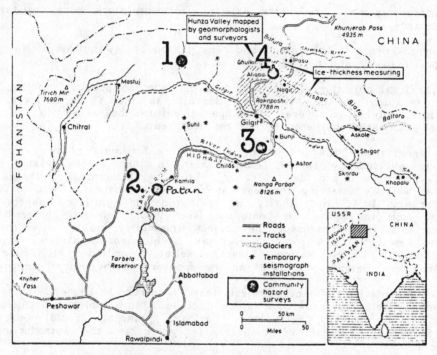

FIG. 2.   MAP INDICATING THE LOCATION OF THE FOUR MAIN
CASE STUDY LOCATIONS.

1.   BARCULTI – Yasin Valley   3.   GHORAHABAD – Indus Valley
2.   PATAN   – Indus Valley   4.   GANESH – Hunza Valley

A work method evolved during the preparations for the expedition and in our period of field study which involved five processes:-

Examination of previous writing on the subject, and
historical records of previous disasters,
Testing the seismic resistance of buildings,
Visual observation,
Social Surveys and
Interviews with local personnel.

Table I indicates the way that methods were applied in the study.

Due to some highly fortuitious circumstances it was possible for the

TASKS TO BE UNDERTAKEN	STUDY TECHNIQUES	Interviews with Local Personnel	Social Surveys	Visual Observation	Testing of Seismic Resistance of Building	Examination of Historical Records
SOCIAL/ CULTURAL/ ECONOMIC CONTEXT	Analysis of:					
	Social needs/priorities	√	√	√		√
	Economic situation	√	√	√		√
	Awareness/Adaptation to hazards	√	√	√		√
BUILDING/ SETTLEMENT CONTEXT	Analysis of:					
	House form	√		√		√
	Settlement pattern	√		√		√
	Building materials/ techniques	√		√		√
	Resistance of buildings to hazards	√		√	√	√
	Resistance of settlement to hazards			√		√
HAZARD CONTEXT	Analysis of:					
	History of hazard occurrence	√	√	√		√
	Scale of severity of risk from hazards	√	√	√		√
	Types of local hazards	√	√	√		√

TABLE 1. THE APPLICATION OF RESEARCH METHODS.

group to work closely with a community in the first of the case studies, the village of Barkulti in the Yasin Valley. Through Israr ud Din, Professor of Social Geography in the University of Peshawar, an undergraduate student of Geology, Shabaz Khan, was located who lived in this community, with his family being one of the leading families. This provided the opportunity for access in order to undertake a social survey. Frances D'Souza organised this study, which has been described in her research paper on social and economic analyses of Barkulti in Yasin with reference to the local perception of risk from natural hazards; see page 289.

It is worth pointing out that this social survey would not have been possible without a female team member, since access to homes was not permitted for the male members of the group if women were present; and in some instances not even when they were absent. The only exception to this rule occurred in the Ismaili communities which were located in the Hunza Valley where a more relaxed attitude was taken to the strict separation of women from men outside the immediate family. After the first case study Dr. D'Souza had to leave the expedition, and in subsequent case studies we never obtained the optimum research conditions that prevailed in Barkulti. Therefore, to some extent, we are reliant on the detailed research findings from Barkulti for our overall analysis since this was the only sample that was statistically valid in comparison with our more haphazard interviews in other locations which were with specific individuals chosen in a random manner. However, these unscientifically selected discussions in our later studies bore out the broad conclusions that emerged from the Barkulti study.

In parallel with the social survey being undertaken in Barkulti, additional study was made of overall settlement patterns, kinship groups, rights to irrigation channels and land tenure. This work was undertaken by Cliff Moughtin and Israr ud Din. Moughtin's paper is concerned with traditional settlement forms as a response to environmental hazards; see page 307.

The wider question of the geology and geomorphology of the Yasin Valley was examined by Richard Hughes. This was undertaken to try and establish by visual observation the physical determinants of settlements, and their relationship and adaptation to the wide variety of hazards to which local communities were exposed. The research paper by Hughes presented in this volume on page 253 is entitled "Yasin Valley: The analysis of geomorphology and building types".

Robin Spence, David Nash and Richard Hughes obtained funding from the Science Research Council to investigate and classify traditional building techniques and house forms and assess the overall vulnerability. This work, carried out with the assistance of Andrew Coburn, involved detailed observation and measurement of buildings, interviews with home occupiers, builders and craftsmen. This analysis assisted in the first case study with in-situ vibration testing of the structures. These testing techniques, recently developed by the Building Research Establishment, were used to help understand the manner in which the structure behaves in an integrated manner under vibrated loading. Essentially the vibration testing consisted of the simulation (in a greatly reduced scale) of certain types of seismic force, with a measurement of the resistance in the building structure to them.

When the expedition was planned it was anticipated that these vibration studies would be made in each of the case study localities, but this was not possible since it was found that a pre-requisite for such testing was the establishment of close rapport with a community, which occurred in Barkulti but not in the other study locations. Therefore, testing only took place in one house in Barkulti. We are aware of the risk of basing conclusions on this single sample, and in future research it will be necessary to widen the testing process to include a range of different types of building and forms of construction. However, one vital lesson from the testing in Barkulti is the recognition that it is not easy to persuade a conservative village to allow an outside group to mount equipment on their house and this will only occur once confidence has been established and perhaps, more important, the family in question are aware of the full aims and context of this testing. The findings from this study of building structures are the subject of a paper in this volume entitled: "The construction and vulnerability to earthquakes of some building types in the northern areas of Pakistan", see page 228.

The second case study was undertaken in Patan on the Indus gorge, the site of the most serious recent earthquake in the area which occurred in December 1974. The primary reason for this study was to try and determine whether builders or craftsmen who undertook the disaster reconstruction had learned anything from the visual aid of past building failure. With the very helpful assistance of the Kohistan Development Board (KDB) the group were able to make extensive observations of earthquake reconstruction in a number of localities in the affected area. This study was particularly important since it revealed the radically different approach of the 'official building' undertaken by the staff of the KDB in contrast to that being built in the traditional manner by local builders and craftsmen. This study has resulted in a paper by Ian Davis in this volume entitled: "Analysis of recovery and reconstruction following the 1974 Patan earthquake", see page 323.

The third case study in Ghorahabad was chosen to enable the team to examine a different type of local architecture which resulted from the higher altitude of the settlement from the other three case studies. This settlement, situated about 3,000 m above sea level, was high above the Indus Valley. At this altitude, close to the tree line, the availability of timber had resulted in buildings with a greater volume of timber in their construction (a potential asset in providing seismic resistance). For this study Dieter Illi from Zurich joined the team, and brought to the group his extensive knowledge of the region and experience of the traditional house types and settlement patterns of the area. (See Figs. 4 and 5.)

The final study was made of Ganesh, an old settlement in the Hunza Valley, close to the historic Altit and Baltit forts. These two structures (two of the major buildings of the entire region) contained examples of traditional building techniques. Richard Hughes made an outline feasibility study of the Baltit fort structure to determine whether it can be conserved to arrest its current structural deterioration. His separate report is titled: "Baltit Fort - Its present state and the need for a conservation policy".

In addition to the four main case studies, various individual members of the group undertook further studies, these have been listed at the conclusion of Appendix 2. A particularly important study was the observation by Richard Hughes and David Nash of a gigantic mud-slide which occurred near Gupis on the Gilgit River. It appears that a glacier may have collapsed on a lake beneath the ice cover which caused a mud-slide.

FIG. 3.   NEW BUILDING IN PATAN,   FOLLOWING THE 1974 EARTHQUAKE.

The housing in the background, and in the lower photograph forms part of the linear bazaar which has been rebuilt since the Earthquake in an unsafe site and in an unsafe form of construction. The 'modern' building on the left-hand side of the top photograph belongs to the Kohistan Development Board built to aseismic standards.

FIG. 4. GHOHARABAD. THE THREE BASIC TYPES OF TRADITIONAL BUILDING.
On the left-hand side a timber granary, in the centre a fort, and on the right-hand side typical housing.

210

FIG. 5.   HOUSING IN KHURTLOT NEAR GHOHARABAD.

This indicates the sandwich form of construction of
heavy timber baulks interspersed with dry masonry.
The timber members act as a framing device and
they also serve to hold the unbonded stones in position.   Simple housing joints are used to join the
timber baulks, without any evidence of pinning.

This slid down the mountain in regular 'pulses' of mud and debris producing at its base a natural dam across the Gilgit River. The effect of this was to form a new lake which gradually rose to inundate a village. Some families were displaced and our group were able to monitor their response to their situation, as well as to observe the manner in which relief assistance was administered by the District Commissioners Office in Gilgit.

Dieter Illi was able to add to the group's work with his documentation of the consequence of a fierce fire which had caused devastation to a village in Gayal in the Darel valley. Although the group had become aware of the fire-risk hazard in these settlements where buildings were built in timber and where water was a scarce commodity, they had not had first-hand experience of this prior to this study.

PROJECT FINDINGS

The detailed findings of the project can be found in the research papers, to which I have already referred. However, in order to draw out the salient results, the following summary has been written which lists both findings relative to our research method as well as those concerning content. Some are general conclusions with application outside Pakistan to similar hazard prone environments, whilst others are specific to the northern areas of Pakistan. In setting out these findings, it is necessary to add the proviso that despite our efforts we were not always able to check or verify a given set of conclusions from one context with another, to the extent that we would have liked. Therefore, although we are confident in the general validity of the findings (which are supported from our analysis of work elsewhere and our study of similar contexts outside Pakistan), a better description would be 'working hypotheses' pending much more detailed study and further testing.

The findings are set out in two parts; firstly, a consideration of ten main lessons concerning the method of analysis, and secondly, the specific project findings relative to the vulnerability of local dwellings.

PART I   Method of Analysis

1.1   Scope of Investigation

The analysis of hazard resistant structures must be a part of a total examination of the values, local perceptions and priorities of a given community. Technical improvements can be introduced to increase the earthquake resistance of both existing and new housing, and to improve their general performance. However, such changes must be within the economic capacity of the occupants. It is therefore extremely difficult to introduce major improvements into existing housing; however, post-disaster reconstruction provides a possible opportunity to introduce such changes, as well as new construction. The full implications of the need to see this topic within a wide frame of reference will be considered in the remainder of these findings. They affect a diversity of issues concerning the type of personnel needed for this kind of study, governmental linkages, and local involvement; see also 2.2.

1.2 The availability of data

The group were handicapped in not having adequate information to enable them to fully undertake the study. In order to work efficiently it was found to be essential to have certain key items of information prior to starting field work. Minimum needs include large scale maps to identify individual houses (which can of course be produced from aerial photography) population figures, and social/economic data. If the latter items are unavailable, they can be obtained by a survey from interviews and standard sampling techniques, but obviously this is a time consuming process and erodes the research time needed to cover the central issues under investigation. Therefore, the group concluded that the lack of maps/aerial photography/social economic data was so important as a pre-requisite that they would not advise any group to embark on these types of studies without this essential information.

Inevitably in any developing country there will be deficiencies in maintaining accurate current data, or updating the physical mapping of settlements etc. This lack of information is an expected fact of life, but it is also within the capacity of most countries to collect this data. In many instances this is already available to the armed forces but restricted for security reasons.

The consequence of this lack of data in our project meant that an excessive percentage of time was spent in mapping the settlements and this information had to be gathered in parallel with undertaking social surveys whereas this information is a pre-requisite to interviewing local residents on their perceptions and priorities. Data is clearly needed in advance of social surveys to help obtain a statistically valid sample, keying in vital information on kinship patterns and linkages, land holdings etc.

Our experiences indicate that the following data is necessary, and ideally should be available prior to a research project:-

All Available Records of Past Disasters
(locations/impact/damage surveys/consequences/magnitude/frequency etc).

Any Available Maps of Hazards
Vulnerability analysis/risk assessment.

Maps of Settlements
On a scale to identify land holdings and individual houses.

Aerial Photographs

Social/Economic Data - Census figures on population/age, spectrum/ occupations/economic status/children at school etc.

Health Data
Life expectancy/mortality/health, nutritional status etc.

Agricultural Data
Forms of agriculture/production figures/crop yields etc.

Local Development Plans
Knowledge of current policies for the overall development of the given area, even if such plans have not been implemented.

Local Building Traditions
Any information on local building patterns – contractors/building costs/ constructional techniques etc.

Information on Public Building Provision
Plans/structural details etc. of governmental buildings.

1.3 Governmental Linkages

The availability of base-line data and settlement maps inevitably raises the issue of linkages between a project of this nature and the role of national and local governments. It is possible that some societies may permit teams to collect data and obtain detailed maps/photographs through normal commercial channels. But even if this is possible, the main source of detailed information on a community will normally lie within governmental departments. Even if they don't possess the necessary data, they will certainly have established channels to obtain information. This led our team to the conclusion that a pre-requisite for this kind of project was the close co-operation from the earliest stages with a host government. This observation is not to deny the helpful co-operation provided to us by central and local governmental officials for our project. But essentially the motivation for this project came from our research team, rather than from the Pakistan government, and given this fact they provided very helpful support.

To summarise, it is envisaged that an ideal formula for this type of project would be for a government to instigate and partially fund the study in close collaboration with the community in question, deploy resources to obtain advance data, and provide a substantial proportion of the project team. This would lead to their key role which would be to utilise the findings as a constituent part of their overall developmental objectives for the area in question. The value of this full co-operation would be extensive, and the only potential risk would be the possibility of arbitrary political constraints being exercised to the detriment of a project's integrity. Some safeguards might need to be retained to offset this risk.

Without active governmental involvement and commitment, any study of this topic, however accurate and thorough will stand minimal chance of either influencing general government policies on mitigation planning, or specific plans for the risk reduction of a given locality.

1.4 Local Participation

Elaborating on the linkage identified above, close contact with local personnel is considered to be crucial to the success of any project concerning risk reduction. This will apply to a variety of personnel within six categories:-

Officials in relevant central government ministries
(i.e., planning, building, agriculture, emergency planning etc.)

Professional/academic counterparts in the disciplines of Engineering, Seismology, Architecture, Planning, Anthropology, Agriculture, Geomorphology, etc. (such personnel would form a substantial proportion of the project team).

Officials in local government concerned with local development issues.

Non-Governmental Agencies   In some countries various bodies may be very effective institutions who can assist in a research project leading to implementation (i.e., agricultural co-operatives; religious groups, local voluntary agencies etc).

The local private sector - local builders, suppliers of materials, craftsmen.

The local community - approached through their leaders, or influential members.

1.5   Reconnaissance

Given the expense of putting a multi-disciplinary team into any area for this type of analysis it is crucial for one or two members to make a detailed reconnaissance visit to all proposed project locations.   These individuals would need to be well versed in all aspects of the proposed study and they would need to be accompanied with local personnel fluent in local language/dialects.

The tasks of this mission would relate to the basic ingredients of the analysis, asking a range of pertinent questions:-

Social Aspects   Are the local community likely to be receptive to such a study, will official permission be forthcoming?

Political   How do the local case studies relate to the overall development plans for the area?

Buildings/Settlements   Are the types of construction and siting typical for this type of study ?   Who are the builders, and when do they build?

Hazards   Are these typical?   Ideally various interlocking types of hazard will be considered.   Is data available - (i.e., within local memories) on hazard frequency?

1.6   Timing

Obviously our project was timed to coincide with the overall Karakoram expedition.   However, it became apparent that a much longer period of time was required to undertake this kind of work.   This was partly due to the reasons already stated in 1.2 above, relative to the lack of basic data and the time needed to collect this information.   But in addition there is the time required to establish and develop good relationships within a village community.

The question of timing is also a very critical issue in terms of when to undertake a project.   We came to believe that this would also need to be considered in a reconnaissance visit as outlined in 1.5 above. The factors to consider include:-

When would be the best timing for a project given local climate (monsoon periods etc.), building season, agricultural cycle (i.e., not to arrive during the busy planting or harvesting period), religious factors (i.e., not to visit during Ramadhan as occurred with our expedition).

1.7   The raising of local expectations

We were at pains to let local communities know that our studies were broadly based with general academic intentions, rather than as a prelude to local projects.   We also pointed out that our work was not being done on behalf of the Pakistan government as part of any development programme. Yet despite such assurances, many villages assumed that out of our work something tangible would eventually emerge in terms of the arrival of new houses to replace ones they now knew to be unsafe in form or siting.

We had always recognised the risk of raising false hopes by our presence and type of enquiry and this may well have occurred as one of the most serious negative consequences of our presence in various communities. However, against this we may have also assisted some families to have a greater awareness of the hazards that threaten their settlements.

The possibility of raising false hopes was inevitable, given the lack of a practical application of our work.   This serves to reinforce the finding 1.3 regarding the value of governmental linkages.

1.8   Project Personnel

For the entire project team one of the most valuable aspects of the study was the inter and multi-disciplinary approach.   We concluded that such teams require personnel in certain key disciplines with the minimum of one research member from each group:

Sociology/Anthropology/Epidemiology
Architecture/Planning/Building Technology
Engineering Seismology/Structural Behaviour of Buildings
Geology/Geomorphology
Agriculture/Rural Economics

In addition some of the team will ideally have had prior experience of vulnerability analysis, or disaster assistance.   In the Karakoram project there were clear benefits in having two members present with this experience which helped form a frame of reference for the structure of the overall project.

We regretted not having the particular insights of an agriculturalist in our team, (preferably one with a good working knowledge of rural economics).   Such a person would have assisted in simple cost-benefit studies considering the cost of erosion/flood damage relative to local mitigation measures.

Previous reference has been made to the need to consider this subject within a wider frame of reference than that of a limited technical enquiry, a serious weakness of many past studies of housing vulnerability.   Therefore the various disciplines suggested above will result in a team able to consider the social, cultural, economic and technical aspects of the subject. However, a further vital dimension for all participants concerns the broad issue of development.   We were very fortunate to have a team where all the members shared an understanding of underlying developmental issues. I would suggest that this awareness, informed with an understanding of the probable impact of measures to address poverty and reduce risk is a further need in any participant in this type of research.

1.9 Construction of Buildings

We found that through careful observation it was possible to broadly determine how a building was constructed. However, it was necessary to see the process in action and in any future project it would be sensible to include funding to pay for local builders to construct a small building (or a portion of a building).

1.10 Methodology for planning, fieldwork and analysis

One of the project team, David Nash, has developed a check-list for this type of research on the basis of our experience:-

(a) Desk Study   Define preliminary objectives; collect background information (maps, photographs, literature); make initial contacts; Redefine objectives; plan reconnaissance.

(b) Reconnaissance   Visit possible locations; meet previously arranged contacts; contact relevant organisations and collect relevant data; decide project location; initiate large scale mapping/photography.

(c) Detailed Planning   Redefine objectives; analyse large scale maps/ photographs (geomorphology, water, energy, transport....); plan detailed field work.

(d) Fieldwork   Several closely integrated projects (e.g., socio-economic analysis, health, nutrition, resources, hazards, building materials/types....); further contact with relevant organisations and individuals.

(e) Post-analysis   Initial post-fieldwork analysis by whole team (undertaken by whole team together); detailed individual analysis of data; second post-fieldwork analysis by whole team (to review papers, reports etc.).

PART 2   Broad Research Conclusions

The following are the main conclusions we reached during the project. As has already been stated the detailed research conclusions are contained within the individual papers by members of our project. In addition the outline findings of our group's work have been described in two overall descriptions of the project (4) and (5).

2.1 Constraints and opportunities for the reduction of risk

We came to recognise that a consideration of the reduction of vulnerability of unsafe houses should be considered in three very different conditions:

Existing Buildings
New Buildings, of two types, firstly all new construction, and secondly post-disaster reconstruction.

217

CATEGORY	MODIFICATION PROCESS	CONSTRAINTS FOR MODIFICATION PROGRAMME	OPPORTUNITIES FOR MODIFICATION PROGRAMME
EXISTING BUILDINGS	Strengthen vulnerable existing houses by modification programmes	A lack of financial resources; A lack of urgency on the part of officials or building occupants; The social upheaval of making any changes to existing houses; The lack of any 'standard building', each building being a 'one-off' design problem requiring skills in appropriate modifications.	1. Very limited for the houses of poor families, much greater for those with higher incomes; 2. Greater opportunities may occur for the introduction of a safe 'core' i.e., a four-poster bed with a heavy top structure to be placed within a house to protect people sleeping from asphyxiation from a collapsing heavy roof. Such measures are relatively inexpensive and are capable of being built by local carpenters. 3. A major problem is the vast proportion of vulnerable housing stock – too large for available resources; 4. Good opportunities for public buildings, especially those involving public assembly, i.e., schools, dispensaries, mosques, etc.
NEW BUILDINGS — All new buildings in hazard prone areas	Modify existing skills, techniques and certain materials for safe construction, design and siting	1. Basic conservation of traditional societies to any change; 2. Additional costs of hazard resistant construction; 3. A lack of teachers to explain basic principles to local builders.	1. Good, if there is the support of public authorities, if there are appropriate techniques, teachers and teaching materials available and institutions to implement projects at a local level.
POST DISASTER RECONSTRUCTION — New buildings	ditto	The prevailing mentality of most reconstruction authorities and relief agencies that suggest that reconstruction should consist of (a) temporary 'stop gap' housing; (b) a radical change of technology; (c) a contractor based activity that is apt to ignore local manpower and skills and the opportunity for training programmes for use at all levels	Very good for three reasons: 1. There is the visual aid of building failure, people are open to change; 2. There is an availability of cash as well as expertise from assisting groups. 3. There is the political will to make a positive action. However, it has to be recognised that all of these three opportunities are 'declining assets over time' and this emphasises the need for rapid action.
Damaged Buildings	Strengthen damaged houses with hazard resistant improvements incorporated into repairs.		

TABLE 2. CONSTRAINTS AND OPPORTUNITIES FOR THE REDUCTION OF RISK.

Table 2 summarises our findings.

There are some important implications from the ideas expressed in Table 2 and they may help to shape future analysis of the vulnerability of unsafe housing. If a primary concern to reduce the maximum number of deaths or injuries is adopted, then this could indicate a major deployment of resources of cash and manpower to existing buildings, since this is obviously the major area of risk. However, as is noted in Table 2 there are significant difficulties in this sector relative to social factors as well as the very high financial costs. Our group therefore concluded that the focus of energy/research etc relative to existing buildings should be sharply focussed on very simple measures capable of being undertaken at virtually zero or bare minimum cost for materials, although there could be a high labour input.

Our research was based on an unwritten assumption that the Pakistan government might be able to instigate some risk reduction programme to improve the safety of the existing housing stock. Therefore our pre-occupation was with houses built in a traditional manner. If we had seen matters a little more clearly, and recognised where the government could make a decisive impact, we would have probably placed greater emphasis on an examination of the vulnerability of new forms of construction which were beginning to emerge (see Section 2.2 below).

To summarise, intervention to reduce risks is advised as follows:-

Existing Buildings   Community based measures for all houses. Government measures to public buildings, schools/dispensaries.

New Buildings   Training programmes, and governmental subsidies to house builders to encourage them to build in a safe manner. This form of subsidy scheme could be on the lines of the very sensible policy adopted by the Pakistan government following the Patan Earthquake. This has been described in:- "Analysis of Recovery and Reconstruction following the 1974 Patan earthquake"; see page 323.

Therefore the focus of future research interest could very usefully be directed towards the two broad questions: what can be done to improve the safety of existing buildings at minimal cost where there are minimal building skills, and secondly the need to draw new construction into the centre of concern.

2.2 Modern Improved Construction

The use of modern building construction such as reinforced concrete construction and new building forms has resulted in the construction of many buildings whose performance in earthquakes is likely to be inferior to the traditional construction it has replaced. Very basic training is required at a village level for local masons and carpenters in building techniques. There is a need for the relevant authorities to recognise the fundamental importance of training (in lieu of the policy of providing safe, well built housing) in reducing the vulnerability of new building. This is particularly vital in view of the migration from the area of skilled building craftsmen to work in the major cities of Pakistan, or even further afield in the Gulf States.

FIG. 6. FOLLOWING THE 1974 PATAN EARTHQUAKE A UNESCO MISSION VISITED THE AREA, AND REPORTED ON THE POSSIBLE SEISMIC RESIST-ANCE OF LOCAL BUILDINGS, GIVEN THE INDEPENDENTLY SUPPORTD ROOF AND THE TIMBER/STONE SANDWICH CONSTRUCTION OF THE WALLING. ONE OF THE OBJECTIVES OF THE HOUSING AND HAZARDS PROJECT WAS TO TEST THIS CLAIM. CREDIT: PROFESSOR N N AMBRASEYS, UNESCO.

NATURAL HAZARDS EXPERIENCED IN THE KARAKORAM (In approximate order of local priority)	FREQUENCY OF HAZARD RECURRENCE	WARNING OF HAZARD	IMPACT ON LOCAL COMMUNITY							CAUSES OF DEATHS OR INJURIES
			Loss or damage to agriculture crops	and animals	Loss or damage to buildings/possessions	Physical hardship	Damage to health (illness, malnutrition)	Physical injuries	Death	
Failure of water supply (for both drinking and field irrigation)	Can occur every winter in some areas	1-2 months	✓	✓		✓	✓		Only in extreme sitns.	No evidence of deaths
Erosion of land (by action of water, wind and frost)	Likely to be a continuing process	Local knowledge exists of where it is occurring except from flood action	✓			✓	Very rarely			
Flooding: Slow impact	Generalisations impossible but communities living beside rivers are likely to have experienced flooding within the past 4-5 years	24-36 hours for slow impact	✓	✓	✓	✓	✓			No evidence of deaths
Flash floods		No warning for flash floods	✓	✓	✓	✓	✓	✓	✓	Drowning from injuries caused by impact of debris
Rockfalls Landfalls Mudslides (all could be induced by earthquakes	A continuous process for small scale rockfalls; major landslides occur less frequently, say 10 year intervals	No warning for rockfalls, but very slight warning (2 mins) following the sound of landslide in distance	✓	✓	✓	✓		✓	✓	From the collapse of houses or asphyxiation from being buried
Earthquakes	Minor tremors occur frequently but major earthquakes occur approx at 5-10 year intervals	No warning other than foreshocks of earthquake	✓ Inside buildings	✓	✓	✓		✓	✓	

TABLE 3.   RESPONSE TO NATURAL

RELATION-SHIP OF DEATHS/ INJURIES TO BUILDING COLLAPSE OR DAMAGE	THE NATURE OF THE VULNERA-BILITY OF HOUSES AND SETTLEMENTS TO SPECIFIC HAZARDS	REMOVAL OF RISK		REDUCTION OF RISK	
		Possible Solution	Difficulties in implementing this solution	Feasible mitigation measures in this context	Difficulties in implementing these measures
No relation-	If failure is total, results in the eventual abandonment of settlements	Build more water channels Reduce the local population by relocation	Excessive cost and in some instances the supply is fully used No obvious alternative	Inject finance to improve water supply as a priority issue	Cost and lack of expertise
No relation-	Can result in the abandonment of settlements adjacent to cliff edges	Build protect-ive river walls/channels	Excessive cost	Rural education for farmers to protect their own lands with simple measures; improve siting of all new buildings.	Lack of expertise
Marginal relation-ships	Location of settlement in low lying ground adj-acent to rivers or on the banks of steep narrow river valleys	Flood control measures relocate homes/settlements at risk	Social up-heaval; excessive cost	In some in-stances low investment flood control measures may be feasible; protective walls can be built	Lack of expertise; cost
Marginal relation-ship	Location of settlements at the base of steep slopes particularly below alluvial slopes	Relocate homes/settlements at risk	Social upheaval; excessive cost	Improve the siting of all new buildings; Encourage the relocation of the most vul-nerable houses in each comm-unity	Cost
Primary cause of casual-ties	Lack of resis-tance of the construction or shape of houses relative to earth-quake forces; dangerous siting on steep slopes subject to earth-quake induced landslides or rockfalls	Relocate homes/settlements at risk; modify houses where they are vulnerable	Social upheaval; excessive cost	Improve the siting of all new buildings; Encourage by education and cash incen-tives the building of aseismic houses	The strength of local building traditions; lack of expertise; cost

HAZARDS IN THE KARAKORAM REGION.

There is evidence of rising aspirations amongst the residents of traditional housing for new homes built in modern materials, with the addition
of guest houses. The vast majority of all new building has departed
from traditional forms and construction. (Fig. 3)

2.3 Earthquake Resistance of Traditional Housing

No traditional houses were examined in any of the areas studied which
had been specifically built to resist earthquakes. Therefore, it would
appear that the assumption that traditional housing in Pakistan has been
built in response to earthquake risk does not appear to be the case.
However, as has been observed in our study of the earthquake resistance
of traditional housing, we found that there was a degree of potential seismic
resistance and relatively simple and inexpensive measures could improve
the safety of such structures. (Fig. 6)

2.4 The perception of hazards by local families is well defined, and frequently in contrast to that of external personnel

In Table 3 the hazards in a given locality have been identified in
an approximate descending order of priority. This chart aims to describe
the response to these hazards, with opportunities to remove or reduce them.
This chart presents a general example since the detail is variable, for
example in some communities the risk of fire (particularly following a
recent fire) would probably assume first priority.

2.5 Hazardous Sites for Houses

In order to secure house sites adjacent to families land, many people
have moved to highly vulnerable locations. They are aware of the vulnerability of their houses to such frequent hazards as rockfalls, landslides
and floods but less aware of earthquake risk with their long return period.

2.6 Local Priorities

The provision of water for irrigation as well as the availability of
land for cultivation were always found to be more dominant concerns than
housing and the effects of natural hazards.

2.7 Availability of Timber for Housing

There is evidence of serious deforestation caused by population pressures
and commercial exploitation of forests. There is an urgent need for an
expansion of forestry and planting policies. One effect of this depletion
is the reduction in supplies of firewood and supplies of timber for building.

2.8 Public Health and Housing

Analysis of traditional housing and settlement plans has identified
significant public health deficiencies. The study of health needs in Yasin
and the educational survey in Ghorahabad and Hunza revealed the need
for education at village levels in hygiene, public health and nutrition.

2.9 Hazard Mitigation Measures

(i) Community Based Activities. If expertise is made available, as
well as cash incentives, it is possible that community based

mitigation measures could be devised to cope with certain hazards, e.g., local communities could build walls to arrest flood damage, water tanks for fire protection, etc.

(ii)   Official Actions.   There is a clear need for officials to advise local communities on safe sites for new buildings, as well as encourage the relocation of parts of settlements which are particularly vulnerable.   We believe that some check needs to be made in each settlement of the safety of public buildings with multiple occupation – particularly schools, mosques and village dispensaries.   This examination must include both the siting of such structures as well as their form and method of construction.

2.10   Developmental Context

The process of vulnerability analysis of poor communities inevitably reveals a wide diversity of causes and effects.   As our research papers, and this summary have indicated some are technical deficiencies in the way buildings are constructed, whilst other limitations relate to cultural or environmental factors.   This introduces the context of development into the analysis, and it is apparent that the areas we studied are no different from other countries where members of our team have previously worked. The similarity being in that there are always political issues that need to be considered in parallel with technical factors.   However, it was not our brief to address the political elements of risk reduction so I merely wish to observe that the political dimensions will encompass such questions as the deployment of governmental resources, the issue of land ownership, local development plans and legislation for the control of deforestation and land-use planning relative to flood risk in the valleys.

In our rather limited discussions with public officials at both local and central governmental levels, we were made aware that some of these topics are already being addressed.   However, the remoteness of the areas under study, as well as the sparseness of population and the scale of severity of the hazards has made this a difficult problem to address. From our limited perspective, it appeared that the work of the Kohistan Development Board is an example for other areas of an integrated government development programme containing elements concerning the reduction of risk.

2.11   Summary Listing of Levels of Risks and the Probability of Damage

1. | The Hazard, Natural Phenomenon | Magnitude
                                            Frequency
                                            Location
                                            Length of Warning

2. | The Settlement Location | Density of spacing of dwellings
                                     Road widths
                                     Specific site with reference
                                        to potential
                                     Deforestation

3. | Types of Buildings |

Age of Building
Quality of Construction
Level of Maintenance
Form of Individual Building
Extent of Hazard Resistant
  Construction
Occupancy
Building Materials
Varieties of Building, types
  in a given location

4. | Social/Cultural/Economic Variables |

Perception of Risk
Economic level of occupants
Patterns of land ownership
Attitude to hazards
Values and Priorities of
  local families
Occupations with reference
  to the settlement

OVERALL CONCLUSION

As we neared the end of the project we became progressively more impressed with the astonishing ability of the local families (perhaps better described as 'survival artists') to cope with the various hazards that continually threatened not only their lives, but also their tenuous relationship with the landscape. Many faced rockfalls on one side of their homes, and on the other erosion risks or even flooding, and virtually the entire region was subject to earthquakes and water supply problems. However 'cope' is the best word to describe their toehold on the land. Only rarely did we see examples of families adapting their homes or living patterns to achieve better protection from natural hazards. When we questioned them why they paid so little regard to these risks they replied that in their eyes there were far greater worries than any of these hazards. In fact, one house owner shook his head in total mystification as to why we should come all the way from England to consider such a strange matter. Local priorities were more to do with their children's education (or lack of it), their health and the lack of medicine in the dispensary, the difficulty in selling their crops for a decent price, which is related to the problems of access to available markets given the very inaccessible terrain.

So vulnerability to natural hazards has to be recognised within the cultural and economic context of poverty and deprivation. And as with our own society long-term logical action is frequently overwhelmed by short-term expediency. Nevertheless, despite this sobering reality, we all came to share the view that risks could be significantly reduced by imaginative and sensitive attention to the issues we have tried to identify.

The project team are encouraged that subsequent to the expedition similar investigations to the Karakoram study are currently being undertaken in the Yasin Valley to expand on the initial social/economic study undertaken by Frances D'Souza. In addition, work has been undertaken in Southern Italy by members of the same team that worked in the Karakoram region. Two closely related projects have recently been approved, a study of the social/cultural and economic aspects of improving traditional housing in

Turkey. This project is linked to research already in progress concerned with the behaviour of traditional housing within seismic areas, including Turkey. Finally, the United Nations Centre for Human Settlements (UNCHS) has recently initiated a major international project in four disaster prone countries concerned with the teaching of safe building techniques.

An important direct result of the expedition has been the establishment, under the joint auspices of the Intermediate Technology Development Group (ITDG) and the International Disaster Institute (IDI) of a panel on the Protection of Buildings and Settlements from Environmental Hazards. This is an international forum for the promotion of risk reduction policies and the exchange of information. Details of membership can be obtained from the author.

From these (and other) developments it is likely that our knowledge of these problems will rapidly expand and that the research approach of the Karakoram project described and criticised in this paper will be progressively refined into an effective working tool for both the analysis of vulnerability and the implementation of risk reduction measures.

In conclusion, it is our hope that our work, with all its obvious limitations may help to stimulate other research teams, governments and funding bodies to address this frequently neglected subject.

APPENDIX 1

Advisers to the Housing and Natural Hazards Group

In the field of engineering seismology, David Dowrick of Ove Arup and Partners, Dr R F Stevens and Dr Keith Eaton of Building Research Establishment and Professor N N Ambraseys of the Department of Civil Engineering, Imperial College.

In the field of settlement planning within marginal economies, Professor Emeritus Otto Koenigsberger of the Development Planning Unit, University College.

In the field of anthropology of shelter and settlements, Paul Oliver of the Department of Architecture, Oxford Polytechnic.

In addition to these principal advisers, the project team have benefited from helpful advice from Mrs Janet Pott, an architect with experience of local architecture in Northern Pakistan; Stuart Lewis, Department of Architecture, Oxford Polytechnic; Professor Karl Jettmar of the South Asian Institute, University of Heidelberg; Dr Schuylar Jones, Pitt Rivers Museum, Oxford; and Dr Robert Muir Wood of the University of Cambridge who contributed to the group's work whilst in Pakistan, and has subsequently written a useful critique of the group's work[4]. The late Jim Bishop was part of the team in its early planning stages, and provided very useful assistance in the early design of the project.

APPENDIX 2

PROJECT ACTIVITIES

Visit to Quetta

A visit was made to examine the housing and public buildings constructed after the 1935 Earthquake which devastated the town. (Project Team – Hughes/ Nash/Spence).

Case Study No. 1, Barculti, Yasin Valley

An examination of a village – its rural economy, social structure, priorities, house forms and construction and settlement patterns. In addition an analysis of the resistance of housing to natural hazards and the local analysis of such risks. A detailed social survey was carried out by Frances D'Souza, and structural tests were made of existing housing. An examination of health status was made by Helen Massil. (Project Team – D'Souza/Israr ud Din/Hughes/Moughtin/Nash/Coburn/Spence/Massil).

Case Study No. 2, Patan, Indus Valley

This period of study was to explore the reconstruction activities of the Kohistan Development Board (KDB) and specifically to:

(a)   examine structures which survived earthquake;

(b)   ascertain whether the surviving population had made any changes in their house construction after the earthquake in the reconstruction phase.

(Project Team – D'Souza/Israr ud Din/Davis/Nash/Hughes/Coburn).

Case Study No. 3, Ghorahabad, Indus Valley

This visit was to examine a very different community from either Barculti or Patan in that Ghorahabad is a very ancient fortified town with close knit housing, as well as being an upper altitude village built primarily out of timber from adjacent forests. The study was similar in emphasis to Barculti with additional studies of the historic buildings within Ghorahabad not previously documented. (Project Team – Davis/Hughes/Nash/ Goodchild/Illi).

Case Study No. 4, Ganesh, Hunza Valley

This final study of a community was made in Hunza to verify whether previous studies in Barculti/Patan/ Ghorahabad were typical of broad trends. As with the study of Ghorahabad some historic structures were found and these were carefully measured and documented. Research methods included observation, measurement, documentation, and detailed interviews with residents of the town. (Project Team – Davis/Goodchild/Hughes/Nash/Holmes/ Illi).

Miscellaneous Studies

Observation of Mud Slide in Gupis and its Social Consequences - Nash and Hughes were able to make first hand observations of the mud slide which temporarily blocked the Gizar River. These measurements were passed to the Geomorphology Group for further detailed analysis. A social analysis was made by D'Souza of the surviving displaced families from the two villages that were inundated in the back-up flood that occurred behind the natural dam. (Project Team - Nash/Hughes/D'Souza).

Visit to Lahore

Following torrential rains there has been extensive flooding in the Lahore vicinity. The group visited the affected area and assessed damage to housing and the physical infrastructure. (Project Team - Davis/Hughes/Nash/Goodchild).

Examination of Distribution of house types throughout the entire Karakoram Region - Swat/Cunjurab; Chitral/Skardu, etc. (Israr ud Din).

Derivation of House Forms/Construction Impact of Yasin/Gupis Earthquake of 1972. (Illi).

Study of House Forms and Settlement in Aliabad Hunza (Davis).

Study of House Forms and Settlement in Chilas and Tangir Valley (Israr ud Din/Davis/Nash/Hughes).

Preliminary Survey of Baltit Fort on behalf of Professor Karl Jettmar, South Asia Institute: Im Neuen Heimer Feld, 330 Heidelberg, Germany, and the International Council of Monuments and Sites (ICOMOS). (Hughes)

Study of Educational Resources in Northern Areas (Ghorahabad, Gilgit, Hunza) (Goodchild).

Study of Gayal, Darel Valley, village fire (Illi).

REFERENCES

1. YUZUGULLU, O., (1980). State of the Art Report. Earthquake Resistant Rural Structures. 7th World Conference on Earthquake Engineering, Istanbul.

2. THE URBAN EDGE, (1981). The Urban Disaster. Preparing for the Worst. The Urban Edge, Vol. 5, No. 3, March 1981.

3. ARYA, A. S., BOEN, T., GRANDORI, G., BENEDETTI, D. et al., (1980). Basic Concepts of Seismic Codes, Vol. 2 Non Engineered Construction. The International Association for Earthquake Engineering, Tokyo.

4. MUIR WOOD, R., (1981). Hard Times in the Mountains, New Scientist, 14 May 1981, pp 414 - 417.

5. DAVIS, I., (1982). Safe Houses in the Karakoram, Geographical Magazine, January 1982, Vol. LIV, No. 1, pp 30 - 39.

The construction and vulnerability to earthquakes of some building types in the northern areas of Pakistan

A.W. Coburn*, R. E. Hughes**, D. Illi***, D. F. T. Nash†, and R. J. S. Spence*

* University of Cambridge, ** Ove Arup and Partners, London,
***Zurich University, Switzerland, † Bristol University, U.K.

ABSTRACT

This paper assembles what was learnt of the earthquake-resistance of the buildings in the Northern Areas during the International Karakoram Project and puts this in the context of other published work on the subject. The building technologies of the area are described, the available information concerning the seismicity of the area is presented and discussed, the earthquake resistance of the buildings as understood from earthquake records, field observation and laboratory studies is considered: and finally some proposals for modification of buildings to improve their earthquake resistance are presented.

INTRODUCTION

This paper draws together what was learnt of the earthquake resistance of the buildings in the Northern Areas during the International Karakoram Project 1980, and puts this in the context of other published work on this subject. Earlier papers (1)(2)(3) have set out the theoretical background to this study, and other papers in this volume, e.g., (4)(5) contain more detailed treatment of some of the matters discussed here.

The paper is in four sections: first, the building technologies of the area are briefly described; secondly, the existing information relating to the seismicity of the area is presented and discussed; thirdly, the earthquake-resistance of the buildings, as understood from earthquake records and field observation is considered; and finally some proposals for modification of buildings to improve their earthquake-resistance are presented.

The study results from fieldwork during the period July to September 1980, which concentrated on the four settlements: Barculti, Ghor, Ganesh, and Patan (Fig. 1) and on observations elsewhere in less detail. Conclusions reached therefore are specific to the building types of these places and their applicability elsewhere would await confirmation from similar detailed local studies.

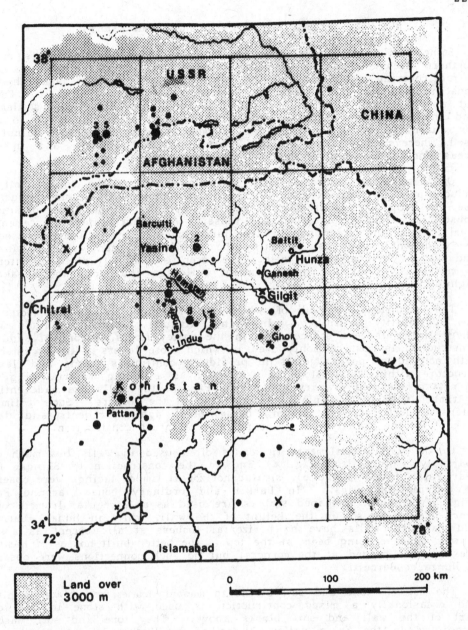

EARTHQUAKE EPICENTRES

- •   M = 4.0 - 4.9
- •   M = 5.0 - 5.9
- ●   M = 6.0 - 6.9
- X   Magnitude unclassified

FIG. 1. THE NORTHERN AREAS OF PAKISTAN SHOWING RECORDED SEISMIC ACTIVITY THIS CENTURY (AFTER QUITTMEYER AND JACOB (7)).

BUILDING TECHNOLOGIES OF THE NORTHERN AREAS

Stone, soil and timber are the traditional building materials of the Northern Areas. Soil is used for blocks, for mortars, plasters and as a roof covering material. In the mountain areas it often lacks the fine constituents which give dried mud cohesion, plasticity and strength. Rough-cut and river-bed stone is the usual walling material in most settlements and in the valleys there is generally sufficient free-lying stone to hand for quarrying to be unnecessary. Timber (deodar pine and walnut) is used wherever available for roof spans, internal columns and in some areas for reinforcement of the wall structure.

Of these three materials, only stone is abundant; the use of soil can reduce precious agricultural land, while timber suitable for structural use has long been so scarce in most valleys that timber from demolished houses is carefully preserved for reuse. All other materials, cement and steel, nails and ironmongery, lime and bricks are virtually unused in the traditional buildings, reflecting the high cost of manufacture, transport and even of the craftsmen's tools. The recent availability of these materials is marked by their use in Government buildings and along the Karakoram Highway.

The traditional method of walling construction is much affected by the availability of timber. In higher altitude villages, such as Ghor, where timber is less scarce than in the valley floors, a form of sandwich construction is most commonly used in which close layers of timber interlock with notched or pegged joints to provide a continuous strong and flexible structure within which stone is used only as a filling material (Fig. 2). In certain elevated granary structures (Fig. 3) the walls are made entirely of timber with scribed planks interlocking at the corners. Such a timber-intensive form of construction, being proof against rodents and damp, is kept for the granaries and was not seen in house construction.

In the valley settlements timber is often used in walls but much more sparingly. In Baltit (Fig. 4) and in the construction of Shingri forts near Patan (Figs. 5 and 6), similar horizontal timber lacings were observed at different spacings. In Darkot and ordinary houses around Patan (Fig. 7), thinner, unshaped timbers are used as an irregular framed course every 60 cm or so up the height of the wall, while in Hunza, Hamaran and Barculti, walls have no horizontal timbers at all, except as a form of wall plate or ring beam at the top. The better built houses of Hamaran are actually pegged at the corners, but no such connections were observed in Hunza or Barculti.

The use of stone and soil blocks in masonry varies from place to place, and occasionally a mixed construction is used with stone in the lower part of the wall and soil blocks above. The stone and soil masonry techniques in the Yasin valley, described by Hughes (this volume), were found throughout the area of study.

Most houses of the area are single storey with a flat, earth-topped roof. The roof-covering consists of a thick layer of beaten soil, resting on a layer of straw or brushwood and supported by unshaped timber rafters at 20 - 30 cm centres. Heavy timber beams support the rafters, bearing on perimeter walls or internal timber columns, the usual span being around 3 metres. One example of this type of roof construction is the baipash of the Yasin and Hamran valleys. The structure of this house-type with its prominent smoke-hole is shown in Fig. 8. In Patan, the smoke-hole

FIG. 2. TIMBER SANDWICH CONSTRUCTION AT GHOR.

FIG. 3. TIMBER GRANARY AT GHOR.

FIG. 4. TIMBER-LACED WALL CONSTRUCTION AT BALTIT.

FIG. 5. TIMBER-LACED WALL CONSTRUCTION NEAR PATAN.

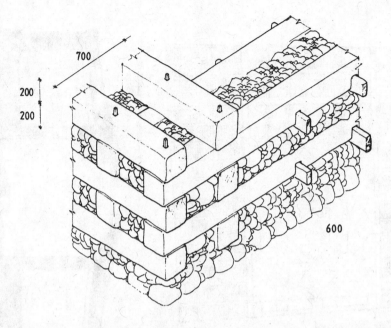

FIG. 6.   SHINGRI TOWER WALL CONSTRUCTION NEAR PATAN.

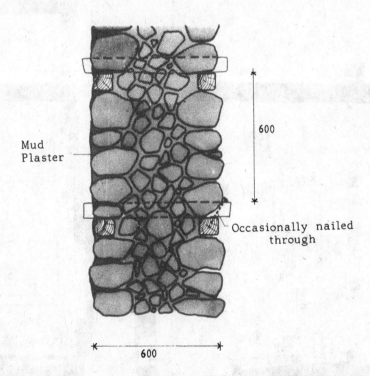

FIG. 7.   DOMESTIC WALL CONSTRUCTION AT PATAN AND DARKOT.
(dimensions in millimetres)

234

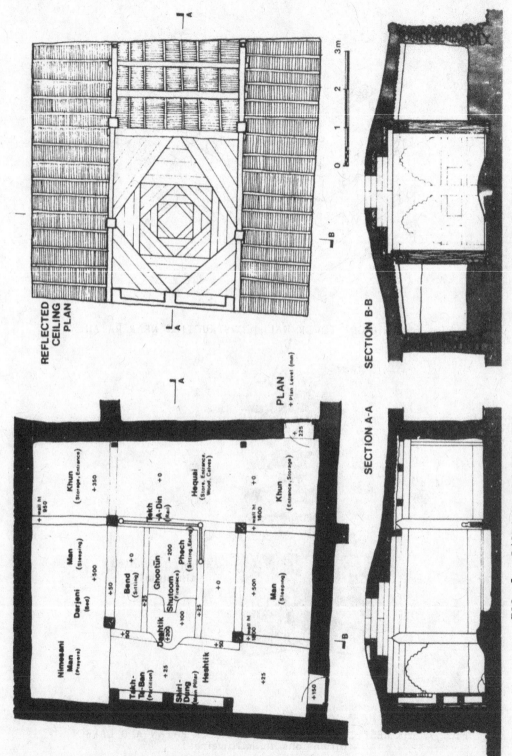

REFLECTED CEILING PLAN

PLAN
+ Plan Level (mm)

SECTION A-A

SECTION B-B

FIG. 8. BAIPASH CONSTRUCTION OF THE YASIN AND HAMARAN VALLEYS

is reduced to a small, functional hole 20 cm diameter, sometimes with a chimney built over it. The houses are cut into the terrace and the rafters supported at one end by the hillside itself and at the other by main beams running parallel to the hillside, bearing on cross-walls and internal columns (Fig. 9).

FIG. 9. TERRACES OF HOUSES AT PATAN.

The traditional forms of construction account for the large majority of buildings in the valleys visited, but other forms of construction are beginning to be used, for some particular structures and for the wealthier house builder. Government buildings, such as schools, dispensaries, administrative buildings, are built using modern construction techniques; walls are of concrete block or even clay brick near the Karakoram Highway, but more usually of squared stone blocks elsewhere, jointed with lime or cement mortar, and with concrete floors and pitched corrugated iron roofs.

At Patan and Dobair Bazaar, substantially rebuilt after the 1974 earthquake when new materials had been made available by the Karakoram Highway, a mixture of construction techniques can be seen with both concrete blocks and reinforced concrete frames beginning to appear alongside the traditional forms. The use of these materials is still unestablished and indicates little understanding of the differences between the old and new materials (Fig. 10). Undoubtedly the increased accessibility of the area, the prestige of using modern materials, and the loss of craftsmen familiar with traditional materials to better paid jobs elsewhere, coupled with the growing scarcity of timber will accelerate the trend towards the use of modern materials and technologies, but for the present and for some time to come the houses in the less accessible valleys will continue to be built in local materials and traditional techniques. It is the earthquake vulnerability of these houses which is the main concern of this paper.

FIG. 10. TIMBER FRAMED HOUSES AND REINFORCED CONCRETE FRAMED
HOUSES AT DOBAIR BAZAAR IN KOHISTAN.

SEISMICITY

A detailed study was made of the seismicity of the area bounded by latitudes 33° and 38° N and 71° and 76° E, within which all the settlements studied lie. Historical earthquake records in this area are few, and are mainly concentrated on the centres of colonial administration. Of the historical earthquakes listed by Ambraseys (6) and Quittmeyer and Jacob (7) before 1914 only those at Swat (1909) and Gilgit (1871) have centres within this area. In each case Quittmeyer and Jacob (7) estimate an epicentral intensity (I_o) not greater than 8. The modern seismic record also lists moderate events at Yasin (1943, $M_s = 6.8$), Tangir/Darel (1972, $M_s = 6.2$), Patan (1974, $M_s = 6.2$) and Darel (1981, $M_s = 5.9$). It is not likely that any other comparably large earthquakes in the area during the last 100 years have been unrecorded.

Teleseismic data are available from 1914 onwards, and the given epicentres of all events within $34 - 38^\circ$ N and $72 - 76^\circ$ E, as relocated by Quittmeyer and Jacob (7) are shown in Fig. 1. Using the Quittmeyer and Jacob (7) list, the recurrence of earthquakes of different magnitudes has been studied for the larger 5° x 5° area $33 - 38^\circ$N and $71 - 76^\circ$ E, between 1962, when accurate magnitude determinations became available, and 1975. Because of the detectability threshold in this area, there are few recorded $M_s < 4.5$ events; there is an abundance of events in the range $4.5 < M_s < 5.5$; and only 5 events of $M_s > 6.0$ of which the largest has $M_s = 6.4$. The spatial distribution of these earthquakes (Fig. 1) is not uniform; there appears to be a concentration of seismicity in the Southern Pamirs and on an axis from Swat/Patan to Yasin/Ishkuman.

The conventional cumulative plot of M against log N (8) for the 5° x 5° gives a straight line (Fig. 11), from which the relationship

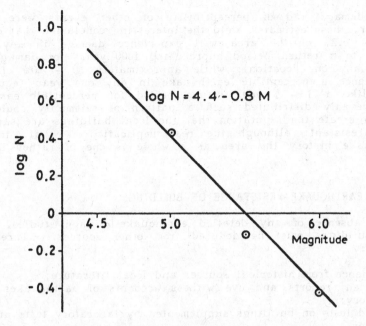

FIG. 11. SEISMICITY OF THE AREA, 33 - 38°N, 71 - 76°E
(5° x 5°), FROM 1962 to 1975 (AFTER QUITTMEYER AND JACOB (7)).

$$\log N = 4.4 - 0.8\,M_s \qquad\qquad (1)$$

is deduced, where N is the number of earthquakes per year with magnitude greater than or equal to M_s occurring in the whole 5° x 5° area. If compared with the data given by Kaila and Narain (9), seismicity deduced from the coefficients in this equation is not very high, considerably below the average for the Himalayan belt, and lower than that of the Mediterranean area, Turkey or Iran; on the seismic activity map they propose, this area is one of relatively low seismicity by comparison with adjacent areas to the north west and to the east. The return periods for earthquakes occurring in the whole 5° x 5° area deduced from equation (1) are

$M_s \geq 5.0$ 0.4 years
$M_s \geq 6.0$ 2.5 years
$M_s \geq 6.5$ 6.4 years (by a small extrapolation)

Because of the short period over which records are available, and the unreliability of much of the data, these return periods are only very tentative.

It is also of interest to attempt an approximate assessment of the frequency with which a damaging (intensity 7+) or a perceptible (intensity 3+) event may be expected to be experienced in any one location. In the absence of clear evidence of the locations of causative faults, the seismicity will have to be assumed uniform. The radii of damage and of perceptibility for one earthquake, that of Patan 1974 are approximately known (6) and some individual accounts of damage in the Yasin 1943 and Tangir/Darel 1972 earthquakes were obtained; from this data the likely radii of damage and of perceptibility of other events were estimated. Put together, these estimates yield the interesting conclusion that on average only 0.1 - 0.2% of the area will experience damage in any one year (equivalent to a return period approaching 1000 years for damaging earthquakes in any one location); while approximately 30 - 40% of the area may experience a perceptible earthquake in any one year (equivalent to something like a 3 - 5 year return period for perceptible earthquakes). Even if unevenly distributed such a pattern of seismicity could well be expected to create the reputation that the local buildings are satisfactorily earthquake-resistant; although the real implication is that, in spite of its earthquake history the area as a whole is one of rather low seismic hazard.

EARTHQUAKE-RESISTANCE OF BUILDINGS

In the absence of any detailed earthquake damage studies, resistance of local buildings must be deduced from other sources. Three possible sources are:

1. Evidence from historical sources and local literature;
2. Written reports and eye witness accounts of earthquakes in living memory;
3. Field tests on buildings supplemented by laboratory tests and calculations.

The evidence from three of these sources is briefly reviewed.

1. Historical evidence

The literature of the Northern Areas in the English language contains very little information about its earthquake history, and only three references are known. Clark (10) describes the shaking of Baltit fort by an earthquake on 6th January 1951; the building 'swayed and creaked' but no damage was done either to the fort or, it must be presumed, to the surrounding settlement so it is clear that this was not a major earthquake. Hussam-ul-Mulk and Staley (11) and Edelberg (12) discuss traditional forms of building further west in Chitral and Nuristan respectively. Edelberg, in describing the timber-framed houses of Nuristan claims that their construction techniques are a response to the "frequent earthquakes of this unquiet zone". Hussam-ul-Mulk and Staley describe the Chitrali baipash which unlike those of Yasin, has its stone walls strengthened by horizontal timbers; and they report a Chitrali belief that this type of walling is particularly resistant to earthquakes. There are, however, no reports of individual earthquakes until recent times, so it is impossible to say what degree of shaking these types of building may have been able to survive.

2. Reports of recent earthquakes

Damage reports of any sort are only available for three of the earthquakes which have been instrumentally located in the area during the last 50 years, (Fig. 1) For two of these, at Qurkalti Bar (1943) and Tangir/Darel (1972) written reports do not exist, but some eye witness accounts were obtained; the third, at Patan (1974) was the subject of a UNESCO report (13).

The instrumentally located epicentre of the Qurkalti Bar (1943) earthquake is near a side valley branching from the Yasin Valley; because of its high altitude this valley has little permanent population, and is mainly used for summer pastures. In the nearest hamlet of Qurkalti it was reported that 90% of the houses (of baipash type) had been destroyed in the earthquake; there was no memory of serious damage to buildings in the nearest settlements in the Yasin valley, Barculti and Yasin, even though these were only 25 km distant from the located epicentre. It may be that the earthquake was not as large as its calculated magnitude (M_s = 6.8) suggests; or alternatively, in common with other earthquakes of the area, it may be that the severe ground shaking was very localised.

The Tangir/Darel earthquake of 1972, with its long succession of aftershocks (see Hughes, this volume, for listing) was much more damaging. Jackson (14) has relocated these shocks within a radius of about 15 km, and at a depth of about 10 km. In the Tangir valley itself, it was reported that 82 were killed and 400 made homeless; some buildings were heavily damaged in Gupis about 40 km to the north, and damage was reported even in Ghor, 100 km to the east. A visit was made to the village of Hamaran, in the valley of this name, close to the epicentre. Here it was reported that 80% of the houses collapsed, and only 4 houses with minimal damage could be repaired. These houses were also of the baipash form already described, and the owner of one of the surviving houses attributed its strength to the craftsmen from Gilgit who built it, three generations ago, jointing the roof timbers with wooden nails and also using many iron nails. Another informant, a carpenter, attributed the strength to the wooden ring-beam which rests on top of the walls and is pegged at the corners. Many deaths were caused by the collapse of the stone walls (even where the roof did not collapse) and also by boulders falling from the mountain.

The Patan earthquake of 1974 had a magnitude M_s = 6.2 and caused around 700 deaths and 4,000 injuries. Damage to buildings was caused both by the earthquake and by subsequent rockfalls, to which the terrace-type houses of the area are particularly vulnerable. The UNESCO (13) report distinguishes two types of dwelling using local materials: in both cases the walls were coursed with timbers of 0.4 to 0.6 m centres as described earlier, but in the first type the walls were non-bearing and the roof was independently supported on timber columns, while in the second case the walls were bearing. The first type was said to have withstood the shaking much better, even though one or more of the non-bearing walls had collapsed. Those walls made of angular blocks were also reported to have withstood the shaking better than those of rounded or semi-rounded boulders. The report also notes that Shingris suffered no damage, and that damage to the few buildings of modern construction in the area was slight or moderate and mainly due either to obvious construction faults or foundation failures.

In September 1981, subsequent to the field study a further severe earthquake (M_s = 5.9) occurred, affecting the Darel and Tangir valleys. As yet no detailed report has been received, but newspaper reports (Pakistan Times, 16 and 17.9.81) put the death toll at over 200 with over 2,000 injured out of a total population of 25,000 in the 15 miles radius area affected in the two valleys. Once again it must be assumed that the collapse of houses of traditional construction was mainly responsible for these casualties.

Taken together, these few isolated reports do tend to confirm that there is some general awareness of the problem of earthquakes in the way buildings are built, and that some strong construction techniques do exist. But the evidence certainly does not justify the conclusion that the people of the area have solved the problem of building their houses to resist earthquakes. The casualty figures in the three most recent earthquakes were large; and even though the modern buildings in Patan were not particularly well constructed, the UNESCO report (13) notes that they performed significantly better than the houses of traditional construction in the 1974 earthquake.

3. In-situ vibration tests

As part of the investigation of the dynamic behaviour of the traditional buildings in the area, some in-situ vibration studies were undertaken. Naturally, these could only be small amplitude tests, but it was hoped that they would indicate the initial vibration mode during an earthquake, so that conclusions could be drawn about the buildings' vulnerability.

The Building Research Station has recently carried out a number of studies of the dynamic sway characteristics of buildings using eccentric-mass vibrators to produce a steady-state sinusoidal excitation, Jeary and Sparks (15). A similar technique was adopted for the field studies of traditional buildings using the test set up shown diagrammatically in Fig. 12. With the vibrator fixed to the roof three types

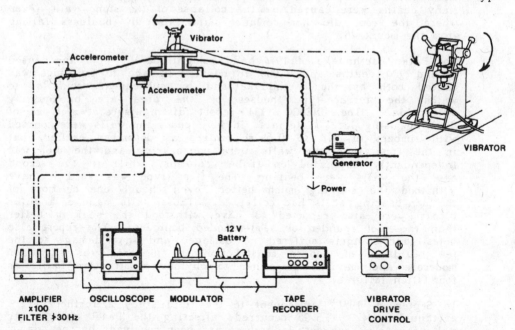

FIG. 12.   EXPERIMENTAL EQUIPMENT USED FOR VIBRATION TESTS.

of tests could be carried out readily:

i) a frequency sweep in which the response of the structure was measured with an accelerometer; when it was subjected to different frequencies of forced excitation it would give its natural frequencies of vibration.

ii) with the vibrator exciting the structure at its natural frequency, the accelerometer could be used to measure the response of different parts of the structure and thus investigate the mode shape.

iii) the natural decay of steady-state vibration when the vibrator was stopped suddenly could be monitored to measure the modal damping.

A typical set of results from the tests on the baipash described in section 2 are shown in Figs. 13 and 14. The detailed studies carried

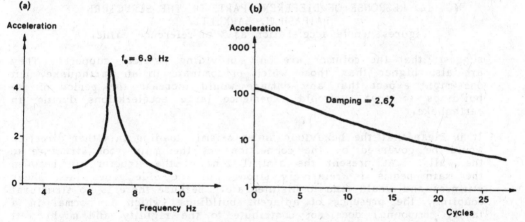

(a) frequency sweep   (b) decay at resonance

FIG.13. RESULTS OF VIBRATION TESTS ON BAIPASH H, BARKULTI, NORTH-SOUTH DIRECTION.

out are described elsewhere (Nash and Spence, to be published), but some general observations and conclusions may be drawn from these.

The natural frequencies of the first mode of vibration of the single-storey baipash buildings tested were in the range 7 - 10 Hz in both directions compared to 10 - 20 Hz for 1.5 to 2.5 metre high free-standing walls swaying about their base. Damping values measured were 2 - 4% of critical damping for the buildings and 1 - 2% for the walls alone. The accelerometer readings taken around the structure in the steady state (Fig. 14) showed that the roof structure had insufficient in-plane stiffness for the longitudinal walls to restrain the vibration, and the first mode involved sway of the central roof structure resisted by the transverse walls cantilevering about their bases.

These natural frequencies are considerably higher than the natural frequency of the roof structure cantilevering on rigid columns, which

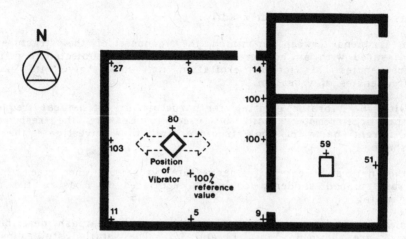

FIG. 14. RESPONSE OF DIFFERENT PARTS OF THE STRUCTURE OF
BAIPASH H, BARKULTI.
Figures denote acceleration as % of reference value.

suggests that the columns are only providing vertical support. They
are also higher than those which predominate in an earthquake, but
one might expect that any damage would increase the period of the
buildings so that it would experience large accelerations during an
earthquake.

It is clear that the behaviour under seismic loading in either direction
would be governed by the connections of the main roof structure to
the walls. At present the central zone of the structure in between
the main beams is relatively strong but the side zones lack shear
stiffness and would tend to disintegrate. Because there is no structural
coupling, the presence of adjacent buildings (which is normal in a
family compound) does not contribute to the rigidity and might well
increase damage as different buildings in the compound vibrated
independently.

4. Laboratory 'rocking-column' studies

The observation made in the UNESCO report (13), that buildings with
roofs supported independently of the walls performed relatively well
in the Patan earthquake led to speculation about the nature and
response of the residual structural system consisting of roof, roof-
beams and columns which would remain after the collapse of the
surrounding walls or the separation of the roof from the walls. In
the absence of the lateral restraint provided by the walls, might
the size of the columns be sufficient to provide stability, even if
they were unconnected to the beams at the top and base? Would this
structural system begin to rock under the increasing base acceleration,
and what would be the characteristics of the subsequent motion?
How much movement would be needed to bring about collapse?

Theoretical and experimental studies of an idealised structural system
with these characteristics were carried out by Stephens and Whitfield
(16) and Evans (17), final year students at Bristol University Dept.
of Civil Engineering and Cambridge University Engineering Dept.

respectively, with the following general conclusions:

1. Beyond a certain base acceleration which depends on the geometry and mass of the system, column rocking will occur, with the effect of limiting the roof acceleration and hence also the forces induced in the structure.

2. There is a maximum absolute acceleration which is a geometrical property of the structure, and independent of the base accleration.

3. Collapse of the structure will occur as a result of displacement of the tops of the columns beyond the point at which the point of loading is vertically above the point of support: this can occur under a base amplitude not much in excess of that needed to initiate rocking.

4. Stability is increased by introducing capping pieces which effectively increase the column width: this also increases the limiting roof acceleration value.

5. There will be considerable loss of energy through the impact between the columns, the beams and the ground, which will create a degree of damping.

The results are interesting, but because of the high degree of idealisation of the system, their application to the prediction of the performance of actual baipash structures is uncertain; in particular it is not known how the interaction of vibration in two horizontal directions and the vertical direction would affect the observed patterns of vibration; whether the friction generated in the laboratory models would also be achievable in the real structure; and what would be the effects of the large 'imperfections' - in verticality, flatness of surfaces and so on - in the real structure. Nevertheless it is reasonable to conclude that a well constructed column-beam system supporting a roof in the manner of the baipash would provide some residual structural support, even after the collapse of the walls, and that it might delay or even prevent the collapse of the roof itself.
Moreover, by careful consideration of the elements of this residual structure, and making use of the method of structural analysis developed, it is possible that this residual strength could be considerably increased without any significant change in the form of the structure. This is considered further in the next section.

IMPROVING THE EARTHQUAKE-RESISTANCE OF TRADITIONAL BUILDINGS

It is generally recognised that it is technically and economically impossible to design buildings which are strong enough to resist the largest possible earthquake shaking without damage; and that more modest criteria for earthquake-resistance have to be established which take account of both the economic possibilities, and the degree of importance of the structure. For low-cost rural housing, prevention of casualties is usually agreed to be the primary concern, and in structural terms this effectively means prevention of collapse. The International Association for Earthquake Engineering (18) have proposed the following, rather severe criteria. "In the event of the probable maximum earthquake intensity the building should not suffer total or partial collapse, or suffer irreparable damage which would require demolishing and rebuilding".

Razani (19) has made a special study of the problem of unreinforced masonry and adobe low-cost housing in Iran, a country with strong similarities to northern Pakistan in terms of climate, materials and living patterns. He proposes the following more realistic criteria for such constructions:

1. No structural damage under earthquakes of local intensity less than or equal to 7 (MM scale);

2. No roof collapse should occur under any moderate or major earthquake having local intensity less than or equal to 9 (MM scale).

To be useful in design, these criteria would need to be modified, since the intensities referred to are only defined in terms of the statistical performance of unstrengthened buildings in earthquakes; nevertheless they provide a good basis for the study of field performance of prototype modified houses.

To achieve the central requirement of preventing roof collapse, generally considered to be the chief cause of casualties in arid areas where thick heavy roofs are used, Razani proposes the following design and construction principles:

1. The roof of the building should remain monolithic with sufficient in-plane rigidity during and after earthquakes;

2. The vertical load-carrying system of the structure should survive the earthquake without its function being impaired, and

3. The lateral load-resisting system of that structure during the main earthquake and succeeding shocks, until the structure can be safely repaired, should retain enough residual capacity to resist safely the lateral forces due to lateral instability, wind loads and future earthquake lateral loads.

In the type of structure common in the northern area of Pakistan, however, wall collapse is a very significant cause of casualties. According to the informant in Hamaran "most people were killed sleeping in the apsidal niches, and buried under stones, even where the roof was still in place". Collapse of walls was also noted in the UNESCO (13) Patan report. Because of the custom of using the spaces adjacent to the walls for sleeping, the most vulnerable activity, strengthening of the walls to resist collapse is therefore as important as roof collapse.

There is an infinite variety of ways in which these four design objectives might be achieved, but the range of available options in the northern areas is in fact limited by the following factors:

1. Very low incomes, which means that only modest, if any, cash outlay is possible;

2. Continued decline in the availability of timber, the traditional material best suited to earthquake-resistant construction;

3. Increased availability, via the Karakoram Highway, of manufactured building materials such as cement and steel;

4. Decline in the number of local building craftsmen, because of better

paid opportunites elsewhere;

5. The continuation of a style of life which is associated with a particular building form, the baipash.

Nevertheless, within these limits, it is considered that some modifications could be made that would considerably improve the earthquake-resistance of dwellings of the traditional type.

1. <u>Walls</u> - The small height/width ratio of walls would ensure good stability if the integrity of the wall was not destroyed. To enable these random rubble masonry walls to act monolithically in earth-quakes, they should have

    -   adequate foundations, at least 0.3 m deep, to prevent subsidence and softening under the wall;

    -   some form of horizontal coursing, to tie the walls from front to back and lengthways, and also to create corner bonding;

    -   a wall plate which can act to spread the loads superimposed on the wall from roof beams, and to tie the walls together at junctions and corners;

    -   mortar of sufficient adhesive strength to prevent the displacement of individual stones during shaking.

    Traditionally the horizontal coursing has been provided by timber, most effectively in the Shingri construction and the construction of Ghor. In normal modern masonry practice, it is achieved by the use of horizontal mesh reinforcement in the mortar courses, or by horizontal reinforced concrete beams. An 'intermediate technology' suitable for the northern areas might be thin concrete or cement mortar bands (50 - 75 mm thick) reinforced by wire mesh, at vertical spacings not greater than the wall thickness. The wall plate could also be either of timber or of reinforced concrete or cement mortar. The general masonry mortar need not be of very high strength: soil mortar with the addition of a small percentage of cement or other stabilizer would be adequate.

2. <u>Roof and supporting structure</u> - The traditional roof structures of the area, with their heavy main beams supported in wide timber columns independently of the walls are potentially earthquake-resistant, but need some modification in order to meet the Razani principles. The roof structure has only very limited in-plane stiffness: this could be considerably improved by the addition of a second transverse set of primary beams (Fig. 15) and by the connecting together of all primary and secondary roof beams and wall plates to limit the tendency to disintegration; the diagonal members in the lantern structure, if firmly connected to the primary beams by pegs or nails could provide adequate diagonal bracing. Where not provided by walls the vertical support to the roof beams is provided by timber columns, resting at their base on timber sole plates or on the ground. These column bases would be better sunk into the ground, and the load carried by a stone pad, (Fig. 15) with stones surrounding them, to provide an extra degree of stability and to eliminate lateral drift.

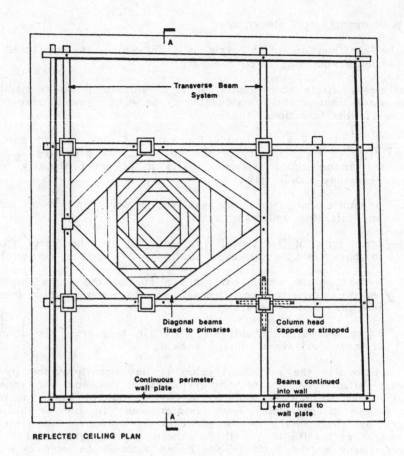

Transverse Beam System

Diagonal beams fixed to primaries

Column head capped or strapped

Continuous perimeter wall plate

Beams continued into wall and fixed to wall plate

REFLECTED CEILING PLAN

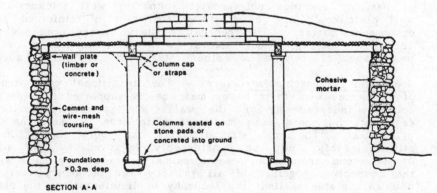

Wall plate (timber or concrete)

Column cap or straps

Cohesive mortar

Cement and wire-mesh coursing

Columns seated on stone pads or concreted into ground

Foundations >0.3m deep

SECTION A-A

FIG. 15. STRUCTURAL MODIFICATIONS TO BAIPASH FOR IMPROVED EARTHQUAKE-RESISTANCE.

The residual lateral strength, in the event of wall collapse, is provided by the connection of the beams to the columns on which they rest. In the baipash structures studied, the beam rested directly on top of the column with no connection of any sort between them. This was the system studied by Stephens and Whitfield (16) who concluded that one way to improve the lateral stability would be the provision of column caps (Fig. 16a); however, these would need to be rigidly connected to the top of the column to be fully effective, which would be difficult to achieve in practice. An alternative method of improving the lateral stability would be by the provision of metal straps or brackets at the top of the 4 main columns as shown in Fig. 16b. One advantage of such a system would be that the brackets could be attached to existing structures to strengthen them as a part of any general upgrading programme. The question of implementation is discussed further below.

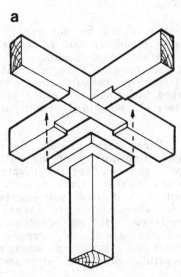 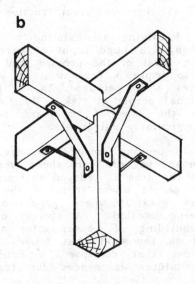

(a) provision of a column capital   (b) addition of metal brackets
       in construction

FIG. 16.   IMPROVEMENTS TO LATERAL STABILITY OF COLUMN HEADS.

IMPROVING THE EARTHQUAKE-RESISTANCE OF BUILDINGS IN NEW MATERIALS

Pakistan already has, in the Quetta code (20), a model for regulations and building methods appropriate to modern materials. The provisions of this code, coupled with techniques advocated in the new IAEE guidelines (18) are appropriate for the new government buildings currently planned and under construction in the area. From those buildings studied it appeared that there was a need for better supervision on the application of these codes, in particular in relation to the position and size of openings, spacing and layout of walls, and site control. A special problem is the growth along the Karakoram Highway of private construction using modern materials, concrete blocks and reinforced concrete especially, where

it is clear that there is little understanding of the techniques needed
for an effective use of these materials. Situations where these materials
have been used for reconstruction after partial collapse in the 1974 earthquake, or for extensions or infill in conjunction with older materials are
especially vulnerable, and probably have less earthquake-resistance than
many of the traditional buildings. This will be discussed further below.

IMPLEMENTATION

A programme to reduce the overall vulnerability of the housing in the
province could be achieved by:

a) modifying existing buildings;

b) improving the construction techniques in new buildings, or

c) improved reconstruction after earthquakes.

Modifying all existing buildings, even if it could be achieved, would
require a huge input of external money and resources out of scale with
the level of the problem and its priority against other needs. It would
also imply a programme of work importing skills and materials into the
area with perhaps minimal contribution from, or consultation with the
actual people at risk. Modification to existing buildings is mainly justified
in the case of important public or historical buildings which appear
particularly vulnerable.

Improving construction techniques is a longer term process dependent
on the housebuilders themselves. The overall vulnerability of the buildings
in the area is reduced as houses are replaced, and requires the involvement
of people where there is the greatest housing market, i.e., opportunities
are greatest where populations are expanding and new settlements are
being created. A special case of this is where there is large scale
rebuilding of houses after a major earthquake. In the reconstruction
it is important that houses are constructed to make them stronger than
those that collapsed. External involvement in improving construction
techniques is needed for training and possibly provision of materials.

1. Reducing the vulnerability of buildings in traditional materials

    Assessing the viability of outside initiatives to help communities
    reduce their housing vulnerability, along the lines proposed by
    Cuny (21), was one of the group's principle aims. The evolution
    of that methodology is dealt with by Davis (4).

    In many areas (for example the tribal areas around Patan) the
    traditional isolation and suspicion of external influence, the deep
    religious feelings and above all the perceived priorities of hazard
    are restraints to any outside initiatives. In these areas an external
    aid programme which attempted to show how to reduce earthquake
    risk, especially in isolation from the other hazards and hardships,
    would seem extremely unlikely to succeed.

    Improving the earthquake-resistance of buildings at a community
    level depends on the availability of money and willingness, both
    by the Government or other aid-giver to mount the programme and
    by the individual concerned to implement it. It depends on the

level of awareness and concern about the hazard shown by both. As shown above, the hazard from larger earthquakes is not as high as in many other places in the world, and its perception as a major hazard by the communities studied was one of far less importance than water shortage, land erosion, flooding, etc.

The extreme marginality of many of the villages means that spare resources are not available to enable the individual to deal with even his most important priorities. Improvements in traditional building, as outlined above, can be made as cheap as possible by using materials carefully and with more care in the construction techniques but improving the seismic resistance of a building will usually represent an additional labour and materials cost. One of the conclusions of the group as a whole is that the reduction of vulnerability to many hazards can only be successfully brought about when the community has some surplus resources of its own to help deal with the problem. Programmes to teach improved building techniques, safe siting of buildings etc., must then be seen in the context of an overall economic development strategy.

A development strategy already in hand is the Government-sponsored irrigation schemes of the Kohistan Development Board (KDB) and the new settlements which are built as a result may well be an appropriate case for a successful housing improvement project.

2.   Reducing the vulnerability of buildings in modern materials

There is already evidence that the Karakoram Highway has made new building materials more available in towns along its route, to the few who can afford them. Wherever the money is available new buildings and extensions are being made from cement blocks or reinforced concrete in preference to the traditional construction, not least due to the prestige associated with the new materials. These new methods, more reliant on machinery and new skills, are also encouraging contractor-style building rather than the owner-built houses of the villages in the mountains. The lack of building experience in these materials means that the under-standing of construction techniques is very poor, resulting in domestic buildings which often appear to have less earthquake-resistance than the traditional structures. The fact that steel and cement are available, affordable by some and in demand means that there is a potential to produce buildings which would be significantly safer against the hazard of earthquake than the existing building stock. A programme to teach builders an under-standing of the best construction techniques in these newer materials could be of vital importance if, as seems likely, they become more and more common as a building style throughout the area.

Public buildings, for example schools, dispensaries, mosques, often Government provided, are recognised as requiring higher levels of earthquake-resistance as well as priority siting to reduce risk from flooding and rockfall. If they are to be built to a satisfactory standard by local contractors a programme of skill training is urgently needed and could, as is happening in the KDB construction programme, lead to their percolation through to the private building industry. The setting up of a building college in Gilgit might be an important way of achieving formal training standards for builders in the Northern Areas and stands a good chance of success, as new building work increases, in keeping skilled builders in the North rather than draining away to the plains or the Middle East.

CONCLUSION

The bulk of the housing stock will however probably remain traditional, owner-built structures for many years to come. In order to reduce their vulnerability to earthquakes what is needed is the overall improvement of the way of life of the housebuilder, enabling him to spend more care and resources in housing himself. By understanding where and how this occurs, by knowing where and when it is important to people to improve the resistance of their houses and by proposing simple and efficient modifications of minimum cost to the housebuilder, the process of reducing earthquake vulnerability can be part of the overall rise in living standards of the people in the Northern Areas.

ACKNOWLEDGEMENTS

The work described in this paper was a part of the work programme of the Housing and Natural Hazards Group of the International Karakoram Project. It was mainly funded by a grant from the Science Research Council. The authors are also grateful to the Royal Geographical Society for providing administrative and logistical support, to Ove Arup and Partners for financial and technical assistance and to Tectronix Ltd for loan of equipment. The work in Barkulti was made possible by the warm-hearted hospitality of Shabaz Khan and his family. The assistance in a variety of ways of Prof. Khan Tahirkheli and Prof. Israr-ud-Din of Peshawar University, Mohamed Farooq of the Survey of Pakistan, and Dr. Abul Farah of the Geological Survey of Pakistan is also gratefully acknowledged. Finally, the project benefited greatly from the support of the Advisors to the Housing and Natural Hazards Group, Prof. O. Koenigsberger, Prof. N. N. Ambraseys, Dr. R. F. Stevens, and D. J. Dowrick.

REFERENCES

1) R. J. S. SPENCE and A. W. COBURN, (1980). "Traditional housing in seismic areas", Proceedings of the International Karakoram Project, Vol. 1, pp 253-264.

2) D. F. T. NASH and R. J. S. SPENCE, (1980). "Experimental studies of the effect of earthquakes on small adobe and masonry buildings", Proceedings of the International Karakoram Project, Vol. 1, pp 245-252.

3) R. E. HUGHES, (1980). "The analysis of local building materials and building techniques", Proceedings of the International Karakoram Project, Vol. 1, pp 336-346.

4) I. DAVIS, (1981). "A critical review of the work method and findings of the Housing and Natural Hazards Group", Proceedings of the International Karakoram Project, Vol. 2, pp 200-227.

5) R. E. HUGHES, (1981). "Yasin Valley: the analysis of geomorphology and building types", Proceedings of the International Karakoram Project, Vol. 2, pp 253-288.

6) N. N. AMBRASEYS, G. LENSEN and A. MOINFAR, (1975). The Patan Earthquake of 28 December, 1974. UNESCO, Paris, 1975.

7) R. C. QUITTMEYER and R. A. JACOB, (1979). "Historical and modern seismicity of Pakistan, Afghanistan, Northwestern India and South-eastern Iran", Bulletin of the Seismological Society of America, Vol. 69, No. 3, pp 773 - 823.

8) B GUTENBERG and C. F. RICHTER, (1954). Seismicity of the earth and associated phenomena, Princeton University Press, New Jersey.

9) K. L. KAILA and HARI NARAIN, (1971). "A new approach for preparation of quantitative seismicity maps as applied to Alpide Belt-Sunda Arc and adjoining areas", Bulletin of the Seismological Society of America, Vol. 61, No. 5, pp 1275 - 1291.

10) J. D. CLARK, (1956). Hunza: Lost Kingdom of the Himalayas, Funk and Wagnalls, New York.

11) S. HUSSAM-UL-MULK and J. STALEY, (1968). "Houses in Chitral: traditional design and function", Folklore, Vol. 79, pp 92 - 110.

12) L. EDELBERG, (1970). "The traditional architecture of Nuristan and its preservation" in Cultures of the Hindu Kush, ed. K. Jettmar, Steiner, Wiesbaden.

13) UNESCO, (1975). The Patan Earthquake of 28 December, 1974, Paris.

14) J. A. JACKSON and G. YIELDING, (1981), "Source studies of the Hamran (1972.9.3), Darel (1981.9.12) and Patan (1974.12.28) earth-quakes in Kohistan, Pakistan, Vol. 2, these proceedings, pp 170-184.

15) A. P. JEARY and B. SPARKS, (1977). Some observations on the dynamic sway characteristics of concrete structures, CP7/78, Building Research Establishment, Watford, U.K.

16) R. V. STEPHENS and M. H. WHITFIELD, (1981). A study of earthquake-induced rockings in relation to low-cost housing, Bristol University Department of Civil Engineering.

17) N. M. S. EVANS, (1981). Seismic effects on monolithic columns in unreinforced masonry and adobe low-cost housing, Part II Project, Cambridge University Engineering Department.

18) INTERNATIONAL ASSOCIATION FOR EARTHQUAKE ENGINEERING, (1980). Guidelines for earthquake-resistant non-engineered construction.

19) R. RAZANI, (1978). "Seismic protection of unreinforced masonry and adobe low-cost housing in less developed countries - policy issues and criteria", Conference on Disasters and Small Dwellings, Oxford Polytechnic, U.K.

20)   QUETTA MUNICIPALITY, (1976).   Building Code.

21)   F. C. CUNY, (1978).   "Scenario for a housing improvement programme in disaster-prone areas", Conference on Disasters and the Small Dwelling, Oxford Polytechnic, U.K.

Yasin Valley:
The analysis of geomorphology and building types

R.E. Hughes – Ove Arup & Partners, London

ABSTRACT

The author's previous paper in these proceedings attempted to define a methodology for the study of buildings. This involves the examination of building construction, behaviour and decay with all aspects being related to the environment.

A second theme showed that structural deformations and material decay if not remedied by "maintenance", "restoration" or "conservation" inevitably lead to abandonment or collapse of buildings and this may be of more significance to the people than the less frequent, though more catastrophic, events such as earthquakes and floods.

The intention of this paper is to show how the methodology was applied to the study of the Yasin Valley. It firstly concentrates on the geomorphology and relates this to the inhabitants' awareness of, and response to, natural hazards. The buildings are then examined in detail; materials, construction, structural performance and decay. Methods for improving the houses are then discussed.

Of all the field studies undertaken by the author, this is considered the most important as there is a real and practical application, in conjunction with the UNDP development proposals, for benefiting the inhabitants.

INTRODUCTION

Looking down on a valley pockmarked from centuries of environmental upheavals, the most striking aspect of the Yasin Valley settlements is their vulnerability. The small brown box-like structures set amongst a quilt of green fields stand out in valiant defiance of their powerful enemies – the harsh realities of the climate and virulent land forming processes. How can the houses be anything but a temporary feature in a continuously changing landscape? see Figs. 1 and 2.

The houses are subject to the vagaries of the environment, for example flooding, rockfalls and earthquakes. Due to their method of construction they undergo structural deformation and the building materials are subject

254

FIG. 1.   LOOKING WEST ACROSS THE LOW ANGLE OUTWASH FANS ON EITHER SIDE OF THE VALLEY AT UPPER BARKULTI.   In the foreground the channel is somewhat protected by flood walls.   In the background are the high angle scree slopes and high level terraces.

FIG. 2.   LOOKING SOUTH EAST ACROSS THE EXTENSIVELY FORMED LOW TERRACE AND FLOOD PLAINS, SOUTH OF BARKULTI.

to rampant decay.   Not surprisingly the oldest house may have been built less than 100 years ago; see Figs. 3 and 4.

FIG. 3.   TYPICAL WALL BUILDING TECHNIQUES OF A YASIN BAIPASH.

FIG. 4.   DETAILS OF A TYPICAL INTERNAL ROOF CONSTRUCTION.

The inhabitants, of course, have their own understanding of these topics and, of necessity, judge how to cope with them. In fact, the houses are not built to last a long time - the idea of a predetermined design life is a somewhat alien concept. These issues are discussed in detail by Frances D'Souza (1). Housing is considered to be well down in the inhabitants' perception of 'priorities' and improvement requirements. Houses are built to serve immediate needs - to provide a place to sleep, to store food and to shelter in during the harsh winter. Interestingly, the same people build a totally different type of house to serve different needs: the rectangular flat-roofed Baipash in the main valley and the random rubble conical-roofed hut in the high summer pastures. The former are the permanent homes occupied all year round, while the latter are used by part of the family while their animals are summer grazing.

A house is not considered a status symbol to be kept in good order or to have an increasing monetary value. A house may stand empty for years while simply out of favour with the family. Rich and poor live in the same type of houses, built by the same methods and with the same materials. Any difference in house dimensions are due to the family size and availability of building labour.

With housing being of little importance to the occupants, the research may seem somewhat irrelevant. However, with a developing shortage of land, traditional houses are being built in positions of increasing vulnerability and their structural performance is proving increasingly inadequate. In addition, the traditional house is slowly being replaced, at unnecessary expense, and often with inferior materials. This is surely a retrogressive development. The inhabitants are also known to be at risk from catastrophic events and it is to be hoped that during any proposed economic improvement, for example as outlined by the United Nations Development Programme (UNDP), house and site modifications can be made to lessen the impact of these natural hazards. Furthermore, any earth-quake-resistant features may prove to be of great interest to other countries. Lastly there is a need to record the Yasin structural types for the interest of architectural historians.

1. BACKGROUND INFORMATION - GENERAL DESCRIPTION OF THE YASIN VALLEY.

The Yasin Valley is located against the very northern boundary of Pakistan, nestling in amongst the high peaks of the Karakoram Mountains (see Fig. 5). To the east is the Ishkuman Valley and to the west is the Chitral Valley. The northern limit of the valley is the watershed between Pakistan and Afghanistan and this forms the Darkot Pass leading over to the Wakhan Corridor. The valley then runs southwards to form a tributary of the Ghizar River. This river in turn flows southeast for approximately 150 km. through Gilgit to meet the main Indus River at Jaglot.

The valley along its present habitable length is approximately 54 km. with the northern half flowing south to Yasin village and the bottom half flowing south-southeast to Gupis, the confluence point with the Ghizar River. The width of the valley varies considerably - the result of the rock structure and the valley's geomorphological development. Due to granitic outcrops and steep scree slopes, the southern entrance to the valley is constricted and is approximately 450 - 800 m. wide. Here the fast flowing river is in a narrow, deep and rugged gorge.

257

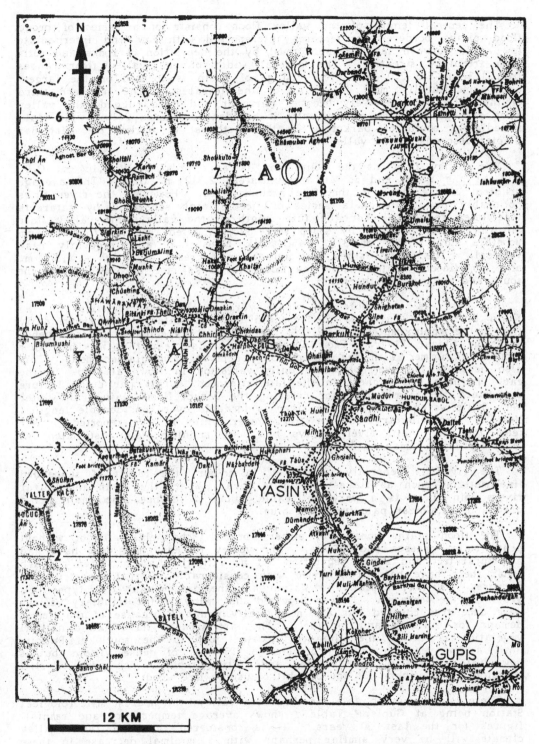

FIG. 5.   GENERAL MAP OF THE YASIN VALLEY 1934 AND PRESENT.

The central length of the valley floor consists of a series of broad flood plains bordered by terraces interspersed with low and intermediate low angle outwash fans, these formed by the deposition of rocks and soils from tributary streams. The general width of the valley floor ranges from 1.3 km. to 2.0 km. The village of Yasin is located on the largest and agriculturally most important of the terraces. The northern third of the valley progressively becomes narrower and more rugged, the valley floor averaging 650 m. width, until it broadens out as a 2 km. wide plain immediately south of Darkot. To the northwest of Darkot the East Bar Glacier descends to the valley floor.

The valley floor at the south end is at approximately 2,164 m. O.D. and at the north end is at approximately 2,712 m. O.D. The average gradient of the river is therefore around 1:100. However, the river flows noticeably down through a series of flood plains separated by cataracts (see later section). For the full length of the valley, the valley floor is dramatically bounded by steep scree slopes, often reaching the maximum possible slope angle of 37°, and rock faces that are frequently near vertical. The valley sides rise up to spur crests of 4,720 - 4,870 m. and immediately behind these the mountain peaks rise to 5,490 m. in the south and 6,400 m. in the north.

No data is available on the hydrological characteristics of the Yasin River. At Barandos the research team estimated the maximum daily flow to be approximately 60 $m^3.s^{-1}$, half of that expected at its confluence with the Ghizar River. From observations it is clear that maximum flow occurs in summer with the water being derived from glacier melt. This also accounts for the noticeable daily variation in flow rate; low in the early morning and highest in the late afternoon. This particular feature determined the team's time for crossing the Darkot flood plain. In winter very little water flows and the river is fordable over considerable lengths. Most of the tributary streams freeze over while it was reported to our team that the main river always has some running water, sufficient at least for daily human needs.

The river is generally very grey in colour, the result of suspended fine material such as flakes of Biotite Mica. Where the Yasin and Ghizar rivers join there is a very noticeable colour difference, the Yasin being grey and the Ghizar pale turquoise blue.

The flood plains consist of either abraded channels, with or without peaty soils, grasses and mosses. The valley terraces may once have had a natural tree and shrub vegetation but as with the former barren outwash fans, these have been transformed with considerable effort for agricultural purposes. Willow, Poplar and Walnut are the predominant tree types now grown on the terraces. It was reported that many of the tributary valleys were once tree covered but are now totally denuded because of the timber demand for building and for fuel. The last area to have had a natural vegetation is called the 'Jungle', located just to the south of Darkot. Until a few years ago this consisted of 150 ha. of dense willows but is now just a mass of tree stumps and rotting roots. The area is still occasionally scavenged for the odd bit of usable fire wood.

There is no climatic data for the Yasin Valley, the nearest Meteorological Station being at Gupis. Table 1 shows average temperature and rainfall figures for the last 30 years. It is considered that the Yasin Valley climate will be very similar perhaps with a minimal decrease in these figures in a northwards direction.

	Temperature		Rainfall
	Max. °C	Min. °C	Mean total mm.
January	3.6	-5.4	7.2
February	5.9	-3.0	11.52
March	12.1	+2.0	16.08
April	17.0	6.9	24.48
May	21.4	10.4	36.48
June	28.6	16.0	9.84
July	31.7	19.4	13.92
August	31.0	17.9	14.88
September	26.5	13.8	6.98
October	37.2	7.6	6.72
November	12.2	+1.3	3.84
December	22.8	-3.7	9.84
		TOTAL:	116.78 mm.

TABLE 1: CLIMATIC DATA FOR GUPIS (average for 30 years)
(2,164 m. O.D.)

Data collected from the Gilgit Meteorological Station

2. GEOLOGY AND SEISMICITY OF THE YASIN VALLEY

The geological structure seen along either side of the Yasin Valley is relatively clear, having fresh rock surfaces, the result of active physical weathering processes. However, the rocks have not been studied in detail because of the valley's remoteness and the difficulty of reaching the rock exposures high up on the valley sides above the great scree slopes. When compared with adjacent valleys, it is found to be less affected by regional metamorphism, and it is considered that the valley will become extremely important, providing a 'Type' section across the 'Northern Megashear'.

During the expedition, there was insufficient time available to map the geology of the Yasin Valley, the following account is based on fieldwork by Matsushito and Huzita (2). From the observations made during the expedition, their five fold division may be too simplified for use in the field, for example in accounting for differential weathering and the formation of the geomorphological features; see later section.

The Yasin, Ishkuman and Hunza Valleys lie astride the Northern Megashear. This represents the major unconformable boundary of the Eurasian Tectonic Plate, to the north, and the Kohistan Island Arc formations to the south. The latter are flysch deposits formed to the north of the Indian Tectonic Plate. Originally the Indian subcontinent was attached to a continental mass comprising Antarctica, Australia, Africa and India. Approximately 80 million years ago these individual land masses separated, with India moving northwards, so reducing the width of the Tethys Ocean to the north. This Ocean became the depository of arenaceous and agrillaceous sediments and also of an assemblage of volcanic rocks. When the Indian Plate, Island Arc deposits and Asian Plate collided, some 15 million years ago, two subduction zones were created, namely the Northern Megashear and the Main Mantle Thrust. These subduction zones are where the rock suite to the south was forced below the rock suite to the north. Accompanying this, folding, faulting and metamorphism occurred along with volcanic and seismic activity; all these processes accounting for the Karakoram Mountain Complex.

In broad terms the Yasin Valley shows a two fold division, the Euro Asian Continental Plate consisting of the Darkot rock suite and the Kohistan Island Arc deposits consisting of the Greenstone Complex and Yasin sequence. Running approximately east-west just to the north of Yasin village is the Northern Megashear thrust zone. This overall sequence is made more complex by the intrusion of two batholithic granite masses. With continuous mountain uplift and valley erosion these are now exposed and have their own typical land forms. Figure 6 shows a sketch of the geology of the Yasin Valley with the five major rock groups and the main thrust zone. A detailed description of the rock types encountered in the valley is given in Appendix A.

The existence of the Northern Megashear across the Yasin Valley, a predicted 4 cm. movement per year, and the location of the Main Mantle Thrust at a considerable depth account for the whole region being considered seismically active. The inhabitants reported to the research team that minor tremors are frequent. However, it is interesting to note that no major shocks are known to have occurred beneath the main valley with the inevitable devastating results. Table 2 shows the location and recorded data of two earthquakes that have occurred recently: Qurkulti Bar, 1943 (a tributary valley of the Yasin), and in the Darel Valley, 1972, southwest of Gupis. Recently, September 1981, a devastating earthquake took place southwest of Gilgit. A separate report is being prepared on this event.

3. GEOMORPHOLOGY

The land forms of the Yasin Valley are typical of the Karakoram region having recently been formed and by the same processes. The weathering is taking place on recently glaciated valleys that are also being continuously uplifted. This has produced features remarkable for their freshness and completeness of form - each recognised with a textbook clarity.

The Yasin Valley is characteristically glaciated, having a typical 'U' shape and other features such as hanging valleys, polished rock surfaces, and moraine deposits. Very little down cutting into the valley floor has subsequently taken place so the overall appearance is dramatically different from, for example, the Indus Valley. Here the river incision may be more than 2,000 m. and is the result of continuous mountain uplift and river down cutting. The oldest recognised glaciation in the Yasin Valley is estimated to have taken place as little as 50,000 years ago, recognised as the high level terrace and perhaps equating to the T_3 terrace of the Hunza Valley; see Derbyshire et al, these proceedings.

The maps of the area and also satellite photographs show that for this part of the Karakoram Mountains the south end of the valley approximately coincides with the present southern limit of active glaciers. Many

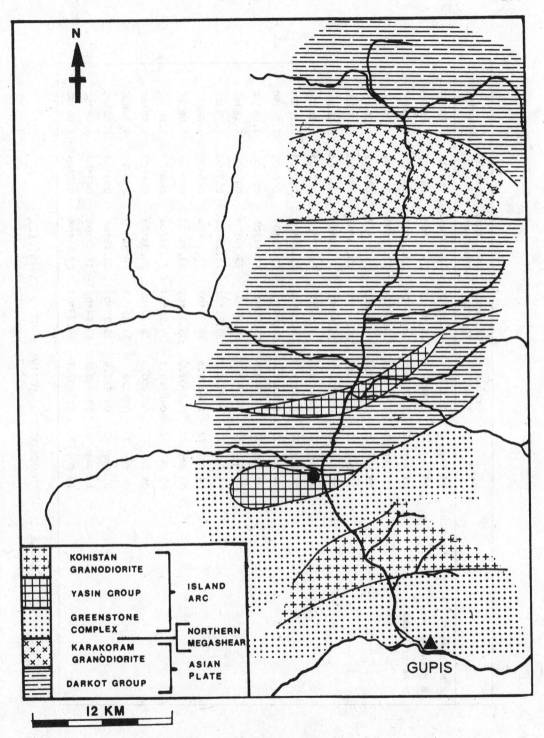

KOHISTAN
GRANODIORITE

YASIN GROUP

GREENSTONE
COMPLEX

KARAKORAM
GRANODIORITE

DARKOT GROUP

ISLAND
ARC

NORTHERN
MEGASHEAR

ASIAN
PLATE

12 KM

GUPIS

FIG. 6.   SKETCH OF THE GEOLOGY OF THE YASIN VALLEY.

	Date	Origin Time	Latitude (Deg*km)	Longitude (Deg*km)	Depth (km)	Magnitude
A Main Shock: No after shocks	24 Sep 43	113130.2	36.38*08	73.59*06	067*12	6.8SPAS
B Main Shock:	03 Sep 72	164829.5	35.94*02	73.33*02	045*03	6.2SPEI
After Shocks:	03 Sep 72	170922.5	35.99*04	73.34*04	062*09	4.7BISC
	03 Sep 72	174620.2	35.94*04	73.22*05	062*07	4.6BISC
	03 Sep 72	203220.	36.0 *03	73.5 *12	049*22	5.2BISC
	03 Sep 72	230353.6	35.96*02	73.24*02	046*03	5.6BISC
	04 Sep 72	005026.7	35.97*02	73.25*03	050*04	5.3BISC
	04 Sep 72	012352.4	35.92*02	73.37*03	053*04	5.2BISC
	04 Sep 72	023619.3	35.98*02	73.38*03	049*03	5.1BISC
	04 Sep 72	035123.0	35.97*02	73.32*03	050*03	5.1BISC
	04 Sep 72	103544.9	35.79*03	73.41*05	054.05	4.8BISC
	04 Sep 72	133752.3	35.89*02	73.30*03	041*04	5.1BISC
	04 Sep 72-	134220.7	35.91*02	73.35*02	057*03	5.7BISC
	04 Sep 72	224409.3	35.95*03	73.51*04	050*05	4.8BISC
	05 Sep 72	030802.3	35.78*07	73.26*07	059*09	4.6BISC
	05 Sep 72	040729.1	35.95*04	73.39*04	049*07	4.8BISC
	05 Sep 72	091400.3	35.87*02	73.33*03	058*03	5.1BISC
	06 Sep 72	025128.	32.49* 3	78.51* 5	14.10	5.7SMOS
	07 Sep 72	042417.4	35.97*03	73.42*04	054*04	4.8BISC
	17 Sep 72	173749.9	35.94*02	73.42*04	049*04	5.4BISC
	18 Sep 72	214202.1	35.78*03	73.51*04	057*05	4.8BISC

TABLE 2: SEISMOLOGICAL ACTIVITY NEAR YASIN. (2)

fossil glaciers are seen in the Kuh Ghizar Mountains to the south of Gupis. Northwards up the Yasin Valley, the frequency and size of glaciers increases, there is a tendency for them to be associated with the larger tributary catchments and, up to the northern watershed, for them to be north facing. At the north end of the valley, west of Darkot, the East Ghamu Bar Glacier descends to the valley floor. The inhabitants report that every seven years this glacier surges out of its valley onto the flood plain.

The geomorphological features that result from deposition in the valley since its last glaciation were mapped during the 1980 field research programme; a sample of these are shown in Figure 7 and are described below.

The valley floor is bordered by steep slopes, a mixture of rock faces (30° - 90°) and high angle scree slopes that rest near the natural angle of repose for angular rock debris (37°). Here the debris results from a mechanical weathering and with gravity being the sole means of transportation, the screes form below. The finer material, sands and silts, works its way down between the coarser rock and somewhat cements the whole mass together. It is estimated that 57% of the valley sides are covered with these scree slopes with those on the west side being slightly bigger; here we are thinking of scree slopes that may be more than 500 m. high. There is also a noticeable difference in percentage cover for each side of the valley, with the west side having 72% covered with scree slopes compared with 42% for the east side. This noticeable difference is due to the valley orientation with the west side experiencing longer daily periods of sunshine, hence having a greater range of diurnal temperatures, and a resultant greater degree of weathering.

Incised into the rock outcrops and located between the high angle scree slopes there are a total of 43 tributary valleys of varying size, out of which there issue 33 fan-shaped outwash deposits. Due to the way these have spread over the valley floor, they form an almost continuous zone along either side of the valley and are of all sorts of shapes, sizes and angles of repose. The deposited debris is normally water transported and water worn, with stones tending to be subrounded to round. A gradation of material is observed with the boulders and cobbles being towards the issue point and the gravels, and occasionally sands, being washed to the fan boundary. Also it is noted that some of the steeper fans consist of uniformly massive angular boulders and it is considered that these resulted from a catastrophic surge of unweathered material where there has been no 'free' water transportation. These may have resulted from an action similar to the flow at Games, and observed by Hughes and Nash.

Here outwash features have been categorised into types based on their angle of repose; see Fig. 7. Low Angle Fans are resting at 2° - 8° while Intermediate Fans are at 9° - 15°. Flood Plains are roughly horizontal features. The fan features cover a total area of 2,600 ha. which is 45% (approx.) of the total valley floor. There are 20 Low Angle Fans of 1,800 ha. total, each varying from 15 - 215 ha. They are evenly distributed either side of and along the valley. There are 13 Intermediate Angle Fans, covering a total of 800 ha., varying from 12 to 145 ha. each. They are evenly distributed along the valley but are noticeably larger on the east side, averaging 94 ha., cf. 41 ha. for the west side. These should be compared with the 3 ha. Games mud flow fan (above water level only) that was observed during its formation.

On the valley floor there is a series of 12 active Flood Plains normally containing abraded river channels. They cover a total of 1,900 ha. and result from silting up of the valley behind naturally formed dams (see below). Associated with some of the flood plains are Terraces, these indicating previous blockages, periods of land uplift and possibly dam breakage. Any resulting from uplift are too old to be dated by observation, giving an idea of the rate and extent of vertical movement. The largest of these is at Yasin Village where two terraces are recognised. In total low terraces 1 - 20 m. above the river level cover about 1,000 ha., and the middle terraces 20 - 100 m. above river level cover about 450 ha.

There is also one terrace recognised high up on the valley side but this is now little more than an infrequent narrow band of soil columns precariously clinging to rock faces. This is considered to be the remnant of a glacial moraine equating with terrace T_3 in the Hunza Valley.

Other features seen on the valley floor include two rockfalls to the south of Yasin Village. These are better described as mountainside collapses covering in total more than 150 ha. Each has flowed from recognisable scars, like waves of water, across the valley to meet in the middle as 30 m. high ridges and mounds. To the north of Yasin Village rising out of a low terrace are sandy hills up to 50 m. high. Although these were not examined it is considered that they are former glacier moraines associated with the Yasin Village middle level terrace.

From the descriptions above it is noticeable that all the outwash features and the two rock falls are of major size and that many, rather than fringing the valley sides, have in fact reached the centre. Indeed 10 Low Angle Fans, 6 Intermediate Angle Fans (equally from both sides of the valley) and the two rock falls have participated in the blockage of the main river. These blockages have been permanent long enough to create the 12 upstream flood plains previously described. The fans also account for 1 instance of causing a temporary blockage with no after-effects, and have on 6 occasions considerably altered the course of the river. Out of the 33 outwash fans, only 10 have caused no secondary effect to the valley floor, indicating just how active the land forming processes have been. No evidence was found, such as broken cross valley features or stripped vegetation, to show that the dam type features have recently collapsed sending a catastrophic wave of water cascading down the valley. Indeed the massive silting up behind the blockages strongly supports a permanence to them.

To account for the size and type and to estimate the potential activity of the outwash fans a preliminary morphometric analysis has been carried out. Here several parameters defining the tributary valleys, the erosion basin and material reservoir, are compared with the parameters defining the outwash deposition features. Very good correlations have been found between pairs of variables, including total stream length, drainage density and catchment area and the angle of the fan and the deposit area. From the size of the data bank it was not possible to carry out a more sophisticated analysis.

On each regression graph it has been possible to define three sets of points. The most recognisable is that formed by a cluster of points defining good correlations or typical relationships. Of more interest for this study are the sets that form scatters, showing untypical relationships and indicating that the deposition features are smaller than expected from the size of the tributary system and those showing deposition features

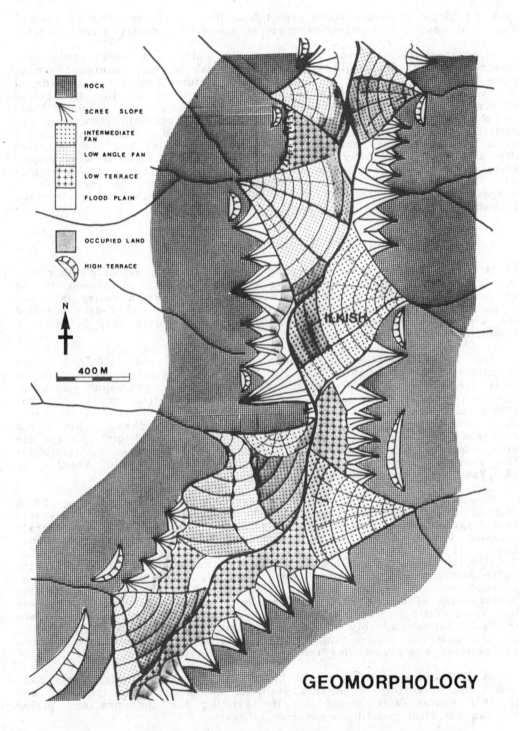

GEOMORPHOLOGY

FIG. 7.   GEOMORPHOLOGICAL FEATURES OF THE YASIN VALLEY.

that are larger than one would expect from the size of the tributary system. Figure 8 shows the distribution of these out-of-the-ordinary sized features.

Since these features may represent both 'under' and 'over' active land forming activity they could represent high risk location points. Firstly, there is a predominance of 'under sized features' on the east side of the valley and this confirms the previous discussion on the distribution of scree slopes related to climatic control on weathering. Secondly there is a good correlation between 'under sized' outwash fans and the distribution of tributary valley glaciers, presumably with this relating to a greater removal action of finer eroded material and the regulated discharge rate imposed by the glacier melt. Lastly there is a very good correlation between the atypical outwash fan and those fans presently with low human exploitation; see Figure 9. It would be reasonable to suggest that these features have been avoided because of the high probability of disaster.

4. SETTLEMENT OF YASIN VALLEY

It was into the actively forming landscape, as previously described, that man settled, possibly first using the valley as a route between trading centres or for seasonal hunting. Occupation in the Yasin Valley is known to go back at least to the 10th century A.D. At this time Bidat controlled the whole of the Gilgit territories and it is reported that he built a Buddhist Monastery in the Yasin (4).

The valley first became Islamic between 1120 and 1160 A.D. when Shamsheer Ishan subjugated Punial, Yasin and Chitral. From then on until the mid 19th century there are numerous, though vague, references to the Yasin, as being either an independent state (for example, in the 1390's under Kharso Khan and the 1430s under Haiden Khan), or as being used as a refuge (for example, in 1790 for Salaiman Shah. This prince of Yasin fled back after he murdered Gori Tham II of Gilgit). There are also references to the princes of Yasin invading neighbouring territories (for example, Gilgit in 1823 and 1840, the latter by Gohar Amen). In 1859 Yasin tribesmen were raiding Punial for spoil.

Between 1842 and 1852 the Gilgit territories, with the backing of the British, came under the control of the Sikhs' Dogra forces. However, raids, skirmishes and rebellions didn't cease until the 1890s. The Dogra's invasion was a reaction to the murder of Raha Bhah Sikander by Gohar Amen. It was Sikander's brother who sought the protection of the Sikhs. By 1860 the Dogra forces were in complete control, with the ruler of Yasin, Malik Amen, having fled to Chitral. Mir Wali was placed in Yasin as governor and it was he who was initially thought responsible for the assassination of George Hayward, the first recorded English man to explore the Yasin Valley. During the latter years of the 19th century, peace in Yasin Valley was only nominal. For example, Malik Amen carried out general guerrilla tactics and on one occasion he massacred men, women and children who were taking refuge in Yasin Fort.

In 1877 following a treaty with the Maharaja of Kashmir, Yasin came under the suzerainty of Aman-ul-Mulk of Chitral. Strife continued and in 1882 Aman-ul-Mulk forced out the existing Mir of Hunza and placed his own son Afzal-ul-Mulk as governor of Yasin.

Reports of these violent incidents reinforce fearsome British ideas of

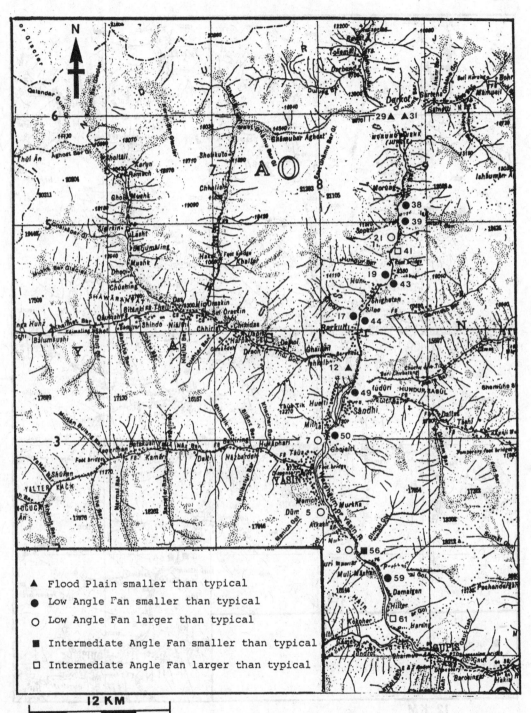

Flood Plain smaller than typical

Low Angle Fan smaller than typical

Low Angle Fan larger than typical

Intermediate Angle Fan smaller than typical

Intermediate Angle Fan larger than typical

12 KM

FIG. 8. YASIN VALLEY, UNTYPICAL ALLUVIAL FANS.

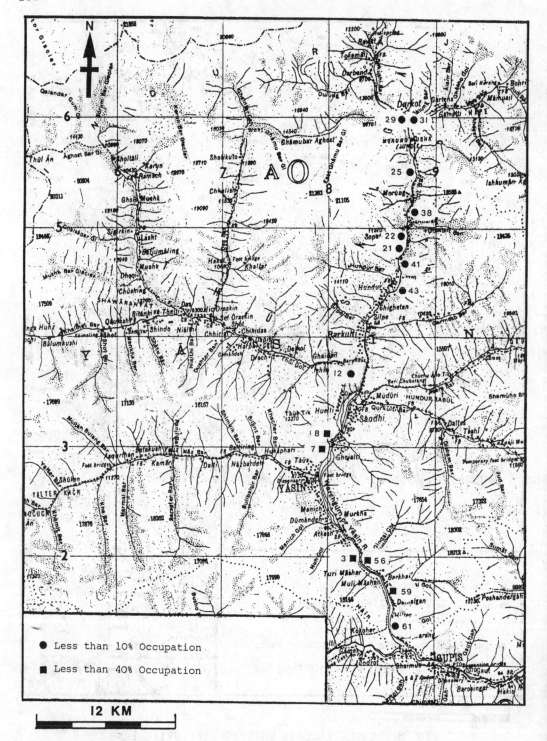

● Less than 10% Occupation

■ Less than 40% Occupation

12 KM

FIG. 9. YASIN VALLEY, POORLY DEVELOPED ALLUVIAL FANS.

the 'Tribal Area'. Yet how different from the Yasin society of today; a more peaceful and hospitable community would be hard to imagine. Clearly pressures, demands and needs have changed. However, echoes of the aggression of the past are still evident among the strongly independent people of the Tangir Valley, south of Gilgit. Here the expedition encountered family and clan feuding. Settlements still rely on the protection afforded by 'Shingri' towers and armed guards stand watch every night.

That the Yasin Valley was once socially similar to Tangir is confirmed by the first British maps (see Figure 10). While it is uncertain how accurate the maps are, for example McNair's map of 1883 shows occupation restricted to the southern half (and may be related to the valley trade routes turning west just to the south of Barkulti), they do demonstrate that each settlement had its own protective tower. From the observations made elsewhere by the expedition team, where such towers still exist it is clear that their location is dictated by defensive and strategic con- siderations. It is only when an overall protection ensures that it is practical for agricultural activity, and for family groups, to expand away from the nucleus.

Often such defensive locations were at vulnerable points for natural hazards and often these positions were also distant from the best agricultural lands. For example, water might have had to be brought in by channels following precarious mountainside routes, or steep outwash fans might have to be cleared of stone, terraced into flattish fields and manured to improve fertility.

The oldest reference to the distribution of Yasin villages is in a report written by George Hayward in March 1870. From the south he records the following ten settlements: Gindai, Dumyal, Yasin, Gujatti, Sandi, Barandos, Hundoor, Dariara, Maskh and Darkote. The India Office survey of 1883 names 11 settlements in the Yasin Valley of which 6 are fortified with towers, by 1921 21 settlements are named and by 1934 there were 30 named settlements. Only Yasin is shown to have a fortified tower – an indication of the decreasing emphasis on defence. This change and expansion of settlement distribution may in part merely reflect better quality mapping but interestingly it happens to correspond with known increases of the valley's population. While the 1934 mapping at 1:250,000 is somewhat diagrammatic, the named settlement distribution correponds very closely to that observed by the 1981 expedition team; see Table 3.

In 1934 farming probably occupied much the same area as today. This area must approach the maximum valley surface which it has been economically feasible to convert for agricultural purposes. In total, it is estimated that 57% of the valley floor has been utilised by settlement and agriculture. It is thought possible, given a high degree of expenditure on civil engineering works, that a further 8% of the land surface could be reclaimed.

Some 51% of the flood plains, amounting to 1,100 ha., are utilised with most being set aside for grazing. Despite the threat of flooding, large tracts of the flood plain are now being used for growing crops, for example at Barandos. All the low terraces are farmed and indeed all the intermediate terraces used to be farmed, although 11% now lie barren due to the failure to maintain a water supply channel just to the north of Yasin Village.

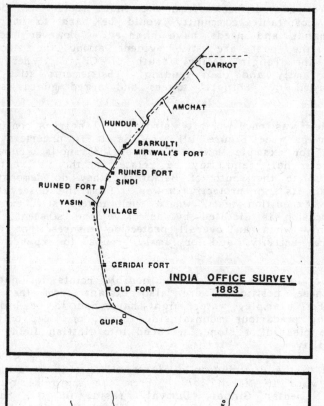

DARKOT

AMCHAT

HUNDUR

BARKULTI
MIR WALI'S FORT

RUINED FORT
SINDI

RUINED FORT

YASIN

VILLAGE

GERIDAI FORT

OLD FORT

GUPIS

INDIA OFFICE SURVEY
1883

N

20 KM

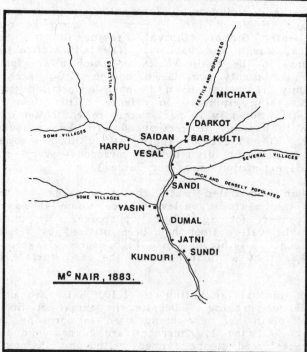

MICHATA

NO VILLAGES

FERTILE AND POPULATED

DARKOT

SAIDAN   BAR KULTI

SOME VILLAGES

HARPU   VESAL

SEVERAL VILLAGES

SANDI

RICH AND DENSELY POPULATED

SOME VILLAGES

YASIN

DUMAL

JATNI   SUNDI

KUNDURI

M^c NAIR, 1883.

FIG. 10.   SURVEYS OF THE YASIN VALLEY 1883.

Nos.	1934 + Present	1921	1883	1870	Geomorpho-logical feature
1	Darkot	Darkot	Darkot	Darkot	L
2L	Morung			Muskh	L
3R	Umaleg	Amalchat	Amchat	Dariara	L
4L	Sopatingdas	-	-	-	L
5L	Tirsit	-	-	-	L
6R	Ilkish	-	-	-	M
7R	Burakot	Burakot	-	-	M
8L	Hundur	Hundur	Hundur	Hundoor	L
9R	Shighetan	-	-	-	M
10R	Silpe	-	-	-	M
11R	(does exist)	Kurkulti	-	-	L
12R	Barkulti	Barkulti	Barkulti	-	L
13L	Barkulti	Barkulti	-	-	L + T
14L		(Mir Wali's Fort)	(Mir Wali's Fort)	-	-
15L	Barandos	Barandos	-	Barandos	M + FP
16L	Dar?	Hundar	-	-	T
17R	Muduri	Munduri	Ruined Fort	-	Rock
18L	Huelta	Kualti	-	-	T
19R	Sandhi	Sandhi	Sandi	Sandi	T + L
20L	Mith	-	-	-	T
21R	Ghojalta	Ghijalti	-	Gujalti	L
22L	Taus	P A R T O F Y A S I N			T
23L	Yasin	Yasin	Yasin	Yasin	T
24L	Manich	-	Village	-	T
25R	Murkha	-	-	-	L
26L	Dumandeh	Doman	Dumal	Dymyal	L
27L	Atkashi	-	Village	-	T
28L	Nuh	Nuh	-	-	L
29R		Jatni	Jatni	-	M
30R	Gindai	Flindal	Geridai	Gindai	M
31L	Turi Mashar	-	-	-	L
32R	Barkhai	-	-	-	L
33L	Muli Mashar	-	-	-	H
34R	-	Kunduri	Kunduri	-	L
35R		Sundri	Sundi	-	L
36R	Damalgan		Old Fort	-	M
37R	Hitter				M
38R	Sili Harang				M

KEY: L or R = Left or Right bank of river   T = Terrace
      L   = Low Angle Outwash Fan   FP = Flood Plain
      M   = Middle "   "   "

TABLE 3: HISTORY OF YASIN PLACE NAMES COMBINED.

The low and intermediate outwash fans used by the inhabitants are characterised by radial fields following the contours of the land form. Associated particularly with the lower angle fans is a strip of barren ground originating at the issue point and descending to the main river. This is the active zone of water flow, out of the tributary valleys. The width of the channel depends on the hydrology of the tributary valley and the degree of incision into the fan. For example, the barren portion of the fan at Darkot is broad and flat and with each year the issuing stream changes its course. Approximately 58% of the low angle fans are utilised, amounting to some 1,039 ha. Slightly more land on the east side of the valley is farmed, about 66% of the total 1,039 ha. Only 27% of the intermediate outwash fans are used amounting to about 215 ha., approximately 70% of this being on the east side of the valley.

It is considered possible to further increase the agricultural coverage of the valley but only by 10% for the low angle fans and by 21% for the intermediate fans. The latter figure reflects the long standing awareness by the people of the fact that these intermediate fans were less terraceable, more vulnerable and harder to farm. Here the reclaiming of 21% of the land form would mean a great deal of effort requiring outside expert help.

It is considered possible with minimal effort to increase flood plain farming by 5%. This process has already been taking place at Barandos. It is important that the reclamation be accompanied by engineering works.

5.   AWARENESS OF NATURAL HAZARDS

The Games 'debris flow' observed by the expedition during its dramatic activity suggests that the Yasin outwash fans (previously described) may also have formed rapidly particularly those consisting of large angular boulders. Frequent references to similar features in the Indus and Hunza Valleys lend support to this view. The occurrence of such catastrophic events raises the question of whether further activity can be expected in the Yasin Valley and whether there is a high risk to the inhabitants.

It is clear that low frequency events with a long term impact over a large area do occur, and have done so recently. Firstly, in 1943, a devastating earthquake of magnitude 8 took place in the tributary valley of Qurqulti Bar. Only one house in the village of Daltas survived. After the main shock, rockfalls and landslides continued for 8 days and nights, according to local reports, killing 2,000/3,000 cattle. For a period of about 3 months, the inhabitants moved down into the main valley and then returned to begin the rebuilding process. This took around three years to complete. The Darel earthquake did not directly affect Yasin but due to family and clan relationships there was an inevitable impact. In the main Yasin Valley the inhabitants cannot recall any damaging earthquakes, but report that ground tremors are very frequent. Likewise, seismographs do not record any major shocks during the last 80 years.

To the immediate west of Darkot Village the East Ghamu Bar Glacier is reported by the inhabitants to surge on a seven yearly cycle but without any effect on the valley farming.

Flooding is recognised as the biggest problem of the low frequency events and the inhabitants reported to our team memorable floods at Darkot, Morung, Umarsat, Tilsit and Barandos. The last major floods, at Darkot and Barandos (September 1978), were responsible for the removal of fields and for the deposition of up to 4 m. of stone debris. At Darkot, 80 houses and a school were washed away. Here it was reported that deforestation in the tributary valleys may have been a contributory factor permitting an increased rapid run-off and transportation of stone debris.

No evidence has been found for the recent formation of low or intermediate outwash fans. No abandoned or destroyed villages were found on such features, indicating recent catastrophic events. However, historical maps do indicate that three named settlements no longer exist (Jatni, Kunduri and Sundri). While it is possible that it is just the names that have changed, their previous map locations do not conform to the present settlement distribution; one of these is on an intermediate type fan and two are on low angle type fans. Further work would be necessary to confirm 'recent' geomorphological activity.

From the fieldwork, it is very clear that the inhabitants are aware of these major land forming processes. The Yasin Valley is known to have been occupied for many hundreds of years but it is only during the last century that a change in the settlement pattern has taken place, through population growth and a need to expand agricultural practices. The following two paragraphs argue that the inhabitants have been aware of moving from safer onto more vulnerable land forms.

Historical information demonstrates that the oldest settlements, those recorded by George Haywood in 1870 and by McNair in 1883, are on terraces and low angle outwash fans, these being the most controllable features. Exceptions are where "old forts" are shown and these are on intermediate angled fans. This supports the theory that the original settlements were related to defensible positions. The risk of occupying low angle fans is acknowledged by the practice of leaving a zone of land unoccupied where outwash debris may be expected to flow. At such places it was noted that dry stone walls are built in an attempt to safely channel any water and debris load.

It is only since 1921 that any villages have been built on intermediate outwash fans and it is only since the publication of the latest maps (1940s) that farming has spread onto the foot of the high angle scree slopes, the latter surely being the most precarious land form type. There are still a few low angle and many intermediate fans that remain largely devoid of settlements, the exception being new small clearances, said by the local inhabitants to have been taken over recently by poor immigrants.

It is again only since 1921 that attempts have been made to reclaim and settle flood plains, for example at Barandos. Here it has already been noted that this reclamation work was lost in a flood of 1978. Even the present attempts to reclaim this land are fatalistically regarded by local families as unlikely to be successful for long. However, despite this pessimistic view, the team felt that future large-scale river channelling and land raising could considerably modify the problem.

The inhabitants are, in fact, much more concerned about high frequency, short duration natural events, that have an immediate and localised impact. Rockfalls and soil erosion interrupting the every day activities assume more importance than the threat of an unpredictable major disaster. From verbal reports rockfalls are the most common natural event as indeed boulder strewn fields seem to confirm. Particular problem areas were noted at Morung Jungle, Yasin, Hilter, Sili Harang, Marick, Atkash, Gindai and Barandos. Lumps of rock falling off mountain cliffs and channelled down scree slopes cause the greatest problems. Here one should note that just south of Atkash, mountains on both sides of the valley have at some time collapsed, so blocking the valley. But at Barandos the rock falls result from boulders becoming detached from the high level terrace and in the spring a guard is posted at night to warn of impending falls. At Morung the rockfalls are associated with high velocity winds. During the first week of April, according to local tradition, a "male" wind blows across the valley from east to west and during the second week a "female" wind takes her turn, blowing from west to east. These winds are capable of moving a 2 kg. stone down one valley side and all the way up to the other! The frequency of this 'mating' event makes travel to Darkot impossible for short periods.

Soil erosion is a seasonal event related to melting snows in the spring and glacier melting in the late summer. Such events produce the greatest fear because the subsequent loss of livestock and damage to the land and foodstocks can result in famine. At Barandos the main river is eroding its banks and the foundations of the buildings, for example the school, are being undermined despite a protection scheme implemented in 1979. At Upper Barkulti several fields have recently been lost because the stream descending the outwash fan made a short cut to the main river.

Due to the vulnerability of fields and crops the water level is monitored by an elected representative of each village, to check flow rates and for overtopping effects. However, rainfall is normally very low (see Figure 5) and generally more problems result from shortage than its abundance. For example, at Darkot, it was reported that it was not untypical for walls to fall down during a storm. However, this appeared to cause little concern as the walls could be rebuilt the following day. While 19 out of 43 tributary valleys have permanent glaciers, guaranteeing a summer supply of water, this may be insufficient for farming when water is most needed in the spring during major crop growth.

Winter-spring snow avalanches are more of a problem at the north end of the valley. At Umalsit 10 or more may take place each year, occasionally resulting in fatalities.

6 THE BUILDINGS OF THE YASIN VALLEY AND THEIR PERFORMANCE

Compared with the magnitude and complexity of the previously described environmental upheavals the houses of the Yasin Valley seem frail. However, for rich and poor the structures are homes and how they are built, how they behave and decay are all significant. There is clearly a great difference between the best and worst house performance recognised by the construction and state of repair. These may be of more significance to safe living than the external influences such as falling boulders.

The traditional house and by far the most common type found in the valley today is based on the Baipash, described in detail by Pott (4). This is a square to rectangular single storey structure with a slightly domed roof. It is built by the poorer families as a single unit or as multiple units by the richer ones. Figure 11 shows one such historically important complex settlement. Inside each main room there is normally a series of six carved columns supporting a complex timber roof (Figures 12, 13 and 14). The columns and smaller timber sleeper beams divide the Baipash into a series of areas each having a well defined social

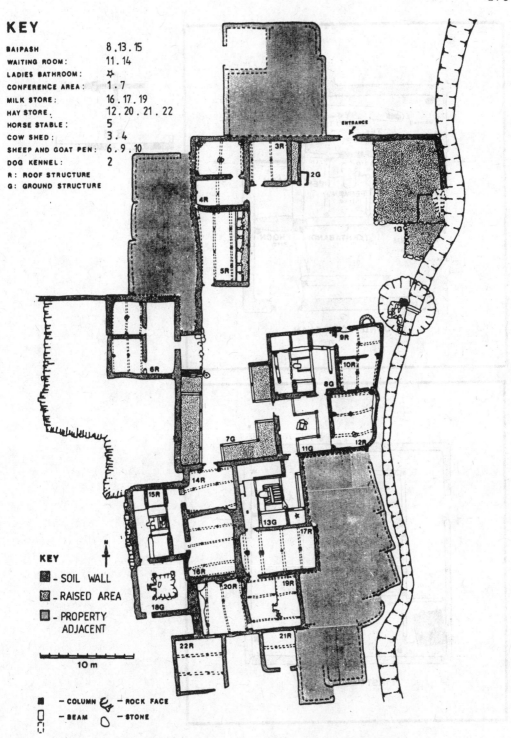

KEY

BAIPASH	8 . 13 . 15
WAITING ROOM :	11 . 14
LADIES BATHROOM :	✿
CONFERENCE AREA :	1 . 7
MILK STORE :	16 . 17 . 19
HAY STORE .	12 . 20 . 21 . 22
HORSE STABLE :	5
COW SHED :	3 . 4
SHEEP AND GOAT PEN :	6 . 9 . 10
DOG KENNEL :	2
R : ROOF STRUCTURE	
G : GROUND STRUCTURE	

KEY

▨ - SOIL WALL

▨ - RAISED AREA

▨ - PROPERTY ADJACENT

10 m

■ - COLUMN   ⬮ - ROCK FACE

▯ - BEAM   ⬭ - STONE

FIG. 11.   PLAN OF GHULBASHIRAY SETTLEMENT.

276

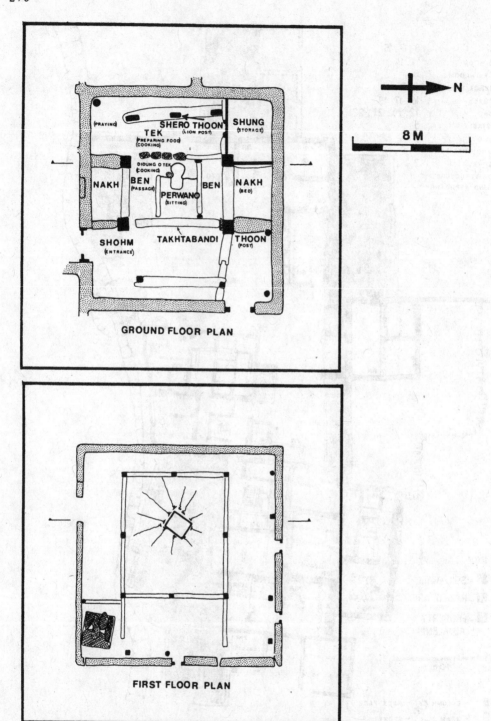

GROUND FLOOR PLAN

FIRST FLOOR PLAN

FIG. 12. GROUND FLOOR AND FIRST FLOOR PLANS OF THE
MAIN GHULBASHIRAY SETTLEMENT.

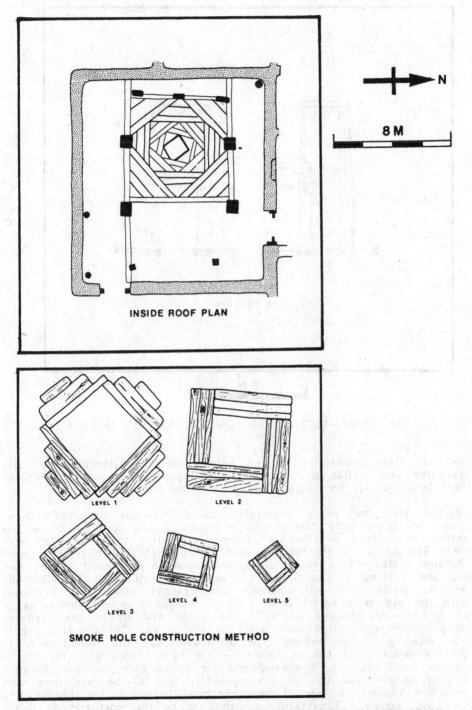

INSIDE ROOF PLAN

N

8 M

LEVEL 1

LEVEL 2

LEVEL 3

LEVEL 4

LEVEL 5

SMOKE HOLE CONSTRUCTION METHOD

FIG. 13.   INSIDE ROOF PLAN AND SMOKE HOLE CONSTRUCTION
METHOD GHULBASHIRAY BAIPASH.

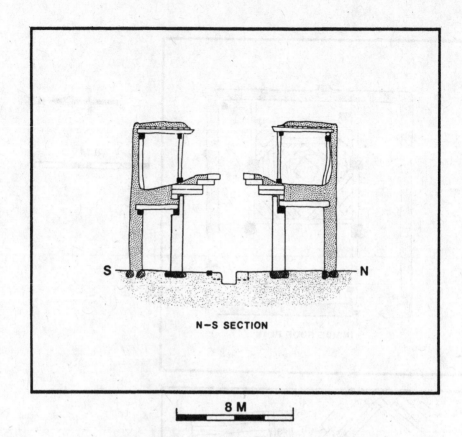

N—S SECTION

8 M

FIG. 14.   CROSS-SECTION THROUGH GHULBASHIRAY BAIPASH.

function.   Ancillary rooms, stores and animal pens are attached, normally more irregular with a flat insubstantial roof supported on unworked timbers. Here there is frequently considerable sagging of the roof joists.

Over the last few years, partially as a response by the younger generations requiring more family independence, a new form of house has been introduced.   This is a single storey rectangular structure often with a verandah and similar to the more western or colonial type buildings found much further south.   This type, along with new shops, guest houses and mosques, are tending to adopt new building techniques.   These include very regular walls built with ashlar stone blocks placed in horizontal rows with the aid of a spirit level.   They also have cement floors, large windows and modern hinged doors.   The roofs are single span trusses supporting pitched corrugated sheet steel panels or are flat and incorporate polythene sheeting.   The modern type of house, say the inhabitants, is very cold in winter and hot in summer – the exact opposite of the traditional house, and that which is required for comfortable living.   Because they do not require the use of new specialist skills, because they cost much less and appear to be of more practical benefit to the inhabitants, the traditional houses will certainly continue to be the most common house type in the foreseeable future.   The following pages describe the engineering of these in detail.

Building materials

Soil is the most widely used material and it is used in several different ways. When very wet, with a moisture content nearly reaching the liquid limit, it is used to form 'freebuild' monolithic walls. These are normally very irregular and often more than 0.75 m. thick. They often have substantial cavities where the soil has not been placed properly. Soil is also made into bricks. The older walls and more elaborate graveyard walls use the uniform fine yellow/brown silts off the high level terrace; this normally collected from the foot of the scree slopes along the northern half of the valley. This soil is always seen to be well mixed and compacted. Recent buildings, particularly of the poorer families, use the soil found at hand, on the lower terraces and flood plains. This is a grey, mixed graded, micaceous silt containing sand and sub angular gravel. This soil was seen being mixed and moulded into bricks clearly in an inferior manner. Such bricks show the soil to have been poorly puddled and placed too dry in the mould, or mixed too wet and hence poorly compacted.

Soil is also used as a mortar and here it has very inadequate properties such as tensile strength. Also it is used as a roof covering where it is placed as a 0.25 m. layer and the surface is compacted by the builder's feet or rammers.

Another important structural material is stone, and throughout the Yasin that chosen is picked from the immediate vicinity. No evidence was found of quarrying stone or transporting it from outcrops of more suitable material. At Barandos and Barkulti the distribution of rock types that were used was determined to be related to the site's geomorphological feature type. These results confirmed observations made throughout the valley. On the flood plains and low angle scree slopes round and subround boulders are used. On the terraces, round to angular pebbles predominate. Where the mountains are granitic intrusions, angular blocks of stone are brought down from adjacent scree slopes. Thinner slabs of stone are used in the immediate vicinity of eroded schists, slates and tabular dolomitic limestones. Only recently have rocks been shaped into rectangular ashlar blocks.

Wood is another important structural material. Walnut is the most prized type of wood and is used for the essential inside structural columns and beams. Its life expectancy is very considerable and we were informed that the walnut timbers are recycled many times. Pine and willow were used for timber lacing the walls of Yasin Fort and also at the north end of the valley in house walls. Wood is becoming rare and expensive, due to the exploitation of timber for firewood, and so every bit is used. Seasonal prunings now provide beams, laths and twig/leaf roof matting.

Building techniques

For obvious reasons it was not possible to examine the foundations of many structures. However, where soil erosion permitted, where new structures were being constructed and from local information, a few comments are possible. Walls usually have no foundations, the top soil is simply stripped away and the base of the wall is placed directly on the subsoil. For soil walls, the bottom two courses are typically stone normally brought up to just above ground level. On the flood plains the stone base may be extended above ground level to form a plinth. Here the stone is used to isolate the wall from the ground moisture and not to spread the wall load.   Occasionally the stubs of old walls are re-used for new ones, particularly where the rebuilding is less substantial.   House walls are normally constructed with a combination of soil and stone.   Occasionally wood is used to lace the walls together.

No differences were found in building methods and the quality of construction between the houses of the various social classes.   Personal choice, time and availability of materials seem to dictate the technique.   House walls are generally single storey ranging in height from 1.75 m. to 2.5 m.   It is very noticeable that towards the north end of the valley there is a reduction in the height of the structures and it was reported that this is due to the problem of snow accumulation in the more exposed locations.

Many wall construction methods were recorded during the field research and these are described below, and should be read in conjunction with Figure 15.   Interestingly they are totally different from those noted elsewhere, for example in the Hunza Valley.

(a)   Here the wall is built to the full height or the bottom three-quarters with angular to semi-angular boulders or tabular slabs.   They are placed in a random pattern, placed as layers embedded in soil mortar, or placed in layers without mortar (see Fig. 15, A and B).

(b)   Here round and semi-round boulders or cobbles are used for the full height of the wall or just the bottom half.   They are embedded closely together with a little soil mortar or are spaced out in 50 – 50 proportion.   Or the stones are placed in layers without mortar (see Fig. C).   Occasionally it is noted that there is a grading of the stones, starting with boulders at the base and becoming pebbles towards the top.   At the north end of the valley these types of walls are laced together with lateral and cross timbers (see Fig. 15 G).

(c)   A modern technique is the use of rectangular granitic blocks of worked stone built to form a facade.   Normally side and rear walls are built with traditional methods (see Fig. 15 D).

(d)   Another favoured technique particularly used where there are no large stones, is the use of small cobbles.   These are basically laid in unmortared horizontal rows and occasionally in rows following ground undulations.   Many variations of this technique exist (see Fig. 15 F):

   (i)   Stones placed in groups of one, two, three or four rows separated by up to 50 mm. thick layers of soil or slates with a little mortar.

   (ii)   Flattish pebbles or slabby stones placed diagonally.   Two three and four rows are produced to create a herringbone effect.   Such bands are separated by soil or slaty stone string courses.

   (iii)   Techniques (i) and (ii) are often combined to further enhance the complexity of the ornamental work.

(e)   Where there is a poor stone supply or where there is a great deal of free time, soil bricks are used.   The bricks are usually 300 x 300 x 100 mm.   They are used for the full height of the wall,

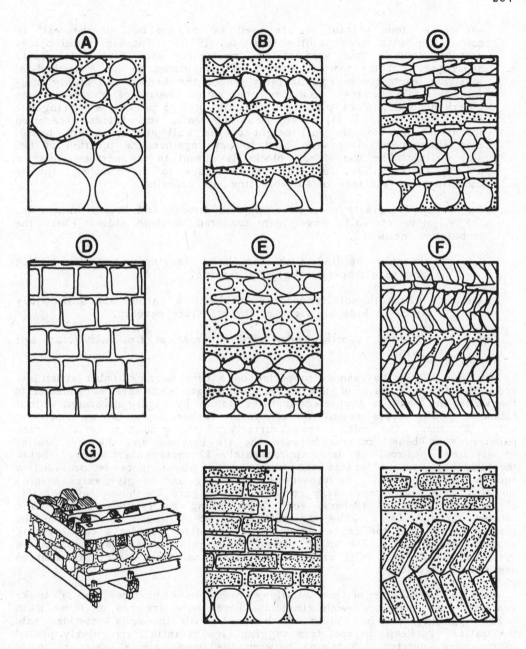

FIG. 15. WALL CONSTRUCTION METHODS, YASIN VALLEY.

set on a stone plinth, or are used as the top half of the wall in conjunction with stone plinths (see Fig. 15 H). The individual bricks are set with a thin skin of soil mortar or each course is widely separated. Thick mortar layers are considered to be a method of levelling each course, particularly where the bricks are highly mal-formed. On several occasions the bottom metre of the wall was observed to be bricks placed on the diagonal to produce a herringbone effect (see Fig. 15 I). On one stone wall, soil bricks were used for the quoins for the full height of the wall and on another house, thin sheaves of slate were used to key together the junction of two soil walls. The use of soil blocks is absent in the northern quarter of the Yasin Valley, and this may be due to the harsher climatic conditions, or the lack of suitable fine silt deposits.

Due to the generally poor embedment of stones and bricks in mortar, the majority of walls have been rendered on both sides. Here the techniques include:

(i)   Roughly applied with the hand leaving a coarse surface and with finger marks still showing.

(ii)   A smooth soil render, applied with a pallet and sufficiently thick to hide all signs of the structure beneath.

(iii)   A thick smooth render with a finish of fine silty clay soil paint.

The roof of the Baipash is supported on a very complex timber structure. From the front to back of the house, two large beams are embedded into the walls and each is supported at three points by highly decorated walnut columns. The cross-section of all these timbers may be as much as 400 x 400 mm. The columns bear directly on the ground or on flat stone pads. Other beams connect between the six columns and the side walls, so dividing the roof up into approximately 12 rectangular areas. Below the central area is located the fire and so above there is provided a square air/light vent. To provide this opening and to give extra height, the roof structure becomes very complex. As Figure 13 shows, this consists of several layers of timbers, each layer being successively smaller and rotated through 45°. The construction is so arranged to maximize the timber lengths and the degree of overlap, this also provides a cantilevering support mechanism. The other eleven rectangular areas each support a system of joists and split branch laths. None of these structural members are pinned together.

The roof structures of the ancillary rooms have no need for a smoke vent and are therefore much simpler. Here there are two or three main beams spanning the full width of the room with the ends embedded into the walls. To keep the roof from sagging, insubstantial, irregularly placed columns are inserted. Spanning between the beams are a series of joists at approximately 300 mm. centres and over these are placed a continuous cover of split branch laths. Sometimes these are placed at 45° to the joists to give a herringbone effect. On all the roofs above the laths is a matting of dried leaves and twigs, although recently polythene sheeting has been substituted for this. Above this is a compacted layer of soil, up to 300 mm. thick.

Structural deformation and material decay

Since the walls and columns have little or no foundations, it is to be expected that many defects result. However, the subsoils immediately beneath are either granular and medium dense to dense or are cohesive and are stiff to very stiff. Both types of soil are therefore capable of taking loads of up to 250 KN/m^2, without noticeable signs of settlement. From typical house forms it has been possible to calculate the load transfer onto the ground. A soil wall including roof loading is about 45 KN/m^2 and a stone wall with roof loading is around 70 KN/m^2. The roof load taken by each of the six main columns onto the ground is approximately 120 KN/m^2. From these calculations no large uniform or differential settlements should be expected and this is confirmed by observations. No structural cracks were noted that would indicate hogging or dishing of the walls or columns. No records were made of cracking or crushing of the soil wall materials below the bearing point of the beam-wall connection. The roof load transfer through the beam onto the wall is approximately 90 KN/m^2 and with the dry compressive strength of the Yasin soil being in the order of 2.5 MN/m^2 such a behavioural fault should not occur.

Occasionally the main beams were observed to be sagging and extra minor columns having been wedged below, quite often around the edge of the walls. This problem is caused by the decay of the beam, the addition of the soil to the roof, or the beam having too long an unsupported length. No observations were made of beams or joists having been displaced by structural movements or earthquake shaking.

The buildings shown in Figures 3 and 4 have little or no diagonal bracing, but no cases were recorded of leaning structures. Here the mass and the geometry of the structures, the stiffness of the walls, the roof beam-wall junction and the small ornamental capitals provide any necessary resistance. Those structures built on outwash fans or indeed scree slopes, likewise showed no signs of leaning over or walls moving apart. These defects would have indicated slow moving landslides, soil creep or soil consolidation.

Various structural defects were noted that resulted from poor construction techniques related to a poor understanding of material behaviour. Firstly there were often near vertical cracks down through poorly staggered joints. Bulging walls were commonly seen where the inside and outside faces of the wall had separated from a loose rubble infill. In walls using semi-round boulders, these stones have often been placed in an unbalanced position, often falling out when the bond with soil mortar weakens. Where stones have been placed without mortar small movements and cracking have taken place when more material is loaded onto the wall mass from above. Vertical cracks were often seen where straight joints exist between walls of different construction phases, where major repairs have been made or where there is junction between two different building techniques. Another common defect noted during the fieldwork was the inadequate wooden lintels above doors and windows, causing the walls above to sag. The last major defect noted on the Baipash house was caused by the total lack of a roof overhang and the provision of an eaves drip trench. With the slightly arched roof, the rain water run-off is concentrated at the wall corners and down the faces of the wall. Here the effect is two-fold: firstly, it leads generally to the removal of the soil render and mortar and, secondly, it softens the ground leading to the settlement of the walls, particularly at the corners.

Building material decay was a most noticeable feature of the Yasin Baipash. Here the main problem is the use of soil, a poor appreciation of its qualities and inadequate maintenance measures. The other major structural materials, granite boulders and walnut columns, do decay, but at a rate too slow to present a problem. Willow laths and the leaf/twig matting rapidly decay, but here it was noted that the smoky micro environment tends to act as a preservative.

A normal feature of most Yasin Valley houses is erosion and soil spalling at the base of walls, the result of capillary rise of water and the efflorescing of salts. This is caused by one of several mechanisms:

a. Flood irrigation in adjacent fields, particularly those fields sloping down to the houses.

b. The inundation of rapidly flowing water - on the flood plains in spring and summer.

c. The ponding of rain water and animal urine in hollows around the base of the walls.

d. Water storage tanks built indoors adjacent to walls. Here there is an increase in the micro humidity causing salt efflorescence and loss of compressive strength.

e. The rubbing away of soil and small stones by animals.

f. The use of polythene sheeting beneath the soil roof cover, allowing water to pond and slow down the rate of roof drying. It may also induce timber to decay.

Another major problem noted with the use of soil is its high linear shrinkage value - approximately 5%. Little effort is put into reducing the soil pasticity by the addition of animal dung, straw or grog. Hence wherever soil is used there are cracks. Firstly the soil render becomes crazed and subsequently falls off. Then the soil mortar shrinks away from stones, soil bricks or wooden frames and the structure progressively weakens. Where additional soil is used to repair the existing walls this does not bond properly and falls off. Although it has not been tested it is certain that the use of Yasin soils for walls speeds up internal mineral decomposition. Biotite mica breaks down into chlorite and Potassium Feldspars to become Kaolinite, a clay mineral.

Improvement techniques for the Baipash of Yasin

It is possible to suggest many methods to counteract most of the defects noted above. These will improve the building performance under ordinary conditions and would certainly provide better protection against some of the natural hazards; the techniques are within the capabilities of the inhabitants and can be done with existing skills without extra material costs:

a. All major wooden roof members should be pinned together. Capital details should be made to serve as diagonal bracing.

b. The bottom level of timbers in the smoke vent should have a greater bearing or overhang length.

c. The major wooden roof members should pass through the wall thickness and overhang the external wall face by a minimum of 0.3 m. Upon this should be placed a roof type apron that would keep rain and snow from running down the wall face. An eaves drip trench filled with granular material should be built around the walls to facilitate water drainage. Water run-off from the roof should be channelled from the roof by drainage pipes.

d. Columns should be sunk into the ground in a 'post pit'. This feature consists of a hole up to twice the diameter of the column, and of a similar depth. In the bottom should be placed a large flat stone to help spread the column load. The sides of the hole around the column should be packed with stone (not soil). This helps to spread the load, and provides resistance when a horizontal force is placed on the column.

e. Foundations should be taken deeper, so that the wall bearing is taken below the soil that is subject to frequent softening, particularly at the corners. The ground around the walls should be formed to slope away so avoiding water collection at the base of the walls and to deflect away rain splashes.

f. Wall building techniques should be improved:

(i)   No dry stone walling should take place;

(ii)   No wall should be a single thickness of stone and no round boulders should be used;

(iii)   All walls should have the bottom metre constructed with squared stone blocks with carefully staggered joints;

(iv)   At frequent intervals stones should pass through the wall thickness to provide keying. At corners quoins should be provided, and the use of slate in the junction would mean that the quoins could effectively be up to one metre long;

(v)   Soil bricks should be better compacted and used with thin layers of soil mortar. The soil bricks should be wetted before being placed;

(vi)   Walls that reduce in thickness towards the top should be vertical on the outside and slope outwards on the inside. The resulting change of centre of gravity would mean that the walls would tend to fall outwards when shaken.

(vii)   There should be an improvement in the soil renders. Firstly they should be made thicker and composed with a finer quality soil. The walls should be maintained on a very regular basis and for example the render near ground level should be patched at frequent intervals. Lastly there should be better methods for bonding the render onto the wall face.

(viii) Animals should be kept away from the face of the walls.

(ix)   The addition of a small percentage of one of the many available soil stabilising agents, to be chosen after mechanical and chemical index tests on the soils. This would improve the durability of soil bricks, soil mortar and also inside the building to help reduce the dusting problem of the soil floors.

CONCLUDING REMARKS

The inhabitants of the Yasin Valley are aware of all dramatic land forming processes and in the past settlement positions have shown a remarkable sensitivity to such potential disasters. However, with recent pressure on land, due to population growth, there is a disregard for safety by expansion into higher risk areas with the inevitable increase of vulnerability. This paper has defined such areas by an analysis of settlement distribution and by examination of geomorphological features. Without immense fieldwork and analytical effort, considerable amounts of money, large engineering works and disruption to the established life style, it would be impossible to reduce the probability of such catastrophic events. Even here one can question whether such ideas are naive when faced with the incredibly destructive-constructive energy released by the environment.

However, there are many simple techniques by which the impact of such events could be lessened including local modification of building sites, improved building techniques (for new houses), engineering protection for individual fields and irrigation channels and river/flood plain revetments. How such improvements could be introduced is more difficult to envisage particularly to the existing situation. Perhaps improved education and greater awareness to the causes of the hazards may be the key, reinforced by regular training schemes and the development of warning systems. How far these alien ideas would be acceptable is questionable considering that the inhabitants regard housing as a low priority.

APPENDIX A

The succession and distribution of rocks in the Yasin Valley

Darkot Group (renamed in 1981 as the Minipin and Baltit Series)

Origin: These are the oldest rocks found in the Yasin Valley and are generally attributed to the Carboniferous Period. They originated as limestones and mudstones that formed a more than 5,000 m. thick sequence fringing the Eurasian Continent.

Rock types: In the Upper Yasin Valley a two-fold division is recognised though further east in the Hunza Valley a triple division has been proposed. The oldest part of the Darkot Group is observed from Sandhi to Umalsit and being nearest to the Megashear is also the most metamorphosed. The rocks consist of mica slates, quartzites and conglomerates with distinct carbonate beds that range from limestones to marbles. Two red sandstone beds form distinctive features. South of Umalsit the black slates are penetrated by numerous acid veins and partly changed to hornfels.

To the north of Darkot, separated from the lower sequence by a major intrusion, is a suite of semi to medium crystalline limestones, fine to medium grained and thinly to thickly bedded, intercollated with black slates and phyllites. This was the original 'Type Section' naming the group.

Structure: It is considered that these rocks form a complex anticline trending NE-SW across the Yasin Valley. This fold is probably pushed up by the massive granitic intrusion just south of Darkot (see below). The rocks to the north of Darkot Village dip approximately 30°N as measured on the south facing limestone scarp faces. Between Umalsit and Yasin, forming the south limb, the dip of the rocks ranges from 60°N to 60°S, often with the rocks standing vertically. However, looking at the mountains as a whole, the dip is slightly southwards and has been overfolded near Yasin Village though without any repeat of the stratigraphy.

The Green Stone Complex

Origin: This sequence to be found over the southern half of the valley is confirmed as being part of Kohistan Island Arc formation. Fossils date the complex to the Cretaceous Period.

Rock type: The name of Green Stone Complex was given in the field to a homogeneous assemblage of rocks dominantly green in colour and consisting primarily of volcanics interbedded with sedimentary deposits now heavily metamorphosed. Rock types include basalts, tuffs, greenschists (with extensive hornblends, epidote and talc), serpentonites, dolerites, quartzites and marbles. These have been injected by a massive granite and by veins of diorite.

Yasin Group

Origin: This is a group of rocks 800 - 1,000 m. thick that extend either side of Yasin Village. They sit uncomfortably upon the Green Stone Complex and are considered to be a suite of flyschoid rocks and the last remnant of the shrinking Tethys Ocean and formed just before the continental collision.

Rock types: The overall recognition of the group is due to the red and pale grey striping observed on the mountainsides. The assemblage consists of mixed sedimentary limestones, conglomerates, shales and sandstones. Some of these have been subsequently metamorphosed into slates, phyllites and phyllitic schists. These are interbedded with volcanics ranging from Basalts, Andersites, Tuffs and Agglomerates. Dykes and Sills are present.

Island Arc Structure

The main strike direction of the Green Stone Complex and Yasin Group is East-West and can readily be traced through to the Hunza Valley. The dip is generally near vertical but near the granitic intrusion the Green Stone Complex schists are extremely contorted. The Yasin Group only occurs over a short strike length and this is considered to be the result of the megashear thrust faulting or erosion of overfolding noted just west of Yasin Village. Here the dip is seen to be 60° SE with the material slightly folded over the older Darkot formations.

Intrusions

Two major intrusions exist in the Yasin Valley and both are post tectonic forming discordant bodies and having sharp contact with the country rocks. These in turn are intruded by varying sized acid and basic veins, sills and dykes.

Morung intrusion: Between Darkot and Umalsit outcrops, an outlier of the muscovite biotite gneiss.   It is uncertain as to whether this is part of the Karakoram Granodiorite that forms the east-west mountain chain to the north.   It has been dated to about 8.6 million years B.C. (Pliocene Age).   It is pale grey, medium grained, sub porphyritic and generally massive.   Accordion type vertical jointing is present, parallel to the gneissosity.   Pegmatites, aplite and quartz with subordinate mafic veins, sills and dykes are common secondary intrusions.

Damalgan intrusion: Intruded into the Green Stone Complex south of Yasin Village occurs an outlier of Hornblende biotite-orthogranite.   It is uncertain as to whether this is part of the main Kohistan Granodiorite found to the south of the Ghizar River.   It has intruded along schistose planes.

BIBLIOGRAPHY

(1) D'SOUZA, F., The Socio-Economic Cost of Planning for Hazards.   An Analysis of Barkulti Village, Yasin, Northern Pakistan. Proceedings International Karakoram Expedition (1983), Vol. 2, pp 289-306.

(2) MATSUSHITO, SUSUMU and HUZITA, KAZUO. Geology of the Karakoram and Hindu Kush. Kyoto University (1965).

(3) HASSNAIM, F. M. Gilgit: The Northern Gate of India.   Sterling Publishers PVT Ltd., New Delhi (1978).

(4) POTT, J., Houses in Chitral: West Pakistan. Architectural Association Journal London, Vol. 890 (1965), pp 246 - 248.

The socio-economic cost of planning for hazards.
An analysis of Barkulti village, Yasin, Northern Pakistan

F. D'Souza – International Disaster Institute, London

ABSTRACT

Local adaptations to mitigate the worst effects of hazards depend largely on two factors: if the threat itself is sufficiently frequent and damaging and if there are, at the same time, adequate resources to afford the cost of preparedness. A socio-economic survey of Barkulti village indicated that the most frequent and destructive hazards were floods, land slips and both shortage and low fertility of land. The seismic risk is not generally perceived as immediate and housing does not reflect any great concern for appropriate building techniques. Since the greater part of each community surveyed follows a subsistence/irregular income economy, with only a few wealthy households able to engage in substantial trade, effective disaster mitigation is too costly for the majority to implement.

INTRODUCTION

The 1980 International Karakoram Project included a group of scientists whose predominant aim was to investigate local adaptation to earthquakes with particular regard to building techniques and materials and the siting of settlements. It was felt that a social and/or economic context was essential and it was hoped that the results of this particular team's work would have a wider application in the general field of low cost adaptations, including housing improvements, to natural hazards.

The Karakoram mountains and the northern areas of Pakistan in general represent one of the most highly seismic zones in the world and it was thought that there would be clear local adaptations to protect communities from the worst effects of earthquakes. Travellers' tales and reports from previous administrators, for example, suggested that design of houses offered significant resistance to earthquakes.

It was decided at the outset that every effort would be made to work as a team, as far as possible, amongst a community small enough in scale to be thoroughly documented. It was particularly hoped that the group would be able to work in either the Yasin or Ishkuman valleys since neither of these areas had ever been adequately investigated, although some studies had already been carried out in the Chitral region to the west of

Yasin valley.   It would be of extreme interest to compare housing designs between the two areas and to relate any differences to such factors as the availability of local building resources, terrain and the perception of hazard.

BACKGROUND

The rationale underlying the socio-economic assessment of communities living in hazardous environments is very simple: any form of disaster preparedness or protection will only be undertaken if -

i)   The threat itself is sufficiently real in terms of perception, frequency and magnitude of damage.

ii)   There are sufficient resources for people to afford disaster prepared- ness or mitigation, including any special building techniques.

It therefore seemed both logical and feasible given the short time available to conduct a survey to determine the following factors:

i)   The nature, type and occurrence of hazards in the region.

ii)   The economic options available for choices to be made.

iii)   The priorities, locally held, in both the short and longer term.

iv)   The traditional mechanisms within the society by which hazard was either reduced, or the risk spread, or the damage minimised or loss compensated for.

That disasters of one kind or another happened was not in doubt, since the impetus for the Karakoram project as a whole was to examine land and life under the more or less constant strain of earthquakes.   The housing and hazards group had a sound reason to concentrate on building structures and materials since the seismological observations following the 1974 Pattan earthquake (Ambraseys et al. 1975) suggested that these high risk communities may have developed earthquake resistant building styles and techniques.   However, these adaptations seen in Chitral to the west and Hunza to the east had not as yet been recorded for the Yasin valley and obviously this aroused scientific curiosity; if such functional designs were present then they could be recorded in great detail - if absent then it would be a genuine scientific endeavour to find out why.

Due to the kindness and hospitality of the various people involved in the project at all levels, the housing group was able to fulfil their original intentions of working in the Yasin valley and within three days of the group assembling a camp was set up in Barkulti village in mid-Yasin valley.*

---

* Mr. Shabaz Begale, a geology student at the University of Peshawar, acted as our interpreter and the team set up camp in his parents' orchard in Barkulti.   We thank them all for their immense hospitality.

FIG. 1. TRANSPORTING WOOD FOR FUEL.

FIG. 2. TYPICAL HOUSE IN BARKULTI SHOWING VERANDAH.

METHODS

Having established a presence in the community and explained precisely the nature and aims of the project to the local people, a preliminary survey of Barkulti was made. Barkulti village consists of five settlements spread on either side of the main river valley and each of these settlements is very roughly associated with a named clan. A grid reference super-imposed on a sketch map of the various settlements enabled a representative sample of households to be selected for subsequent interviewing. Interviews were carried out through an interpreter recruited locally.

A reference interview format was prepared but interviews tended to be informal and discursive. Notes made at the time were then transcribed at the end of each day and any omissions noted to follow up in the follow-ing day. Every effort was made to interview the household head or heads to get the requisite information. However, two strategies were used if such persons were unavailable – either because they were in the high summer pastures or working in more distant fields – a return visit was made or close relations or neighbours were used as informants. As a last resort, if members of a preselected household were absent, the closest neighbouring household was interviewed instead. Verification of data thus obtained was relatively easy in this open and friendly community where such vital statistics as number of fields and animals and occupation was common knowledge. Often interviews were held in any one 'baipash'** with half the neighbourhood present and there was no opportunity or indeed desire to falsify information. In addition to the research interviews described above, several dozen informal discussions were held with Barkulti villagers all of which helped to build up a more detailed picture of the community and thus confirm the results of the interviews. Finally, following a disastrous flood affecting the two villages of Games and Khalti in the Gupis district to the south west of Yasin, family heads of fourteen households were interviewed on the consequences of the floods and their immediate needs.

The first task was to establish patterns of leadership, household structure, income and wealth distribution. Particular attention was given to the whole question of hazard and individual or group efforts to protect and survive. Any difference following on this initial survey either between settlements or even between individual households in any one settlement could be related to the results gained by the engineers and architects on buildings. During the Yasin fieldwork period a total of seventy four families were interviewed representing approximately 660 persons out of a total population of about 2,500.

CONTEXT

The most common and destructive disasters in this area included flooding in the spring and early summer, when snow melt from the surrounding mountains is at its most rapid: Landslips, spontaneous or earthquake induced, which can and do destroy fields, crops and even settlements; and which also block rivers with subsequent severe flooding. Other, so-called creeping, hazards include land erosion and a growing scarcity of

---

** Urdu word describing the communal room signified by the hearth.

wood for fuel and housing construction. The rapid population growth*** in the valleys of the Northern areas has resulted in pressure on land which in turn has several effects, one of them being a fairly regular exodus of young and older men to the larger towns as wage labourers; and secondly a decreasing fertility of the land due to the fact that it is no longer possible to leave fields fallow for any one season. Finally, the community is made more vulnerable due to the poor amenities such as communications with larger marketing areas and inadequate health and education provision.

Until very recently some of the larger valleys in this region engaged in substantial amounts of trading; in particular as recently as the 1930's and 1940's the Yasin valley provided wheat for a large number of the surrounding areas. However, today probably about half of the community has a subsistence economy and it is only the more wealthy families who engage in trade and sale of surplus produce to any great degree.

The peoples of the Karakoram mountains are the descendants of ancient tribes, previously hostile to one another, and much of the traditional kind of building and location of housing reflects the former need for defence, rather than overt concern with the effects of natural disasters.

(a) Houses and Compounds

Settlements consist of groups of houses all of which, with the exception of one settlement, Ghulbashiray, are single storey. The complex of houses representing one extended household are more often than not enclosed by a compound wall within which are the orchards. 'New' and 'old' houses can be roughly distinguished not only by type of construction since later arrivals in the valley, presumably having less need for stringent defence, tend to live in houses which are somewhat isolated on the fringe of the cluster of compounds and generally do not have compound walls. Most households except the very poor have more than one house and wealthier families can have up to six houses. The choice as to which house is used depends on factors such as seasonal variation. Thus, in summer, one family of a household group will live in an outlying house in order to watch over fields and ensure that stray cattle do not wander into the fields and destroy crops. In winter the tendency is to choose an inner house better protected from the bitter cold by compound walls and adjacent to the animal byres. All the older types of house are standard – having a main room the 'baipash' where fires are maintained and all cooking, eating and living take place; leading off this communal room is the "milk store" which is primarily the women's domain for storing, preserving and preparing food. In summer families sleep on their porches, but in winter entire households and sometimes animals too, sleep in the stalls surrounding the 'baipash'. Today all but the poorest families have what are called "guest houses", modern single or double roomed houses constructed away from the compounds and intended for both visitors and perhaps newly married sons of the family.

Since relatively little status is attached to traditional houses

*** A 1931 Census records a total population of 8,000 for the Yasin valley and approximately 1,000 Barkulti villagers. Today the respective figures are nearer to 20,000 and 2,500.

themselves and possessions are few, there is great freedom in moving from one to another. The destruction of a house due to earthquake or flood appears less important than even the smallest damage to fields. Many houses have been abandoned rather than reclaimed and modern ones built with increasing use of stoves instead of open fires.

(b) Social Organisation

The Muslim community of Barkulti belongs to the Ismaili sect but in spite of the relative freedom afforded women as compared to the Shi'ites it is a male dominated society. Inheritance of land, its produce and other assets passes from father to eldest son, the younger sons having a varying share in the produce of land rather than fields. This avoids the uneconomic practice of 'parcelling' land to its eventual depletion. Women are betrothed between twelve and sixteen years (or even earlier in the case of wealthy and powerful families who wish to create favourable alliances). After the marriage ceremony has taken place it is customary for the girl to spend a further year in her father's house before moving permanently to her husband's village and household where she usually serves a long apprenticeship under the tutelage of her mother-in-law or the senior wife of the household head. Men are allowed up to four wives but even second wives were very rare (two cases) amongst the Barkulti people. However, divorce, i.e. sending a women back to her parents' home is less rare and commonly arises due to "incompatible temperament" or infertility. Re-marriage on the death of a first or even second wife is the rule. In the event that the husband dies, a woman will return to her father's household if she is childless or young enough to re-marry. If there are children, however, especially male children, the woman will remain in her late husband's household.

Households usually consist of extended families, that is often up to three generations of males with their wives, sons, daughters-in-law and grand and great grandchildren. It was reported that younger sons customarily used to leave the natal home and with wife and children travel to adjacent or even distant valleys to colonise virgin land. People say that this is no longer possible due to overpopulation and a great dearth of cultivable land and the more usual practice nowadays is for young men to seek jobs in towns further south or, if money is available learn a trade or study for a profession.

(c) Division of Labour

The division of labour between men and women has important economic consequences particularly because of the marked seasonal variation in agricultural work resulting in greater leisure for men than women during the winter months. Men plough, plant and harvest crops, supervise animal breeding and husbandry; mill grain for flour, construct and maintain irrigation channels and when possible have outside employment. Crafts such as weaving are traditionally a male activity. Women's tasks include all care of young children, all house-hold affairs such as cooking, preparing milk products; preservation of fruits and vegetables for the winter and oil production from the apricot kernels. All women will plant and tend their own vegetable gardens as well as weed the main crop fields. Women are also responsible for fruit gathering and tending fruit trees; supervising the grazing and milking of animals, (mainly goats and sheep), chaffing grain, making clothes and bed linens and, finally, the daily collection of water which can involve a long and slippery journey to the main river, the only source of water in the freezing conditions of mid-winter.

(d) Climate

The general environment can be described as highly mountainous, with widely distributed and reasonably fertile valleys in which communities carry out a predominantly mixed farming economy consisting of crops, vegetables, animal husbandry and horticulture. The valleys themselves are remote and often difficult of access and the climate is an extreme one of intense heat during the summer and bitter cold from the end of November right through until April. Seasonal variation in agriculture is marked in that from about May to the beginning of October at least half of any community will move to the high summer pastures in order to graze their cattle, leaving the other half of the community to harvest the crops in the lower valleys.

(e) Irrigation

Rainfall in this land full of water is scanty – approximately four inches per year and tends to fall, sometimes torrentially, in the summer months – rainfall is feared as it can wash away top soil at an alarming rate and also damage young crops. Fields are watered by means of irrigation: essentially there are three types of channels main, village and field networks, each tended by different sectors of the valley and village community – the engineering skill and labour involved requires a cohesive society since irrigation has a clear role in social and economic life.

Briefly, glacial waters from the main mountain rivers are trained to serve the valley each village having its own take off channel and each group of fields will also have a network of much smaller channels. Village allocations are controlled by one individual who is elected each year by the village men and the appointment is approved by the village elders – that is a group of respected men from the whole valley who meet to discuss and agree on crucial decisions such as irrigation channel maintenance and repairs, the building of new channels for the creation of new fields on steep mountain slopes; a procedure which normally requires dynamite for blasting. The channels, particularly the main ones can run for many miles (12 miles is not unusual) and are intricately built around almost vertical mountain sides. Because they are exposed and run for such great distances they are open to damage from even the most minor rockfalls and landslips and require constant vigilance – a dangerous and strenuous task. Sluice gates are of stone slabs packed with mud and fibrous materials or wooden shutters neatly fitting into wooden slots on either side.

It will be obvious that since survival depends on adequate watering of fields, the task of hacking out new fields from the stony, bare and steep hill and mountain sides difficult enough in itself, is only possible if there is a higher and relatively nearby water source.

(f) Agriculture

The agricultural season begins as soon as the winter snow has melted and since the growing season is short (April-September) every effort is made to reap two harvests. The staple crop is wheat but buckwheat, maize, barley and more recently soya beans are also cultivated. All women have vegetable gardens and are also responsible for the gathering and preservation of fruits such as apricots, mulberries, apples and to a much lesser extent pears. Potatoes are a relatively recently introduced crop (1920's) and other vegetables regularly grown include spinach, beans, carrots and herbs. Animals, that is sheep, goats, chickens and for the wealthier oxen, donkeys, cows and yak are kept as work and milk animals, meat for feasting and perhaps most importantly as capital on the hoof. In summer the staple diet consists of chapatti, tea and stewed vegetables; in the winter there is far greater reliance on dried and preserved vegetables and animal products since animals are kept in nearby sheds for easy milking and slaughter. The summer pasture people live almost entirely on milk products which is considered to be a great treat. A variety of dishes consisting of milk at different stages of fermentation are prepared and enjoyed.

The major imports are necessities such as tea, some rice, sugar and fuel. Many households are forced to buy in wheat towards the end of winter when their previous year's stocks are exhausted. Exports include grains if there is a surplus, dried apricots, apricot kernel oil and the occasional animal which is more often sold due to need than as part of any wider and regular economic activity. In those households which have sufficient members for some to engage in crafts, handmade rugs are sent to Gilgit for sale at the markets.

RESULTS

What emerged was a picture of overall marginality with only a very few wealthy households with sufficient resources to make long term plans for future survival, and, thus, also to have - to a certain extent - disaster preparedness. The greatest concern expressed time and again was first of all, shortages and/or loss of land, and secondly a fear of poor harvests (see Table 1). Famine is remembered and the last serious

Hazard Type	Percentage expressing fear of Hazard
Flood	35.5%
Landslip	9.7%
Land Shortage	9.7%
Famine/food shortage	6.4%
Disease (human and animal)	6.4%
Earthquake	3.2%
Rockfalls	3.2%
Thunder and lightning (i.e. damage to crops)	3.2%

Suffered flood damage in previous 3 years	22.6%

TABLE 1:- VULNERABILITY EXPRESSED BY BARKULTI VILLAGERS.

food shortage occurred twelve years ago when people were forced to eat "grasses and snakes". By contrast, the last destructive earthquake occurred over 35 years ago in 1943 and although vividly remembered by older members of the community is not perceived as a major hazard.

There are, of course, substantial differences in wealth and not all households live in constant fear of insufficient food for the winter months. But the wealthy families are apparently few in number (a total of three households in the sample of seventy four can be considered to be extremely secure) and wealth is not, institutionally at least, distributed throughout the community by way of payment for services or patron-client relationships.

Larger households are nearly always wealthier than smaller ones and the association between the two factors of size and income can be demonstrated and to some extent analysed.

Table 2 illustrates the range in both household size and the numbers of fields and animals per household.

Settlements		$\overline{X}$	Median	Range
Barendas	Household size	11.1	7	5-40
	No. of fields	10.8	8	4-30
	No. of animals	34.3	25	6-115
Aduhar	Household size	9.5	7	3-29
	No. of fields	6.0	5	1-15
	No. of animals	29.6	23	1-157
Silpe	Household size	9.0	9	6-14
	No. of fields	8.8	9.5	4-13
	No. of animals	41.7	42	33-51
Ghulbashiray	Household size	9.3	7.5	7-18
	No. of fields	7.3	5.5	2-14
	No. of animals	42.7	22	11-146
Harayoudeh	Household size	5.9	5	4-9
	No. of fields	5.1	4	2-11
	No. of animals	14.9	15	5-24

Total number of households = 74

TABLE 2:- HOUSEHOLD SIZE AND ASSETS, BARKULTI, YASIN.

Given the division of labour between men and women and also the necessary splitting up of the community during the summer grazing period in high and distant valleys, the ideal family is one which can diversify its economy and gain an income from several different types of activity. This normally means having sufficient manpower for men to do the heavy agricultural work and spare men who can perhaps take up employment temporarily or permanently.

An example of a varied and fail-safe economic strategy is seen in one of the wealthiest families in Barendas which has a total of forty members. This large extended family results from the teaming up of three brothers several decades ago. One brother is the local school teacher, the second is the agricultural head and the third and eldest is the family head now too old to work in the fields, but clearly having the authority role and overall responsibility for weighty family issues such as forward planning of agricultural policy, education of male children and grand-children and marriage alliances. The five sons of the eldest brother, now all grown up and all but one married, are employed as follows; the eldest is an engineer working in the hydro-electric project in Hunza, the second son is studying to be a geologist at the University of Peshawar, the third was about to become a shop owner and the remaining two are apprenticed to the agricultural head of the family - the uncle - in order to learn land management. The children of the other brothers are either being educated in preparation for a university career or work in the fields providing each year a substantial surplus which is sold at markets in Gilgit. At the same time the family can rely on cash income from the members of the household who are gainfully employed.

Having a diverse source of income and increasing opportunity to produce surplus food enable this particular family to plan ahead in such matters as education, crop policy, land reclamation schemes and because of the leisure available to some of the older members, indulge in crafts such as rug weaving and carving; the products of which are again sold for cash in markets to the south. At the other end of the range is the small farming unit which is definitely not beautiful in Yasin and has none of the opportunities implicit for the wealthy.

The assets of families in Harayoudeh are less than half those in other settlements and it would seem that they travel on a vicious cycle of poverty and even destitution. Small families in Harayoudeh and other settlements could never hope to produce a surplus let alone have the capital resources to release one or more member for employment or even secondary education since young men are needed to work in the fields. Theoretically males should be able to supplement household income by taking temporary employment as labourers in the agriculturally slack winter months. However, all construction work on roads and buildings in the towns comes to a halt in winter. Families such as these have extreme difficulty in stretching their grain through to the next harvest and are often forced to sell animals from an already depleted herd in order to buy essentials such as extra wheat, fuel, tea and sugar.

Smaller households show other clear signs of poverty. For example, poor families usually had only one house, which was generally small and often lacked a milk store. The houses were dilapidated, in one case a family of five was living in a household, the outer wall of which was crumbling very badly, and in another house a stream had sprung up within the house itself, which is an indication of the unsuitability of the land for housing construction. The people in this particular settlement suffered

FIG. 3. MEADOWS
IN HIGH SUMMER
PASTURES, QURKULTI

FIG. 4. PATHWAY BETWEEN
COMPOUNDS IN BARENDAS.

markedly from ill health as seen, for example, by a large number of skin
infections, a lack of teeth, and a very much higher proportion of feeble-
minded individuals and deaf mutes.

Infant mortality was predictably very much higher in the extremely
poor settlement of Harayoudeh than any other area and this is a typical
consequence of poverty. Table 3 illustrates the number of recorded infant

	Number of recorded births	Number of recorded deaths (0 - 3 years)	$\overline{X}$ Number of births/family	Infant Mortality as a % of live births
Barendas	37	2	6.2	5.4%
Aduhar	90	35	7.5	38.8%
Silpe	60	18	5	30.0%
Ghulbashiray	29	6	5.8	20.6%
Harayoudeh	22	12	3.6	54.5%

Table 3:- INFANT MORTALITY RATES

deaths, that is up to three years of age, as a percentage of the total
number of births. Again, Barendas which shows by far the greatest number
of richer households has an extremely low infant mortality rate (5.4%),
Harayoudeh on the other hand shows a 54.4% infant mortality rate and
relatively low birth rate which partly accounts for the small household
size. Finally it is significant that Harayoudeh settlement was located
far (approximately half a mile) from the main river and winter water
collection presented yet another heavy and dangerous task for women.

There is a third category of household employing a slightly different
strategy for survival and these are the families which rely almost entirely
on a cash income and usually refers to families which have a hereditary
craft such as jewellery. Their goods are either sold in main markets
or to local Yasin inhabitants for cash and foodstuffs can then be bought.
However, even these households which also tend to be small will have
one or, at most, two fields which can be tended by the male members alone
or perhaps with help from neighbours or nearby relations for payment
in cash or kind. Aduhar has more households of this type than any of
the other four settlements and one typical family in this category consists
of three members, a young man, his wife and one young child; he is a
shop owner who sells general household items, soap, matches, material
as well as some produce such as eggs from his own chickens. He has
one field which yields a small amount of wheat and the grain when
harvested is threshed by his neighbour in return for eggs. His wife keeps
a vegetable garden and one goat for milk.

Finally there are those households which have, as it were, acknowledged
the downward spiral of their poverty and where males have elected to
take up employment at the expense of remaining at home to tend the fields.
This job is left to the women who, in turn, rely to some extent on help
from neighbours and kin if available. In these households, again

predictably small and poverty stricken, the hope, obviously, is to survive
on cash alone and in at least one case an old woman had leased her
two remaining fields to her wealthier neighbour and cultivated a vegetable
garden only. Some grain was obtained as payment for the field leasing
arrangement and cash was sent by her two sons who had temporary summer
employment in Gilgit.

Table 4 shows the proportion in each settlement having either –

(i) a subsistence economy – that is except for tea and sugar, relying
almost entirely on home food production or

(ii) an income from the sale of surplus produce or

(iii) where the greatest reliance is put on cash income from
employment.

	Subsistence	Surplus Produce and/or Outside Income	
Barendas	22.2%	61.1%	50%
Aduhar	27.2%	22.7%	72.7%
Silpe	16.6%	50%	33.3%
Ghulbashiray	16.6%	66.6%	33.3%
Harayoudeh	45.5%	9.1%	45.5%

Total number of households = 74

TABLE 4: PERCENTAGE OF HOUSEHOLDS HAVING A PREDOMINANTLY SUBSISTENCE
ECONOMY OR SURPLUS PRODUCE AND/OR CASH INCOME FROM NON-AGRICULTURAL SOURCES

Thus Harayoudeh shows almost half its households following a subsistence
economy as compared to only 22% in Barendas, a wealthier and better
sited settlement. By contrast only 9% of Harayoudeh households have
sufficient surplus produce to sell at all and again nearly half rely on
an outside income to the detriment of efficient agricultural management.
Barendas, correspondingly, shows over 60% of its households having a
surplus produce and in addition up to 50% also have a cash income from
jobs.

Ghulbashiray is a relatively small settlement (about 25 extended households) and is also the oldest group and possibly because of its long
standing presence in the region with all the attendant opportunity to claim
fertile, well-drained fields and other resources, it is also rich. The
same kind of pattern is discernible in Silpe but probably for different
reasons. Silpe is a spread out settlement to the north west of
Barendas – and according to local people is younger in settlement pattern
than Ghulbashiray but it has one very clear advantage of being sited
near to a natural and abundant spring which provides clear water for
drinking and precludes the necessity for labour intensive and vulnerable
irrigation channels. This source of water has allowed the Silpe inhabitants
to be relatively independent of other villages and the people tend to

express a certain kind of superiority concerning the quality of their produce and especially of their animals – which according to them do not suffer from liver fluke. The relative economic security of these two settlements, Silpe and Ghulbashiray, is reflected in the low proportion of households showing a subsistence strategy and the much higher proportions having either or both surplus produce or employment benefits.

During the fieldwork period much attention was given to the opportunities for distributing wealth. Although the local people reported a kind of patron-client system for the past, whereby poorer people were employed as servants by the richer, this is rare today. Accounts of social organisational principles by previous travellers though admittedly not for Yasin itself, report near feudal systems; for example, in Chitral the poor were vassals in the service of the rich land-owning or warring lords or chieftains and payment was in kind, ensuring long-term loyalty.

However, people do say that 'no-one need starve' in Yasin – meaning that in extremes even the destitute could expect to benefit from the Muslim ethic of sharing and taking care of the less fortunate. This charity was, however, not evident. The really poor did not express any expectation of help from their wealthier kinsmen and in one case a household which had lost its fields in a previous flash flood had subsequently rented two fields from a close relative no longer living in the valley. They were required to pay half the produce of the land as rent which left them severely short of food for the winter and with few prospects of being able to survive in the valley in the long term.

Further evidence for the lack of a system for distributing wealth even within a clan is illustrated in Table 5. The Hilbitingay clan is an ancient

Clan Name	$\overline{X}$ Household Size	$\overline{X}$ Number of fields/household	$\overline{X}$ Number of animals/household
Begale	12.3	12.8	51.3
Hilbitingay	11.7	8.0	45.3
Quazian	13.3	11.0	39.7
Ghulbashiray	9.4	7.0	40.2
Mulay	7.0	6.8	17.4
Harayoudeh	5.5	4.2	13.5
(Hilbitingay in Harayoudeh)	5.0	2.3	12.0

TABLE 5:- DISTRIBUTION OF ASSETS ACCORDING TO CLAN

one and very well established in Barkulti. It ranks among the richer clans and yet those few members who for one reason or another live in Harayoudeh, the poor and dilapidated settlement, are extremely poor and have far less assets than the average for the village as a whole. It would seem that there is no mechanism for ensuring that clan members are maintained to the standard of the majority and little sharing of capital resources.

The extreme vulnerability of villagers in the Northern areas was underlined in the immediate aftermath of a combined mud slip and flood in the Gupis region. The events are described elsewhere (see this volume) but the result was that two villages - Games and Khalti, were severely affected including 357 people requiring emergency supplies out of a total of 500. Table 6 illustrates the type and extent of damage to these small farmers and what is perhaps most interesting is the options they felt open

Total Number of People Affected	500
Total Number of Households Affected	58
Number in Need of Emergency Supplies	357
Number of Households Interviewed	14
Losses: Houses including animal sheds	66
Fields	62
Animals	50*
Trees	> 6500
Main Irrigation Channels lost	9
damaged	3

* Note that the majority of livestock were in the
  high summer pastures above the level of the mud
  flow.

TABLE 6:- THE GAMES/KHALTI FLOOD: LOSSES INCURRED

to them after the disaster. Eleven out of the fourteen households interviewed had lost all their fields. Those fields remaining and having a standing crop would probably also be lost due to severe damage to major irrigation networks. Of these fourteen households only one - an old man and his younger wife and daughter left the area to stay with relatives in the valley beyond the mountain range and even they had returned by the third day. Eleven families stayed in the area and camped in caves or in the open on the steep slopes and two families remained in their half inundated homes in spite of rapidly rising water levels. The majority expressed great dependence on government relief supplies distributed by the army and five families appeared very willing to be relocated even to distant valleys providing that they would also receive the necessary subsidies to farm virgin land. Not one family expressed any intention or plans to stay with relatives or to solicit their help. While it is understandable that under the stressful conditions all the villagers were keen to elicit as much government help as they could it is nevertheless consistent with the prevailing system in Barkulti, that none of the victims actually used their kin networks when disaster struck. The overall conclusion must be that when resources are both scarce and subject to loss, they are restricted to the household and not available to the wider kin group.

DISCUSSION

There are three main ways in which any community might adapt to a hazardous environment in the absence of national plans and that is by making physical, social and/or economic adjustments. In Yasin and Barkulti particularly the results of this short study indicate that only those who have spare resources can hope to protect themselves. For example, physical measures include such procedures as building dams along the river banks which are only ever a temporary measure and again require a labour force and are, thus, an option open only to larger/ wealthier households. The siting of houses themselves and land can be said, as well, to be a physical measure of protection, but there is no possibility that families who find themselves in vulnerable positions are able to resite their houses, let alone their fields. Building techniques are certainly important in attempting to mitigate the worst effects of earth- quakes and very briefly the traditional technique of using wood beams for a frame upon which the heavy mud roofs of houses rest is an appropriate one in that the wooden frame has a "sway factor" which means that the houses won't necessarily collapse at the touch of a tremor. But the wood required to make a house relatively resistant to earthquakes is a scarce commodity; it requires land to cultivate the poplar trees and adequate extra fuel sources for cooking and heating to allow groves to mature. At the moment poorer families are forced to travel many miles in this difficult terrain to get the resinous juniper and sandalwood from the high mountains, wood which is adequate for fuel but certainly not of the right kind for building. So therefore, once again, poorer families are unable to reinforce their houses to make them earthquake-resistant. Attempts to stabilise slopes and river banks require capital resources as do the use of agricultural aids, such as fertilisers and high-yield hybrid crops.

Social adjustments include teaming up, that is where closely related members of families choose to live and work together thereby releasing some members of the agricultural force for outside work, and in fact the wealthier families are very much extended ones which employ just this strategy. The fact that poorer households in Harayoudeh do not take up this option is surprising and, as yet unclear. Given that organisation is predominantly at the household level, there appears to be relatively little potential or impetus for economic co-operation between poor households and not so poor ones. Because of the small size of each household; there is never a sufficient labour force to manage the division of tasks in the summer months. Thus, for example, many of the families in the Harayoudeh settlement are forced to leave their animals during the summer months in the care of other people, a service for which they pay in grain. Because of lack of cash and the need to have all their members working on the land they cannot afford to educate their children beyond primary level. Thus, there is no way in which they can plan for the future.

Richer households on the other hand do not have a subsistence economy and rely on the sale of surplus grain and other products as well as other income. In some cases these families are able to 'play the system' to their own advantage; they will sell their own grain on the open market and buy in government subsidised wheat - a lucrative option available to those who have enough money to share in a transport system.

The opportunities for both avoiding the effects of disasters and for survival in an economic sense, are really only available to large families having a surplus and adequate capital. What happens to the poorest of the poor - very crudely, they very often don't survive in the community and become migrants to urban areas, part of the squatter population living in the slums of larger towns. Their by no means unique problems can be summarised as a combination of the following factors:

1. A lack of surplus produce for trade and a lack of cash capital for entrepreneurial activity.

2. Increasing pressure on land resulting in less produce per hectare; this mainly results from abandoning the traditional practice of leaving fields fallow for between one and three years. The potential for serious food shortages is considerable and to add to this particular threat recent and severe government restrictions on hunting activities, previously a valuable protein source, today necessitates importing goods if malnutrition is to be avoided.

3. A lack of adequate communications especially of vehicles and roads to facilitate marketing of any surplus.

4. A lack of employment opportunities; since road construction work is seasonal and the opportunity arises in the summer which is precisely the time of heaviest agricultural commitment.

5. Little real opportunity, even if the work force is available, to cultivate virgin land, mainly because it is not available; irrigation is an extremely tricky process and it is simply not cost effective to construct channels on unstable slopes where they will be destroyed by even a minor land slip.

6. Increasing use of land which is subject to disasters mostly flooding.

7. In the longer term, extensive logging of the hills, which in turn, makes them more unstable and thus vulnerable to land slip and a concomitant shortage of wood plus a lack of cash to buy kerosene for fuel.

FIG. 5.   TEMPORARY SUMMER SHELTERS IN HIGH SUMMER PASTURES OF QURKULTI.

306

The fact that the majority of people do not construct elaborate or even simple defences against the exigencies of the environment is not at all surprising – nor is it necessarily a special case due to extreme poverty, but rather a matter of priorities and choice. The importance of these kinds of studies lies, perhaps, outside the immediate scope and range of the International Karakoram Project. Its message should be directed more to that huge body of planners and development pundits who are increasingly concerned with disaster preparedness and mitigation. The message is simple – disaster preparedness whether it be adopting earthquake-resistant building techniques or constructing flood defences will be undertaken when development has reached a level sufficient for the majority to choose to allocate a portion of their income to protect themselves from what they consider to be a likely catastrophe.

BIBLIOGRAPHY

AMBRASEYS, N., LENSEN, G. and MOINFAR, A. (1975)
Patan Earthquake of 28 December, 1974. UNESCO. Paris. No. FMR/SC/Geo/75/134.

BARTH, F. (1959)
Political Leadership among Swat Pathans. London School of Economics and Social Science. Monograph on Social Anthropology, 19.

CONWAY, W. M. (1894)
Climbing and Exploration in the Karakorums-Himalayas. T. Fisher Unwin

ILLI, D. (1979)
Archetypal Forms of Construction in the Hindukush Architecture and Urbanisation, 19

JETTMAR, K. (1961)
Ethnological Research in Dardistan 1958. Pro. Amer. Philos. Soc., 105, (No. 1), pp 80 - 97.

JONES, S. (1979)
Men of Influence in Nuristan. A Study of Social Control and Dispute Settlement in Waigal Valley, Afghanistan, Academic Press.

LORIMER, D. L. R. (1935 - 38)
Collected Papers held at the School of Oriental and African Studies, University of London, U.K.

LORIMER, E. O. (1939)
Language Hunting in the Karakoram. George Allen & Unwin.

MACFARLANE, A. (1976)
Resources and Population. A Study of the Gurungs of Nepal. Cambridge Studies in Social Anthropology 12 Cambridge University Press.

UD DIN, I. (1965)
A Social Geography of Chitral State, M.Sc. Thesis, King's College, University of London. 2 volumes.

Barkulti in the Yasin valley: A study of traditional settlement form as a response to environmental hazard

C. Moughtin – Institute of Town Planning, Nottingham University

ABSTRACT

This paper analyses the factors that affect the physical form of the settlement of Barkulti in the Yasin Valley. It pays particular attention to the precaution taken by the community to mitigate environmental hazards.

INTRODUCTION

The first impression on entering Barkulti in the Yasin valley is of a peaceful tree studded parkland sandwiched between two converging masses of forbidding rock forming almost perpendicular walls to a lush valley floor. Along the narrow alluvial plain that borders a treacherous river Yasin, often swollen by the melting snow and ice, the people of Barkulti engage, with nature, in a constant battle for survival. The neat houses, well managed fields, intricate drainage constructions and elegantly manicured landscape belies the hazards faced by a group of people who eke out an existence in what can only be described as being on the margins of the modern world; see Figs. 1 and 2.

The aim of this study is to analyse the physical factors that affect settlement form and the use of land in Barkulti; in particular it sets out to show how far the people of this settlement recognise environmental hazards and make provision for them in terms of settlement layout. This exercise must be viewed as part of the more extensive study undertaken by the Housing and Hazards Group of the International Karakoram Expedition which looked in detail at the wider social and economic factors affecting the response to hazard control and settlement planning together with a study of the engineering qualities of structures in Barkulti.

THE PHYSICAL EVIDENCE

The survey on which the findings of this study are based, consisted of mapping the outline of the settlement and its main features such as roads, rivers and building groups; a land use study of the area within the surveyed settlement boundary; a study of the main drainage channels; an examination of the areas facing environmental hazards; measured drawings of sample building clusters and individual buildings; a photo-

graphic survey and finally discussions with two residents of Barkulti about
the ownership and use of land and the purpose for which the spaces within
the family home are used. Figure 2 shows the result of the physical survey
of the settlement; figure 3 is an analysis of this survey; figure 4 shows
land ownership patterns; figure 5 illustrates the layout of typical housing
clusters; figures 6 and 7 show the form of spaces between the buildings in
a housing cluster and figures 8 to 10 illustrate the form of a typical
house in Barkulti.

HAZARDS IN BARKULTI

The people of Barkulti have lived with hazards for generations and
have learned how to mitigate or manage some of the effects of a hostile
environment; in other respects they have also learned to live and work
within the shadow of possible disasters.   The traditional hazards faced
by the people, not only of Barkulti, but of the region as a whole, include:-
destruction of property and loss of life from earthquakes; destruction of
property, damage to, or complete loss of agricultural land and loss of
life caused by flash flooding; the ever present danger from rockfall; the
loss of agricultural produce through poor harvest and finally the loss
of life, or its impairment through illness.   In addition to these traditional
hazards a new and daunting prospect for this group of people is the growing
attraction of urban life.

Communities like Barkulti may, in the future, face a loss of young
active members leaving behind the aged, infirm and less able bodied to
maintain the elaborate infrastructure so necessary for the control of the
environment.

DESCRIPTION OF BARKULTI

To the south of Barkulti is an area of desolate wasteland; the debris
left from a previous flash flood which flowed from the side valley Ishkaibar
inundating and destroying a once prosperous area.   This wasteland forms
the beginnings of the surveyed area; the northern limit being the bridge
across the Yasin.   The size of the surveyed area is 1,000 metres by
3,500 metres.   The land use for this section of the valley is shown on
figure 2;   the major use being arable land which requires constant
irrigation for cultivation.   The local people distinguish three main soil
types:- stony , clayey and sandy soils running in approximately parallel
lines to the mountainside and the river.   A typical cross section through
the valley would consist of the precipitous rocky mountainside, an area
of fallen scree and debris, a zone of transition consisting of orchards,
cemeteries and wasteland; it is through this zone that the road passes
in the south of Barkulti.   This transition zone is followed by a band
of rocky soil which is very narrow to the south of the settlement; then
comes the main agricultural area of clayey soil turning into the less
productive sandy soil on the margins of the river.

ANALYSIS OF SETTLEMENT FORM

A group of important housing clusters are sited along the zone where
there is a change from stony to clayey soil, placing them strategically
to take advantage of these two fertile land types.   Some of these housing
clusters are the main centres of kin and clan groups; the important ones

FIG. 1. GENERAL VIEWS OF BARKULTI.

310

$$\log N = 4.4 - 0.8\, M_s \qquad (1)$$

where N is the number of earthquakes per year

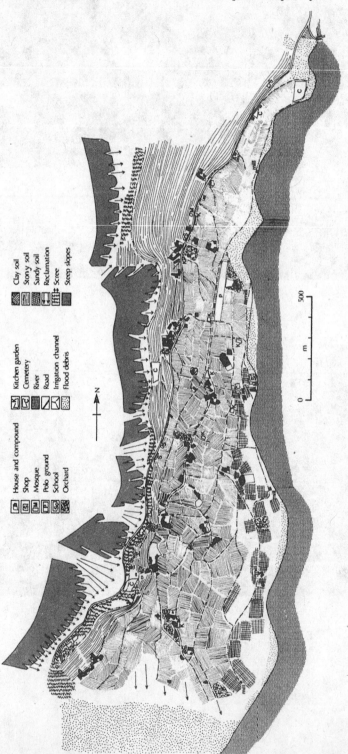

FIG. 2. LAND USE.

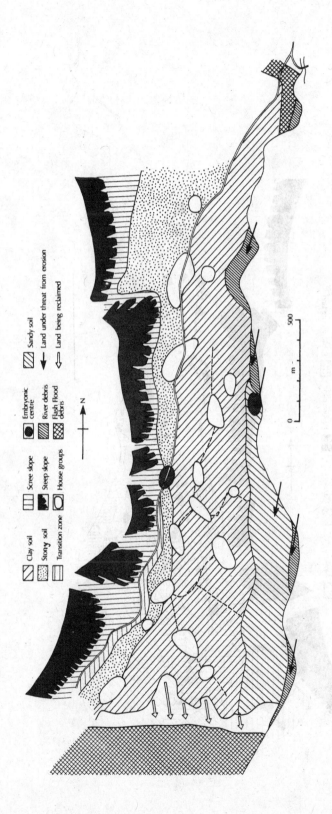

FIG. 3. ENVIRONMENTAL ANALYSIS.

312

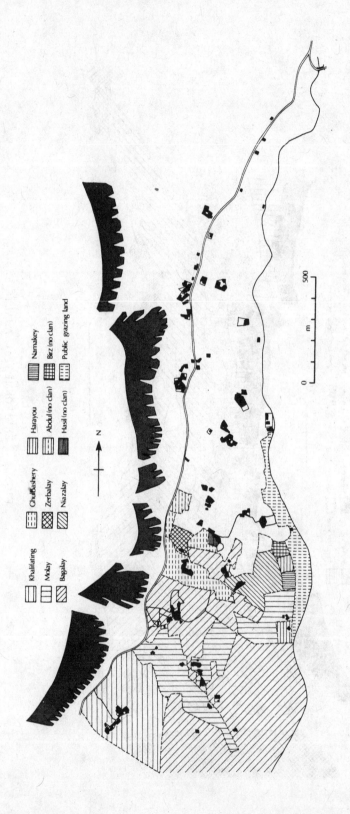

FIG. 4. LAND OWNERSHIP.

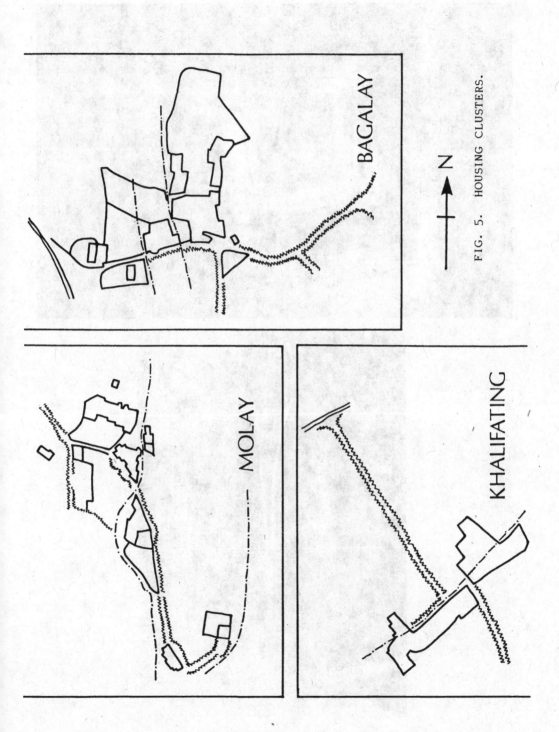

N

FIG. 5. HOUSING CLUSTERS.

BAGALAY

MOLAY

KHALIFATING

314

FIG. 6.   FOOTPATHS.

FIG. 7.   SEMI-PRIVATE SPACE.

BRUSHISKI	CHITRALI	ENGLISH
Havli	Sharan	Courtyard
Mokhen	Moken	Verandah
Kamra	Kamra	Guest room
Buch Bia Koang	Shal Mudi	Cattle shed
Oolha		Store room
Dalah	Joi	Water Channel
BAIPASH	BAIPASH	MAIN ROOM
Khun	Shung	Nursing/wash space
Man	Nakh	Area
Darzheni		Bed
Dung		Main pillar
Bend		Sitting area for men
Phech	Pherwan	Sitting area for young people
Ghotun		
Shutuma		Hearth
Heshtik	Tek	Dancing platform
Takhtabano		Little sideboard
Almari		Big sideboard
Sheridung	Sherothün	Lion Pillar
Nimezheni Man	Nimezheni Nakh	Praying area
Tiss	Kash	Grain Store
Heqai	Shom	Entrance area
Takadini	Bitari	Entrance screen

LEGEND OF FIG. 8

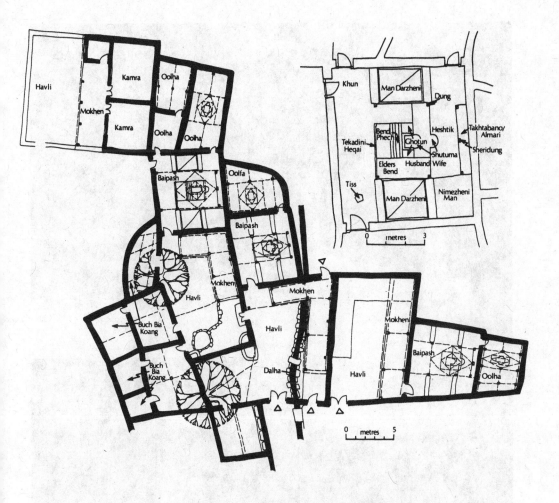

FIG. 8.   HOUSE PLAN OF AKBAR MURAD SHAH.

owning land which runs across the valley floor from mountainside to river-side (see Fig 4). Secondary housing clusters sometimes connected by kinship ties and by pathway to one of the more important clusters are sited further from the western edge of the settlement towards the river; these later housing clusters may have provided homes for the founding clan and have certainly reduced walking distance from home to field so facilitating the use and reclamation of the marginal land close to the river. A further group of important housing clusters follow the largest and most important drainage channel back to the source of irrigation at the river bank to the north of the settlement. This line of the drainage channel is now an important route through the settlement and like the line of soil change was probably one of the features structuring the early shape of Barkulti.

The eastern fringe of the settlement bounding the river, made up of the least fertile sandy soil, is subject to erosion and inundation in places. This zone of the settlement is used for communal grazing purposes, firewood collection and the small isolated farms of the very poorest members of

FIG. 9. THE COURTYARD OF AKBAR MURAD SHAH.

FIG. 10.   THE BAIPASH OF AKBAR MURAD SHAH.

the community; these people are the most vulnerable members of a marginal
community with tiny land holdings of poor quality soil which can be lost
to the river at any time. Along this edge of the settlement are also banks
of river borne debris, a cemetery, the polo ground, a mosque and the
school: all are land uses which are of secondary importance to the
traditional economic life of the community.

To the south of the settlement is the wasteland of debris left from
the last major flash flood. This area is claimed by one of the most
important families in Barkulti, members of the Bagalay clan. They are
at the moment trying to reclaim the land by extending the irrigation
channels through it and planting young trees along the banks of the channel
at 3 or 4 metre intervals. Between the small fields formed by the irrigation
channels the stones and debris are collected into piles and removed; a
long painful process which illustrates vividly the constant battle with
nature that has been fought through the centuries to win some land on
which to secure a living.

The road through Barkulti connecting it by jeep with Gilgit and the
urbanised world follows its own imperative, skirting the scree slopes to
the south of the settlement then following the line of soil change and,
in the main, connecting the important housing clusters. It is along this
latest axis that new shops and mosques are located with the beginnings
of an embryonic centre forming at the mid point along the route through
Barkulti.

TRADITIONAL METHODS OF DEALING WITH HAZARDS

Very little, if any, pre-planning of a traditional nature, in terms
of settlement layout, can mitigate the effects of a major earthquake or
an immense flash flood. In order to contend with this type of hazard
fear has to be suppressed through a process of denial of its existence,
the assumption being that a particular catastrophe is unlikely to happen
to any given individual in his or her lifetime. Precautions, however,
can and are taken to control the effects of minor hazards. Transition
zones are evident in areas of rockfall and in areas liable to erosion or
flooding. In these areas land uses less vital to the economic life of the
community are located. For example, along the south-western boundary
of the settlement there is a transition zone between the rockface or scree
slope and the main part of the settlement: it consists of the road,
cemeteries, wasteland and wooded orchards. Sensibly, the housing clusters
are not built within the area normally affected by rockfalls, being protected
by a barrier of land and in some cases by trees. Similarly, along the
margins of the river are found the less intensively farmed plots of land and
the homes of those too poor to consider an alternative. The location of
the poorer members of the community in this area is additional evidence
for the notion that obtaining a livelihood today is more important to the
occupant of this valley than some possible future danger that may or may
not happen; precautions are taken only for those hazards within the
economic means of those under threat, all other fears are suppressed in
order that day to day existence can continue.

There are signs along the river bank that action has been taken to
retain and strengthen the bank against erosion. Judging by the effects
of a recent change in the path of the river, the traditional engineering
efforts appear to be of only marginal effect. In these circumstances it
is difficult to understand the siting of the village school so close to a

highly vulnerable part of the river bank. Again, this appears to be a quite deliberate choice with economics taking precedence: the school is not placed on safer but valuable agricultural land despite the obvious dangers to the school children.

Crop failure can be mitigated to some extent by large land holdings spanning all soil types, and of course by diversification of effort through ownership of animals and grazing rights in the "high fields" on the rocky mountainside. The main mechanisms, however, for the mitigation of this particular hazard, are social. Short term assistance to families within the Barkulti community is not unknown and connections with members of the clan in other communities and in other valleys provides a safety mechanism for small scale and temporary problems. Some members of the larger family groups obtain employment out of the settlement in the army, the professions, administration or labouring which brings in added income to the group. The chief measure for ensuring family prosperity is, however, a large land holding formed by a united extended family or kin group.

The Bagalay clan of Barkulti consists of 80 people farming 31.4 hectares of land which is 0.4 hectares per person (this does not include the land being reclaimed) (see Fig. 3). The family has recently taken a group decision to remain as a single social and economic unit so that any surplus gained from the land can be used to educate selected young men from the family: this is in effect an investment policy which will bring additional wealth and power to the group. The small nuclear family in Barkulti does not have the security of a large holding nor is it able to make plans to capitalise on the opportunities presented by the urbanisation process. If indeed this type of family is drawn into the modern world it will be because of the loss of land or its inability to feed itself; those forced from the valley in this way will probably join the ranks of the urban poor.

The health hazards with which this paper is concerned are those illnesses caused by poor water supply. The drinking water supply for the settlement is taken from the irrigation channels via settling tanks. The water coming directly from the river or mountainside contains heavy quantities of mica and often traverses areas used as latrines and fields paraded by animals. The people of Barkulti seem totally unaware of this potential hazard and make little use of natural springs in the area. This may be a further case of "denial" so that the time which would be spent travelling to and from distant springs can be more usefully employed in direct economic activities.

HOUSING CLUSTERS AND HOUSE DESIGN

Clan groups remaining together in the same social and economic unit occupy one or more housing clusters. The cluster of houses is connected by a series of semi-private spaces or walkways. These semi-private spaces give access to the more private courtyard or groups of courtyards of the family group of closely related males:- a father, his sons and their dependents or brothers and their dependents. Where the social and economic boundaries between families of the same clan group harden, that is, where the family land has been subdivided, the outward physical appearance of the housing cluster remains the same but the meaning of the spaces between the homes changes and they become semi-public rights of way.

Figure 8 is the house plan of Akbar Murad Shah of the Khalifating clan. This home is occupied by three brothers each of whom has his own Baipash where the nuclear families meet, eat and sleep, each Baipash being entered from a courtyard. In this case, since the brothers and their dependents remain as a single social and economic unit there is one main Baipash where the whole extended family gather. The main Baipash in this home belongs to the eldest brother and is in the oldest part of the building: it is entered through two courtyards. Progression to the inner sanctum leaves on the visitor an impression of entering a series of locks each one becoming more private and less public than the previous one. The spatial rhythm of the architectural composition proceeds from the bright but enclosed valley through a series of shaded semi-public, semi-private spaces to the dark smoky privacy of the Baipash lit only through its opening in the roof.

An analysis of the ability of the structure of the house in Barkulti to withstand the shock of an earthquake has been dealt with elsewhere. The evidence gathered for this paper, however, suggests that a family increases its ability to survive disaster through joint ownership of several properties, particularly if those properties are distributed through the three zones of the settlement. A family, therefore, is more likely to survive if it has property at the foot of the scree slopes, in the main area of the settlement on the clayey soil and on the sandy soil near the river bank. Ownership of property only at the foot of the scree slopes or near the river bank renders such a family susceptible to the effects of even a minor disaster. Those homes built at the foot of the scree are the ones most likely to survive even a large flash flood issuing from the main gorge of the Yasin river: they are built beyond the range of the largest boulder falling from the threatening rockface above and could therefore be regarded as the most secure property in this highly insecure environment. As an extra precaution the Baipash of the house built in this position is sunk 3' to 4' into the hillside at the position of the lion pillar, traditionally the strong point of this type of house construction.

CONCLUSION

The people of Barkulti have, over many generations, learned to live with the possibility of disaster. From the physical evidence gathered for this paper the people of the region appear to practise the defence mechanism described by psychologists as 'denial'. Dangers and hazards over which there is little or no control are ignored or forgotten so that the day to day economic activities which sustain life can proceed. Precautions are taken only against those hazards within the control of the community and for which they have the technology. In this situation the survival of the community, the clan or the extended family, is paramount and on behalf of this it is possible to take social and physical precautions.

ACKNOWLEDGEMENTS

The physical survey of Barkulti was carried out with the assistance of my colleagues in the "Housing and Hazards group" of the International Karakoram Expedition. Many of the ideas contained in this paper have their origin in camp fire discussions during the survey period. I wish, therefore, to record my indebtedness to my associates in the group not only for their labour but also their stimulating ideas.

Analysis of recovery and reconstruction following the 1974 Patan earthquake

I. Davis – Oxford Polytechnic, Oxford

ABSTRACT

The Patan earthquake was the most severe earthquake in Pakistan since the 1935 Quetta disaster. This paper will review some aspects of the relief and reconstruction activities with particular reference to earthquake damage and subsequent reconstruction in Dobair Bazaar and Patan. Detail is provided concerning shelter and housing provision, and in the light of the objectives of the Housing and Natural Hazards Project, the paper discusses changes that may have to be made in construction techniques, thus reducing future vulnerability.

INTRODUCTION

Four questions are central in the study of risk reduction in hazardous areas and were the primary focus of the research work of the Housing and Natural Hazards Group, namely:-

What transpires after a major disaster in a remote rural area and who takes what action with what consequences?
How do societies learn to mitigate the effects of natural hazards?
Do disasters result in rapid learning processes for house builders?
What are the long-term consequences of disaster assistance?

The group were concerned to visit the site of the most serious recent disaster that had occurred in the area with the primary objective of determining what forms of adjustment had followed in terms of either the siting or construction of buildings.

This paper will initially reconstruct the events which transpired after the 1974 earthquake, using material gained from conversations with officials and residents of the area, observations within the locality, and also an examination of the available literature on the subject; (see references 1 - 18). Then in conclusion comment will be made on the broad implications of recovery from the Patan earthquake.

This research in Patan was used to contribute to a study of Shelter after Disaster (the most comprehensive analysis of shelter provision yet undertaken) which has been recently published by the Office of the United Nations Disaster Relief Co-ordinator (19). This study has been based around 13 basic principles concerning the subject of post-disaster shelter and housing. This list of principles has been included in Appendix I of this paper as a checklist of recommendations which may form a reference point to the material in this paper. Therefore, the topics covered in this paper have been matched to these 13 principles in order to facilitate comparison of the events that occurred in Patan to the broader set of principles that emerge from case studies of varied disaster types from a wide range of countries.

PART I

DESCRIPTION OF THE PATAN EARTHQUAKE

Context

Location. The epicentral area consists of 600 sq. kilometres in the Swat, Hazara and Kohistan Districts of the North West Frontier Province of Pakistan. (Subsequently the region was designated as the 'Northern Areas'). (See Fig. 1)

Topography. Very rugged, high peaks ranging from 3657 - 3737 m. above sea-level. Narrow valleys, with a river that meanders widely producing alluvial terraces such as Besham and Patan. The main river is the Indus, which has cut a deep gorge.

Climate. Extreme, winter temperatures below freezing, summer temperatures average 43 °C. Monsoon period late July-early September; little precipitation at low altitudes.

Population. Very sparsely populated, approximately 60,000 in the affected region, with concentrations in the river valleys particularly in Besham, Dobair Bazaar and Patan. Seasonal migration occurs throughout the area.

Local Economy. Subsistence levels. Farming communities producing rice and maize as principal crops, as well as pastural agriculture. The area is undeveloped, with a deficiency of resources.

Traditional Architecture. River smoothed stones or untooled masonry walling, sandwiched between timber levelling members. Roofing is flat, made up of poplar tree trunks overlaid with branches and brushwood finished in mud. However, due to extensive deforestation in the past twenty years, timber has become scarce, and concrete block and reinforced concrete construction is slowly replacing the tradition of the masonry timber sandwich construction.

Earthquake History. The area is characterised by active seismic activity. This has been documented in references (7) and (13). Immediate past earthquakes in the area occurred in Punial/Gilgit in September 1972; and on the Hindu Kush Frontier in July 1974.

Earthquake Impact

Date/Time. 28 December 1974, 1211 hours GMT.

Scale/Severity.   Magnitude: 6.5 (Richter Scale); moderate intensity. The epicentre was in the Dobair valley, and the earthquake originated at a depth of 10-15 km.   No foreshocks were experienced, but extensive aftershocks occurred.

Casualties.   1,000-1,500 dead, but very unreliable figures; there is evidence that both local and overall totals were probably inflated to attract aid.   For example, the Outline Development Plan of March 1975 reports 5,000 deaths and 15,000 injured.   The UNESCO mission suggest a figure of 800-1,500 deaths out of a total population of 60,000. A more realistic figure for the injured is 4,000.   The Army Hospital treated 3,807 patients, this included 2,000 patients treated in the Patan Hospital.

Damage to Property.   Extensive damage was caused to villages lying in a strip 150 km. long running north-south in the Indus Valley.   The villages of Patan and Palas were the most severely damaged.   (See Fig. 1 and the photographs taken immediately after the earthquake: Figs. 2 - 5).

Damage to houses was caused by seismic ground motion as well as from the fall of boulders and landslides from unstable slopes, which are a characteristic feature of the local topography.   (There is a description of the damage to the main villages in reference (13), whilst in reference (7) the damage was reported at 13,000 houses and 900 shops, with 18,000 houses partially damaged.   This latter report claimed that the loss of live-stock was 20,000.   However, the Geological Survey of Pakistan Record (1980), reference (16), reports that only 4-6,000 houses collapsed or were damaged.)

Actions taken for Initial Relief Phase Assistance

By Survivors.   By 4 March, Day 61, the Outline Development Plan reported (17) that 'most of the partly damaged houses have since been repaired by their owners'.

By Local Authorities.   Tents were provided by the Government but no reliable statistics exist on numbers provided, occupancy or method of distri-bution.   Grants were made to surviving families as follows:

500 Rs   (approx. £23) for the owner of a partially damaged house
1,000 Rs (approx. £46) for the owner of a totally damaged house
1,500 Rs (approx. £69) for the owner of a house who was prepared
                        to rebuild according to a Government
                        approved design.

A total of 90 million Rs (£5.2 million) was paid out in compensation for destroyed homes or terraces.

Tents were provided by the Pakistan Army but again, as with local authority provision, no data is available on this provision.

By National Government.   On 15 January, the Prime Minister ordered that all relief funds should be used within the affected area, and asked the Government of the NWFP to produce an outline plan.   This had actually been instigated on 6 January (Day 8) and just three weeks had been allocated for the preparation of this document.   However, the plan took longer to produce, and was discussed by the Federal Government on 4 and 8 March (Days 61 and 65 respectively).

326

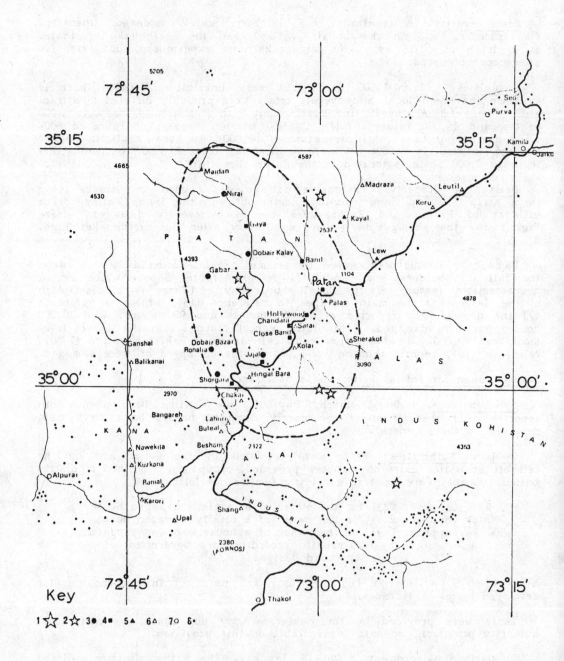

FIG. 1. MAP OF THE AREA AFFECTED BY THE EARTHQUAKE.
(THE LARGER THE STARS, THE GREATER THE DAMAGE CAUSED BY THE
EARTHQUAKE. THE AREA WITHIN THE DOTTED LINE CONTAINED THE
MAJOR DAMAGE TO BUILDINGS).   CREDIT: UNESCO Ref 13.

FIG. 2. RELIEF SUPPLIES BEING DELIVERED BY THE PAKISTAN ARMY
TO EARTHQUAKE SURVIVORS. CREDIT: UNICEF.

FIG. 3. DAMAGE TO PATAN BAZAAR (PHOTOGRAPH FROM THE SAME VIEW-
POINT AS FIG. 7). CREDIT: UNICEF.

Footnote

UNICEF photographs: Wayne Pennington of Lamont-Doherty,
Geological Observatory.

328

FIG. 4.   CHECK POINT CLOSE TO BANDAR.   THIS AREA WAS DAMAGED
BY FALLING ROCKS AND A SCREE FLOW.   PHOTOGRAPH:
AMBRASEYS, UNESCO.

FIG. 5.   PATAN.   TENTS SET UP BELOW THE BAZAAR.   THIS
PHOTOGRAPH WAS TAKEN FROM A SIMILAR VIEWPOINT
TO FIG. 10.   PHOTOGRAPH: AMBRASEYS, UNESCO.

Footnote

Photographs on this page were taken approximately four
weeks after the earthquake.

By Foreign Governments. Apart from gifts in kind, a total of 420 million Rs was received (£24,705,000). This came from 29 countries, however, four of these countries provided between them no less than 392,177 million Rs. These contributions all came from Arab states as follows:

Kuwait   49,209 million Rs

Libyan Arab Republic - 165,803   "   "

Saudi Arabia   - 98,431   "   "

United Arab Emirates   - 78,734   "   "

            TOTAL   392,177 million Rs

These gifts were not tied to specific projects, and this sum has provided the capital for the Kohistan Development Board to continue its work from 1975 with a target date for completion of 1982. The Kohistan Development Board (KDB) decided to divide the 420 million Rs into two broad categories: (a) 86% for Public Works; (b) 14% for Land Development. Out of the total of 420 million Rs, 90 million Rs was paid out in cash sums as compensation payments to affected families who had suffered damage to their fields, livestock, houses or shops.

PART 2

DETAILED ANALYSIS OF POST-DISASTER SHELTER AND
HOUSING PROVISION

This part of the paper should be read in conjunction with the 13 principles listed in Appendix I.

The Resources of Survivors

Families in the immediate relief period appear to have followed several courses of action:-

(a) They accepted the tents provided by the Government or the Military and erected them as near as possible to their ruined homes: in some cases pitching them on the roof of their damaged home (see Figs. 4 and 5).

(b) They moved temporarily into other homes that they or their relatives owned. These additional homes may have been guest houses, or they were used during the summer grazing period in the upper pastures.

(c) They immediately started work on rebuilding their homes. The cash grants of 500 Rs (£23) which were distributed by the authorities for help in rebuilding homes did not appear to have been used for this purpose in most instances. Interviews indicated that this was sometimes used to purchase guns, or to employ people to help rebuild damaged terrace walls. This is an indication that a higher priority was attached to agricultural recovery than to the rapid reconstruction of homes. However, this priority did not result in the neglect of house rebuilding which was virtually accomplished within months. The Outline Development Plan was published on 29 March 1975 (Day 91). This included the statement 'most of the partly damaged houses have since been repaired by the owners'.

Allocation of Roles for Assisting Groups

This was undertaken initially by the Government Officials of the NWFP, then subsequently after March 1975 by officials of the Kohistan Development Board. No outside governmental staff or volunteers were permitted to enter the affected area.

The Assessment of Needs

This was undertaken by the Pakistani Army who had a significant presence in the area at the time of the earthquake. Certain assumptions were made from damage surveys which were probably false. The assessors calculated that 13,000 homes were totally damaged and 18,000 were partly damaged. On the basis of these figures they suggested that 5,000 had died, and they also made projections of the number of survivors rendered homeless. However, as the UNESCO Mission have observed, reference (13), the existance of perhaps four homes for each household makes it impossible to correlate house damage with casualties.

The Report of the Geological Survey of Pakistan (16) has modified the earlier estimates of damage to 4,000 houses destroyed or damaged. This adjustment indicates the wide margin of error in the initial damage survey.

Evacuation of Survivors

There was no official policy to evacuate the survivors from the earthquake. It appears however that there was a voluntary movement of families from the affected area, either to move in with their families in nearby towns or villages, or more generally to their other houses. (The transhumance system of farming in the region means that families may possess additional houses in the high altitude summer pastures, as well as additional 'guest houses'. However, due to inclement winter weather conditions it is very unlikely that many were able to move up to the higher altitude.)

The Role of Emergency Shelter

In summary, emergency shelter consisted of: (i) Tents; (ii) Improvised shelters made out of house debris; (iii) Short distance evacuation to the homes of relatives or to the families' other home, and (iv) Rapid rebuilding.

The photographs taken by the UNESCO Mission (Figs. 4 and 5) indicate that the tents which had been distributed to the surviving families were placed within the existing damaged towns and villages. Fortunately no attempt appears to have been made by the Pakistan army to establish organised camp sites. Probably, there was no actual policy of dispersal, in all likelihood it took place because of the demand from survivors for their own tent with freedom to put it where they wanted. However the consequence of the method of distribution was positive in terms of emergency rapid reconstruction.

Shelter Strategies

The government followed a simple and direct policy of: (a) immediate relief period in the form of tents, and (b) longer term reconstruction, i.e., grants. Clearly other policies could have been followed such as providing building materials, or grants for such, or building temporary shelter or housing for the surviving community.

Contingency Planning (preparedness)

There is no evidence of any contingency plan for an earthquake to occur in this area.

Reconstruction: The Opportunity for Risk Reduction and Reform

In the Outline Development Plan of March 1975, reference (17), the planners reported: "The ideal solution will be to reconstruct the entire areas. This is, however, both practically and financially impossible." and that the system of cash grants for reconstruction offered Rs 1,500 (£88) for improved construction was to "ensure proper reconstruction, in those cases where the owners show their willingness to reconstruct their homes according to Government approved design."

However, detailed discussions with various officials responsible for reconstruction (the Chief Engineer of the Patan District, Mr Ali Ayub Shokoti, and Brigadier Jonkhan Nadir) revealed that apart from the intention to improve the construction and earthquake resistance of local houses backed up by a system of financial incentives, no provision for advising or training local builders, masons, bricklayers, or carpenters had been made. The broad philosophy behind this stance appeared to be that officials believed that aseismic design was a specialist matter to be undertaken by engineers, and built by established contractors. Despite the notional financial incentive of Rs 1,500 this sum was not related to any specific design proposed to either strengthen existing traditional houses, nor to the housing as built by the KDB in conformity to the Quetta Code (devised after the 1935 earthquake).

In Dobair Bazaar and Patan various small buildlers or masons/ carpenters rebuilt highly unsafe houses and shops using concrete block or concrete block infill within the reinforced concrete frames. This took place during January 1975 and was still in progress in August 1980, although most structures had been rebuilt. This reconstruction has produced some houses that are probably more dangerous than the original; see Figs. 6-11.

Good opportunities existed for education programmes for local builders since qualified engineers were based in the area throughout the reconstruction period.

The Relocation of Settlements

In the town of Patan the authorities decided that the bazaar, which constitutes the existing main street (sited on a terrace with a steep drop on one side of the road) was very dangerously situated. Therefore a planning team (within the month of January) produced a new plan for the bazaar and new housing for the entire community of Patan. This proposal was made because of the damage caused by landslides on housing and the bazaar. The site for the new settlement was a flat area, near the helicopter landing pad, adjacent to the foot suspension bridge across the Indus.

The plan was produced following the receipt of an offer from the Turkish Government to rebuild the entire town. However, the offer was never realised. Two conflicting reasons have been given for this: firstly, some residents of Patan informed us that they were offered the choice of cash grants to rebuild their homes and shops, or to await the Government proposal to relocate the bazaar and houses on the steep slopes. All the interviewees told us that they had no hesitation in accepting the cash

FIG. 6. REBUILT HOUSE IN PATAN BAZAAR.
PHOTOGRAPH TAKEN DURING AUGUST 1980, 5½ YEARS AFTER
THE EARTHQUAKE.

FIG. 7. REBUILT STREET, PATAN BAZAAR.
THE HOUSES ON THE RIGHT HAND SIDE ARE ALL POST-
EARTHQUAKE RECONSTRUCTION.

FIG. 8. HOUSING IN THE PATAN BAZAAR REBUILT IN THE TRADITIONAL MANNER. IN THE FOREGROUND A BUILDING FOR THE KOHISTAN DEVELOPMENT BOARD BUILT OUT OF RENDERED CONCRETE BLOCK. THIS BUILDING CONFORMS TO THE QUETTA ASEISMIC CODE.

FIG. 9. PART OF THE REBUILT PATAN BAZAAR. VERY POORLY CONSTRUCTED REINFORCED CONCRETE BEAMS.

FIG. 10. PATAN.   IN THE BACKGROUND OF THIS PHOTOGRAPH THE
LINEAR FORM OF THE BAZAAR, REBUILT ON ITS PRE-EARTHQUAKE
PLAN.   THE WHITE BUILDINGS ARE ALL OWNED BY THE KOHISTAN
DEVELOPMENT BOARD AND WERE BUILT AFTER THE EARTHQUAKE.

FIG. 11.   RECONSTRUCTED HOUSING IN DOBAIR BAZAAR.   TWO YEARS
AFTER THE EARTHQUAKE THE REBUILT SETTLEMENT WAS DESTROYED IN
A SEVERE FIRE.   THE RECONSTRUCTION AFTER THIS OCCURRED IN CON-
CRETE BLOCK AND CONCRETE FRAME CONSTRUCTION.

grants, which had been of 500 Rs each (£23). The reasons they offered for this decision were their fears that the Government would not fulfil their promise to relocate the township.

The second reason came from discussions with Brigadier Nadir Jan Khan, the Chairman of the Kohistan Development Board. He indicated that initially the Turkish Government promised $20 million for the reconstruction of Patan but later, after the outline plans had been produced, the Turkish Government withdrew the offer.

The Ministry of Fuel, Power and Natural Resources asked the Geological Survey of Pakistan to make a detailed survey of the earthquake affected area. The objectives of these surveys were to 'select sites for the new townships'. These field surveys took place between 1974-75 and 1975-76. The report from these surveys was published in the spring of 1980. To some extent this advice on relocated settlements became something of an academic exercise in view of the virtual completion of all reconstruction long before this report was published.

Perhaps the greatest missed opportunity was that offered by the authorities who could have capitalised on the situation with proposals for the construction of houses in the area which would have satisfied both aseismic criteria, whilst also meeting the demand for 'modern homes'. One possibility, given declining timber supplies, would have been to substitute pre-cast concrete for the timber together with the establishment of simple labour-intensive small plants near the main settlements of Patan and Dobair Bazaar.

Land Use and Land Tenure

Good fertile farming land is very scarce in the area. It is owned by individual families or in some instances by small landowners who let out their fields to tenant farmers who pay 50% of their crop in payment for the use of the land.

The availability of new sites for homes or settlements is closely tied to this issue, since families living at the subsistence level of local farmers must live as close as possible to their fields and this prerequisite overrides the need for safer sites. In addition it is apparent that families normally build on inferior land that cannot be farmed. Such sites are frequently those on steep, vulnerable slopes.

Financing Shelter

This aspect has previously been discussed.

Rising Expectations

The reconstruction of homes in the entire region shows evidence that families are attempting to build 'Pucka' houses. This is in part a response to rising aspirations for better homes, but it is also a response to the shortage of timber for roofing and lateral posts, and timber baulks for the walling of houses. Therefore these realities have stimulated construction in reinforced concrete and concrete block construction. In this process of changing the technology from masonry and timber construction to concrete, it would appear that the vulnerability of the local population has markedly increased. The traditional buildings with their discontinuous roof separated from the perimeter walling, as well as the 'framery' of timber baulks within the masonry wall, possessed certain qualities which, with minor adaptation, could have made them earthquake resistant.

Accountability of Donors to Recipients of Aid

There was no relationship between the foreign government donors and the survivors.

At a local level the Kohistan Development Board have made decisions for people. There has been some limited participation in the process, but most decisions have been made with little assistance from the community in question, and there is no accountability between the KDB and local communities.

PART 3

THE IMPLICATIONS OF THE RECOVERY FROM THE EARTHQUAKE

There are many highly commendable features about the longer term recovery from the Patan earthquake. These became increasingly apparent to the study team during their visit to the region.

Firstly, with regard to the immediate actions of the government, the general policy on relief assistance for shelter provision seems to have been effective, unlike many situations where authorities have attempted to forcibly evacuate the affected population, or congregate them in large campsites. The government appear to have been sensitive to local needs by providing tents which were erected within the ruins of people's homes, thus initiating a rapid process of recovery.

Secondly, the policy of providing grants which incorporated incentives to build in a safe manner was extremely sensible and is a very unusual characteristic of government response to this type of situation. However, the flaw in the policy was the absence of expertise at virtually every level to assist the local community to site their homes in safe locations or modify the construction to an anti-seismic approach. Reasons for the lack of expertise concerned the team in their discussions with various officials of the Kohistan Development Board, and the general opinion we gained from these discussions suggested that the main approach of the government was to place reliance on a totally different system of construction. This was the concrete block type of building which incorporated the measures of the Quetta Code of Construction in Seismic Areas. Naturally there is something of a conflict here between the policy of improving rural housing through incentive grants and the prevailing view of the authorities regarding the need to replace vulnerable traditional housing with improved 'modern building'. It is easy to speculate on what could have been achieved had sensitive local expertise been available, and it is facile in the view of the team to expect major shifts in attitude amongst public authorities to aseismic risk and building techniques. An overall recognition of the need to work within existing building traditions is required and more specifically the development of what we came to think of as a 'barefoot architect or engineer' who would be available to give simple advice on improved construction. Such men would not need to have a degree in civil engineering or construction, they might be drawn from the building industry and would need to have simple knowledge of how to build safely in either traditional or modern techniques. Our observations of the concrete construction in Dobair Bazaar indicated that there is a very strong need for expertise to be available on improving reinforced concrete construction within that town and within the whole region.

To a large extent the entire region has been rebuilt in a very similar form of building to that which persisted prior to the earthquake, and it is unlikely that the risks have been substantially reduced in terms of building construction. It is however likely that the broad developments in infrastructure which have come about through the Kohistan Development Board, including the improvement of terracing, irrigation and some limited forms of flood management may have significant effects on reducing risks from landslides and flood, but we saw little evidence of major reductions in seismic risk. One of the most striking contrasts in the forms of building was that of the official structures built by the Kohistan Development Board and the surrounding traditional buildings (Figs. 8 and 10). There was a large gulf between both structural systems which had its roots in the contrast of attitudes already referred to and also inevitably in the contrast of the sums of money spent on both types of building.

The manner in which the Kohistan Development Board have used the extensive grant funding from the Arabian countries of the Gulf States is very impressive and it appears that long term developments in the region which are still in progress were all triggered by the disaster. This is not a new phenomenon, being observed in many past contexts and it would appear that the benefits from this development are not being confined to the wealthier sections of the community.

In conclusion, it would seem that the Patan region has received long term benefits following the earthquake. The government have taken the issues very seriously and have adopted some wise policies for both relief and reconstruction. However, set against this achievement on the question of aseismic risk, there remains a need to develop an effective policy for the improvement of both traditional housing in terms of siting and construction, as well as the more modern techniques of building which seem equally vulnerable. Such a policy is necessary for three contexts; (i) the improvement of the existing housing stock (this will require simple measures developed at a community level), (ii) for new construction through-out the region, particularly using the example of all public buildings such as dispensaries, schools, etc, which can serve as an educational tool in training local builders. Therefore, there may be a value in such buildings being built in the future more closely within local building traditions, yet incorporating anti-seismic construction methods, and (iii) since the area is subject to very frequent earthquakes of either minor or major dimensions such as the event in 1981 (a further major earthquake has since occurred in the Darel and Tangir Valleys), there is a need for the government to prepare contingency plans, notably absent in this event, so that the policy of grant giving so wisely adopted for improved building in this situation, is augmented by the provision of expertise. This may be in the form of simple publications relative to the different types of construction in the region but more radical measures will also be needed relating to the training of local builders and craftsmen.

We are confident that there are high levels of resourcefulness within the local population and commitment on the part of the governmental officials, and with these two vital ingredients substantial progress can obviously be made in reducing seismic risks which may significantly reduce the possibility of future damage on the scale of the Patan earthquake.

APPENDIX I

PRINCIPLES ON THE PROVISION OF POST-DISASTER SHELTER AND HOUSING (Taken from the UNDRO Study: Shelter after Disaster, 1982 pp 3-4.

1.   Resources of survivors

The primary resource in the provision of post-disaster shelter is the grass-roots motivation of survivors, their friends and families.   Assisting groups can help, but they must avoid duplicating anything best undertaken by survivors themselves.

2.   Allocation of roles for assisting groups

The success of a relief and rehabilitation operation depends on the correct and logical distribution of roles.   Ideally, this allocation should be undertaken by the local authorities, who are best qualified to decide who should do what, when and where.   However, if the local administration is too weak to assume this responsibility, the priority must be to strengthen it.

3.   The assessment of needs

The accurate assessment of survivors' needs is in the short-term more important than a detailed assessment of damage to houses and property. Partial or inaccurate assessments of human needs by assisting groups have been a frequent cause of past failure of relief efforts.

4.   Evacuation of survivors

The compulsory evacuation of disaster survivors can retard the recovery process and cause resentment.   The voluntary movement of survivors, where the choice of venue and return is timed by their own needs, on the other hand, can be a positive asset.   (In the normal course of events some surviving families may seek shelter for the emergency period with friends and relatives living outside the affected area).

5.   The role of emergency shelter

Assisting groups tend to attribute too high a priority to the need for imported shelter as a result of mistaken assumptions regarding the nature, and, in some cases, relevance of emergency shelter.

6   Shelter strategies

Between emergency shelter provision and permanent reconstruction lies a range of intermediate options.   However, the earlier the reconstruction process begins, the lower the ultimate social, economic and capital costs of the disaster.

7.   Contingency planning (preparedness)

Post-disaster needs, including shelter requirements, can be anticipated with some accuracy.   Effective contingency planning can help to reduce distress and homelessness.

8. Reconstruction: the opportunity for risk reduction and reform

A disaster offers opportunities to reduce the risk of future disasters by introducing improved land-use planning, building methods, and building regulations. These preventive measures should be based on hazard, vulnerability and risk analyses, and should be extensively applied to all hazardous areas across the national territory.

9. Relocation of settlements

Despite frequent intentions to move entire villages, towns and cities at risk to safe locations, such plans are rarely feasible. However, at the local level a disaster will reveal the most hazardous sites (i.e., earthquakes, faults, areas subject to repeated flooding, etc.). Partial relocation within the town or city may therefore be both possible and essential.

10. Land use and land tenure

Success in reconstruction is closely linked to the question of land tenure, government land policy, and all aspects of land-use and infrastructure planning.

11. Financing shelter

One of the most important components of a post-disaster shelter programme is its financing system. Outright cash grants are effective in the short-term only, and can create a dependency relationship between survivor and assisting groups. It is far more advantageous for both the individual and the community to participate in the financing of their own shelter programmes, especially permanent reconstruction.

12. Rising expectations

Apart from the tendency of prefabricated, temporary housing to become permanent because of its high initial cost, and in spite of its frequent rejection on socio-cultural grounds, temporary shelter, nevertheless, frequently accelerates the desire for permanent modern housing, well beyond reasonable expectations. It is important for assisting groups not to exacerbate social and economic tensions by such provision where there are widespread and chronic housing shortages among low-income and marginal populations.

13. Accountability of donors to recipients of aid

Since the most effective relief and reconstruction policies result from the participation of survivors in determining and planning their own needs, the successful performance of assisting groups is dependent on their accountability to the recipients of their aid.

APPENDIX II

PATAN CASE STUDY

<u>Interviews were conducted with the following local personnel</u>

Amir, Wali Khan   – Senior Administrative Officer, Kohistan
            Development Board (KDB)
Nadir, Jon Khan, Brigadier – Chairman, KDB

Qamar, Colonel   – Deputy Director, KDB

Shokoti, Ali Ayub   – Executive Engineer, Patan District KDB

together with interviews with local officials and residents in Besham, Dobair Bazaar and Patan.

STUDY METHOD

Analysis of available literature (see reference list)

Interviews with local officials and residents

Interview with N N Ambraseys who was a joint author of reference (13).

Observation of reconstruction, August 1980 (5 years and 8 months after the earthquake).

REFERENCES

1) AHMAD, N., (1976, Nov.). 'Earthquake Hazard Reduction in Pakistan'. Proceedings: CENTO Seminar in Earthquake Hazard Minimisation, Tehran, (CENTO Science Report No. 27) pp 6 - 7.

2) FARAH, Abul., (1976, Nov.). 'The Assessment of Earthquake Hazards in Pakistan'. Ibid.

3) JACKSON, R., (1936). Thirty Seconds at Quetta: The Story of an Earthquake, Evans Brothers, London.

4) KHAN, A. Q., (1976). 'Earthquakes and Aseismic Designs in Pakistan', World Conference on Earthquake Engineering.

5) MIAN, Z.-ud-Din, MAHMOOD, K., WASTI, S. T., (1980, Jan.). 'Building in Natural Hazard Areas of Pakistan'. Proceedings: International Conference on Engineering for Protection from Natural Disaster, Asian Institute of Technology, Bankok, pp 907 - 919.

6) Building Code, Quetta 1937 (various revisions). Public Works Department (PWD).

7) QUITTMEYER, R. C. and JACOB, K. H., (1979, June). 'Historical and and Modern Seismicity of Pakistan, Afghanistan, North Western India and South Eastern Iran'. Bulletin of the Seismological Society of America', Vol. 69, No. 3, pp 773 - 823.

8) WASTI, S. T. and AHMAD, N., (1976, Nov.). 'Improving the Earthquake Resistance of Rural Dwellings in Pakistan', Proceedings: CENTO Seminar in Earthquake Hazard Reduction, Tehran. (CENTO Science Report No. 27), pp 215 - 222.

9) KIBRIA, B., (1979, June). 'Housing and Life Style'. Paper given at Symposium: Architecture and National Development, Karachi.

10) MAHMOOD, D., MIAN, Z. and WASTI, S. T., (1978, Feb.-March). 'Design and Construction Needs for Rural Structures'. Proceedings of International Seminar on Low Cost Farm Structures for Rural Development, Faculty of Engineering, University of Peshawar, pp 97 - 100.

11) MUMTAZ, K. K., (1982). 'The cultural context of the use of Adobe and other earthen buildings'. International Workshop Earthen Buildings in Seismic Areas. (Including a description of Adobe/Masonry buildings in the Swat Region). Conference proceedings Vol. 11, pp 131 - 155, National Science Foundation, Washington.

12) QURESHI, S. A., (1979, June). 'No Cost Building Material'. Paper given at Symposium: Architecture and National Development, Karachi.

13) AMBRASEYS, N., LENSEN, G. and MOINFAR, A., (1975). The Patan Earthquake of 28 December 1974, Pakistan, promotion of the study of natural hazards of geophysical origin. Restricted Technical Report RP/1975-76/2.223.3, UNESCO, Paris.

14) AMBRASEYS, N. 'The Patan, Pakistan earthquake of 28 December 1974', Proceedings: Institute of Civil Engineers, London.

342

15) GHAZNAVI, M. and ASLAM, M.   A note on Patan Earthquake, Hazara-Swat Kohistan NWFP (28 December 1974), Unpublished Report, Geological Survey of Pakistan, undated.

16) KANWAR, S. A. K., (1980).   'The Patan   Earthquake of 28 December 1974 and Reconstruction Programme in the affected areas of Swat, Hazara and Kohistan Districts, North West Frontier Province Pakistan'. Records of the Geological Survey of Pakistan, Vol. 47, Geological Survey of Pakistan, Quetta.

17) GOVERNMENT OF THE NORTH WESTERN FRONTIER PROVINCE, (1975, March). Outline Development Plan for Reconstruction and Rehabilitation of Earthquake Affected Area in Swat and Hazara Districts, Planning and Development Department, Government of NWFP Pakistan.

18) KOHISTAN DEVELOPMENT BOARD (KDB).   Projects at a Glance, Abbottabad, Pakistan, undated.

The vulnerability to fire risk of some building types and traditional settlements with reference to the fire which destroyed 'Gayal-Kot' in July 1981

D. Illi – University of Zurich, Switzerland

ABSTRACT

The Housing and Natural Hazards project was initially conceived to examine earthquake risk, gradually this expanded to include other hazards, and during the expedition it became apparent that fire risk was one of the major hazards facing communities in the area. Moreover due to their localised impact such disasters are rarely noted. This paper is an example of the analysis of housing to fire risk, and resulted from a chance visit by the author to the village of Gayal-Kot.

INTRODUCTION

The destructive force of fire has always been a threat to people living in densely populated villages or towns. In the area of our investigations we noted that most families lived in isolated houses scattered throughout the irrigated land used for planting. The risk of fire was thus reported as an extremely low priority for obvious reasons because it would always be limited to the farm where it started.

The catastrophe of a fire affects the socio-economic group of the extended family (using the term of 'extended family' as identified in Refs. 1, 2 and 3 which explain the family structure of the region). However, the major fear of fire is on account of its damage to livestock, the basic necessity of life and indicator of wealth in the area. If animals die in a fire that occurs during a cold period, then the family retreats to another house they may own in a more isolated area. (It is normal for farmers in these areas to cultivate the ground in different areas and at various altitudes, in upper levels relying on rain for irrigation, in contrast to the complex irrigation systems at lower altitudes.) As an alternative strategy the family may split up to join relatives for a limited period. To survive in the latter situation they will need to use any cash reserves they may possess or they may sell their jewellery as a last resort.

Fires within settlements in the region are frequent disasters and it is important to analyse both their cause and simple preventive measures which can be undertaken.

CAUSE OF FIRES

During earlier research in 30 villages within the Khowar and Shina speaking areas, I came across only four buildings that had suffered fire damage (4). In each of these accidents my informants reported that the houses burnt when the family or responsible person was absent. It never emerged whether the destruction occurred in response to ignition by the sun, by sparks from the glowing hearth or by arson as a deliberate act of revenge by an enemy.

I will attempt to list below the causes of such fires. (Points 1 – 3 were given by my informants. I have added 4 and 5 from my observation of houses under similar conditions in a hot dry climate). It appears from the sparse evidence from such areas as Afghanistan, the Pamir region and Nepal, that in areas where there is a chronic shortage of timber, where men have to travel long distances to collect fire wood, that uncontrolled fires are very unusual. The following appear to be the major causes of fire:

1. The spontaneous ignition of hay or of wood on the roof of the house (see Fig. 1).

2. Careless handling of the hearth fire and of inflammable material such as matches, candles, lamps and other materials.

3. The malice of neighbours or enemies.

4. The spontaneous ignition of the fire wood stored inside or near the house.

5. The short circuit in electric installations inside the houses (where electric power is available as in Chitral and Gilgit).

EXAMPLES OF THE FIRE DAMAGE

I will now describe four places where fire occurred:

FIRE OF THE 'SHINGRJI-TOWER' AT LOSNOT-KOT IN GOR

Losnot-Kot, a once fortified village, suffered rapid decay but it is obvious that only one of its tower-buildings was burnt down. In all other cases the houses deteriorated for a variety of reasons. (5, 6). Of that tower of the 'Shubilae-clan', only the cellar walls remained. Pieces of wood which did not burn completely were salvaged for further use or for fire-wood. The reconstruction of the fortified building, one of the few traditional 'Dardic Area' buildings was not undertaken since its owner lived in open land. It was reported, that the fire broke out at the time when men were conducting evening prayers at the mosque, whilst women were working in the fields. Ignition from the hearth was believed to be the cause. Of the attached buildings, at least three burnt down totally, two of which were reconstructed in the same place. The inhabitants of the buildings across the narrow path were luckily not affected, because of the heavy wind from that direction and due to the preventive measures of the inhabitants.

FIG. 1. MANIKAL, DAREL: A TYPICAL VIEW OF A SERIES OF HOUSES, FACING A STREET. The wooden planks, which shelter the top roof, are an immense fire risk for livestock and for the village with its inhabitants.

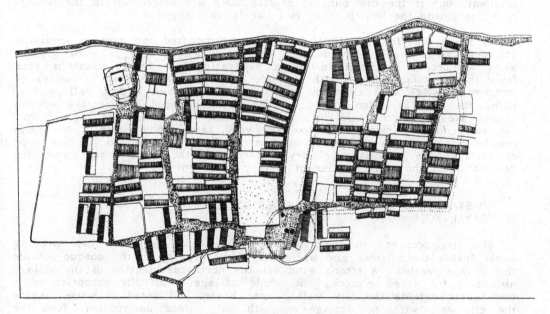

FIG. 2. A PLAN OF MANIKAL-KOT IN THE DAREL. The plan of the ancient site shows the closely packed nature of the buildings.

FIRE OF A HOUSE OF THE FOUNDING FAMILY AT USHNU
IN RICH-VALLEY (CHITRAL)

The outside walls of the main room, constructed in the local technique of superimposed beams (2), stayed in place with the roof partially collapsed.   When I visited the site, the 'coal' of the partially burnt down main structural beams (Sansjir) had just been removed to be used as building material.   The people told me, that these beams had recently been sold to a relative for building. They pointed out that the dimensions were so excessive, that the wood would supply enough material for four new houses. This size could therefore withstand the effects of the fire. It also prevented the roof from collapsing altogether.   Thus the stored grains inside earth-pitches were not affected.   The fire was said to have been discovered so late that it could not be stopped, even though there was a water channel nearby.   Attempts were made to extinguish the fire by throwing earth and water from the nearby channel on the burning roof.

FIRE IN A HOUSE IN MASTUJI (CHITRAL)

In the large complex of a farmer in Mastuji   the main room, the so called 'baipash', suffered from a fire some years back.   The reason was the same as mentioned above.   All neighbouring rooms could be saved from the fire by pouring water on its roofs.

Later, a new addition of the same 'baipash-type' was constructed, the roof of the partially destroyed house was repaired, now serving as a store room for fire wood and for some of the grain.   The owner was happy to have a separate wood store, because "Most fires occur by some little accident, but if the dry bunches of fire wood are stored within the living-room containing the hearth, they can hardly be stopped."

We found that for the people living in the fortified and densely populated villages, fire was a considerable risk.   Fire exposure is even greater when we cross the mountain range separating the inner mountain area from the southern slopes with a more humid climate, where a series of fortified clusters still exist in the Darel, Tangir and Bagrot valleys (5). Here, the hay is often stored on the roof sheltered by massive wooden planks preventing the mud-roofs from erosion.   (North of the high range, the rains in the monsoon period are very rare, and generally the roofs remain dry).   Those roofs pose an immense fire risk, and in the Kot type of village, the quantities of dry material on the roofs could cause the destruction of the entire settlement if they were ignited.

THE EXAMPLE OF THE FIRE WHICH DESTROYED
' GAYAL-KOT'

The fire occurred in August 1980, while people prepared their evening meals inside their homes and while they were praying in the mosque outside the village walls.   A strong wind set the north-east section of the village ablaze.   By seven o'clock, the whole village (with the exception of a few houses outside the city wall), was ablaze.   (I noted 28 houses inside the city-wall with no damage or with only minor destruction, from the total number of 280 houses with an approximate population of 1600 inhabitants).   It was reported, that all the inhabitants and neighbours from a village four miles up the valley tried to stop the fire by throwing mud and dust on the flames.   Men immediately broke the roof planks and

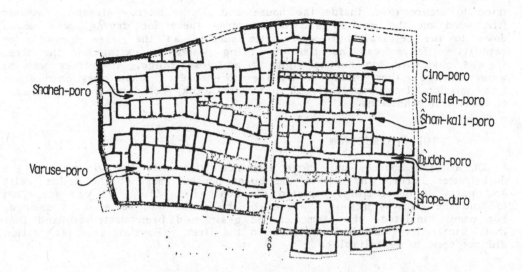

FIG. 3. THE MOST ANCIENT FORTIFIED VILLAGE, 'GAYAL-KOT', DAREL.
This is after the devastating fire of 1980, and gives the location
of streets and houses, indicating the cellar-walls which resisted
the fire. The storage for winter grain, in boxes at the rear
of the house, saved the most important livestock in the houses
of the landowners; the houses of the servants had no personal
stock and no storage facilities inside the houses.

FIG. 4. RUIN AT 'GAYAL-KOT'. The only house inside the walls
of the city, where the roof of the cellar resisted the fire; most probably due to a thicker layer of mud and the large hardwood timber
sections of the supporting structure.

tried to secure them inside the house and in the narrow streets. Second, fire wood on the rooftop and hay, stored there for drying, was moved down to the cellars, but with little success as the fire spread too rapidly. There was no time to bring water to extinguish the fire. It was obviously too hot to fight it at close range, also there was no water storage either inside or outside the houses. When questioned about this omission, local men insisted that water tanks could not be dug because of local soil conditions.

THE DESTRUCTION

Of all the houses, only four remained with only the window and the door-frames burnt, and within the burnt houses, most of the cellar walls and parts of the ceilings resisted the heat (Fig. 4). I was surprised to find the grain stores in the rear of the cellar had suffered no damage, but people insisted, that some of the grain had been destroyed and that their winter livestock had perished in the fire. (However this information did not seem to be reliable).

THE REACTION OF THE COMMUNITY AND THE INTRODUCTION OF PREVENTIVE MEASURES

One and a half months later, almost all the population lived under temporary shelter inside or near the stables, which are situated outside the village near the river. I found only about a dozen of the destroyed houses being rebuilt; the majority of the population was said to live outside the village in new houses. I had the impression, even though denied by the people, that before the fire many of the houses had not been used during the summer and served as shelter in the cold period only, or had been abandoned much earlier. (The actual percentage may vary between 30% and 40%).

Their need for help and priorities were reported by the village chief and additional interviewees as follows:

(a) Most people wanted improvements in land irrigation: The aid should allow for the construction of new channels. The inhabitants would bear the cost of reconstruction of the houses in the 'new style' (with different rooms for men and women, and an addition for guests) in adjacent fields. A majority would thus prefer to migrate to the 'open land' and live in segregated and isolated houses in a similar manner to others within the area.

(b) Most people insisted, that the 'Kot' village was suitable for their living patterns, and given financial help for reconstruction on a family basis, they would rebuild according to the traditional pattern and shape, but using corrugated metal or other fire-proof materials to reduce the risk of fire.

CONCLUSIONS

Given the traditional house with its typical four pillar structure supporting a flat mud roof, and with its traditional open fire, we may consider the risk of fire minimal in isolated settlements, which are becoming more popular at the present time. The following points may help to explain its high fire resistance:

1. The overdimensioning of the pillars, and more importantly of the beams, is essential both for fire resistance and for the time to withstand the heat and thus support the roof. The edges should be cut at 45°.

2. The hay should not be stored on the roof top but aside on a separate platform; the same precaution is necessary for the firewood.

3. In the cluster of the (fortified) town, the roof cannot be sheltered with wooden planks; the roofing should be made with an incombustible material such as stone, corrugated metal, or other incombustible materials.

4. Only a few wooden boxes and other combustible materials and kerosene should be stored within the room.

5. There are two reasons why the form of traditional walling (where timber beams are set within the masonry as 'spacing members') does not improve the overall fire resistance of the structure: The beams nourish a strong fire, and any one of these beams would make the wall collapse if they were to catch fire.

REFERENCES

1) SCHURMANN, F., (1962). "The Moghuls of Afghanistan", ed. K. Jahn and J. R. Krueger, Mouton Co. S-Gravenhage.

2) KUSSMAUL, F., (1965). "Siedlung und Gehoft bei den Tagiken in den Bergtalern Afghanistans", ANTHROPOS Int. Review of Ethnology and Linguistics, Vol. 60, St. Augustin.

3) KUSSMAUL, F., (1965). "Badaxan und seine Tagiken", TRIBUS Veroffentlichungen des Linden-Museums, Vol. 14, pp 11 - 100, Stuttgart.

4) ILLI, D. W., (1978). "Volkerkundlicher Bericht und Reiseaufzeichnungen". Sta. Maria and Zurich.

5) SNOY, P., (1975). "Bagrot. Eine Talschaft im Karakorum", ed. K. Jettmar in 'Bergvolker im Hindukush und Karakorum', Vol. 2, Graz.

6) ILLI, D. W., (1980). "Karakoram Project. Preliminary Report". Zurich.

BIBLIOGRAPHY

EDELBERG, L. and JONES, S., (1979). "Nuristan", Graz.

FRIEDRICH, A., (1955/56). "Wissenschaftliches Tagebuch der deutschen Hindukush-Expedition", MS.

HASSAM-UL-MULK and STALEY, J., (1968). "Houses in Chitral, Traditional Design and Function", FOLKLORE, Vol. 79, pp 92 - 110.

JETTMAR, K., (1975). "Die Religionen des Hindukush", in: 'Religionen der Menschheit', Vol. 4,1., ed. Ch. M. Schroder, Stuttgart, Berlin, Koln, Mainz.

SNOY, P., (1962). "Die Kafiren. Formen der Wirtschaft und der geistigen Kultur", Diss., Frankfurt am Main.

WUTT, K., (1978). "Zur Architektur einiger Hindukush-Taler im Umkreis von Nuristan", Diss., Wien.

Medical aspects of the IKP

Dr D. Giles – Medical Practitioner, Bude

ABSTRACT

The medical problems in Hunza may be catagorized as follows: The
water supply was poor; domestic hygiene was extremely poor; there was
no effective contraception or health education; the acute medical services
offered by the dispensers in the Hunza area, although honestly given
and excellent in concept, were completely unable to deal with the relatively
large number of often complex medical problems that presented themselves
daily at the dispensaries.

On the basis of the disease patterns which we have tabulated it would
seem that the large proportion of the medical problems in Hunza Valley
could be greatly alleviated by the following measures:

1. The provision of piped, filtered water supply.

2. Intensive education perhaps on the basis of the health guards who
   already operate in the area.

3. Effective contraception (this hopefully would follow from health
   education).

4. A resident Doctor and Nurses.

5. The provision of a small hospital and obstetric delivery unit would
   be ideal, but it is recognised that this also would be very expensive.

The main points made above would not, in our opinion, be expensive or
difficult to provide particularly since the framework for the provision
of these services is presently extant in Hunza. We are aware that our
solutions are probably facile and that the Government almost without doubt
has already considered them.   It is our purpose to encourage the
Government or any other Agency to consider no longer but to implement
instantly.

It has been our privilege to work amongst a remarkable people in
an area of outstanding natural beauty. We can only express our gratitude
for this opportunity to serve in some small way both the Expedition and

the people in Hunza.

INTRODUCTION

The Medical team was formed by Dr. Helen Massil and Dr. David Giles whose major task was to look after the health of the expeditionaries and to assist, as far as was practical and consistent with their primary duty, with the medical care of any local people, either at Base Camp or anywhere else where the Expedition might be working.

The team had met and exchanged medical information with each member of the field party and had tried, before the Expedition left the United Kingdom, to form a liaison, as far as possible, with local medical authorities in Pakistan.

Some innoculations and general medical examinations were carried out by the medical team at the Royal Geographical Society shortly before the Expedition left for Pakistan.

WORK LOAD

Not all consultations, either with expeditionaries or with the local people, were recorded although every effort was made to ensure that the vast majority of consultations were documented. Table 1 shows a breakdown of major causes for consultation amongst members of the Project. It will be noted that there was a total of 186 consultations of which almost a third was for intestinal infections. Although there were 9 recorded consultations for infectious hepatitis it was probably considerably higher than this for the seven patients with the complaint. A most gratifying figure is that of 9 consultations on account of trauma. This reflects the extremely careful and responsible attitude of project members when faced with hazardous working conditions.

TABLE 1

EXPEDITIONARIES' HEALTH – CONSULTATION RATES

Gastro–enterestinal infections 51

Respiratory Tract infections 16

Infectious Hepatitis 9

Trauma 9

A number of blood tests were also carried out. Most Project members had a test for haemoglobin and haematocrit on arrival in Pakistan and about 50% were re-tested before they left the field. Unfortunately, the results obtained were hardly representative of the group and certainly did not fit into the physiological pattern expected. It is felt that this is probably due to a degree of morbidity from gastro–intestinal infections and also because the time spent at any given height above that of Base Camp (approximately 9,000 ft.) was not very long. Table 2 shows the major reasons for consultations that took place with Hunzacuts. These tables will be discussed later.

FIG. 1. A PATIENT
WITH HIS FATHER.

FIG. 2. PROF. ZHANG
A CHINESE GLACIOLO-
GIST - INJURED DURING
A FALL ON THE HISPAR
GLACIER BEING TENDED
BY DRS. HELEN MASSIL
AND DAVID GILES WITH
THE HELP OF A LOCAL
DISPENSER.

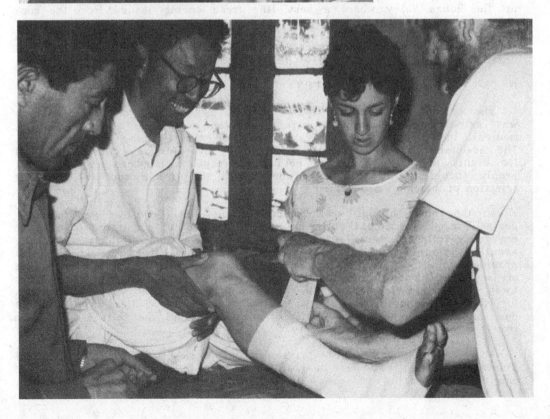

HYGIENE

Despite the best efforts of the administrative and medical teams it was almost impossible to persuade our locally recruited cooks and camp servants at Aliabad that the common house fly was a disease vector and that standard hygiene measures were essential for the well-being of the somewhat tender European gut. Washing facilities and flush lavatories were provided but the basic operation of these wonders of modern plumbing was ill-appreciated by the local people and the maintenance of function and hygiene divulged, eventually, on Expedition members including the medical team and several stalwarts from the geomorphological group.

Our water supply was a constant problem. Early offers from the reconnaissance team of water from the standard valley supplies were quickly rejected on the principle that not only was the water unpleasant to look at, but it had flowed across many fields and through various cisterns and channels, picking up an enormous load of bacterial contamination by the time it reached Aliabad. We were recommended to a spring at Hasanabad which certainly looked and tasted adequate. The presence of many green plants growing around the spring itself was encouraging. Test drinkings by the medical team for three or four days before the arrival of the main group failed to produce significant gastro-intestinal upset and accordingly it was decided on these simple criteria that the water was probably fit to drink. All water for use in hot drinks and other forms of cooking was accepted without treatment but other water was treated with sterilizing tablets.

Further water supplies were eventually obtained from 20/30 kilometres up the Hunza Valey where a very fine fresh spring issued from the rock close to the road. It is clear, now, that there was a quantity of magnesium sulphate dissolved in the water from Hasanabad and that this could have had some minor part in the causation of some gastro-intestinal upset although it is extremely unlikely to have been the major factor.

WORK PATTERN

The medical team worked from a specific medical tent on the site which was extremely well supplied with drugs and instruments and tools, for medical, investigation by many members of the British Drug Industry. The assistance given to the Expedition by this Industry cannot be rated too highly and the thanks of both the Expedition members and the local people should go to all those who helped to provide such a wonderful selection of materia medica.

In the course of our work periods in the Base Camp we were able to devise emergency medical rolls for use in the field, consisting of a basic emergency roll for coping with acute trauma and medical problems, weighing about $5\frac{1}{2}$ kilograms, and a back-up roll weighing 6 kilograms, which would provide further drugs and instruments for coping with the second phase of any medical or traumatic emergency which might be likely to occur away from the Base Camp. We feel we have produced a useful pattern for the equipping of emergency medical teams in the field based on our own experiences. A list of the contents of the rolls is available from the Royal Geographical Society.

We were also extremely privileged to work in the local Aliabad dispensary, which had been a hospital erected by the British in 1928 and originally consisting of some 16 beds and other facilities such as an

FIG. 3.   THE VILLAGE MIDWIFE AND HEALTH GUARD, IN ALIABAD.

FIG. 4.   ON THE WAY TO A HOUSE CALL AT HISPAR.

Out-patients Clinic and a Dressing Station. Unfortunately this had now been somewhat run-down and only one in-patient bed was available, together with the Out-Patient facilities and a rather basic Dressing Station. Staffing this unit was an extremely dedicated and very able Dispenser in the form of Hamayun. Beg for whom our admiration in his attempts to cope with the myriad problems that were brought to him is unbounded; he was assisted by two dressers and a dispensary cook. It was clear that their best efforts were unable to effectively determine most of the complaints that were brought to them and that the supply of drugs and medicines with which they were manfully struggling to control a wide range of disease was inadequate for the job, although it may well have been as much as they could reasonably be expected to handle.

Both Doctors attended at least five days per week each seeing between 30 and 40 patients daily. We were somewhat embarrassed to find that patients were coming from over 100 kilometres away in order to consult us. We confined our attention to the acute and, in our opinion, remedial diseases which we encountered. We tried to refer to the already over-stretched medical authority in Gilgit such conditions as were chronic or long-standing, on the principle that we did not wish to leave behind a legacy of partly treated medical problems.

The Aliabad dispensary also served to help treat some of our Expedition members, particularly those who had been brought back from a scientific sub-camp and whom we had seen during the course of our fairly frequent visits away from the Base Camp to follow up our medical duties further into the field.

LOCAL DISEASE PATTERNS

Table 2 shows that certain medical problems were common in Hunza and of these many were unexpectedly common. A complete break-down of every case is not shown here although it is available from the medical team.

TABLE 2

DISEASE INCIDENCE IN HUNZACUTS

1. Gastro-enterestinal tract:
   Enteritis - total cases - 35 (under 5 years of age - 19 cases).

2. Respiratory Tract:
   Infections - Chronic Bronchitis 27 cases (all over 30).

   Tuberculosis - 15 cases (mostly in the 16/30 age group).

3. Cardio-vascular Disease:
   Heart Disease - 25 cases.

   Heart Disease plus hypertensions - 12 cases.

4. Locomotor Disease:
   101 cases (Male 61, Female 50).

5. Thyroid Disease:
   Total 40 cases (Male 11, Female 29).

6. Ocular Disease:
   Cataract – 24 cases (Male 20, Female 4).

   Infections – 27 cases (Male 16, Female 11).

7. Psychiatric Disease:
   Neurosis – 88 cases (Male 37, Female 51).

   Psychotic Disease – 2 cases (Male 1, Female 1).

8. Anaemias and Deficiencies:

   35 cases (Male 14, Female 21).

Briefly it will be noted that diseases of the gastro-intestinal tract were fairly common amongst local people and what is more, infections were shown to be much commoner at the lower end of the Valley around Aliabad than they were at the upper end of the Valley around Karimabad, following the transit of standard drinking water across fields. Peptic ulcer was extraordinarily common and twice as common in women as in men. There is no question that this finding was surprising to the medical team. Respiratory tract illnesses, including chronic bronchitis, that good old English winter disease, were extraordinarily common. Its incidence was thought to be a reflection of the design and use of houses during the winter months when a small central fire, despite the outlet in the roof, must contaminate the breathed atmosphere very greatly with respiratory irritants such as sulphur dioxide and soot.

Cardio-vascular disease, with or without hypertension, was seen surprisingly often and was infinitely more common than had ever previously been described. In fact, our information prior to arrival in Hunza was that disease of any kind was extremely rare. This certainly has not proved to be the case. However, thyroid disease, which we had expected to find in large quantities, was far less common than our pre-project information would have indicated. We suspect that this is due to the fact that, with improved communication, provision of sea salt with its contained iodine is helping to counteract the natural lack of iodine in the diet of the people of Hunza. We do, however, note that it is twice as common to find thyroid deficiency in females as males and attribute this to lack of mobility of the former. Ocular disease was fairly common although infections were not seen as often as we felt they might have been. Trauma to the eye subsequently causing cataract was seen as a reflection of poor attention during the acute phase and high exposure to a potential trauma in the occupations of the men of Hunza who are engaged in agriculture and mining.

It was unexpected that psychiatric disease would be found. In fact we discovered that neurosis, both depression and anxiety neurosis, were extremely common. We feel that the sudden change of life style from an entirely traditional Hunza style to a partially Western-influenced style due to the opening of the Karakoram Highway, together with the increased pressure on land due to a greatly enlarged population trying to live a mostly subsistent agricultural life, are responsible for these surprising figures.

Many unusual problems were seen such as Pneumoconiosis in a mine worker; Congenital Albinism; Infantile Hirsuitism; Porphyria presenting as abdominal pain; and lastly the curious complaint of "swelling" complained of exclusively by women. We were never able to discover any convincing physical signs to accompany this complaint and subsequent investigations by one of us (H.M.) in other parts of Asia show that this is a common complaint throughout Asia and is due to an hysterical reaction following apparent cursing by some malefactor.

The medical team was particularly privileged to be allowed to visit many local people in their homes. We were usually asked to attend a house where there was said to be one elderly sick patient. We invariably arrived to discover a very large and extremely extended family, all of whom felt obliged to complain of one or another symptom! Despite this, we were most kindly and hospitably received wherever we went and were honoured to be allowed to see so many of the local people in their home environment and to help them with a few of their medical problems.

Pregnancy is not regarded as being an even potentially abnormal state. It would appear to be the norm that most married ladies of reproductive age would have two babies every three years. The infant mortality rate would appear to be around 25%. We were asked to see three cases of puerperal fever which were in extremis. However, we are pleased to report that, as far as we could subsequently ascertain, none of these patients failed to recover. We estimated an annual pregnancy rate of around 3/4,000 and given a population of reproductive age of between 25,000 and 35,000 fact that we only were asked to see about 5 ladies specifically because of their pregnancy is an indication of the local attitude to this condition.

FIG. 5. THE LOCAL MEDICAL DISPENSER, HAMAYUN BEG, WITH HIS FAMILY.

Geomorphology

The geomorphology of the Hunza Valley, Karakoram mountains, Pakistan

A. S. Goudie, D. Brunsden, D. N. Collins, E. Derbyshire
R. I. Ferguson, Z. Hashmet, D. K. C. Jones
F. A. Perrott, M. Said, R. S. Waters and W. B. Whalley
(Authors' affiliations are given in the front of this volume)

ABSTRACT

The Hunza Valley in northern Pakistan shows greater relative relief than any other place on Earth. It cuts through the diverse structural and lithological elements of the Karakoram Mountains. At low altitudes it flows through an area of aridity and of desert vegetation. The slopes in the area are highly unstable, weathering by salt and frost is active, glaciers are active and large, discharge is highly seasonal, channels are subject to marked changes, and sediment and solute loads are high. The overall denudation rate is approximately 5000 t $km^{-2} yr^{-1}$.

INTRODUCTION

The Hunza Valley can lay claim to contain the "steepest place on Earth". The other location that has a comparably great vertical height difference over a comparably short horizontal distance is the Kali Gandaki valley in northern Nepal which rises from around 2470 to 8167 m at the summit of Dhaulagiri I, an elevation difference of 5697 m over 11 km. The Hunza valley for its part rises from about 1850 m (Fig. 1), to the summit of Rakaposhi at 7788 m, a vertical difference of 5938 m over 11 km (Fig. 2).

Such grandeur of relative relief results in long steep slopes of great elevational (Fig. 3) and micro-climatic range, which are variably sculptured by wind, water, ice and mass movement processes. The topographic scale is matched by geomorphological variety, for there exist in close juxtaposition sensitive zones undergoing intense modification by contemporary processes, ranging from the micro-scale to the catastrophic, and areas of at least temporary inactivity or stagnancy of development where the impact of Pleistocene events can still be distinguished. One of the first geographers to visit Hunza, G. N. Curzon, was well aware of the geomorphological uniqueness of the area, writing: "The little state of Hunza alone is said to contain more peaks of over 20,000 feet than there are over 10,000 feet in the entire Alps. The longest glaciers in the globe outside of the Arctic circle pour their frozen cataracts down the riven and tortured hollows of the mountains." He described it as "this great workshop of primaeval forces". (1)

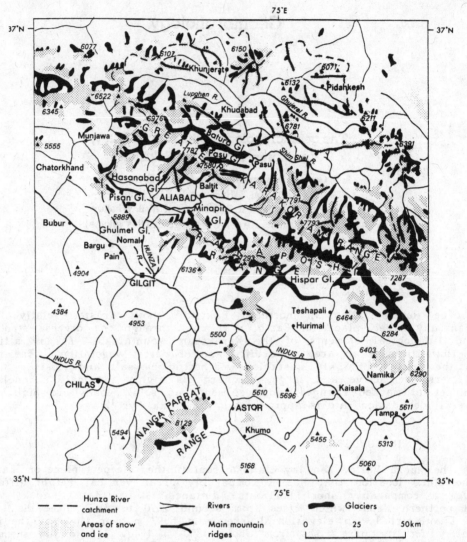

FIG. 1. TOPOGRAPHY, DRAINAGE AND LOCATION MAP OF THE KARAKORAM MOUNTAINS.

The Hunza Valley itself commences on the border between China and Pakistan at the passes of Kilik and Mintaka and can conveniently be divided into Upper (source to Bulchi Das), Middle (Bulchi Das to Chalt) and Lower (Chalt to confluence with Gilgit River) reaches. Although the Upper Valley is winding, it has a general north–south trend until it reaches Bulchi Das, where it turns abruptly westwards to form the Middle Valley. On both flanks of the Upper Valley, tributaries and glaciers join at right angles. These include the Batura, Pasu, Ghulkin and Gulmit glaciers. The mountainous watersheds between these valleys, called <u>mustaghs</u>, are capped with ice and represent the highest ridges of the Hunza basin. On the east side there are only two important tributaries, the Ghujerab and the Shimshal, both of which rise in Chinese Turkestan.

The Upper Hunza is a consequent valley that is transverse to the main structural trend (Fig. 4). When hard rocks such as the Gujhal Dolomite or the Karakoram Granodiorite are present the valley narrows, whereas it tends to widen where it cuts across outcrops of soft rocks such as the

FIG. 2. THE STEEPEST PLACE ON EARTH. View of Rakaposhi, the avalanche slopes, glacier basin and moraines of the Pisan Glacier and cultivated terraces of the Hunza Valley at Pisan village. The vertical relief in this view is greater than that at any other point on the Earth's surface.

Misgar and Pasu Slates and the Girchar Formation.

In its Middle reach, the generally westward oriented Hunza approximately follows the strike of the rock outcrops and main structures (eg, the Northern Megashear Suture; Fig. 4) although local transverse sections occur across the outcrop of resistant lithologies, as above and below the Hindi embayment. This portion of the valley is generally open between wide-spaced <u>mustaghs</u>, and the steeply sloping tributary glaciers rarely approach close to the main river. Although gorge sections occur locally on the discordant sections, the river is generally entrenched up to 150 m in extensive and thick valley floor sedimentary infills (Fig. 5).

The Lower Reach, to the south of Chalt, runs generally transverse to the structural trend. This is particularly obvious in the northern part which is engorged (Fig. 6) where the river crosses the megashear (Fig. 4). To the south, however, the valley progressively opens out so that the relatively broad contemporary channel is slightly incised in a complex assemblage of low terraces and alluvial fans.

362

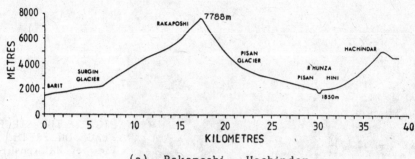

(a) Rakaposhi – Hachindar

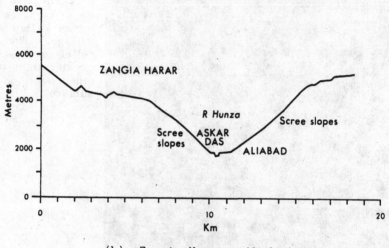

(b) Zangia Harar – Aliabad

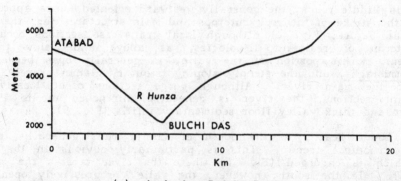

(c) Atabad – Bulchi Das
FIG. 3. CROSS-SECTIONS OF THE HUNZA VALLEY AT THREE LOCATIONS:

A) Nanga Parbat (8125 m)
B) Rakaposhi (7788 m)
C) Indus River
D) Batura Glacier
E) Hispar Valley

FIG. 4.   LANDSAT IMAGERY SHOWING THE ZONE OF OPERATIONS
OF THE INTERNATIONAL KARAKORAM PROJECT.

(Photograph supplied by NASA/USGS EROS Data Center)

FIG. 5.   THE MIDDLE HUNZA VALLEY
Illustrating the way in which accordant, glaciated valley reaches
are wider, infilled with tills, alluvial fan sediments, scree slopes
and fluvial deposits to create irregular but wide terraces and vertical, unstable river cliffs.

GEOLOGY

The Karakoram Mountains are part of an Alpine orogenic belt which
includes the South and Central Pamirs, the Aghil Range, the Kashmir
Himalaya and the eastern Hindu Kush (3).   The maximum width of this
belt, at 72° E, is about 600 km.   At the longitude of the Hunza Valley,
74° - 75° E, the whole Karakoram Range is only 150 km wide, measured
at right angles to the main structural trend, which runs WNW-ESE (4).
The bedrock geology of this segment has been studied by numerous investigators.   Their main findings up to 1964 are summarized by Gansser (5).

The constant strike of the various lithological and structural elements
in the Karakoram allows a subdivision of the range into several distinct
zones, each of which is a more or less well defined structural unit (5).
Gansser divides the Karakoram as a whole into a northern sedimentary
zone, (Tethys Karakoram (6)), a central metamorphic zone with a plutonic
core, and a southern volcanic schist zone.   In the Hunza Karakoram, which
includes elements of all three zones a more detailed subdivision has been
proposed by Desio (3).   This is as follows:

FIG. 6. THE ROCK GORGE OF THE HUNZA RIVER BELOW CHALT.
The rocks here consist of the steeply-dipping green schist complex
and the Bahrain pyroxene granulites of the Kohistan Island Arc
formations.

1. Northern sedimentary zone	4. Chalt green schist zone
2. Axial granite zone	5. Basic intrusive zone
3. Central metamorphic zone	6. Gilgit zone

In this sector, the axial granite batholith separates a northern sequence
of unmetamorphosed sedimentaries from a sequence of metasediments to the
south. The margins of all zones are sharply defined, < ½ km, except for
the southern margin of unit 6. Major northward dipping thrust faults
occur at the boundaries of zones 3 and 4 (Hindi Fault) (sic), 4 and 5
(Chalt Fault) and 5 and 6 (Nomal Fault).

Each of zones 1 - 6 will now be briefly described, following Desio
and Martina (7) and Desio (3)(4). The stratigraphic sequence is summarized
in Table 1 and Figure 7.

1 Northern sedimentary zone

The rocks in this zone decrease in age from north to south. They
consist chiefly of limestones, dolomites and slates, and are generally
unfossiliferous. The Misgar Slate, Kilik Fm., and Gircha Fm. outcrop only

Sedimentary and metamorphic rocks:-

Element	Unit	Lithology	Average Thickness	Zones
a	Hini Fm., ? Eocene	Biotite-feldspar-quartz-graphite schists with garnet and staurolite	?	4
b	Chalt Fm. Cretaceous- ? Eocene	Green sandstone, arenaceous quartzite, 'green schist', with minor limestone and conglomerate. Metamorphosed submarine lavas and tuffs in upper part	ca. 3000 m	4-6
d	Shanoz Conglomerate, Cretaceous- Tertiary?	Conglomerate with marble pebbles and silty cement	ca. 50 m	1
1	Gujhal Dolomite, Triassic - ? Jurassic	Thick bedded dolomite and limestone with minor conglomerate and slate	6000 m	1
	Pasu Slate, (within the Gujhal Dolomite?)	Black to dark grey, mainly arenaceous, occasionally phyllitic, slates with minor limestone and quartzite	?	1
e	Gircha Fm. L. Permian	Arenaceous slates with minor limestone and sandstone	4600 m	1
c	Dumordo Fm. U. Palaeozoic	Marble with garnetiferous calc schist and gneiss	ca. 3000 m?	3
f	Kilik Fm. Palaeozoic	Thick limestone and shales	2500 m	1
g	Misgar Slate, Palaeozoic	Arenaceous slates with minor quartzite and porphyrite sills	5000 m	1

continued next page...

Igneous rocks:-

Element	Unit	Lithology	Average Thickness	Zones
h	Tourmaline granite Pliocene			2-3
i	Hornblende granite Mio-Pliocene			6
j	Axial Karakoram Granodiorite, ? Miocene	Granodiorite, biotite granite and minor hornblende granite		2
k	Peribatholithic plagioclase-gneiss, Neogene	Migmatitic gneiss		2
m	Twar Diorite Cretaceous-Eocene	Basic intrusives - norites and hypersthene diorite		5-6
n	Askore Amphibolite Eocene	Garnet - or epidote - amphibole gneisses		5
p	Basic agmatite Oligocene-Neogene	Strips of basic rock within granite gneiss		5-6

TABLE 1: ROCK SEQUENCE IN THE HUNZA KARAKORAM

368

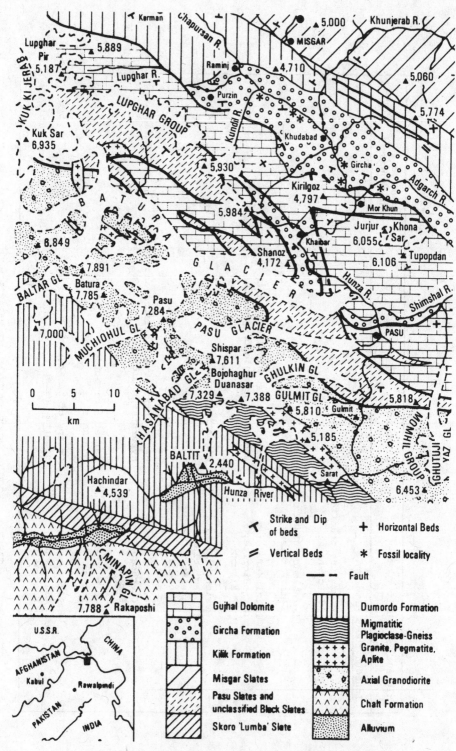

FIG. 7.   GEOLOGICAL MAP OF THE HUNZA VALLEY.

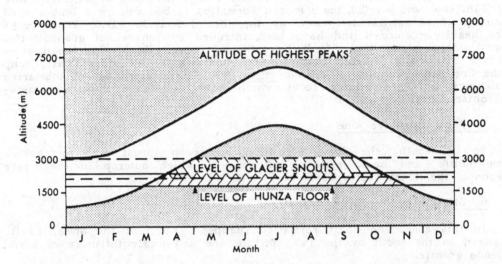

FIG. 8. GRAPH OF THE ELEVATION AND TEMPORAL OCCURRENCE OF FREEZE-THAW CONDITIONS IN THE KARAKORAM MOUNTAINS (ADAPTED FROM (8)).

to the north of the Lupghar Range. The latter is underlain by the thick Gujhal Dolomite, which also outcrops between the Batura and Pasu glacier snouts. The Pasu Slate outcrops on the south side of the Batura, above the lower Pasu Glacier and in the Borit Jheel diffluence col. Its exact stratigraphic position is uncertain because all its contacts with adjacent rocks are faulted (Fig. 7) (7).

2 Axial granite zone

The Axial Karakoram Granodiorite, which forms the core of the Batura Mustagh, outcrops in the headwaters of the Batura, Pasu, Ghulkin, Gulmit and Hasanabad Glaciers, and in the Hunza Valley between Ghulkin and Sarat. It consists of tectonically deformed and recrystallized biotite-grano-diorite, biotite granite and locally, hornblende granite. Radiometric age determinations range from 25 to 8 My (4). The Karakoram Granodiorite is a particularly suitable rock for relative dating of moraines using boulder-weathering criteria, since it is reasonably homogeneous and consistent in its weathering behaviour. It also develops strong, even coatings of desert varnish with increasing surface age.

On its southern side, the axial intrusion is bounded by a narrow belt of migmatitic gneiss; the Peribatholithic Plagioclase Gneiss. Both rock bodies are intruded by tourmaline granite dykes and sills and younger, basic dykes.

3 Central metamorphic zone

This zone comprises the Dumordu Formation, of possible Upper Palaeozoic age (4). It consists primarily of marble with garnetiferous calc schist, and is intensely folded and faulted. Dykes and sills of microdiorite, tourmaline pegmatite and aplite are common.

4 Chalt green schist zone

To the south, across the Hini (Hindi) Fault, Northern Megashear Suture of Tahirkheli and Jan (2), the Dumordu Formation is bounded by a lens-shaped outcrop of metasediments known as the Hini Formation (3). These rocks are heavily deformed and have been intruded by tongues of granodiorite, particularly along their faulted northern contact (9). The headwaters of the Pisan and Minapin Glaciers are underlain by the older Chalt Group (the Greenstone Complex of Ivanac et al, (10)), an assemblage of submarine volcanics and sedimentary rocks which have undergone low-grade meta- morphism (3)(11).

5 Basic intrusive zone

To the south, the Chalt Formation has been thrust over the Askore Amphibolite and basic agmatite of zone 5, which outcrop in the Lower Reach of the Hunza Valley north of Gilgit.

6 Gilgit zone

In this zone, the Chalt Formation reappears south of the Nomal Fault, flanked to the south by the Twar Diorite and a younger intrusion of horn- blende granite.

The distribution of rock types in the Hunza Valley not only influences valley character but also makes it possible to infer past directions of ice movement with considerable confidence, particularly in the Batura-Pasu and Hindi sectors (see later).

The classic view of the geological history of the area, summarized by Desio (3), has recently been reassessed in the light of plate tectonics by Powell (12) and Tahirkheli et al (13). According to this new model (Fig. 4), zones 1 - 3 form part of the Asian landmass, whereas zones 4 - 6 represent the crust of an ancient calc-alkaline island arc (the Kohistan Island Arc) which has been partially subducted under the Eurasian Plate along a major thrust fault, now renamed the Northern Megashear (2). By the late Cretaceous, the northward drift of the Indo-Pakistan continent had resulted in the reduction of the Tethys Ocean to two linear basins on either side of a volcanic island arc. Subduction of oceanic crust under the Kohistan arc generated high pressure metamorphism during the Late Cretaceous-Eocene period. Suturing of the two continents is thought to have occurred in two stages (13): first by underthrusting of the Kohistan Island Arc by the Indo-Pakistan landmass (Early Tertiary) and secondly, by suturing of the island arc to the Asian mainland along the Northern Megashear (Late Tertiary). Deformation and uplift resulting from the continent-continent collision have continued throughout the Neogene and Quaternary.

CLIMATE

The climate of the area, transitional between that of Central Asia and the monsoonal lands of South Asia, varies considerably with latitude, altitude, aspect and localised relief, eg., rain shadows caused by high mountain masses such as Nanga Parbat. Climatic data are few, but reasonably accurate long-term temperature and rainfall records are available for Gilgit (1490 m), and these are summarised in Table 2. The most signifi- cant features are the low overall precipitation (on average just over 130 mm

	Mean Max. (^{o}C)	Mean Min. (^{o}C)	Rainfall (mm)
J	7.9	0.0	6.35
F	11.4	2.7	6.60
M	16.9	7.3	20.32
A	22.3	11.5	24.38
M	27.9	15.2	20.32
J	32.9	19.1	9.40
J	35.3	22.1	9.91
A	34.2	21.7	13.97
S	29.6	17.2	10.16
O	23.2	11.2	6.10
N	16.8	5.2	1.27
D	9.9	1.6	2.80
Year	22.3	11.2	131.58

TABLE 2: TEMPERATURE AND RAINFALL DATA FOR
GILGIT, NORTHERN PAKISTAN.

per year), the large range of mean monthly temperature values, the low winter temperatures, and severe frosts during the period early December - late February (8).

Analysis of the rainfall data for Gilgit Airport for the period August 1979 to July 1980 showed a yearly total of 155.4 mm spread over sixty-two rainy days (Table 3). Thus the average precipitation per rainy day was 2.51 mm. However, it is clear that a relatively small number of days produced appreciable quantities of rain; 66.9 mm of precipitation (43.1% of the annual total) fell on just five days, 39.7 mm of precipitation fell in three consecutive days in May, and the largest daily fall was 18 mm. This tendency for precipitation to be low overall but for there to be a small number of storms each year, sometimes with intense rainfall episodes, has considerable geomorphological significance, especially in terms of the production of debris flows. For example, in August 1974, 10.33 mm of rainfall spread over two days was enough to promote a catastrophic mud-flow in the vicinity of the Batura Glacier (14).

Most of the precipitation is not derived from the Indian monsoon but from depressions moving in from the west during the spring and summer (15). However, on occasion monsoonal disturbances do succeed in extending sufficiently far north so as to enter the area, and when they do precipitation levels can be substantially increased. Great fluctuations

Month	mm	Rainy Days
August 1979	39.8	14
September	1.8	5
October	0	0
November	trace	1
December	4.4	5
January 1980	0.8	1
February	12.7	6
March	5.1	6 (?)
April	10.6	7
May	43.7	8
June	7.6	6
July	27.9	9
Total	154.4	68

From data kindly provided by Meteorological Department, Gilgit Airport

TABLE 3: GILGIT RAINFALL DATA

in regional rainfall totals due to the annual variation in the scale, extent and frequency of disturbances is further complicated because of the localized nature of most storms, marked rain shadow effects, and topographical influences on regional and local wind fields. Thus variability is as important as the more obvious aridity. Although rainfall amounts probably decrease in a general north-eastwards direction along the Hunza Valley as compared with Gilgit, eg., 100 mm at the snout of the Batura Glacier (14), they almost certainly increase in magnitude and intensity with altitude (see later under Hydrology). Measurements of the thickness of the annual layers of ice in the firn basin of the Batura Glacier and the monitoring of the annual discharge of meltwater from the glacier have led Chinese investigators to suggest that precipitation above the regional snow line (here at 4700 - 5300 m above sea-level) may exceed 2000 mm p.a. (16). Hewitt (8), working further to the east near the Biafo Glacier, broadly confirmed this greater precipitation maximum at high altitude, and suggested totals in excess of 1400 mm p.a.. Müller (17), however, has suggested that precipitation totals may decrease again at very high elevations.

Temperature regimes are similarly variable. In the valleys shade temperatures can exceed $40\,^{\circ}C$ on a number of occasions each year and surface temperatures on exposed surfaces with southerly aspect can be

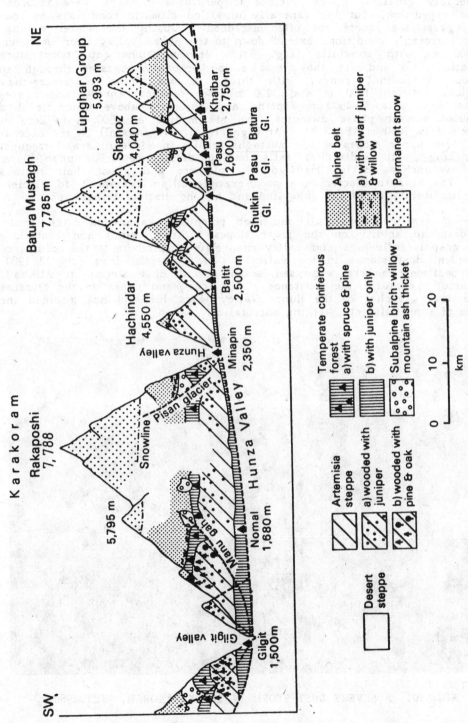

FIG. 9. THE ALTITUDINAL VEGETATION ZONATION OF THE KARAKORAM MOUNTAINS; SEE REFERENCE (15).

considerably greater. Rock surface temperatures of 50°C were recorded by the expedition, but the generally unsettled climatic conditions at that time suggest that these are often exceeded. During the winter months, however, freezing conditions extend down to the Hunza Valley floor (minimum of 1500 m) with snowfalls (Fig. 9). Investigations into temperature fluctuations (8) indicate that frost action probably operates through an enormous altitudinal range, with three to six months of freeze-thaw conditions between 1500 m and 3000 m, two separate three-month phases between 3000 m and 4500 m (spring and autumn), above which level a continuous summer phase dwindles from six months at 4500 m to zero at approximately 7000 m (Fig. 8). At high elevations, snowfall on the exceedingly steep upper slopes of the mustaghs results in widespread and frequent avalanching, with individual falls having displacements of up to 3500 m. These avalanches nourish many of the glaciers and affect their regimes (18). The Minapin Glacier is a good example of an avalanche-fed glacier, for it lies mainly below the snow-line and has no distinct tributaries.

The generally low rainfall and high potential evaporation rates result in widespread aridity on the lower slopes. Strong winds and gusts due to topographic influences and valley channelling contribute to the occurrence of frequent dust storms in the valleys. In six weeks from July 12 1980, the expedition members witnessed seven severe dust storms at Aliabad. Such aridity explains the existence of the true sand dunes on the terraces of the lower reaches of the Hunza Valley (Fig. 10) and has moulded the surface of some tills into striking microrelief.

FIG. 10.   A SEVERE DUST STORM IN THE KARAKORAM, JULY 1980.

VEGETATION

The Hunza Karakoram lies on the northeastern margin of the Northwest Himalaya biogeographic region (19). Its vegetation, which has been described by Pampanini (20) and Paffen et al (15), shows important similarities with that found in Dir, Chitral and the rest of Northern Areas (21) - (26) and strongly reflects the northward increase in aridity.

Throughout northern Pakistan, the vegetation displays conspicuous altitudinal zonation. Dry, subtropical vegetation at lower elevations passes upwards into moister, temperate forest, woodland or scrub. Above the treeline, plant growth is limited by atmospheric cold, intense insolation and night-time reradiation, high ultra-violet intensities, wind stress, extensive snow cover, and in many cases, geomorphological instability (19).

Within each altitudinal zone, the general northwards decrease in precipitation results in a marked reduction in species diversity. Between Nanga Parbat and the Hunza Valley, the total number of vascular plants is reduced by almost half (15) (21). There is also a shift in floristic composition, as taxa typical of areas with summer rains give way first to Mediterranean species and then to taxa which thrive in the arid, continental interior of Central Asia (15)(24). In the Hunza Karakoram, the Central Asian element reaches its greatest abundance in the lowermost, desert-steppe and uppermost, alpine, zones.

A third important factor governing the vegetation distribution is the strong contrast in insolation and moisture stress between dry, south- and east-facing, and moist, north- and west-facing, slopes (15)(19). Other local variations result from differences in geomorphological stability and in the degree of human interference (23)(25)(26). Unstable areas such as talus cones, alluvial fans, young moraines and avalanche chutes, tend to be characterized by sparse, successional communities.

The general altitudinal sequence of vegetation in areas with a winter-spring rainfall maximum is as follows (22)(24):

(1) Dry subtropical broad-leaved forest or steppe with Pistacia, Indian olive (Olea cuspidata) and Artemisia maritima;

(2) Dry holm oak (Quercus ilex) forest;

(3) Dry temperate coniferous forest with cedar and chilghoza pine (Pinus gerardiana), passing up into moist fir-pine forest;

(4) Subalpine birch-conifer forest; and above treeline;

(5) Dwarf juniper-rhododendron scrub and alpine pastures. In Northern Areas, dry juniper forest takes over the lower part of the coniferous forest belt (26).

The Hunza Valley field area is situated at the extreme arid limit of forest. Thus, Pistacia and Olea have their northernmost stands in the Indus Valley, Quercus ilex on the southern flank of Rakaposhi, and Pinus gerardiana in the Chaprot Valley just west of Chalt (15). Fir also drops out north of Nanga Parbat, whereas spruce and mesic pine extend as far as the north side of Rakaposhi. In the eastern Hunza Valley and the Indus headwaters, Juniperus is the sole remaining arboreal taxon (15)(23).

Between Gilgit and Pasu, almost all the main altitudinal zones are still recognizable, especially on north- and west-facing slopes (15). The sequence is as follows (Fig. 9):-

```
                                                     5000 m
              ( Isolated high-altitude plants
              (
              (                                      4500 m
              ( Low-growing mats of Cobresia and
 Alpine zone  ( Polygonum spp. with abundant
              ( bare ground
              (                                      4100 - 4200 m
              (
              ( Willow scrub and recumbent juniper
                                                     3800 - 4000 m

 Subalpine birch-willow-mountain ash thicket
    with a diverse understorey of flowering
    shrubs and herbs
                                                     3400 - 3600 m

 Temperate coniferous forest
    with spruce, mesic pine and juniper
                                                     3300 - 3400 m

 Artemisia steppe with an incomplete ground
    cover of grasses and herbs. Clumps of
    junipers in upper part
                                                     2700 m

 Desert steppe
    Low, sparse vegetation dominated by
    Artemisia, Salsola, Ephedra, globe
    thistles, Capparis spinosa and grasses
```

The altitudinal limits of each vegetation belt increase upvalley, this trend being most pronounced in the case of the lower zones. Above the Minapin Valley, the coniferous forest is represented only by sheltered patches of juniper between 3500 and 3800 m. It is eliminated entirely north of the Ghulkin Valley. On the sunny north flank of the Batura Glacier, wooded Artemisia steppe passes directly into alpine juniper scrub at ca. 4000 m. In general, the vegetation becomes much more open in an upvalley direction, although this trend may partly reflect human interference through grazing and wood cutting. In addition, large areas of the valley floor are now devoted to cultivation or are occupied by thickets of sea buckthorn (Hippophäe rhamnoides) and other thorny shrubs which line the numerous irrigation canals and ditches.

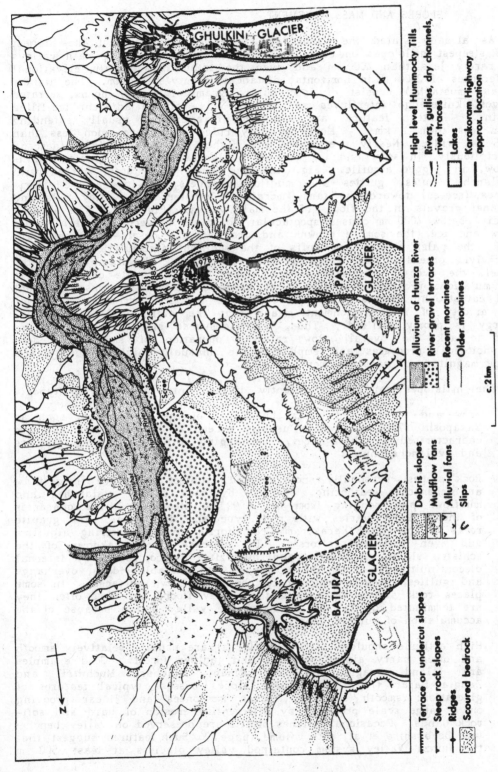

FIG. 11. GEOMORPHOLOGICAL MAP OF PART OF THE UPPER HUNZA VALLEY NEAR PASU TO SHOW THE DISTRIBUTION, EXTENT AND FORM OF MUDFLOW FANS.

Terrace or undercut slopes
Steep rock slopes
Ridges
Scoured bedrock

Debris slopes
Mudflow fans
Alluvial fans
Slips

High level Hummocky Tills
Rivers, gullies, dry channels, river traces
Lakes
Karakoram Highway approx. location

Alluvium of Hunza River
River-gravel terraces
Recent moraines
Older moraines

c. 2 km

GHULKIN GLACIER
PASU GLACIER
BATURA GLACIER

SLOPES AND MASS MOVEMENT

As already stated the Karakoram ranges contain some of the highest and steepest hillslopes on earth. The relative relief of the main valleys is rarely less than 4000 m and even the tributaries contain elevational differences of 2000 m in horizontal distances of only 1 - 2 km. The highest areas (<u>mustaghs</u>) consist of spectacular pyramidal peaks, horns, serrated ridges, knife-edge interfluves and abrupt cols between adjacent ice-filled basins. Plateau features are rare and fragmentary, usually extending over less than 1 km^2. More commonly ridge crests are much less than 0.25 km wide. Nearly everywhere the higher peaks and ridges descend precipitously, in snow and ice covered avalanche slopes to glacier basins below, in jagged aiguilles (Fig. 12) and bare rock faces scarred with avalanche chutes, gullies and mudflow channels, or in massive debris slopes (screes) toward the fans, moraines, low terraces, valley fills and channel gravels in the floors of the main valleys (Fig. 13). The abiding aridity below 4000 m causes spectacular contrasts between the high, white snow and ice, the sombre browns and greys of the desert-varnished bare rock, the paler browns and buffs of the sediment accumulation areas and the vivid greens of the irrigated cultivated land. The magnitude of the relief, the overall steepness of the slopes and the scale of the debris accumulations provide an overwhelming sense of instability, mass movement and catastrophic events, an impression that is only correct in part, for the aridity of the environment means that the high available potential energy is poorly utilized. Thus, the widespread signs of slope failure tend to mask the overall slow rate of operation of small-scale geomorphological processes and emphasise the unusually high frequency and large magnitude of mass movement activity.

The hillslopes can be conveniently divided into the following types:-

(a) Snow- and ice-covered avalanche slopes of the high peaks, such as Rakaposhi (7788 m), which usually exceed 40° in steepness, and are characterised by sheer faces, overhanging cornices, hanging glaciers and avalanche chutes.

(b) Rock slopes, seasonally snow covered, subject to severe freeze-thaw and chemical weathering, scarred by rock falls, avalanches and mudflows and ranging from 40° - 90° in steepness. The character of these slopes varies with rock type and dissection. The granite rocks of the Axial Karakor'am Granodiorite produce towering aiguilles and even rectilinear 'tors'. By contrast, the breakdown of the schists, slates, limestones and quartzites is controlled by internal discontinuities to give rise to jagged and unstable ridges, overhangs and gullies, beneath which occur boulder strewn surfaces. In some places rock slopes descend to the valley floor, but more often they are terminated up to 1000 m above the valley bottom because of the accumulation of surface debris.

(c) High altitude shoulders (2600 - 3600 m) are usually relatively smooth and often carry alpine pastures and juniper forest. Good examples are Jaland Buaghntum above Pisan, Miachar, Muchntril and Zangia Harar (Fig. 3). These slopes possess typical features of glaciation; smooth surfaces, roches moutonées and linear grooving on a large scale produced by differential erosion of hard and soft-rock bands. Occasional patches of till are preserved on valley benches (see Derbyshire et al, this volume, page 496. Such features suggest that the Hunza Valley system contained valley glaciers at least 500 m

FIG. 12. SERRATED ROCK PEAKS AND RIDGES (AIGUILLES) IN THE
UPPER HUNZA RIVER VALLEY, FROM PASU VILLAGE.
Steep scree 'fans' with mudflows descend from the slopes to the
Hunza river terraces and Batura glacier moraines.

FIG. 13. THE
VALLEY OF THE
HUNZA BELOW PASU.
Of particular note
is the enormous
volume of channel
sediment which
occurs in braided
river bar form.
These are probably
lake sediments behind
a natural landslide
dam.

FIG. 14.

a. THE NATURE OF HILL-
SLOPES WHICH GENERATE
MUDFLOW TALUS DEPOSITS,
A STEEP-ANGLE MUDFLOW
SCREE-FAN AND IN THE
FOREGROUND LEVEES OF A
LOW ANGLE MUDFLOW FAN.

b. DETAIL OF THE SUR-
FACE FORM, LEVEES,
COARSE CLAST DISTRIBU-
TION AND CHANNELS OF A
MUDFLOW FAN.

thick during some stage, or stages, of the later Pleistocene. Similar evidence exists down the Indus Valley as far as Shatial and indicates that former glaciers extended at least 100 km downstream from the confluence of the Gilgit and Indus rivers (27). Evidence of the extent of this glaciation can be seen clearly on the Landsat imagery, see page 363.

(d) Scree slopes are probably the most conspicuous and dramatic element of the Karakoram valley landscape (Figs. 5 and 14). They vary in form from simple small rockfall - scree cones to huge compound debris accumulations with slope lengths of up to 1000 m and relative heights of up to 500 - 600 m. Although typically conical in form, overlapping screes can extend 3 - 10 km along the valley sides. Slope angles usually range from 25° - 35°, sometimes increasing to 40°. Surface characteristics reflect both lithology and formative processes and range from smooth, via striped, to complex patterns of stripes, lobes, scallops and festoons; see Brunsden, Jones and Goudie, (this volume page 536(28)).

(e) Mudflow debris cones and fans occur where fed by discharges of sediment and water from narrow gullies and ravines in the steep rock slopes above. These represent an intermediate member of a continuous series of debris accumulation forms ranging from high-angle scree slopes to low-angle alluvial fans. They have extremely rough surfaces (Fig. 14) created by levéed gullies (mudflow tracks) and lobes of coarse debris, with generalized surface slopes in the range 7° - 20°. Mudflows are probably the most important sculpturing process operating on the lower debris slopes and can occasionally be of such enormous size and destructive force as to create lobes of sufficient volume to dam rivers (29). Mudflow fans occur throughout the length of the Hunza Valley, although particularly impressive examples are located in the Upper, near Pasu, (Fig. 11) and Lower Reaches.

(f) Large low-angled (2° - 7°) alluvial fans occur at a number of points where powerful meltwater streams debouch at locations where the main valley is broad enough to allow for their development (e.g. Batura, Pasu, Gulmit, Aliabad, Hindi, Chalt); Figures 11 and 15.

(g) Terrace fragments can be identified throughout the whole length of valley studied. In the Middle and Upper Reaches they are best developed in the narrow sections upstream from Baltit and Sarat. A number of levels can be clearly identified up to 50 m above the present river (Fig. 16), and the terraces appear to be in part lacustrine features produced by the blocking of the main valley by advancing tributary glaciers, although some of the Sarat examples appear to be the product of ponding by a very large landslide. Good examples of low terraces also occur in the Lower Reach below the gorge and appear to be fluvial in origin.

(h) Irregular topography is often developed on moraines in the Middle and Upper Reaches of the Hunza. Their morphology ranges from low irregular mounds to steep-sided knife-edged ridges up to 50 m high (see Figs. 11 and 15).

(i) Valley fills. Although the whole length of the Hunza Valley is floored by unconsolidated sedimentary deposits (d-h) above, particularly in the wider sections, the Middle Reach is unusual for it contains

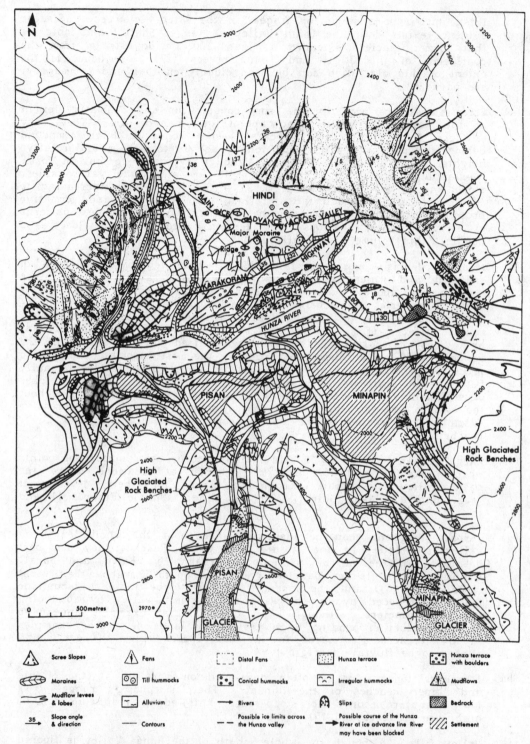

FIG. 15. GEOMORPHOLOGICAL MAP OF THE VALLEY OF THE HUNZA AT HINDI, INCLUDING THE HINDI EMBAYMENT AND MORAINE, THE VALLEY INFILL AND THE MORAINE SEQUENCES OF THE MINAPIN AND PISAN GLACIERS.

FIG. 16.   TERRACE LEVELS IN THE ROCK GORGE OF THE HUNZA
AT MUHAMMADABAD, 8 KM UPSTREAM FROM BALTIT.
The terraces appear to be lacustrine deposits produced by ice-damming
of the Hunza by advancing tributary glaciers.   The age is unknown but
probably refers to the last main glacial advance since they are older
than the oldest known man-made features of the area, including rock
carvings.

particularly thick sequences (up to 150 m) over virtually the whole
distance from Sikandarabad, in the west, to Baltit at the confluence
of the Hispar River (see Figs. 3 and 6).   The surfaces of these fills
usually slope gently (c. 7°) towards the centre of the valley, except
where overlain by young moraines (see Figs. 11 and 15) and are
deeply trenched by the Hunza River.   The sediments appear to consist
of complex interdigitated assortments of tills, outwash and matrix-
supported mudflow deposits, all overlying an irregular sub-drift
rockhead.   Why such thick fill sequences should exist in this section
remains unresolved.   One possible explanation is that bedrock lithological variations have controlled both valley width and the depth
of sub-glacial erosion, so that irregularities in plan are synchronous
with undulations in the rockhead thalweg.   As a consequence, thick
fills would have accumulated in 'basin' reaches and be separated
by engorged sections.   An alternative explanation is that sediment
accumulation occurred upstream of temporary dams created by the
repeated advance of steep-gradient tributary glaciers, including
the Minapin, Pisan, Gulmet and Hasanabad.   Irrespective of their
origin, the Hunza River has become deeply incised into these very
young sediments to produce unstable near-vertical cliffs, up to 150 m
high, which periodically fail in spectacular debris falls (Fig. 17).

FIG. 17. A RECORD OF THE SPECTACULAR ROCK FALL
OF AUGUST, 1981, IN THE UNCONSOLIDATED SEDIMENTS
OF THE HUNZA VALLEY AT MIACHAR, 4 KM FROM HINDI.

Evidence of mass movement is ubiquitous, as is to be expected in mountainous terrain, although activity is not uniform but varies in character and geomorphological significance with respect to elevation, lithology and local relative relief. On the lower ground in the valleys (<3000 m) slope movements are inhibited by the general aridity, but potentially unstable steep slopes abound. However, slope failure grows in both absolute and relative importance with altitude as precipitation and weathering activity increase. The most common types of failure are rockfalls (including toppling failures), debris falls and avalanches, and result in the widespread production of scree cones. Avalanching is the single most important mechanism at high elevations (>4500 m) but its influence also extends well downslope during late winter and spring to modify bedrock and scree slopes down to 2500 m. Rock falls (Fig. 18) and rock slides occur

FIG. 18. RECENT ROCKFALL IN THE HUNZA VALLEY NEAR CHALT WHICH IS VIEWED FROM THE KARAKORAM HIGHWAY.

throughout the attitudinal range, although they appear to be most significant between 2500 m and 4500 m, while debris falls are mainly concentrated below 2500 m in the zone of large scale valley sediment storage. Large falls tend to be produced by three mechanisms:

(1) the undercutting of fill slopes by rivers, especially the Hunza River; (see Fig. 17)

(2) pressure-release failure in over-steepened rockslopes following the retreat of ice-tongues, a particularly important mechanism in this area because of the rapid fluctuations in glacier size (see Goudie, Jones and Brunsden, this volume, page 411, (30), and

(3)   instability caused by earthquake shocks.

Small failures, down to individual blocks, are widespread and common, and account for the great thickness of glacier surface moraines and the massive development of talus slopes, many of which are dangerous to traverse because of falling debris. The scale of activity is explicable in terms of the generally very steep slopes, the fractured nature of the bedrock and the impact of salt weathering and freeze-thaw activity, processes that are extremely potent when acting alone but virulent when operating in combination. As a consequence, the rock slopes are generally shattered and unstable, to the extent that small showers of rain and gusts of wind can result in dislodgement, the latter often disturbing the surface of screes below and contributing to duststorm generation.

The other important mass movement mechanism is debris flow, where surface debris becomes mobilised by rainfall or snowmelt to produce waves of slurry which pass down gullies and across fans to create the characteristic levee-flanked channel and terminal lobe fans (see Fig. 14 and 15) (29). Evidence for such activity is widespread and includes,

(1)   striped scree slopes;

(2)   mudflow fans veined by mudflow tracks;

(3)   sediment infillings in gullies;

(4)   large depositional lobes in valley bottoms; and

(5)   extensive accumulations of matrix-supported sediments within the Hunza Valley infills.

Field observations indicate that small storms can generate large numbers of flows, particularly if preceded by a lengthy dry spell during which loose debris accumulates in the gullies. Larger, catastrophic mudflows can be created either by intense storms or by sudden discharges of glacial meltwater. Examples of the former include the large mudflow which blocked the Hunza River at Batura on August 14 1974, following a rainfall of 10.3 mm in two days, and a similar occurrence at Shishkat (Vol1 pg 76) that resulted in the destruction of the newly completed 8-span Friendship Bridge (Table 4; see later under Channel Processes and Forms). The latter mechanism is more unusual but equally destructive, the best example being the catastrophic mudflow at nearby Gupis on the Ghizar River. Here pulses of meltwater from a high debris-covered glacier relic, or rock glacier, generated a series of lobes over a 36-hour period spanning 26 - 27 July 1980. These descended to the main valley floor at speeds of up to 60 km/hr, producing a depositional lobe up to 30 m thick which dammed the river to create a lake 3.5 km long which did not overtop the dam until 12.00 p.m. on 27 July.

The significance of mudflow activity reflects the sporadic nature of rainfall, the generally low infiltration rates and the ample supply of loose surface debris. Mudflows have small clay (3 - 10%) and low water contents, usually 10 - 30%, and therefore can form under relatively arid conditions, whereas mudslides usually do not; cf. the wetter Nepal Himalayas (31). The magnitude and frequency of occurrence undoubtedly varies over space and time in response to lithological, geomorphological and climatic controls. Mudflow activity is probably at a maximum following widespread deglaciation, as is testified by the important contribution of mudflow deposits in the thick valley fills. Under present conditions, mudflows

Type	Recurrence in years	Recovery in years	Status
1: Glacial Landforms of major glaciations			
(a) Valley infills	1:30-50,000	10,000	0.3-0.2 Characteristic
(b) Moraines	1:30-50,000	10,000	0.3-0.2 Characteristic
(c) Overall glacial valley forms	1:30-50,000	10,000	0.3-0.2 Characteristic
(d) Detailed glacial slope forms	1:30-50,000	5,000	0.2-0.1 Characteristic
2: Minor, eg, 20, glacial advances			
(a) Moraines	$1:10 - 10^3$	$10 - 10^3$	1.0 Transient
(b) Detailed slope forms	$1:10 - 10^3$	$10^2 - 10^3$	10-1 Transient
3: Mass Movement			
(a) Large deep seated failures	1:50 (in valley as a whole)	1000	20 Transient
(b) Large Mudflows			
(i) Track	1:5-10	$10^2 - 10^3$	10-100 Transient
(ii) Lake	1:5-10	10^3	200-100 Transient
(c) Small debris flows	1:1-10	10^2	100-10 Transient
(d) Small rock falls	1:1	$10 - 10^2$	10-100 Transient

Thus hillslope events are seen to occur more frequently than the landscape can recover and the slope forms are therefore in a state of transience, typical of frequent disturbance and continuous adjustment. It is possible for the evidence of major glaciations to be almost entirely removed before the next advance and for the hillslopes to be converted into forms typical of frequent mass movement events or to stable forms characteristic of the environment. The Hunza slopes are, at present, still recovering from (or displaying) the effects of the last major advance but the evidence is not everywhere very strong and some detailed slope forms are now recovered from the effect of glaciation.

TABLE 4: ESTIMATES OF RECURRENCE AND RECOVERY TIMES OF LANDFORMING EVENTS.

continue to be an important process on the lower slopes (< 3000 m), particularly where the natural surface has been disturbed by Man. Nevertheless, their scale and overall significance appear to have declined since the last glacial maximum. While large, catastrophic, flows still occur in the Hunza Valley, with a probable frequency in the valley as a whole of 1 in 5 - 10 years, it is the ubiquitous small flows that are most conspicuous, although their overall contribution to the downslope movement of debris (in absolute terms) may be less than subjective assessments based on visual inspection suggest. This is because the recovery times of surface modification and desert varnish development following an event are larger than the frequency of occurrence. Flow tracks may continue to appear fresh for relatively lengthy periods. For example, very fresh tracks examined on a fan near Hindi in September 1980 were clearly visible on an air-photograph taken in 1965.

Apart from a variety of movements on scree slopes, such as shallow sliding, grain flows and shallow slumping, the only other types of mass movement noted were deep seated failures. These are poorly developed because of the aridity, and only two, albeit extremely large, examples were identified in the main valley, at Hakuchar and Sarat. In both cases undercutting by the river had been partly responsible for failure. In the first instance water infiltrating from irrigation leets has resulted in successive slumping of fill materials, while at Sarat complex and large-scale failures have continued on both sides of the valley following a major movement in 1858; this was a bedding plane slip where the rock dips steeply into the river.

Finally, four important general aspects of slope evolution should be emphasised:

There is abundant evidence that the Hunza Valley has been filled with an enormous glacier system which extended deep into the Indus Gorge to the south of Gilgit (27). Thus the lower rock slopes have been profoundly influenced by the history of glaciation.

The repeated rapid advances of the large, steep gradient, tributary glaciers to the main valley floor have considerably influenced slope form by;

(1)   damming the Hunza River to cause sedimentation,

(2)   deflecting the Hunza River against the opposite valley side, e.g., the Hindi embayment; see Fig. 15

(3)   creating complex irregular terrains of moraines up to 60 m high with slopes of as much as 55 - 56°, which in any other landscape would be regarded as major relief features.

These glacial reliefs have suffered "post-glacial" modifications by weathering, fluvial and mass movement activity. Enormous accumulations of unconsolidated sediment have been formed as slope and valley fills. As these depositional features are very young, it is clear that modification must have been extremely rapid, particularly during deglaciation, with slope movements clearly supplying sediment more rapidly than the Hunza River could remove it. The continuing scale of glacial activity and pace of mass movement clearly go a long way towards explaining the high sediment loads of the present rivers.

High magnitude, relatively low frequency geomorphological events are important in both short and medium term landscape evolution and should be regarded as the current formative events. Little of the landscape is, however, of any great antiquity. While the overall pattern of relief has essentially remained the same for at least 10^6 years, the effects of tectonic instability, glaciation and mass movement have ensured that the extant landforms and deposits are essentially very young, probably ranging up to 10^4 years at most, with the possible exception of perched fill remnants on high shoulders and cols. In such a dynamic environment, the large-scale, or catastrophic, events (mudflow, glacial surge, earthquake, flood surge, etc.) are extremely important in topographic sculpturing. However, although these are the formative events, the impact of an individual event may persist in the landscape for as long as in other terrains, for the Karakoram slopes can be characterised by a suite of forms typical of frequent disturbance, continuous adjustment and rapid transmission times of debris to the valley stores beneath (Tables 4 and 5).

Location	Hazard	Date or frequency	Source
1. LANDSLIDES			
a) Phungurh, Ghammesar	Landslide and lake	1858	1
b) Nanga Parbat	Rockfall	?	1
c) 80 miles N. of Gilgit	Landslide	1 : 2 months	1
d) 45 miles N. of Gilgit	Landslide, Rockfall	Winter thaw	1
e) Hakuchar	Landslide	1950's	4
2. MUDFLOWS			
a) Batura	Mudflow blocked Hunza	1972 and 1975	3
b) Batura	Mudflow and dam	1974	3
c) Shishkat	Mudflow destroyed 134' Friendship Bridge	1976	3
d) Gupis	Mudflow dams Ghizar River	1980	4
e) Nilt-Aliabad	Mudflow closed Karakoram highway 9 times in 1980	1980	4
3. GLACIER MOVEMENTS CAUSING DAMS, LAKES AND FLOODING			
a) Minapin Glacier	Rapid advance	1889-1892	2
b) Yangutsa	Surge	1901	2
c) Hasanabad	Rapid advance	1903	2
d) Shimshal valley glaciers - various advances, dams and floods. Includes Yangil or Malangutti glacier	Rapid advances etc.	1884 1893 1905 1906 1927 1928 1959	2
e) Batura Glacier	Rapid advance	c. 1770 1925? (1885-1930) 1974	3 3 3
4. FLOODS			
a) Batura Glacier and Hunza	30 m bridge damaged by meltwater	1972	3
b) Batura Glacier and Hunza	8 m bridge damaged	1973	3
c) Rapid river migration at Batura outlet	30 m flood	1973	3
d) Hunza	10.3 mm rain in 2 days	1975	3
e) Hunza Valley	Flooding caused by frozen culverts	1978	1
f) Ghulkin Glacier	Rapid change in outwash location destroyed Karakoram Highway	1980	4
5. EARTHQUAKE			
a) Patan Earthquake 140 km S. W. Gilgit	2000 dead in Kohistan	28.12.74	1

References: 1. Far East Ec. Rev. 14.10.79   4. Karakoram Project, 1980 (these proceedings)
2. Mason, Rec. Geol. Surv, India. 1930. (34)   5. Finsterwalder, 1959 (57)
3. Batura Investigation Team, 1976 (49)

TABLE 5: A SELECTION OF NATURAL HAZARDS CAUSED BY HIGH MAGNITUDE EVENTS IN THE HUNZA AND KARAKORAM MOUNTAINS.

WEATHERING

As will be indicated later, analyses of river-water chemistry indicate that in the catchment of the Hunza, the rate of chemical denudation is of the order of $70\,t\,km^{-2}yr^{-1}$, approximately two times the world average. While some of this may be explained by the existence of some relatively soluble dolomites and limestones within its watershed, this can only provide a partial explanation. It is plain that chemical weathering is an active process in spite of the low precipitation levels over the lower parts of the basin and the low air temperatures over the higher parts of the basin. As Whalley et al (32) point out, rocks show signs of chemically caused disintegration and weathering rind formation even at above 4000 m, while below that altitude desert varnish becomes more common, especially below 3000 m.

One possible explanation for this high rate of chemical denudation and the extensive effects of chemical weathering is provided by the temperature observations undertaken by Whalley et al (32). They found that rock surface temperature could reach high levels even at high altitudes. They concluded: "Rock surface temperatures can reach $10^{\circ}C$ or more above air temperatures even at 4000 m and above. Such surface temperatures may easily reach $30^{\circ}C$ under sunny conditions and $20^{\circ}C$ under cloud. When coupled with moisture, either from cloud or precipitation, conditions very favourable to chemical weathering of rocks may be achieved in the summer".

However, the cause of rock breakdown is not restricted to chemical processes. Three main mechanical processes also play a role: salt, frost, and unloading. Many rock surfaces are covered by a salt efflorescence and where this is so, rock disintegration is evident. The older glacial tills on the north side of the Hindi embayment, the post-1920 tills and rock fall debris of the Hasanabad valley, and the post-1954 till of the Minapin valley, are all extensively broken down, as are some farm walls. X-ray diffraction studies of the efflorescences (33) indicate that they consist predominantly of hexahydrate ($MgSO_4.6H_2O$) and gypsum ($Ca_2SO_4.2H_2O$).

Frost may be effective in the area, even affecting the floor of the Hunza Valley in the winter months (Fig. 9). Its effects in cracking rock and producing some of the coarse angular debris that makes up the well developed scree slopes of the area, are assisted by the fact that glacial excavation of major troughs has led to the development of dilatation joints along major valleys as a consequence of unloading.

THE GLACIERS: PROCESSES AND FORMS

The Karakoram contains some of the longest glaciers outside the polar regions (Table 6). There are more than 100 of them that are 10 km or more long. All the glaciers investigated are of alpine type in which maxima both of snowfall and ablation occur in the summer half year. However, this simple statement conceals considerable diversity. The earliest classification of the Karakoram glaciers was based on morphology (34), in which longitudinal glaciers, controlled by the roughly east-west structural trend, constitute the dominant group and include the longest glaciers, such as the Hispar, while the shorter, transverse glaciers make up the second group. An alternative classification (35) cf. (18) was based on the morphology of the accumulation basin of the glaciers, and consisted of basin reservoir, incised reservoir, and avalanche types, the latter

Glacier	Approximate lengths (km)	Approximate elevation of snout (m)
Hispar	61	3000
Batura	59	2448
Barpu (Hopar)	25	2561
Pasu	24	2550
Ghulkin	18	2534
Hasanabad	17	2223
Ghutulgi Yaz	15	3323
Minapin	13	2150-2350
Pisan	11	2070-2420/2450
Ghulmet	7	3050

TABLE 6: SOME GLACIERS OF THE HUNZA VALLEY AND TRIBUTARIES

lacking tributary reservoirs. The incised reservoir and avalanche types are dominant in the Karakoram, with narrow, high gradient accumulation zones.

The Karakoram alpine glaciers are amongst the steepest in the world and their termini are the lowest in the Himalaya (Table 6). They extend through a wide range of climatic environments and so are glaciologically complex. The high source areas (sometimes extending above 7000 m) have a negative mean annual temperature with perennially frozen ground, but the glacier snouts reach into subtropical desert conditions. For example, the mean annual temperature near the terminus of the Batura Glacier is about $9\,^{\circ}C$ with a summer diurnal maximum of over $34\,^{\circ}C$ and a minimum of $-10\,^{\circ}C$ (36).

The mean annual precipitation near the terminus is less than 100 mm with a summer daily maximum of 15 mm. Equilibrium lines in the region lie in the range 4800 to 5400 m (37), where mean annual temperatures are negative, the mean air temperature at the equilibrium line of the Batura Glacier being $-5\,^{\circ}C$ (38), rising to $0\,^{\circ}C$ as low as 4200 m above sea-level.

On the basis of detailed measurements made on the Batura Glacier, ice in accumulation zones above about 4500 m is below $-10\,^{\circ}C$ and so is of cold metamorphic type (39), ice forming predominantly by infiltration and congelation is indicated by well developed depth hoar. Negative ice temperatures prevail in the seasonally variable upper ice layer as low as 3000 m on the Batura Glacier, reaching $0\,^{\circ}C$ only 20 m above the snout; 2650 m (37). However, many of the glaciers are of considerable thickness, with the Minapin and Hasanabad glaciers being rather thinner. The substantial accumulation and ablation totals result in generally rather high activity glaciers. A figure for total annual ablation on bare ice of 1841 mm/yr has been given for the snout of the Batura Glacier, the ablation season lasting 315 days (40). The characteristic cover of coarse rockfall debris on the lower parts of many ice tongues acts as an effective insulator. It is known that only 3 mm of debris cover reduces the annual ablation figure by 25%. Thus the thicker (> 1m) mantles, such as that covering the lowermost 3 km of the Batura ice tongue, reduce it by over 75% (37). In terms of the heat balance, radiation is by far the most important ablation source, accounting for between 80% and 90% of the total, with turbulent eddy transfer and latent heat of condensation accounting for the rest. The maximum radiation balance reached over $27.9\,MWm^{-2}$ per day on the Batura Glacier in August 1974 (40). Over 80% of the total heat loss is due to melting with less than 20% due to evaporation and convection. It follows that high summer radiation levels, combined with steep, unvegetated slopes, constitute a major control of glacier ablation patterns and resultant ice morphology in the ablation zone. They result in the well developed 'ablation valleys' between lateral moraines and glacier margins remarked upon by Mason and many other early workers. Well developed ice pinnacles, typical of clean glacier surfaces under high radiation ablation, are seen on many glaciers including the Minapin, Ghulmet, Pasu and Batura.

Given their thickness and relatively high activity indices, these glaciers have average flow rates of between 100 and 1000 m/yr, a measured figure for the Batura, just below the equilibrium line, being 517 m/yr; Table 7 (14). Their great thicknesses also imply that, over a majority of the ice tongues, their deeper parts are at, or close to $0\,^{\circ}C$, and so behave

General	Hewitt (1969)	240–460
Baltoro	de Filippi (1910)	299–612
Batura	Shi and Wang (1980)	1000
Batura	Batura Glacier Investigation Group (1976)	517.5 37–115
Biafo	Suizu (1977)	187
Chogo Lungma	Workman (1905)	296–366
Ghulkin	Pillewizer (1957)	117
Hasanabad	Pillewizer (1957)	329–449
Kutiah	Caputo (1964)	80–216
Minapin	Schneider (1969)	350–645
Pasu	Pillewizer (1957)	157

TABLE 7: THE VELOCITIES OF SELECTED HUNZA GLACIERS (m/year)

like temperate glaciers. Although supraglacial water can be seen ponded in summer on some glacier surfaces (e.g., the Pasu Glacier and parts of the upper Ghulkin Glacier below the ice fall), the bulk of the water in the glacier system is englacial and subglacial and exits by means of major subglacial tunnels in the case of the Batura, Pasu, Ghulkin, Gulmit, Hasanabad, Minapin and Pisan glaciers. Migration of these major subglacial rivers, as a result of thermo-erosion, ice collapse and subglacial debris removal, results in shifts in the proglacial stream. The outlet river of the Batura Glacier shifted a distance of 1 km in 1973 from the northern to the southern edge of the terminus (41). A smaller but equally sudden shift occurred at the snout of the Ghulkin Glacier in July 1980. Both ice collapse and debris removal are indicated in this case, the line of the subglacial tunnel being expressed on the surface of the ice upstream of the terminal area by a sinuous depression.

In the late nineteenth and early twentieth centuries, the glaciers in the Karakoram were generally advancing, followed by predominant retreat in the period 1930 - 1960. Since that time, both advance and retreat have been recorded in different parts of the Hunza Karakoram (see Goudie, Jones and Brunsden, this volume (30)). In their study of glacier behaviour since 1812 A.D., Mayewski et al (42) found that advancing glaciers flow most commonly eastward, southeastward, northeastward and northward and that advance is rare in glaciers flowing south and west. The glaciers considered here flow between east and north (Batura, Pasu, Ghulkin, Ghulmit), northward (Minapin and Pisan) and between east and south (Hasanabad). These authors tentatively suggest a positive correlation between glacier advance and a more vigorous monsoonal flow, assuming a precipitation-glacier advance response time of between 2.5 and 25 years. Their correlation is shown graphically in Figure 19. This throughput time may be too short for the larger glaciers, the lag between cold decades (as determined by tree-ring width) and advances of the Batura Glacier having been estimated at 50 - 70 years by Shi and Liu (36). Zhang (43) found that the general curve for Karakoram glacier fluctuations in the twentieth century is similar to that of the European Alps (Fig. 20) although with longer response times. Mason in 1914 (44) reported an advance of the Hasanabad Glacier of about 100 feet (30 m) per day during part of 1890, rates approaching those of one of the best known surging glaciers; 50 - 100 m/day (Medvezhiy Glacier, USSR: (45)). A number of glaciers have since been regarded as surging in the Karakoram region (46) and, of these, the Hasanabad Glacier is the best documented (see Goudie, Jones and Brunsden, this volume (30)). Zhang (43) suggests that the cyclic response time of the Minapin and Hasanabad glaciers (~ 20 - 30 years, latest rapid advance 75 years) is much greater than in true surging glaciers elsewhere in the world, although it should be noted that the records of this phenomenon are relatively short lived; cf. (47) page 284.

Kinematic waves have been recognised on the Batura Glacier, but its recent behaviour has not been reported to include surging (14). Rapid advances also characterise the Ghulkin Glacier, but the rates (e.g., 420 m between 1885 and 1913; or 15 m/yr: (43)) do not approach the values for surging glaciers (Fig. 20). We conclude that although periodic rapid advances have occurred in some of the glaciers, true surging behaviour has not been clearly demonstrated. It should, however, be noted that remnants of lateral moraine with gradients noticeably steeper than those of the present glacier surfaces, occur adjacent to rock-steps along both the Ghulkin and the Pasu glaciers, so that past behaviour of surging type cannot be ruled out. The concave lower reach of the Ghulkin, sunk 20 - 30 m below fresh lateral moraines, matches the appearance of some post-surge ice tongues. Trimlines and stranded avalanche cones of talus

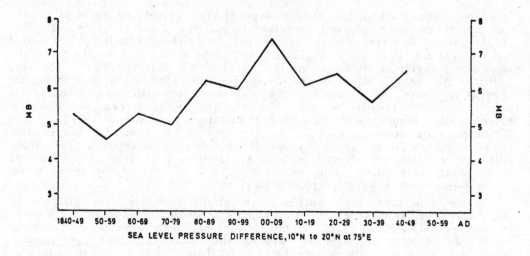

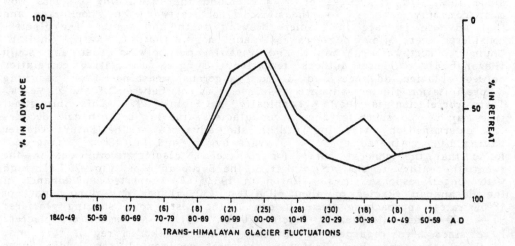

FIG. 19. SUGGESTED CORRELATION BETWEEN THE SEA LEVEL
PRESSURE DIFFERENCE 10°N – 20°N AT 75°E (AND THEREFORE
THE VIGOUR OF THE MONSOONAL FLOW) AND TRANS-HIMALAYA
GLACIER FLUCTUATIONS 1850 – 1960; (FROM (42)).

and ice hang above the upper reaches of the Ghulkin near the ice fall,
indicating rapid thinning of the order of 30 m or more in the fairly recent
past (Fig. 21).   At the same time, the convoluted medial moraines so
characteristic of surging glaciers in Alaska and elsewhere are unknown
in the Hunza Karakoram, perhaps because of the importance of block flow
in these glaciers.

The combination of a core of isothermal ice yielding high sliding
velocities, abundant englacial and subglacial meltwater, and very steep
longitudinal gradients results in some of the highest glacial erosion rates
on earth.   Ice thickness measurements show that all the major glaciers
in this region are deeply entrenched in rock troughs, and the high sliding
rates have produced polished and moutonnéed bedrock surfaces where recent
retreat has been marked, e.g., lower Pasu Glacier.   The efficacy of sub-

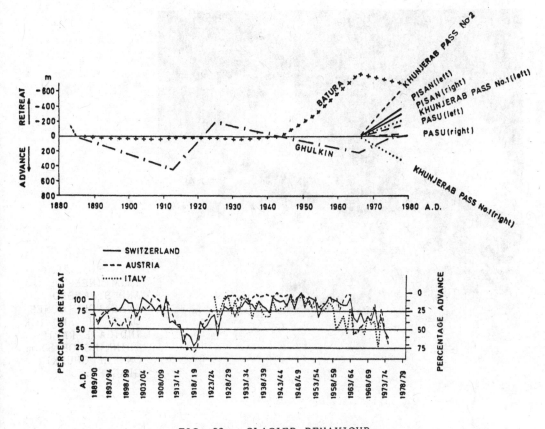

FIG. 20.   GLACIER BEHAVIOUR.

(a)   Advance and retreat of the Batura and Ghulkin Glaciers
      (1800 - 1980) and some data for other glaciers since 1974
      (Zhang, (43)).

(b)   Glacier termini behaviour in the European Alps for com-
      parison with the Karakoram data.

aerial weathering and the maintenance of basal ice at the pressure melting
point in the lower reaches combine to produce a predominantly supraglacial
and englacial debris load which contains much unmodified talus and is
widely modified by meltwater action.   Lodgement till is thus thin and
localized around margins of the Hunza glaciers, but basal sliding is
demonstrated by well developed fluted moraine and small push moraines
in the foreland of the Pasu Glacier between the 1925 and the 1980 ice
fronts.   Moraines are a composite of meltout, slide and push processes,
and the larger end-moraines have a characteristic cone form as a result,
radially incised by old meltwater channels floored by steep (15 - 18°)
straths of coarse torrential gravels.   This end-moraine morphology is a
long-standing one, for the late Pleistocene moraines of the Ghulmit Glacier
have this form.

Evidence demonstrating that the Karakoram glaciers have relatively
recently advanced across the Hunza Valley and diverted and ponded the
main drainage, includes abundant outcrops of finely-bedded lacustrine

FIG. 21.
TRIMLINES
AND STRANDED
AVALANCHE CONES
OF TALUS AND ICE
INDICATE RAPID
THINNING OF 80 m
ON THE GHULKIN
GLACIER.

silts well above river level, moraine remnants, and also the recollections of villagers, such as those at Pasu, recorded in the early scientific literature (44). The embayment at Hindi provides a fine example of this complex interaction of glaciers and river preserved in the geomorphology and the sedimentary record (Fig. 15). An expanded ice-foot made up of the extended Minapin and Pisan glaciers obviously maintained a position over 1 km to the north of the present Hunza River for some time as indicated by a series of substantial arcuate end moraines. The river must have been diverted around this ice tongue, for channel-fill sediments are preserved in cliff sections above the present river at the western end of the embayment. A series of retreat positions can be inferred from end-moraine remnants but slope processes rapidly filled in the depressions between the moraines, so that only the larger now pierce the fill as inliers. This mudflow/debris flow surface is preserved on both sides of the Hunza River and underlies Minapin village. With the establishment of the present channel of the Hunza, which is entrenched c. 150 m into the valley fill, this surface is, in its turn, being dissected by gully extension and suffering encroachment by high angle debris fans along the mountain foot.

In conclusion then, the processes and forms are those to be expected from high activity, high gradient, low latitude valley glaciers, namely deep erosional incision, rapid modification of deglaciated surfaces, well developed moutonnée, fluted tills, large lateral and end-moraines made up of abundant granular tills of meltout and slide origin with coarse glaciofluvial gravels, and only localized thin lodgement tills.

HYDROLOGY

The Karakoram Mountains lie almost entirely within the headwaters of the Indus river and contribute about 25% of its total flow from about 15% of the catchment area. Like the rivers of Kashmir and the Punjab that provide most of the rest of the flow of the Indus, the Karakoram headwaters have a strongly seasonal regime with a summer peak, but one that is due to snow and ice melt rather than monsoon rainfall. The seasonal, short-term and year to year fluctuations of river flow in the Karakoram all appear distinctively alpine in character.

The water balance equation

Runoff = Precipitation - Evapotranspiration - Storage gain

cannot yet be quantified for Karakoram river basins. The Pakistan Water and Power Development Authority, WAPDA, maintains three gauging stations near Gilgit in the western Karakoram and one near Skardu in the east. River discharges are thus known to fair accuracy but monthly and annual runoff depths calculated from these data (Fig. 22) represent outputs averaged over large and heterogeneous areas within which the spatial distribution of precipitation is unknown. WAPDA rainfall data for Gilgit and Karimabad and Chinese measurements below the snout of the Batura glacier (49) show annual totals of only about 100 mm in the valleys which are climatically arid in summer with potential evaporation averaging over 10 mm d^{-1} (WAPDA data for Skardu). Since annual runoff past these sites averages 300 - 1000 mm, precipitation must increase rapidly with elevation (50). This conclusion is supported by mountaineers' reports of the frequency and depth of summer snowfalls and more quantitatively by two observations by the Chinese Batura Investigation Group (49)(51):

1 More than twice as much precipitation fell at 3400 m above sea-level by the Batura Glacier in the summer of 1974 than at base camp 900 m lower down.

2 Annual layers in crevasses near the equilibrium line of the glacier indicate net accumulation of 1030 - 1250 mm water equivalent.

Beyond this, neither the vertical variation in accumulation (and ablation) nor any regional trends are known for, regrettably, there have been no detailed mass balance studies of Karakoram glaciers and no systematic

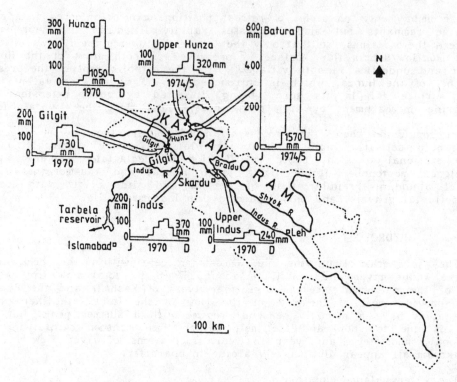

FIG. 22.   SEASONAL STREAMFLOW REGIME AND ANNUAL RUNOFF IN THE
KARAKORAM MOUNTAINS AND ADJACENT PARTS OF THE INDUS BASIN;
DATA FROM (48)(49).

snow sampling for oxygen or hydrogen isotope analysis.   In the absence
of mass balance measurements the storage term in the water balance here
representing glacier growth or shrinkage is also unquantified.   Its possible
significance is discussed later.

Figure 22 shows total annual runoff and its monthly distribution at
four gauging stations for 1970, a year of approximately average runoff
(48) and for the Batura and Upper Hunza rivers in 1974/75 (49).   Discharge
begins to rise in March at the lower stations as snowmelt begins but not
until May in the upper Hunza and upper Indus.   Between 40% and 70%
of the total runoff is in July and August, when discharges are 15 to 40
times those in March.   The timing of this peak suggests an icemelt origin
and observation confirms that all significant tributaries of the Hunza river
are fed by glaciers.   Spatial differences in average runoff appear to
correlate with differences in the percentage of each catchment covered
by permanent snow and ice.   For example, the discharge of the Upper
Hunza river draining 5000 km^2 on the relatively low mountains north of
the main Karakoram chain and without major glaciers is doubled by the
confluence of the short torrent draining the 300 km^2 Batura glacier and
its surrounding peaks.   From here to Gilgit the Hunza river is swollen
by meltwater from a succession of major glacier systems to give a mean
runoff for the whole 13,000 km^2 catchment of about 900 mm yr^{-1}. The similar sized, but less heavily glacierised basin of the Gilgit river to the
west, produces 20% less streamflow.   Further east, the Upper Indus drains
four times the area of the combined Gilgit and Hunza rivers but has only

about 30% more discharge because the undoubtedly high runoff from the great glaciers of the central and eastern Karakoram in the Braldu and upper Shyok catchments is offset by very low runoff in Ladakh and the Tibetan plateau.

Streamflow also varies considerably from year to year. For example, runoff from the entire Karakoram region (upper Indus catchment plus Gilgit and Hunza rivers) was 390 mm in 1970 but 540 mm in 1973. Differences in the annual flow of the Indus further downstream have been related to winter snow cover as measured on satellite images (52) but this can only be a partial explanation. Runoff from the sub-alpine zone south of the Karakoram proper has a late spring snowmelt peak that ought to be bigger after an unusually snowy winter, but glacier-fed rivers normally have smaller discharges in the summer following the snowy winter because the lower and more slowly retreating snow line means a smaller ablation area (53). Annual variations in runoff from the Karakoram proper are primarily due to differences in July and August streamflow (e.g., Batura catchment: 840 mm in 1974, 1230 mm in 1975) and these must represent changes in the storage term of the water balance i.e. fluctuations of glacier mass balance. These may well depend less on winter snowfall than on weather conditions and ablation rates in the summer. The Chinese Batura Investigation Group of 1976 (49) found a close correlation between discharge in the Batura river and air temperature and even the much bigger Hunza river near Gilgit exhibits characteristically alpine fluctuations of flow

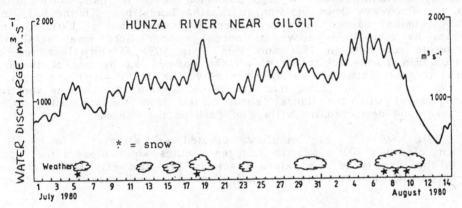

FIG. 23. DIURNAL DISCHARGE CYCLE AND WEATHER-RELATED RECESSIONS OF THE HUNZA RIVER NEAR GILGIT, SUMMER 1980 BASED ON UNPUBLISHED WAPDA DATA AND IKP OBSERVATIONS.

in summer (Fig. 23). A clear but considerably lagged diurnal cycle is interrupted by sharp recessions when snowfall or prolonged cloud cover halts glacier ablation. A sunny summer thus gives high runoff at the expense of glacier storage. Year to year fluctuations of this kind may well apply over sufficiently extensive areas to affect regional runoff. The general 20th century movements of Karakoram glaciers (54)(30) must also have affected runoff but no attempt has been made to quantify this.

These major seasonal variations in the discharge of the Hunza are not the only variations to be considered in any treatment of the hydrology of the area for from time to time, the river's flow is affected by the creation of major natural dams as a consequence of either glacial, mudflow or landslide blocking (55).

In the 1850's for example, following the Phungurh landslide near Sarat (see above and Fig. 13), a catastrophic flood-wave came down the Indus, raising the river level at Attock by some 9 m in less than 10 hours (55). The discharge may well have been over 1 million $m^3.s^{-1}$. Belcher (56) wrote (p. 223):

> "All accounts agree that the late stoppage was of the river of Hoonza, about a day's journey above the fort and town of that name, and four or five days' journey northward of Gilgit. It was caused by the subsidence of a mountain side called "Phungurh" on the left or Nuggur bank, from the action of rain and snow above, and of the stream below, in the winter of 1858. The vast fragments of rock, and earth, and trees and drift wood, dammed up the narrow bed of the river for six months; when the waters swollen from the melting of the snow in the mountains burst a sudden passage in August, and destroyed many forests and villages and lands of the Hoonza and Gilgit districts."

Advancing glaciers have also created dams and subsequent floods from time to time (Table 5). In 1884 an ice-dam burst in the Shimshal valley, a northern tributary of the Hunza and led to a 3 m rise in river level that caused considerable devastation at Ganesh and Baltit. This was followed by a similar event in 1893 and then again in 1905. The latter sent a 9 m floodwave down the Hunza, causing landslips. In the following year the Shimshal caused an even bigger flood than that of 1905, raising the Hunza by over 15 m above its normal summer flood level at Chalt. Other events occurred in 1927 and 1928. In 1959 Finsterwalder records (57) a sudden burst took place of a lake dammed up by one of the big glaciers in the Shimshal Valley (probably the Mulungutti or Yazgil). The flood caused by the burst had a depth of around 30 m at the junction of the Shimshal with the Hunza, about 40 km from the assumed position of the lake, and destroyed the village of Pasu on the Hunza.

In 1974 (49) a big mudflow created an alluvial fan with dimensions of 300 x 200 x 4 m in just ten minutes which blocked the Hunza river in the vicinity of the Batura Glacier. The dam only lasted for one hour but must have created a small floodwave and increased sediment yield.

CHANNEL PROCESSES AND FORMS

The rivers of the Karakoram convey huge quantities of meltwater and suspended sediment from glacier and valley-side source areas. That they also erode their beds is testified by the unsurpassed height difference between mountain tops and valley floors, and the very existence of rivers crossing the axis of the Karakoram despite rapid and continuing tectonic uplift.

Spectacular rock gorges exist where the rivers have incised across the generally WNW-ESE trend of the more resistant lithologies. Examples in the western Karakoram include the lower Shimshal; the Hunza below Shishkat, above Minapin, and below Chalt; and the Indus above the Gilgit confluence. Nevertheless the Hunza river, at least, is rock-walled over comparatively short distances only (Fig. 24). A much greater length is confined by high alluvial/colluvial terraces and valley fills, presumably

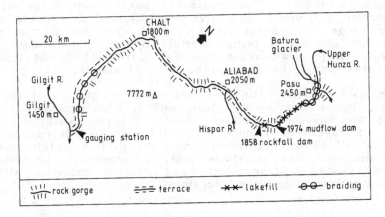

FIG. 24. CHANNEL TYPES OF THE LOWER HUNZA RIVER.

created by tectonically-induced incision, to which Schneider (6) ascribes a late Quaternary age; and in places the river is not closely confined at all.

Some indication of the hydraulic properties of the Hunza can be derived from stream-gauging measurements (WAPDA, unpublished) from a terrace-confined section near Gilgit. The bed material is coarse, giving a high friction factor that tends to reduce water velocity but this is offset by the steep slope (about 0.006 here and averaging 0.008 for the 130 km from Batura to Gilgit, according to an altimeter traverse), high peak discharges (over 2000 m^3s^{-1}), and relatively narrow deep channel (summer width c. 100 m, mean depth 4 – 5 m). At peak flow the mean velocity reaches 5 m^3s^{-1} with a bed shear stress of about 250 Nm^{-2} which, according to Shields' function, is sufficient to entrain boulders 250 mm in diameter. Since these conditions obtain for many days each year there is no difficulty in explaining the existence of a coarse alluvial bed even in rock-walled reaches of the river.

High energy channels are generally braided, but along the Hunza extensive braiding is restricted to the 15 km of relatively wide valley of the Upper Reach, from the confluence of the Batura past the Shimshal and Pasu (Fig. 24). The other relatively unconfined stretch, above the Gilgit confluence (Lower Reach), has a gently sinuous pattern with longi-tudinal bars along about half the length of the river. The rock gorges of the lower Shimshal, lower Hispar, and Hunza below Shishkat also have braided-channel alluvial floors. Elsewhere channels are single, apart from minor bars and rock islands, and either straight or gently meandering-as would be expected from Parker's pattern criterion (58) in view of the high Froude number and low width/depth ratio of these coarse-bedload channels.

Several minor channel changes had occurred by 1980 compared to 1965 air photographs, including one neck cutoff of a confined meander. The major change has been the creation in 1974, and subsequent filling, of a temporary lake some 12 km long behind the huge debris flow that dammed the Hunza near Shishkat and destroyed the newly-completed Friendship road bridge. The dam has been degraded to bouldery rapids dropping 20 m in less than 1 km, and the lake has already largely been filled by a sediment wedge with a mean surface slope declining downstream from

0.003 to less than 0.001. The proximal deposits are of gravelly braid-plain character but distal ones are unconsolidated fine sands and silts. The entire wedge is just submerged at peak flow but it appears that a single low-flow channel is being re-established so that in due course the lakefill will form a low terrace like that visible above the 1858 rockfall dam near Sarat some 15 km downstream; Figs. 13 and 16.

The rapidity of lake filling testifies to the magnitude of the sediment load in the Hunza (discussed below), and to the importance of sediment storage in the Karakoram valleys despite their overwhelmingly erosional character. Storage over a much longer timespan is indicated by the high Hunza valley fill already mentioned; but although this contains some coarse alluvium, and some fine lacustrine beds, it appears to be predominantly composed of weathered fanglomerate and mudflow deposits, although in a few places incorporating old end moraines.

EROSION RATES FOR THE HUNZA RIVER CATCHMENT

The highly peaked discharge regime of the Hunza River, together with the high velocities that the river attains, means that it has a substantial ability to transport sediment. Moreover, there are a number of factors which ensure that large quantities of material are delivered to the trunk stream for evacuation from the area. These include:

(a)   the glaciated nature of the catchment

(b)   the limited vegetation cover

(c)   the steep relief

(d)   the fractured and contorted nature of the bedrock

(e)   the efficacy of frost and salt weathering

(f)   the presence of an easily eroded store of Pleistocene debris

(g)   the frequency and magnitude of landslides, mudflows, avalanches, etc.

The only practical way to assess the denudation rate in such a large, heterogeneous, and inaccessible region as the Karakoram is through measure-ments of sediment concentrations in rivers. Concentration in kg m^{-3} x discharge in $m^3 s^{-1}$ gives the instantaneous rate of removal of sediment from the drainage basin which can be integrated over time as discharge and concentration vary. Since the late 1960's WAPDA have collected water samples for suspended sediment analysis at one to two week intervals at several gauging stations primarily for estimation of sediment loads to the Tarbela Reservoir whose useful lifespan depends on the rate of sedimentation. The load of the Indus below its confluence with the Gilgit and Hunza rivers has been estimated as 155 Mt y^{-1} (59).

Estimates such as this based on suspended sediment rating curves are notoriously unreliable (60). Double logarithmic sediment ratings for Karakoram gauging stations are statistically significant but the scatter about them and thus the potential error in estimating daily sediment loads is considerable. Sixty-four measurements in the Hunza river during four weeks of July/August 1980 showed no simple relationship between concentra-

tion and discharge: rather, sediment concentrations show flushing and
depletion effects familiar from other environments but rather unexpected
in the Karakoram where loose unvegetated sediment is abundant. Re-analysis
of WAPDA data shows that flushing and depletion occur seasonally too
(Fig. 25). This is consistent with the tapping of new sediment sources

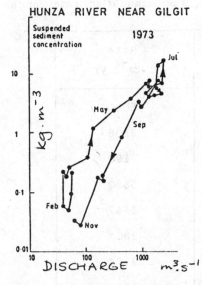

FIG. 25. RATING CURVE OF SUSPENDED SEDI-
MENT CONCENTRATIONS AGAINST DISCHARGE FOR
THE HUNZA RIVER NEAR GILGIT (DATA FROM
WAPDA, 1973). Clockwise loop indicates sea-
sonal flushing and depletion of sediment
supply.

as the snowline rises from valley floors in the spring to mountain walls
and glacier surfaces in summer, sub-glacial drainage systems are
re-activated and unvegetated floodplain surfaces are inundated.

The mean annual load of the Hunza river is estimated by WAPDA (59)
as 61 Mt although calculations for individual years show a range from
< 30 to >100 Mt according to runoff and type of sediment rating used (61).
Between 50% and 75% of the annual load is transported in July and August
(Fig. 26) when sediment concentrations can exceed 15 $kg\,m^{-3}$ (15,000 mg L^{-1})

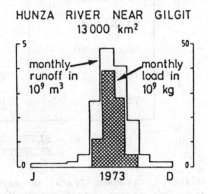

FIG. 26. SEASONAL REGIME OF RUNOFF AND
SUSPENDED SEDIMENT LOAD (ESTIMATED BY
DISCHARGE RATING CURVE) FOR THE HUNZA
RIVER NEAR GILGIT (DATA FROM WAPDA, 1973).
Scales match at a concentration of 10 $kg\,m^{-3}$
(10,000 mg L^{-1}).

corresponding to daily loads in excess of 2 Mt of predominantly silt sized
suspended sediment. This annual load is equivalent to a denudation rate
of about 5000 t km^{-2} yr^{-1} for the 13,000 km^{-2} catchment and would result

in uniform surface lowering by almost 2 mm yr^{-1}, which for this size of catchment is probably surpassed only in the loess region of northern China (62) (see Table 8). Since both runoff rates and sediment concentrations are somewhat lower in other Karakoram rivers, probably because of a lower percentage of glacier cover, the Hunza and Gilgit rivers contribute almost as much sediment as the entire eastern Karakoram (upper Indus) which is four times more extensive. The denudation rate for the

River	Basin area (km^2)	Rate $(t\ km^{-2}\ yr^{-1})$
Hunza	13000	5000
Lo	190000	7308
King Ho	56930	7190
Wei Ho	29880	5880
Tamur (Nepal)	5680	5147
Yellow	673000	2804
Kosi	62000	2774
Hun	1740	1966
Yeyo	370	1895
Ganges	1730000	1040 - 1518
Mekong	810000	1200
Brahmaputra	666000	1090
Irrawaddy	410000	850
Si-Kiang	400000	660
Shuliogou	1364	567
Indus	960000	420
Yangtze	1175000	234

TABLE 8: RATES OF MECHANICAL DENUDATION IN SELECTED ASIAN RIVERS.

entire Karakoram tract is nevertheless over 1,000 t $km^{-2}yr^{-1}$ (0.4 mm y^{-1}) and the region contributes 40% of the load of the Indus from only 15% of the area. This denudation rate is very high by world standards (Table 9) despite there being no significant contribution from accelerated soil erosion.

River	Period	Mean Annual suspended sediments 10^6 metric tonnes	Observed mean annual maximum concentration ppm	Mean annual discharge ($10^6 m^3$)	Sediment/ Discharge Ratio by weight (%)	Area (km^2)	Sediment Removal tons/km^2/yr
HUNZA (Dainyor)	1966-75	63.23	15052	11965	0.51	13157	4805
GILGIT (Gilgit)	1963-72	13.57	4458	9041	0.15	12095	1122
INDUS (near Kachural)	1970-75	87.09	6277	30220	0.29	112664	773
SHYOK (Yugo)	1973-75	33.57	5100	9769	0.29	33670	997

TABLE 9: SEDIMENT YIELDS OF THE HUNZA AND NEIGHBOURING RIVERS (CALCULATED FROM DATA SUPPLIED BY WAPDA)

The sediment yields just discussed omit the chemical component of denudation. Measurements of river water conductivity in summer 1980 suggest diurnally varying total dissolved solids concentrations in the range of 0.02 - 0.06 kg m^{-3} in major pro-glacial streams. The glacial source of most of the summer runoff in the Hunza river is apparent in slight diurnal conductivity fluctuations but the general level is approximately double that in pro-glacial streams (TDS between 0.07 and 0.09). The river water is probably enriched partly by solution/desorption from the abundant silt and clay in suspension and partly by irrigation return water and groundwater which, although small in volume, are often very saline (TDS concentration about 0.3 kg m^{-3}). If atmospheric and biologic inputs are deemed negligible (precipitation is mainly snow and vegetation cover is sparse), the rate of chemical denudation is probably of order 90 t km^{-2} yr^{-1} which, although two orders of magnitude less than mechanical denudation, is still three times the world average.

CONCLUSION

The Hunza Valley is one of the most active geomorphological environments on Earth. It displays a greater degree of relief than found anywhere else; its long steep slopes are characterised by the operation of frequent mudflows, rock falls, etc.; weathering (both chemical and physical) greatly exceeds the world norm; its fast moving and rapidly fluctuating glaciers are numerous and long; its rivers carry immense loads and are prone to catastrophic floods; and frequent dust storms have imposed an aeolian stamp on the landscape. Its fascination for the geomorphologist is heightened by the fact that besides being the scene of such active processes at the present day, in the past it has been affected by major climatic changes, the imprint of which can still be seen.

REFERENCES

1) CURZON, G. N., (1896). 'The Pamirs and the source of the Oxus', Geographical Journal 8, pp 15 - 54, 97 - 119, 239 - 263.

2) TAHIRKHELI, R. A. K. and JAN, M. Q., (1979). 'A preliminary geo-logical map of Kohistan and the adjoining areas, N. Pakistan', Fig. 1 in

eds. Tahirkheli, R. A. K. and JAN, M. Q., (1979), Geology of Kohistan, Karakoram Himalaya, Northern Pakistan, Geol. Bull., Univ. Peshawar (Spec. Issue) (11)1, 187 pp.

3) DESIO, A., (1974), 'Karakoram Mountains', Geol. Soc. Lond. Spec. Paper (4), pp 254 - 267.

4) DESIO, A., (1979), 'Geologic evolution of the Karakorum'. In Farah, A. and DeJong, K. A., eds., Geodynamics of Pakistan, pp 111 - 124, Geological Survey of Pakistan, Quetta.

5) GANSSER, A., Geology of the Himalayas, Interscience Publishers, London, 289 pp.

6) SCHNEIDER, H-J., (1957). 'Tectonics and magmatism in the NW-Karakorum', Geologische Rundschau (46), pp 426 - 476. (in German)

7) DESIO, A. and MARTINA, E., (1972). 'Geology of the Upper Hunza Valley, Karakorum, West Pakistan'. Bull. Soc. Geol. Ital. (91), pp 283 - 314.

8) HEWITT, K., (1968). 'The freeze-thaw environment of the Karakoram Himalaya', Canadian Geographer 12, pp 85 - 98.

9) POWELL, C. McA., and VERNON, R. H., (1979). 'Growth and rotation history of garnet porphyroblasts with inclusion spirals in a Karakoram schist', Tectonophysics (54), pp 25 - 43.

10) IVANAC, J. F., TRAVES, D. M. and KING, D., (1956). 'The geology of the NW portion of the Gilgit Agency', Rec. Geol. Surv. Pakistan (8)2, pp 3 - 27.

11) TAHIRKHELI, R. A. K., (1979). 'Geology of Kohistan and adjoining Eurasian and Indo-Pakistan continents, Pakistan'. In eds., Tahirkheli, R. A. K. and Jan, M. Q., Geology of Kohistan, Karakoram Himalaya, Northern Pakistan. Geol. Bull., Univ. Peshawar (Spec. Issue) (11)1, pp 1 - 30.

12) POWELL, C. McA., (1979). 'A speculative tectonic history of Pakistan and surroundings: some constraints from the Indian Ocean'. In Farah, A. and DeJong, K. A. (Eds.) Geodynamics of Pakistan, pp 5 - 24, Geological Survey of Pakistan, Quetta.

13) TAHIRKHELI, R. A. K., MATTAUER, M., PROUST, F. and TAPPONIER, P., (1979). 'The India Eurasia suture zone in Northern Pakistan: synthesis and interpretation of recent data at plate scale', In Farah, A. and DeJong, K. A., (Eds.), Geodynamics of Pakistan, pp 125 - 130, Geological Survey of Pakistan, Quetta, 361 pp.

14) THE BATURA GLACIER INVESTIGATION GROUP, (1979). Investigation Report on the Batura Glacier in the Karakoram Mountains, The Islamic Republic of Pakistan (1974-1975). Beijing, Engineering Headquarters, the Batura Glacier Investigation Group of Karakoram Highway, 123 pp.

15) PAFFEN, K. H., PILLEWIZER, W. and SCHNEIDER, H-J, (1956). 'Investigations in the Hunza Karakorum', Erdkunde (10) pp 1-3. (in German)

16) SHI YAFENG and WANG WENYING, (1980). 'Research on snow cover in China and the avalanche phenomena of Batura Glacier in Pakistan', Journal of Glaciology (26), pp 25 - 30.

17) MÜLLER, F., (1958). 'Eight months of glacier and soil research in the Everest region', The Mountain World (1958-59), pp 191 - 208. Swiss Foundn. for Alpine Res. BM. Sci. Mus., SPRI, Cambr. UL.

18) WASHBURN, A. L., (1939). 'Karakorum glaciology', Amer. Jour. of Science 237, pp 138 - 146.

19) MANI, M. S., (1978). Ecology and phytogeography of high altitude plants of the Northwest Himalaya, Oxford and IBH Publishing Co., New Delhi, 205 pp.

20) PAMPANINI, R., (1930). 'The flora of the Karakorum'. In Spedizione Italiana De Filippi vell' Himalaja, Caracorum, etc., Ser. II, Vol. X-XI, Bologna. (in Italian)

21) TROLL, C., (1939). 'The vegetation cover of Nanga Parbat. Explanation of the vegetation map of the Nanga Parbat Group 1:50,000', Wiss Veroff. Mus Dt. Landerk zu Lpz. (NF7).

22) CHAMPION, Sir H. G., SETH, S. K. and KHATTAK, G. M. (1965). Forest Types of Pakistan, Pakistan Forest Institute, Peshawar, MS, 214 pp.

23) WEBSTER, G. L. and NASIR, E.D., (1965). 'The vegetation and flora of the Hushe Valley (Karakoram Range, Pakistan)', Pakistan Journal of Forestry (15), pp 201 - 211.

24) ZELLER, W. and BEG, A. R., (1969). 'Review of some vegetation types of Dir and Chitral', UNDP-FAO Pakistan National Forestry Research and Training Project Report (11), Ecological Papers 1968-1969, pp 9 - 25.

25) BEG, A. R. and BAKHSH, I., (1974). 'Vegetation of scree slopes in Chitral Gol', Pakistan Journal of Forestry (24), pp 393 - 402.

26) SHEIKH, M. I. and ALEEM, A., (1975). 'Forests and forestry in Northern Areas', Pakistan Journal of Forestry (25), pp 197 - 235 (I) and 296 - 324 (II).

27) DESIO, A. and OROMBELLI, G., (1971). 'Preliminary note on the presence of a large valley glacier in the middle Indus Valley (Pakistan) during the Pleistocene', Atti della Accademia Nazionale dei Lincei 51, pp 387 - 392. (in Italian)

28) BRUNSDEN, D., JONES, D. K. C. and GOUDIE, A. S., (this volume). 'Particle size distribution on the debris slopes of the Hunza Valley'.

29) BRUNSDEN D., JONES, D. K. C., WHALLEY, W. B. and GOUDIE, A. S., (in preparation). 'Debris flows of the Karakoram Mountains'.

30) GOUDIE, A. S., JONES, D. K. C., BRUNSDEN, D., (this volume). 'Recent fluctuations in some glaciers of the western Karakoram Mountains, Hunza, Pakistan'.

31) BRUNSDEN, D., JONES, D. K. C., MARTIN, R. and DOORNKAMP, J.L, (1981). 'The geomorphological character of part of the low Himalaya of eastern Nepal', Zeitschrift für Geomorphologie Supplementband 37, pp 25 - 72.

32) WHALLEY, W. B., MCGREEVY, J. P. and FERGUSON, R. I., (this volume). 'Rock temperature observations and chemical weathering in the Hunza region, Karakoram: preliminary data'.

33) GOUDIE, A. S., (this volume) 'Salt efflorescences and salt weathering in the Hunza Valley, Karakoram Mountains, Pakistan'.

34) MASON, K., (1930). 'The glaciers of the Karakoram and neighbourhood', Records, Geological Survey of India 63, pp 214 - 278.

35) VISSER, P. C. and VISSER HOOFT J. Karakoram, 1938.

36) LIU GUANGYUAN, (1980). 'The climate of the Batura Glacier and its adjacent areas'. In Professional Papers on the Batura Glacier, Karakoram Mountains, Shi Yafeng (ed.) Chapter 7, pp 99 - 110 (in Chinese). Beijing, Science Press.

37) SHI YAFENG, (1979). 'Some achievements on mountain glacier researches in China'. Lanzhou Institute of Glaciology and Cryopedology, Academia Sinica, 29 pp.

38) ZHANG XIANGSONG, CHEN JIANMING, XIE ZICHU and ZHANG JINHUA, (1980). 'General features of the Batura Glacier'. In Professional Papers on the Batura Glacier, Karakoram Mountains, Shi Yafeng (ed.), Chapter 2, pp 8 - 27 (in Chinese). Beijing, Science Press.

39) SHUMSKII, P. A., (1965). Principles of Structural Glaciology. Transl. d. Kraus, New York, Dover Pub.

40) ZHANG JINHUA and BAI ZHONGYUAN, (1980). 'The Surface ablation and its variation of the Batura Glacier'. In Shi Yafeng (ed.), op. cit., pp 83 - 98 (in Chinese).

41) ZHANG XIANGSONG, SHI YAFENG and CAI XIANGXING, (1980). 'The migrating sub-glacial channel of the Batura Glacier and the tendency of the new channel'. In Shi Yafeng (ed.), op. cit., pp 153 - 165 (in Chinese).

42) MAYEWSKI, P. A., PREGENT, G. P., JESCHKE, P. A. and AHMAD, N., (1980). 'Himalayan and trans-Himalayan glacier fluctuations and the south Asian monsoon record', Arctic and Alpine Research 12, pp 171 - 182.

43) ZHANG XIANGSONG, (1980). 'Recent variations in the glacial termini along the Karakorum Highway', Acta Geographica Sinica 35, pp 147 - 160.

44) MASON, K., (1914). 'Examination of certain glacier snouts of Hunza and Nagar', Records, Survey of India 6, pp 49 - 51.

45) DOLGUSHIN, L. D. and OSIPOVA, G. B., (1973). 'The regime of a surging glacier', I.A.H.S. 107, pp 1150 - 1159.

46) HEWITT, K., (1969). 'Glacier surges in the Karakoram Himalaya (Central Asia)', Can. J. Earth Sci. 6, pp 1009 - 1018.

47) PATERSON, W. S. B., (1981). The Physics of Glaciers (2nd. ed.), Oxford, Pergamon, 380 pp.

48) WAPDA, (1970). Annual Report 1970, v-1, Pakistan Water and Power Development Authority, Lahore, 55p.

49) BATURA INVESTIGATION GROUP, (1976). Investigation report on the Batura Glacier in the Karakoram Mountains, the Islamic Republic of Pakistan (1974 - 1975): Peking, 51 pp.

50) FLOHN, H., (1974). 'Contribution to a comparative meteorology of mountain areas', in Ives, J. D. and Barry, R. G. (Eds.), Arctic and alpine environments, Methuen, pp 55 - 71.

51) BATURA INVESTIGATION GROUP, (1979). 'The Batura Glacier in the Karakoram mountains and its variations', Scientia Sinica 22, pp 958 - 974.

52) GLOERSON, P. and SALOMONSON, V. V., (1975). 'Satellites - new global observing techniques for ice and snow', J. Glaciology 15, pp 373 - 389.

53) YOUNG, G. J., (1978). 'Relations between mass balance and meteorological variables on Peyto Glacier, Alberta, 1967 - 74', Zeitschrift für Glazialgeo. Gletscherkunde 13, pp 111 - 125.

54) MAYEWSKI, P. A. and JESCHKE, P. A., (1979). 'Himalayan and Trans-Himalayan glacier fluctuations since AD 1812', Arctic and Alpine Research 11, pp 267 - 287.

55) MASON, K., (1929). 'The representation of glaciated regions on maps of the Survey of India', Professional Paper, 25, Geological Survey of India, 18 pp.

56) BELCHER, J., (1859). Letter addressed to R. H. Davies, Esquire, Secretary to the Government of the Punjab and its dependencies. Journal of the Asiatic Society of Bengal 28, pp 219 - 228.

57) FINSTERWALDER, R., (1959). 'Recent German expeditions to Batura Mustagh and Rakaposhi', Journal of Glaciology 3, p 787.

58) PARKER, G., (1976). 'On the cause and characteristic scales of meandering and braiding in rivers', Journal of Fluid Mechanics 76, pp 457 - 480.

59) WAPDA, (1979). Tarbela Dam project: sedimentation base survey September - October 1979: Pakistan Water and Power Development Agency, Lahore, 55 pp.

60) WALLING, D. E., (1978). 'Reliability considerations in the evaluation and analysis of river loads', Zeitschrift für Geomorphologie Supplement Band 29, pp 29 - 42.

410

61)   FERGUSON, R. I., (1982), (this volume).   'Sediment load of the Hunza River', page 581.

62)   WALLING, D. E., (1981).   'Yellow river which never runs clear', Geographical Magazine 53, pp 568 – 575.

Recent fluctuations in some glaciers of the Western Karakoram mountains, Hunza, Pakistan

Andrew S. Goudie – Oxford University
David K. C. Jones – London School of Economics
Denys Brunsden – King's College, London University

ABSTRACT

During the International Karakoram Project, 1980, the snout positions of the Minapin, Pisan, Ghulmet, Hasanabad, Ghulkin, Pasu and Batura glaciers were surveyed, and their positions related to previous investigations over the last century. They have shown considerable fluctuations, including major advances at the turn of the century, and general retreat from the 1930s to the 1970s.

INTRODUCTION

The Karakoram Mountains possibly possess more ice cover than any other extra-polar mountain system and the valley glaciers include some of the longest in the world outside the polar regions. Estimates of ice cover range from 28% to 37% (see references 1 and 2). The local name, Mustaghs, (ice mountains) is used to describe the crests of entire mountain systems and in the Greater Karakoram Range these include the Batura (Fig. 1), Hispar, Panmah, Baltoro, Rimo, Siachen, and Saser Mustaghs. Further north lie the Lupgha and Ghujerab Mustaghs and to the south those of Rakaposhi (Fig. 1), Haramosh, Masherbrum and Saltoro in the Lesser Karakoram Range. The mustaghs may be further subdivided into smaller ice field areas usually called groups (eg, the Pasu group, Fig. 1).

The total area of glaciers in this region is approximately 13,600 to 15,000 km^2 (2) (3) with a snow line ranging from 5,100 m in the south to 5,600 m in the main ranges (4) (5). In the north the snow line lies at 4,700 – 5,300 m (6).

Because of the extremely steep relief, there are numerous ice falls, heavy crevassing and massive debris loads, so that many of the glaciers have been observed to move by a blockschollen motion. Although slow

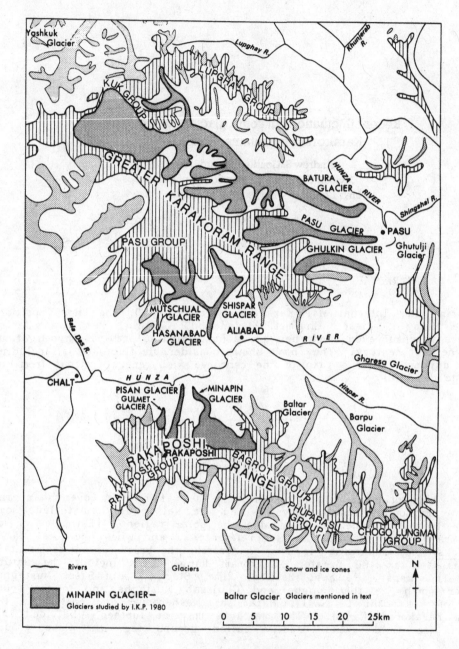

FIG. 1.   LOCATIONS OF GLACIERS STUDIED BY THE GEOMORPHOLOGISTS
OF THE IKP, 1980

advances and retreats have been noted, reports of catastrophic advances or surges have made the Karakoram glaciers notorious. For this reason the snouts of them have been the subject of a great deal of interest, particularly those which Mason (7) termed 'threatening glaciers'.

In the past this interest was partly related to the effect that such advances had on the local communities in terms of flooding, landslide activity, the disruption of the irrigation channels upon which most of the population depend, and the relative prosperity of villages. Today, there is increasing concern that rapid glacial advances may cause temporary damming of the Karakoram rivers and thereby threaten the safety of recently built highways and reservoirs, such as that impounded by the Tarbela dam. In addition, records of glacial movements are seen to be of scientific importance because they are associated with, and are indicators of, environmental change (8) (9) (10). As the existing historical record of such glacier fluctuations in the Himalaya and Karakoram regions has been considerably augmented over the last 30 years, and in places now extends back over 150 years, so it is becoming possible to prepare regional syntheses of the composite record and thereby reveal the dominant nature of retreat, advance and standstill regimes in relation to particular regional environments. Unfortunately, the record is often fragmentary and disparate, with cartographic and other evidence that appears, at least at first sight, to be incompatible. The purposes of this paper are to make a collection of all the available historical data for the Hunza valley of the Karakorams, and to add to this record the position of seven valley glacier snouts surveyed by the International Karakoram Project in 1980. The results not only provide evidence of local and regional patterns, but also provide a data base of value for future investigations in the area.

Regional syntheses relevant to the Karakoram have been made by Mason (7) (8), based on his own surveys and those of the officers of the Geological Survey of India, especially Hayden (11), and by Visser and Visser-Hooft (12), who reported that in the period 1900 to 1910 most of the glaciers were advancing, but from 1910 to 1920 were mainly retreating. More recently, Mercer (9) carried out a general survey of the literature on 50 glaciers in the region, 31 of which had previously been discussed by Mason and Visser. Mercer was able to show that the glaciers could be grouped into three categories on the basis of movement patterns: those with steady advance characteristics which were also the longer glaciers; those displaying cyclic advances being those with short, steep courses; and those experiencing catastrophic advances, which he believed to be the result of earthquake activity. The most comprehensive study to date is by Mayewski and Jeschke (10), who produced regional syntheses for all the Himalayan-Karakoram environments. Their results will be discussed later in the conclusions to this paper. Finally, Chinese scientists (13) from the Batura Glacier Investigation Group of the Karakoram Highway Engineering Headquarters surveyed fourteen glaciers and paid particular attention to the so-called "surging" glaciers (14).

THE 1980 SURVEY

As part of the IKP 80 scientific programme, the snout positions of seven glaciers were mapped at varying levels of detail: the Minapin, Pisan and Ghulmet glaciers which drape the steep northern flanks of Mount Rakaposhi; the Hasanabad Glacier, renowned for the rapidity of its movements, which descends the southern slopes of the Pasu Group; and three eastward moving glaciers, the Ghulkin, Pasu and Batura.

All these glaciers occupy tributary valleys of the Hunza with snout positions 0.3 to 7.2 km from the main river, and were selected for study because of reasons of proximity to the Expedition Base Camp at Aliabad, accessibility, the existence of historical records, or that their forward movement posed a direct threat to the recently constructed Karakoram Highway.

Field surveys were undertaken at a scale of 1:2500 using planetable and telescopic alidade in combination with a Hewlett Packard 3805A Distance Meter (see Fig. 2) mounted on a WILD 11431A Tribrach (Model GDF 6)

FIG. 2.   THE USE OF THE HEWLETT PACKARD 3805A DISTANCE METER
BY MEMBERS OF THE IKP, 1980

and tripod, and sighting onto a triple prism target.   In addition, snout positions for all glaciers except the Hasanabad were obtained from 1:30,000 scale air photography flown in 1965 and provided by the Pakistan Airforce.   Both sets of snout locations were plotted on maps either of 1:30,000 scale, and made by air-photo interpretation, or of approximate 1:10,000 scale created by the transposition of air-photo detail onto photographically enlarged versions of the 1:50,000 scale map of part of the Middle Hunza valley produced by the German-Austrian Himalaya-Karakoram Expedition of 1954 using terrestrial stereophotogrammetric methods (15).

THE MINAPIN GLACIER

The Minapin Glacier lies on the south side of the Hunza River, descending steeply from the north flank of the Rakaposhi Range at a col between Rakaposhi peak (7788 m) and Minapin peak (7196 m).   Its course can be divided into two distinct units: an upper basin approximately 10 km across, and a lower sinuous valley track enclosed by steep mountain walls. Overall it is approximately 13 km long, and has one of the lowest snout altitudes in the entire Himalaya-Karakoram range, with an elevation that has ranged between 2150 and 2350 m during the period of observations (1889 - 1980). The glacier is fed by avalanches which originate from cornices and ice cliffs along the entire 10 km long headwall. A full description of the morphology of this valley is given by MacBryde (16).

For two km above the snout the irregular surface of the ice is almost completely covered by moraine (Fig. 3) and is bounded by two high, steep-

FIG. 3. THE DEBRIS COVERED LOWER PORTION OF THE MINAPIN GLACIER IN AUGUST 1980

sided lateral moraines which converge below the present ice-front to enclose a deep rock-cut gorge. This gorge has, in part, controlled the form and position of the glacier tongue for the past century (Fig. 4).

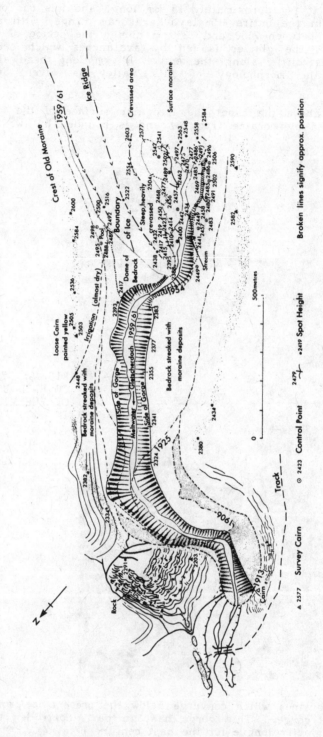

FIG. 4. SNOUT POSITIONS OF THE MINAPIN GLACIER IN RELATION TO THE TOPOGRAPHY
OF THE GORGE
(see also page 419)

Recordings of the snout position of the Minapin Glacier have been frequently made in the past due to its accessibility from the main Hunza and Nagir routes (Table 1). Indeed in 1931 the Editor of the Himalayan Journal appealed "to every traveller to record the position of this glacier's snout" and to date approximately fifteen such observations have been made so that the Minapin has one of the best catalogued of all Karakoram glacier histories. The advance and retreat positions of this glacier are summarised in Figure 5, which shows that since 1889, when Ahmed Ali Khan made the first reconnaissance map, the glacier first advanced by approximately 1700 m until 1913, but since then seems to have been in a general state of retreat though there are some gaps in the record (see (3) (11) (15) (17) (18) and (19)). At some time prior to August 1978 (13), but after the Cambridge expedition's visit of 1961-1962 (16), there must have been a slight advance, for the Chinese survey reported (14) that the ice tongue terminus was at + 2350 m, at least 150 m forward of its 1961 position and approximately equal to that recorded in 1954. The glacier does not seem to have been accurately surveyed during this time and no evidence of this advance has been identified either on the 1965 Pakistan air force air photographs, or by members of the International Karakoram Project. In July 1980 the glacier snout was still less advanced than it had been in 1961. It appeared to have generally thinned by 10 - 40 m (Fig. 6), and had retreated, especially on its eastern side, by 50 - 60 m. The snout was divided into two lobes by a moraine-blanketed outcrop of smooth buff coloured rock, as reported by the Cambridge expedition. The lowest kilometer or so of glacier ice was covered in moraine and the whole snout appeared to be in a state of decay. The 1980 position of the snout, when plotted on the graph of advance and retreat (Fig. 7), appears to continue the retreat phase of the previous sixty years, with the result that the ice front is in a broadly comparable position to that when it was first observed in 1889.

These records, therefore, cover a complete cycle of short term advance and retreat, superimposed on a general secular pattern of overall retreat; see Figures 5 and 7. Mason (8) speculated that if reliance could be placed on the early observations, and assuming that the linear retreat rate from 1913 to 1930 were continued, then the Minapin could have a cycle with a periodicity of roughly 48 years (24 of advance and 24 of retreat), a further assumption being that if this glacier was periodic then it would begin to advance again at the end of the 1930's. Mason was obviously concerned for the safety of the people at Minapin, for in 1935 he stated "forty-two years have now passed since this glacier last started to come forward. It is of the upmost importance that it should be examined yearly now that it is so decadent" (7). It is now clear that the retreat phase has been much longer lasting than he anticipated, and that if there is a periodicity in the movement of this glacier, then it is of a longer duration than he estimated, for the degraded nature of the terminal portion of the glacier as recorded in 1980 suggests that an advance is unlikely for a few years yet.

THE PISAN GLACIER

The Pisan Glacier descends towards the Hunza River only about two km to the west of Minapin. It is easily visible from the valley below and is even more accessible than its neighbour. Yet, surprisingly, there are only two mentions in the literature of the existence of this glacier and no comments of any substance. The survey reported here is, therefore, the first on record and must form the basis of future work.

OBSERVER	DATE	DETAILS OF SNOUT
AHMED ALI KHAN (8)	AUG - SEPT 1889	2750 m FROM SITE OF FUTURE BRIDGE CROSSING ON MINAPIN RAVINE.
CONWAY (17)	1892	ADVANCE OF 100 m. ROUGH SKETCH
KHANSAHIB ABDUL GAFFAR (8)	1893	ADVANCE OF 1100 m
HAYDEN (11)	1906	ADVANCE OF 275 m. MARKS MADE ON VALLEY WALLS. CAIRNS. PHOTOGRAPHS
MASON (18)	1913	ADVANCE OF 200 m. FRESH MARKS MADE. PHOTOGRAPHS. "AN IMMENSE PACK OF BLACK ICE"
VISSER (19)	1925	RETREAT OF 600-700 m. "AN INSIGNIFICANT NARROW STRIP OF ICE BURIED BENEATH RUBBISH." ROCK GORGE EXPOSED
TODD (8)	1929	PROBABLY A 100 m RETREAT (ESTIMATE ONLY)
TODD (7)	1930	CONSIDERABLY MORE MELTING REPORTED. 300 m RETREAT SINCE 1925
WOOLDRIDGE (7)	1932	5-10 m RETREAT FROM 1930 CAIRN. "A MISERABLE TONGUE"
SHIPTON AND MOTT (16)	1939	CONTINUED RETREAT ESTIMATED FROM SHIPTON AND MOTT MAP PUBLISHED IN 1950
GERMAN KARAKORAM EXPEDITION (15)	1954	c 750 m RETREAT SINCE 1925
GERMAN KARAKORAM EXPEDITION (3)	1959	c 40-100 m RETREAT SINCE 1954
CAMBRIDGE EXPEDITION (16)	1961/1962	c 200 m RETREAT SINCE 1954
PAKISTAN AIR FORCE	1965	AIR PHOTOGRAPH SHOWS CONTINUED SHRINKAGE
ZHANG XIANGSONG (13)	1978	ADVANCE OF 150 METRES COMPARED WITH 1961 AND REPORTED TO BE NEARLY AT 1954 POSITION
INTERNATIONAL KARAKORAM PROJECT	1980	NORTHERN LOBE RETREATED 100 m AT FURTHEST POINT SOUTH OF SNOUT, 10 m BETWEEN THE TWO LOBES, AND 0-15 m ON SOUTHERN LOBE COMPARED TO 1961

TABLE 1 OBSERVATIONS ON THE MINAPIN GLACIER

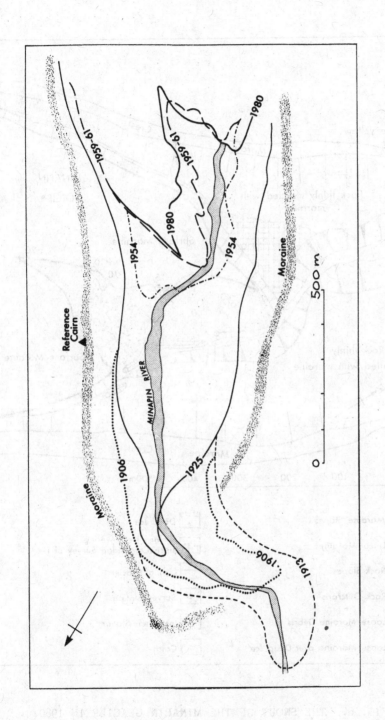

FIG. 5. SUMMARY DIAGRAM OF THE SNOUT POSITIONS OF THE MINAPIN GLACIER

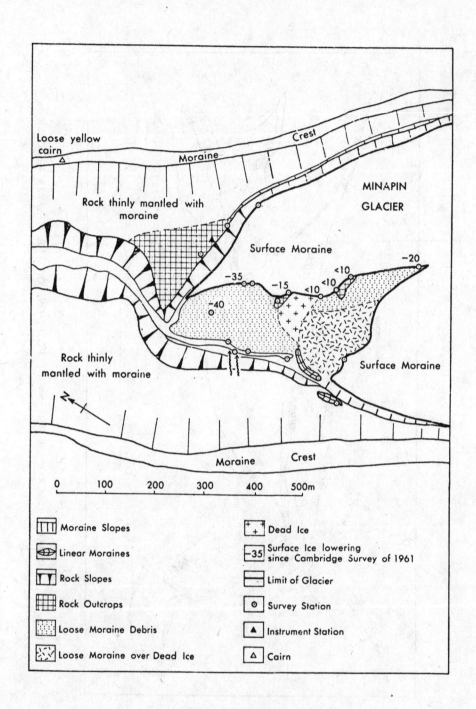

Loose yellow cairn

Crest

Moraine

MINAPIN

GLACIER

Rock thinly mantled with moraine

Surface Moraine

−20

−35

<10

<10

−15

<10

−40

Rock thinly mantled with moraine

Surface Moraine

N

Moraine   Crest

0   100   200   300   400   500m

	Moraine Slopes		Dead Ice
	Linear Moraines	−35	Surface Ice lowering since Cambridge Survey of 1961
	Rock Slopes		Limit of Glacier
	Rock Outcrops		Survey Station
	Loose Moraine Debris		Instrument Station
	Loose Moraine over Dead Ice		Cairn

FIG. 6.   THE SNOUT OF THE MINAPIN GLACIER IN 1980

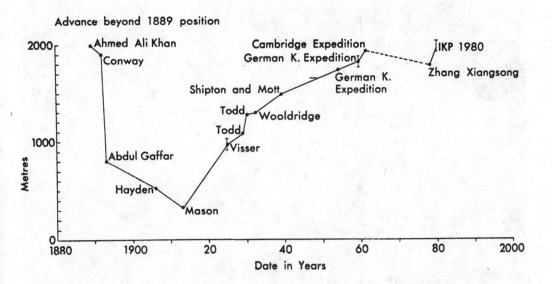

Advance beyond 1889 position

FIG. 7.   GRAPH OF CHANGES IN THE POSITION OF THE MINAPIN SNOUT
SINCE 1889

This glacier must rank as one of the steepest in the Karakoram region and possibly on the Earth as a whole (Fig. 8).   The headwalls that supply avalanche debris descend from the peak of Rakaposhi itself (7788 m).   The glacier proper appears to start at about 5300 m and descends in a remarkably straight deep trough to approximately 2400 m over a horizontal distance of 10 km.   This approximates a mean slope of about 29%.   The last two km are sinuous and the whole of the valley track is enclosed between massive and impressive lateral moraines.   The lowest part of the glacier lies between four moraines that curve to the east and tower above the village of Pisan (Fig. 9).   The snout is covered in morainic debris and is obviously rapidly retreating, for it is divided into two lobes by a meandering meltwater stream (Figs. 9 and 10) the easternmost lobe descending to a lower elevation than the western.

The previous records for this glacier are only approximate.   The termini shown on the 1:10000 topographical map of Pakistan (1966) based on 1965 air photography, were at heights of 2070 and 2020 m respectively.   In August 1978 Zhang Xiangsong (13) reported considerable retreat to elevations of 2420 and 2450 m, a distance retreat of 300 – 370 m since 1966.   An earlier survey in 1954 by Paffen et al. (15) is indistinct but shows a position rather higher than that shown on the 1965 photograph, namely near 2380 m (Fig. 9).

The present survey (Fig. 9) used a yellow painted cairn erected by the Chinese Investigation Group on the crest of the curved moraine above Pisan village.   This cairn, which is both substantial and prominent, should form the reference point for all future work.   Unfortunately, the correct position of this cairn, with respect to other topographic surveys, is unknown

FIG. 8. THE MORAINES OF THE PISAN GLACIER IN AUGUST 1980

as are the positions of the Chinese observations with respect to it. In 1980 the eastern lobe was 717 m (horizontal distance) from the cairn, and 148.086 m above it, and the western lobe 987 m and 166.467 m respectively. In order to compute an approximate elevation, the distances were plotted on the 1954 map of Paffen et al. (15). The resulting figure of 2400 m crudely agrees with the figures produced by the Chinese investigations of 1978.

THE GHULMET GLACIER

Like the Minapin and the Pisan, the Ghulmet Glacier (which is approximately 7 km long) descends from the avalanche slopes of Rakaposhi toward the Hunza river with an elevation of range from approximately 5400 m to 3000 m. A series of very fresh, unvarnished moraines separated from the 1980 snout by a broad polished rock apron (Fig. 11), suggests that the glacier has been beyond its present position by as much as 1000 m, probably within the last century, (see Conway's map of 1892, Fig. 12). There have, however, been no detailed observations and available historical data are restricted to an early photograph and published survey maps. A photograph taken in 1923 (Fig. 13), kindly provided by Mr. S. A. Bucknell, with a copy deposited in the Royal Geographical Society, shows the glacier to be slightly in advance of its present position (Fig. 14) but to be remarkably similar in outline, thickness and fresh white appearance. It is not debris covered on its lower portion (c.f. the Pisan and Minapin), and seems to be actively maintaining its position. The survey maps by

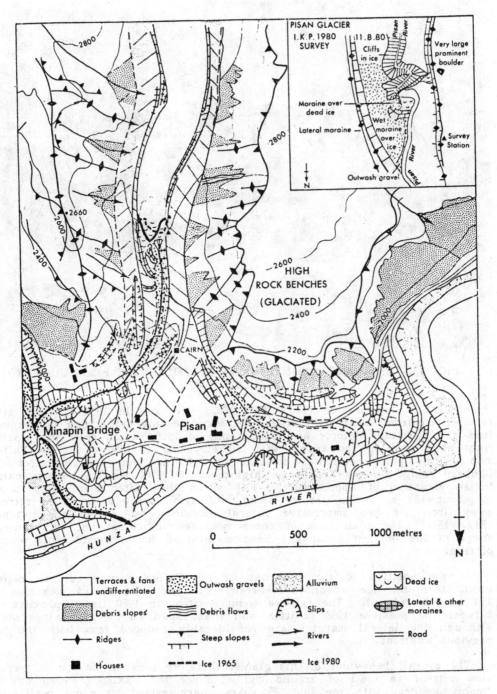

FIG. 9.   GEOMORPHOLOGICAL MAP OF THE PISAN GLACIER ILLUSTRATING
THE SNOUT POSITION IN 1980

FIG. 10. THE SNOUT AND MELTWATER STREAM OF THE PISAN GLACIER
IN AUGUST 1980

the American Map Service (1:250,000, 1962; prepared in 1954) do not give reliable ice-front locations. The first accurate information can be derived from the Pakistan Airforce air photographs of 1965, which show the glacier front at a position comparable with an elevation of 2950 m on the Paffen map, this at a horizontal distance of 3.25 km from the centre of the Ghulmet bridge on the Karakoram highway. By 1980 the ice had retreated to an elevation of approximately 3000 m on the Paffen map, a distance of about 120 m. The glacier snout had also narrowed so as to retreat away from its two impressive lateral moraines by a similar distance (Fig. 15). The plan form of these moraines indicates that the general shape of the Ghulmet snout has been maintained for a considerable period of time.

If the position of the ice-front on the Paffen survey can be relied upon, then the ice advanced between 1954 and 1965 by a maximum of 250 m and widened. The extreme snout position in 1980 lies approximately between the possible 1954 position and that shown on the 1965 air photos, although the lateral margins are considerably reduced from both the two previous locations.

The overall behaviour of this glacier therefore seems to be one of overall slow retreat (a total of around 650 m) since the maximum shown by the fresh moraines, with the last 25 years characterised by minor oscillation about the same general location. There are no signs on this glacier of the decay associated with either the Pisan or Minapin. This is strange in view of their adjacent positions on the north slopes of Mount Rakaposhi.

FIG. 11. THE SNOUT OF THE GHULMET GLACIER IN AUGUST 1980

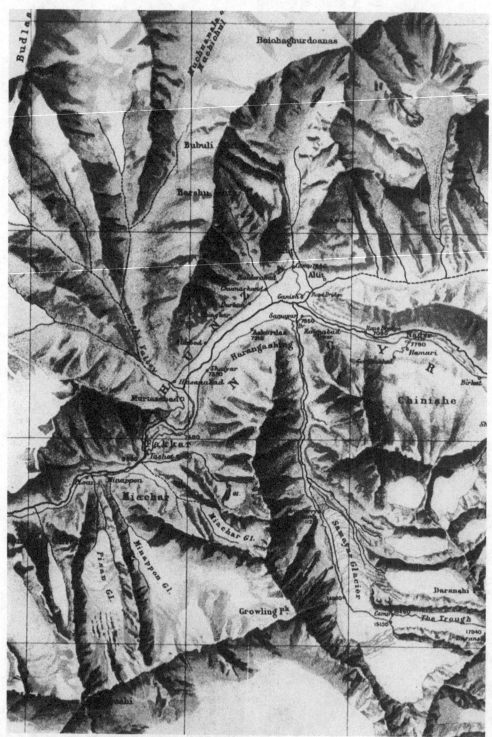

FIG. 12. CONWAY'S MAP OF HUNZA IN 1892

FIG. 13. THE GHULMET
GLACIER WITH RAKAPOSHI
IN THE BACKGROUND; as
photographed by Melaram
in 1923

A possible explanation is that the Ghulmet glacier is not enclosed in a long valley trough and so does not descend to such low elevations. By contrast it is a short tongue which emerges from an upper basin with a limited catchment. It is, therefore, more likely to reflect very short-term high altitude accumulation/ablation conditions rather than the dominantly ablation conditions of the lower slopes. In fact, the Ghulmet does not seem to have extended to low elevations for a very long period, because there is virtually no morainic debris on the valley sides below the snout. Instead the valley floor consists of a sheet of outwash torrent gravels flanked by rock and scree slopes.

THE HASANABAD GLACIER

The Hasanabad Glacier system drains two enormous basins located on the flanks of the peaks of the Pasu Group, which rise to between 7286 and 7777 m. The western basin is called Mutschual or Muchiohul, the eastern

428

FIG. 14. THE GHULMET GLACIER WITH RAKAPOSHI IN THE BACKGROUND; AS PHOTOGRAPHED BY THE IKP IN 1980.

429

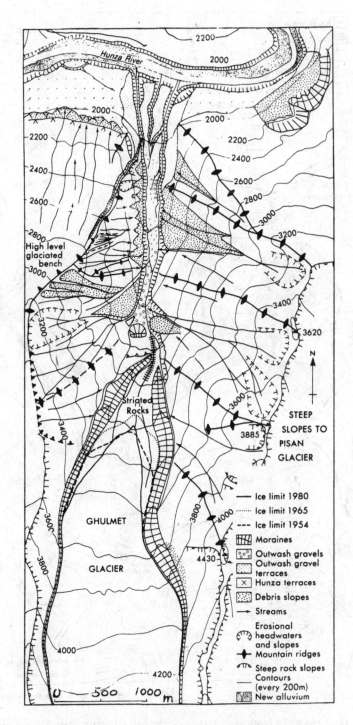

FIG. 15. GEOMORPHOLOGICAL MAP OF THE GHULMET GLACIER SHOWING
THE ICE LIMITS OF 1980, 1965 and 1954.

430

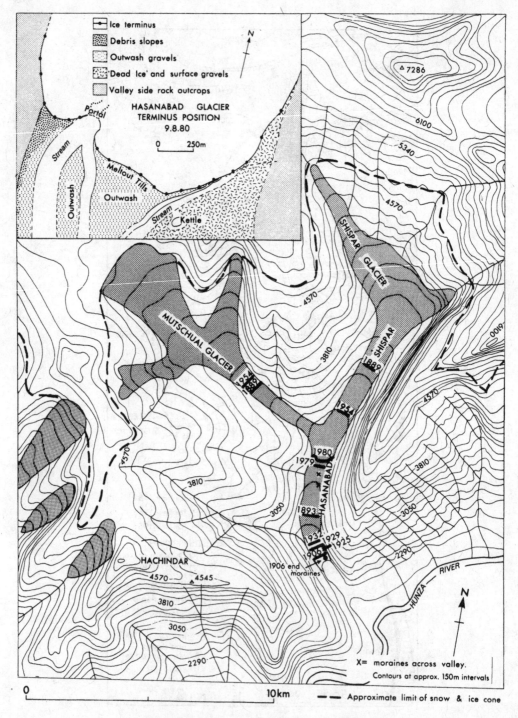

Ice terminus
Debris slopes
Outwash gravels
'Dead Ice' and surface gravels
Valley side rock outcrops

HASANABAD GLACIER
TERMINUS POSITION
9.8.80

0   250m

Portal

Stream

Meltout Tills

Outwash

Outwash

Stream

Kettle

N

△ 7286

6100

5340

4570

SHISPAR GLACIER

MUTSCHUAL GLACIER

4570

3810

SHISPAR
1889

1951

6100

1980
1979

4570

3810

HASANABAD

3050

1893

3810

1932 1929 1925

3050

1906
1905

2290

HACHINDAR

HUNZA — RIVER

4570   4545

3810

N

3050

X= moraines across valley.

2290   Contours at approx. 150m intervals

1906 end
moraines

0   10km   ▬ ▬ Approximate limit of snow & ice cone

FIG. 16.   VARIATIONS IN THE POSITION OF THE SNOUT OF THE
HASANABAD GLACIER, 1889 – 1980.

Sources:- Ams. India and Pakistan 1:250 000   Baltit Sheet sheet NJ 43-14 Series 0502   and references listed in text.

basin, the Shispar. At present the two ice tongues draining these basins combine 8.75 km from the Hunza river to form the Hasanabad itself. Both arms rise at around 4600 m, and descend to approximately 2750 m, a fall of 1850 m in 12.5 km, equivalent to a mean slope of about 14.8%.

The Hasanabad is of particular interest because it has the reputation of having undergone greater and more rapid advances and retreats than almost any other glacier on earth and, apart from the Minapin, its snout has been known to descend to lower elevations than any other glacier in the Himalayan-Karakoram ranges. Over the period 1889 - 1980 a total of 14 observations have revealed the snout position to be in radically different locations over a horizontal distance of 11 km; see Table 2 and Fig. 16. The two earliest reports of Ahmed Ali Khan in 1889 and Conway in 1892 indicate very limited ice-cover with the glacier restricted to the upper parts of its catchment and separated into a number of independent tributary tongues. The minimal extent of ice as portrayed by Conway (Fig. 12) has remained the subject of controversy ever since. By contrast, a number of observers during the early years of this century noted a major tongue of ice extending to within 4 km of the Hunza. (Figs. 17, 18, 19.) Thus, at some time around 1890, although the precise date is uncertain (7), the glacier must have rapidly advanced between 9.6 and 11.2 km in a matter of months. In fact Hayden (11) reported that it advanced about 9.7 km in two and a half months. The subsequent recorded history suggests a static snout position for about 22 years (1903 - 1925), then a period of great decline lasting for some three decades which culminated in the glacier withdrawing into the two tributary valleys so as to recreate a pattern similar to that reported by Ali Khan in 1889. A further recent phase of rapid re-advance resulted in the subsequent rejoining of the two branches by the end of the 1970's. The scale of these fluctuations in movement is shown in Figures 16 and 20, which are based on a synthesis of the observations in the literature but with care taken to reconcile many of the vague or contradictory statements. For example, the published reports of Hayden in 1907 have been preferred to those of Dr. Arthur Neve in 1895. Similarly, the observation of Zhang Xiangsong in 1980 that the positions of the east and west branches of the Hasanabad as shown on the map of Paffen et al. (published in 1959) indicates retreats of 4.5 km and 7.0 km respectively with regard to the 1929 position, also appears in error, the correct distances being 5.6 km and 8.15 km.

The visit to the glacier in 1980 revealed that the snout lay slightly up-valley from a mass of decaying moraine-covered stagnant ice with developing kettle holes. This suggests a slight advance to beyond its 1980 position and is compatible with the observations made by Chinese investigators when they visited the glacier in 1979 (13). The re-united glacier tongue downstream from the confluence consists of approximately 0.75 km of heavily crevassed and debris-mantled ice (Fig. 21 a) which is terminated by a steep, high and very active snout with a large unstable ice-cave on the western side (Fig. 21 b) from which emerges a large melt-water stream; inset of Fig. 16. The surface moraine varies in colour according to which valley source is supplying debris. The Chinese survey results suggest that the west branch advanced first so as to reach and extend beyond the confluence point, and that the later advance of the eastern ice tongue has resulted in it currently overlapping that of the west branch. Downstream from the ice-front the steep valley walls are relatively clean, although thin skins of till are plastered on to the walls up to a high clear trimline which records the depth of ice associated with the early twentieth century advance. At the furthest point of this advance there is a pronounced moraine which stretches half way across the valley and possesses several degraded kettle hole features. The recently vacated

432

FIG. 17: THE HASANABAD VALLEY DURING THE TWENTIETH CENTURY; IN 1906 (AFTER HAYDEN, 1907)

FIG. 18: THE HASANABAD VALLEY DURING THE TWENTIETH CENTURY; IN 1980.

OBSERVER	DATE	DETAILS OF SNOUT
AHMAD ALI KHAN (8)	1889	9.65 KM FROM HASANABAD RAVINE BRIDGE (EQUIVALENT TO 11.15 KM FROM HUNZA)
CONWAY (17)	1892	ROUGHLY 14.3 KM FROM HUNZA.
ABDUL GAFFAR (8)	1893	SNOUT 4.93 KM FROM HUNZA.
NEVE (8)	1895	REPORTS LOCAL COMMENT THAT ICE HAD ADVANCED 3.2 KM THAT YEAR AND 6.4 TO 8 KM IN PREVIOUS YEAR. THIS PROBABLY REFERS TO MOVEMENTS OF 1892/1893.
HAYDEN (11)	1903	TOLD THAT GLACIER HAD ADVANCED c. 9.7 KM ("A DAY'S MARCH") IN TWO AND A HALF MONTHS.
HAYDEN (11)	1906	SAME AS 1903. SURVEYED.
WORKMANS (20)	1908	SAME AS 1906.
MASON (18)	1913	SAME AS 1906 (3.329 KM FROM HUNZA).
VISSER (19)	1925	VOLUME OF ICE DECREASED. PORTION OF SNOUT WAS DEAD ICE. LENGTH THE SAME AS 1913.
TODD (8)	1929	GLACIER DIRTY. SNOUT 2.4 KM FROM RAVINE BRIDGE AND 366 M UPSTREAM FROM HAYDEN'S MARKS.
MASON (7)	1932	c. 550 M BACK FROM 1906 MAXIMUM (3879 M FROM HUNZA).
PAFFEN ET AL. (15)	1954	GLACIER HAD SPLIT INTO ITS TWO TRIBUTARY VALLEY GLACIERS. MUTSCHUAL AT 11790 M FROM HUNZA, AND SHISPAR 9286 M FROM HUNZA (EQUIVALENT TO 7 KM and 4.5 KM RETREAT SINCE 1929).
ZHANG XIANGSONG (13)	1979	GLACIERS RE-UNITED AND JUST BEYOND CONFLUENCE. MUTSCHUAL 6990 M FROM HUNZA AND SHISPAR 7786 M.
IKP	1980	NEAREST POINT OF SNOUT TO HUNZA = 7200 M.

TABLE 2   OBSERVATIONS ON THE HASANABAD GLACIER.

FIG. 19.   THE HASANABAD VALLEY DURING THE TWENTIETH CENTURY; IN 1980
(THE SNOUT POSITION IS MARKED BY AN X).

436

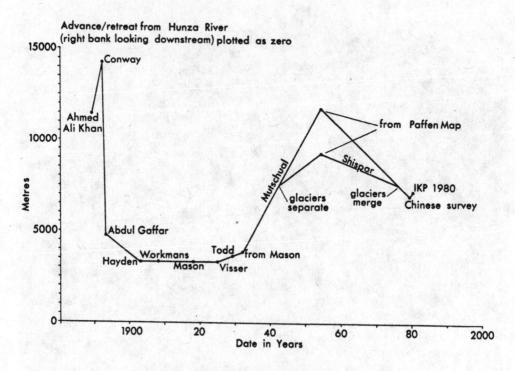

FIG. 20. GRAPH OF CHANGES IN THE POSITION OF THE
HASANABAD GLACIER SNOUT, 1889 - 1980

valley floor contains other debris (including outwash and morainic materials), with two irregular moraine ridges located at 2350 - 2400 m and 2400 - 2500 m elevation (Fig. 22) which probably record periods of relatively stationary activity and surface decay during main retreat phase. Since this retreat the valley walls have responded by unloading, the most dramatic feature being an enormous rockfall cone, 1650 - 1820 m down valley approximately from the 1980 ice-front position, which has succeeded in deflecting the river westwards. A wide variety of slope processes appear to be active, and the glaciated trough form is being modified with extensive boulder scree slopes.

Reconstructions of ice extent (Fig. 20) show that this glacier, although it has exhibited major advance and retreat phases, nevertheless displays a periodic pattern of fluctuation which is similar to that of Minapin (see Fig. 7). Mason (8) argued that such a dramatic and vigorous advance could only be explained as the product of accidental causes, such as earthquakes, or as a consequence of overcoming obstacles in its bed. He likened the Hasanabad to a clinical thermometer with the bulb lying in the mountains and the narrow valley below recording the eccentricities of the upper slopes just as the mercury registers the fevers of a sick man. He argued that secular or periodic variations would be entirely obliterated by such major accidental movements, and predicted a significant retreat between 1935 and 1940, followed by a further advance. This curious confusion of ideas, which expressed itself in his apparent confidence in his ability to predict accidental behaviour, is now recognised as misleading.

FIGS 21 a and b. THE HASANABAD SNOUT IN AUGUST 1980

a) The debris-covered area

b) The main portal

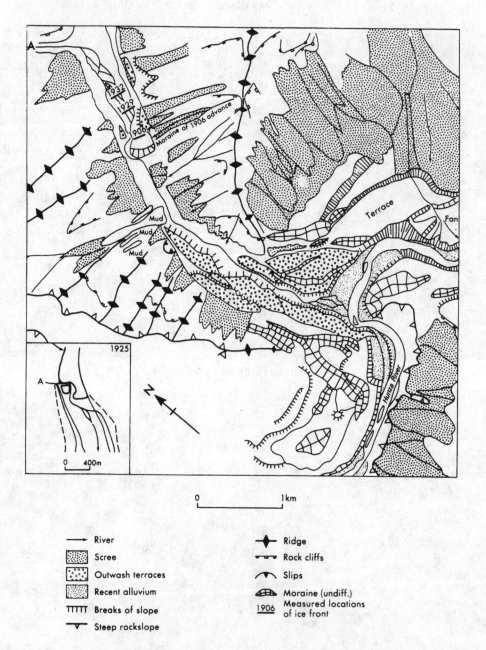

FIG. 22. GEOMORPHOLOGICAL MAP OF THE LOWER HASANABAD VALLEY SHOWING
SNOUT POSITIONS OF THE EARLY TWENTIETH CENTURY.

The fact that the Hasanabad glacier demonstrates a periodic pattern of movement very similar to that of the Minapin, indicates that the major advances and retreats simply reflect the way in which this glacier finds it possible to respond rapidly as and when environmental conditions dictate. Aspect could well be of fundamental importance in this respect, for it should be noted that the Hasanabad is the only glacier studied which is located on the southern slopes of a mountain mass. Visual inspection of small, steep, southward moving glaciers to the east of Aliabad (see Fig. 1) indicate that they too have recently changed in extent by relatively great amounts in proportion to their total length.

THE GHULKIN GLACIER

The Ghulkin (Sasaini) Glacier is the southernmost of the three eastward flowing glaciers that were investigated. It is approximately 17.5 km long with a headwall at up to 5950 m between two peaks that rise to 7331 m and 7613 m. The glacier descends from 5500 m to 2550 m, with a mean slope of 14.6%. Its terminus in 1980 was separated into two parts, a northern tongue and an irregular southern margin which could be further sub-divided into two small lobes separated by a meltwater portal (Fig. 23). It appears to have retained this general form since at least 1966. However, its heavily crevassed snout and numerous small caves testify to large-scale meltwater production. These meltwater flows frequently change in volume and exit position. The changing patterns of debris production and redistribution on the steeply sloping outwash cone (Fig. 24) pose a major hazard to the Karakoram Highway (Fig. 25) and a 200 m section of road was actually destroyed by these means during August 1980.

The main valley section of the glacier is enclosed between steep high moraines, the black debris-laden ice smothered by large quantities of superficial moraine. Ogives and rockfall crevasse patterns are well developed, and there are extremely steep ice falls towards the head. The twin lobes of the snout appear to have an extension in the ice topography for several kms up the glacier, the two ice ridges being separated by a pronounced central depression.

The recent history of the glacier is poorly known since most visitors have merely confined themselves to discussing the difficulties of crossing the outwash spreads and only give crude estimates of the distance of the snout from the Hunza River. Although the precise pattern of movement is unknown, reliable records do exist (Table 3) which indicate that the snout position has fluctuated by 625 m during the period 1885 – 1980 (Fig. 26), with a notable advance between 1913 and 1925 and a further small advance between 1966 and 1978. This again repeats the general pattern observed for the Minapin and Hasanabad glaciers, although the advance early in this century appears to lag by some ten to fifteen years (c.f. Fig. 1 and Figs. 7 and 20). It is significant that the minor advance of the early 1970's appears on all three records. The movements, therefore, appear to be periodic but in this case there is no overall secular retreat. There has been no record of blocking of the Hunza River, and no evidence either of massive retreat. Instead, the snout seems to have experienced limited oscillation around its present position.

THE PASU GLACIER

The Pasu Glacier, like the Ghulkin, rises from a col at just under 6555 m between two peaks of 7286 m and 7613 m. It is fed by a basin

FIG. 23. MAP OF THE TERMINUS OF THE GHULKIN GLACIER IN 1980

approximately 5 km wide, and is about 22.5 km long, descending to a snout elevation of around 2600 m (a mean slope of 17.6%). It is avalanche fed, and falls from steep hanging glaciers via a most beautiful cascade to a long straight, deeply incised valley track.

The 1980 snout (Fig. 27 b and Fig. 28) lies within two small arcuate moraines formed during the retreat stage of the mid-twentieth century (Fig. 27 c). The snout is now sharply pointed and lies on the southern side of the valley. Only the lower part is extensively covered in moraine. On the northern side of the valley there is an extensive area of dead ice covered with glaciofluvial gravels, meltout gravels and meltout till. There are two meltwater outlets, the most westerly having recently extended further up the glacier.

FIG. 24.   THE GHULKIN GLACIER SNOUT AND OUTWASH FAN.   ON THE RIGHT OF THE PICTURE IS THE HUNZA RIVER AND THE KARAKORAM HIGHWAY (NOTE THE BREACHED SECTION).

OBSERVER	DATE	DETAILS OF SNOUT
WOODTHORPE (8)	1885	732 M FROM "LEFT" BANK OF HUNZA. DESCRIBED AS A FEW HUNDRED YARDS WEST OF THE ROAD.
PRICE-WOOD (8)	1907	WITHIN A COUPLE OF HUNDRED YARDS OR LESS OF THE HUNZA.
MASON (8)	1913	300 M FROM LEFT BANK OF HUNZA RIVER. BEGUN TO RETREAT.
SKRINE (8)	1924	SPACE OF AT LEAST HALF A MILE BETWEEN SNOUT AND RIVER.
AFRAZ GUL KHAN AND VISSER	1925	1000 YARDS FROM LEFT BANK OF HUNZA.
PAKISTAN SURVEY (13)	1966	TERMINUS AT 2520 M. NORTHERN TONGUE 500 M FROM LEFT BANK OF HUNZA RIVER. SOUTHERN TONGUES AT 600 M.
CHINESE SURVEY (13)	1974	THEODOLITE SURVEY. SOUTHERN TONGUES AT 2534 M AND 800 M FROM HUNZA. A RETREAT OF 180 M SINCE 1966.
INTERNATIONAL KARAKORAM PROJECT	1980	NORTHERN TONGUE IS MINIMUM OF 418 M FROM HUNZA. THE NORTHERN PORTAL IS 733 M FROM HUNZA. THE INDISTINCT SOUTHERN LOBE VARIES FROM 550 - 600 M FROM HUNZA. THE RESPECTIVE FIGURES TO THE KARAKORAM HIGHWAY (RIVER SIDE) ARE 393 M, 708 M AND 548 - 560 M.

TABLE 3   DETAILS OF THE GHULKIN GLACIER SNOUT POSITION.

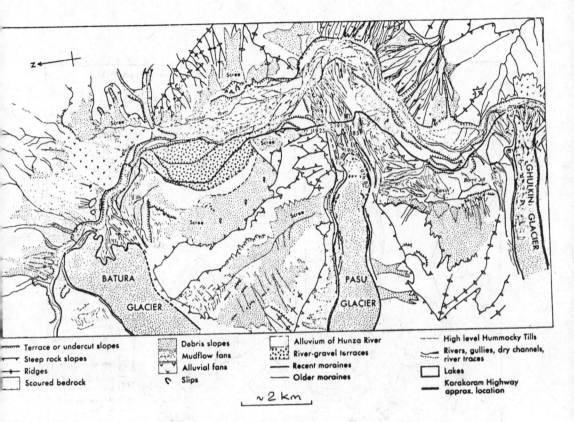

FIG. 25.   GEOMORPHOLOGICAL MAP OF THE GHULKIN GLACIER
IN RELATION TO THE HUNZA VALLEY

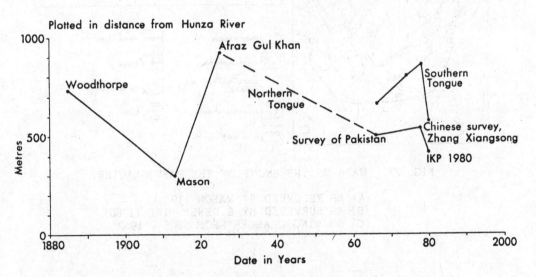

FIG. 26.   GRAPH OF THE CHANGES IN THE POSITION OF THE
GHULKIN GLACIER SNOUT, 1885 – 1980

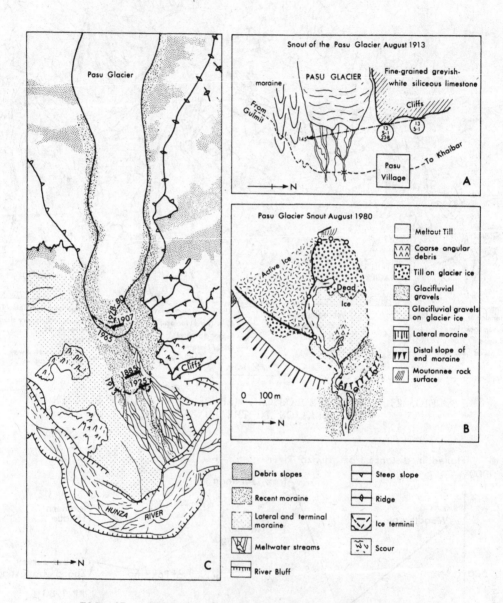

FIG. 27.   MAPS OF THE SNOUT OF THE PASU GLACIER:

(A) AS RECORDED BY MASON (1913)
(B) AS SURVEYED BY E. DERBYSHIRE (1980)
(C) SHOWING CHANGES FROM 1885 – 1980

445

FIG. 28. THE SNOUT OF THE PASU GLACIER IN 1980, WITH THE KARAKORAM HIGHWAY IN THE FOREGROUND AND THE LIMESTONE CLIFF OF FIGURE 15A ON THE RIGHT.

Earlier records (summarised in Table 4 and Fig. 29) show that in the early part of the century there was a rapid advance so as to bring the glacier snout parallel with the main limestone cliffs near the village of Pasu and close to the Jeep track (Fig. 27 a). The area covered by ice at that time (c. 1910 - 1930, see Fig. 27 c) is now a confused zone of moraines, till mounds, and outwash gravels, and there is a conspicuous polished trim line testifying to the former extent of ice. The present outwash forms a broad plain that extends all the way to the Hunza River (Fig. 27 c). The graph of movement (Fig. 29) shows a broadly similar pattern to that recorded for the other glaciers described previously, except that there does not appear to have been a minor advance at the time of the Chinese survey (1978), so that the glacier has now effectively withdrawn to the position it had in the early 1900's. Thus, this glacier also displays the periodic pattern of movement noted for the other glaciers.

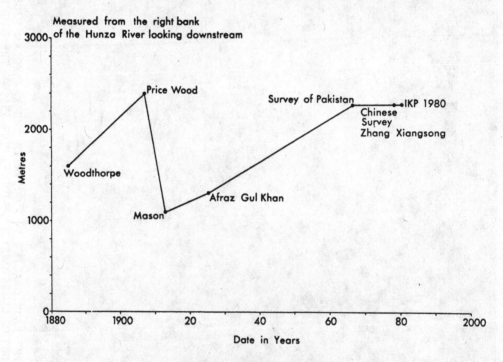

FIG. 29. GRAPH OF THE CHANGES IN THE POSITION OF THE
PASU GLACIER SNOUT, 1885 - 1980

THE BATURA GLACIER

The Batura Glacier is a large west-north-west to east-south-east longitudinal valley glacier and is one of the eight largest glaciers in the middle and low latitudes of the world, with a length in excess of 50 km. The drainage area totals 687 km^2 and the basin contains peaks up to 7795 m. The total area of all the glaciers in the system is 332 km^2, and they occupy 48% of the basin area. The Batura itself has a length of 59 km, an area of 285 km^2, and descends from a maximum elevation of 6200 m on the western ice flow down to the Hunza River at 2516 m (a mean slope of 6.2%). The glacier consists of five main ice flows and over 20 subsidiary tributary glaciers. There are numerous other glaciers which,

OBSERVER	DATE	DETAILS OF SNOUT
WOODTHORPE (7)	1885	ORIGINAL SURVEY SHOWS GLACIER 1.6 KM FROM HUNZA
PRICE-WOOD (7)	1907	SNOUT 2.4 KM FROM HUNZA
MASON (18)	1913	SNOUT IN LINE WITH LIMESTONE CLIFF TO NORTH, SKETCH. 1100 m FROM HUNZA.
AFRAZ GUL KHAN (7)	1925	SNOUT 1326 M FROM HUNZA. SIGNS OF RETREAT.
SURVEY OF PAKISTAN	1966	TERMINUS AT 2550 M ELEVATION. 1150 M FROM JEEP ROAD (WHICH IS AN ESTIMATED 1125 M FROM HUNZA).
CHINESE SURVEY	1978	RETREAT OF ICE CAVE BY 140 M, NORTHERN TONGUE "HEAVY RETREAT", SOUTHERN TONGUE 30 M, ELEVATIONS OF SNOUT 2615 - 2640 M.
INTERNATIONAL KARA-KORAM PROJECT	1980	SURVEY BY E. DERBYSHIRE INDICATES LITTLE CHANGE FROM CHINESE SURVEY.

TABLE 4   DETAILS OF THE PASU GLACIER SNOUT POSITION.

448

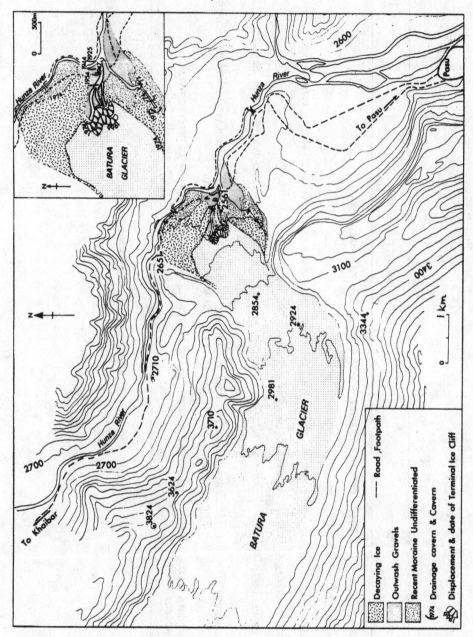

FIG. 30   THE SNOUT AREA OF THE BATURA GLACIER (MODIFIED AFTER THE BATURA GLACIER INVESTIGATION GROUP, 1976).

while not actually part of the Batura, contribute meltwater and debris to it. The south facing slopes of the basin are less heavily glaciated than those that face north, with magnificently developed hanging glaciers. Two thirds of the main glacier is covered with supra-glacial moraine, the last 3 km being totally blanketed. The morainic characteristics show that the combined flows from the western part of the basin exert a dominant influence, and where the flows from the north-facing slopes meet the trunk, pronounced ogive ridges are formed which persist for some distance down the glacier. The basin is notorious for the frequency and vigour of its avalanches (6), descending from a local snowline at approximately 4200 m. This extremely large glacier is considered by the Chinese Batura Glacier Investigation Group of 1976 (14) and Yafeng and Zhang Xiangsong in 1978 (13) to be still active and not simply a residue (relic) of the last glaciation, and that the glacier, in its upper parts, is a cold one, but changes to a temperate one in its lower and middle reaches.

The Batura glacier terminus is both well known and of considerable local significance because it descends so close to the Hunza River (Fig. 30). Historically it is known to have occupied a position either close to, or across, the Hunza River (see Table 5 and Fig. 31), and may

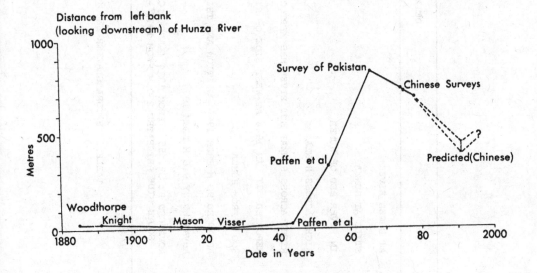

FIG. 31. GRAPH OF CHANGES IN THE POSITION OF THE
BATURA GLACIER FROM 1885 - 1980

even have created a temporary dam in the past, with resultant flood damage to the village of Pasu (18). Since the 1940's, however, the glacier has retreated quite rapidly until 1966. Since that time it has again tended to advance steadily and is at present 700 m from the Hunza River. (Fig. 30). This latter advance caused considerable damage to the Karakoram Highway and led to the intensive study carried out by the Chinese investigation group. They concluded from ice flux, ablation and velocity studies, that the glacier should stop advancing during the period 1991 - 7, with a predicted further advance of 180 m - 240 m before that time, taking it to within 300 m of the Highway. Starting from the 1990's they predict that the glacier will once again suffer decline and this will last for at least 20 to 30 years.

OBSERVER	DATE	DETAILS OF SNOUT
WOODTHORPE	1885	AT HUNZA RIVER
KNIGHT	1891	STILL AT HUNZA
MASON	1913	IN RIVER BED OF HUNZA
VISSER	1925	RIGHT ACROSS HUNZA
AFRAZ GUL KHAN	1925	RIGHT ACROSS HUNZA? BUT RIVER HAS KEPT CHANNEL OPEN.
PAFFEN	1954	REPORTED IT IN 1944 AT WEST SIDE OF RIVER AND IN 1954 3OO M AWAY FROM RIVER
SURVEY OF PAKISTAN	1966	8OO M FROM HUNZA
CHINESE	1974	ADVANCED 9O M FROM 1966 POSITION AND THICKENED BY 15 M BUT DECLINED AT SIDES.
CHINESE	1975	ADVANCED BY 9.6 M FROM 1974
CHINESE	1978	ADVANCED BY 32. 55 M FROM 1974 (LARGE ICE CLIFF IN CENTRE). SMALL ICE CLIFF ON SOUTH SIDE RETREATED 23O M FROM 1975 - 78.

TABLE 5 BATURA GLACIER SNOUT DETAILS.

It is noticeable that the Batura, which is a longitudinal valley glacier of great size and complexity, does not behave in a similar temporal fashion to the much smaller and relatively simple transverse glaciers hitherto described. The phase of rapid advance in the early part of this century displayed by those glaciers did not occur on the Batura, which remained virtually stationary from 1885 to the 1940's (Fig. 31). However, the retreat which occurred in the 1950's and 1960's, and the subsequent advance (both real and predicted) perhaps shows a pattern which is related to larger scale changes in the regional environment.

OTHER GLACIERS

There are numerous other glaciers in the Hunza catchment, the majority of which are short, steep hanging glaciers. There are, however, some very long valley glaciers similar to those described and these include the Baltar, Barpu, Gharesa, Hispar, Ghutulji Yaz, Gul Khwaja Ulwin, Yashkuk Yaz and the Gulmit. The character and evolution of the glaciers are unknown, for they have never been properly surveyed and there appear to be no readily available historical records of their snout positions. There is obviously much scope for further work on the glaciers of this region.

CONCLUSION

The glaciers of the Hunza valley that have been discussed in this paper have undergone considerable change over the last century. The historical evidence is most complete for the short transverse glaciers which descend from the Rakaposhi and Pasu Mustaghs. These appear to have generally advanced or to have been stationary during the early part of the century (Fig. 32). In some cases the advance was extremely rapid. During the

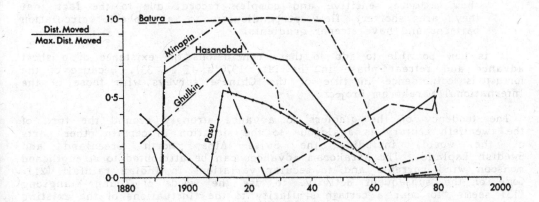

FIG. 32. COMPOSITE GRAPH OF CHANGES IN THE POSITION
OF THE HUNZA GLACIERS SINCE THE 1880's

subsequent period lasting from the 1930's to mid-1970's, the glaciers nearly all displayed patterns dominated by retreat, although some of those studied show a short minor pulse of advance and retreat in the 1960's. In some cases there appears to be a time lag before the dominant movements appear on the glacier record. Although this is partly due to the fragmentary nature of the historical record and the different timing of observations on each of the glaciers, it may reflect differences between the environmental conditions of the basins concerned. The Rakaposhi glaciers recorded here tend to show a long-term retreat. Those from the Pasu and Batura ranges show an oscillation about a generally stationary snout position. Five of the glaciers (Minapin, Pisan, Hasanabad, Ghulkin and Pasu) show similar short-term behaviour, to the extent that in four cases there is a record of slight advance in the late 1970's. It is, therefore, possible to offer the tentative proposal that there may be a regional pattern of glacial behaviour. If this is true, then it is necessary to reject the hypothesis tendered by Mason (8), and followed in the literature by others, that the sometimes rapid advance displayed by some of the glaciers (e.g. the Hasanabad) were merely accidental events caused by such factors as earth-quakes and valley obstruction. On the contrary, they appear to display instead a variable rate response to environmental changes that have affected the whole area. This conclusion is in broad agreement with the comprehen-sive review of Trans-Himalaya glacier activity presented by Mayewski and Jeschke (10); (see also Fig. 33) which suggested that:

a) The dominant regime since 1850 has been retreat.

b) At some time during the period 1870 – 1970 there have been secondary trends of advance and standstill.

c) During the years 1920 to 1940 the number of advancing and stationary termini exceeded the number of retreating termini, suggesting a near reversal in the dominance of the retreat regime for this period.

d) Longitudinal and transverse glaciers demonstrate different fluctuation histories, and

e) The transverse glaciers, which are mainly described in this paper, show a more sensitive and complex record, due to the fact that they are shorter, flow perpendicular to atmospheric circulation patterns and have steeper gradients.

It is now possible to add to these conclusions the existence of a short advance and retreat phase in the late 1970's (Fig. 33), because of the fortuitous coincidence in time of the Chinese surveys with those of the International Karakoram Project.

The tendency of the glaciers to advance strongly around the turn of the twentieth century is analogous to the situation in certain other parts of the world, including the Swiss Alps, north Greenland and Swedish Lapland. The Karakoram advances can be attributed to strengthened monsoon wind currents and to secular variations in Indian rainfall (21). Some of the subsequent activity, to use the words of Zhang Xiangsong (13) seems "to bear a certain similarity to the fluctuations of the existing Alpine glaciers in Europe, with the exception that the time lags a little behind." Given that an increasing number of Alpine glaciers seem to be in a state of advance at the present time it will be valuable to monitor the Karakoram glaciers over the next decade to see whether they behave in a similar manner.

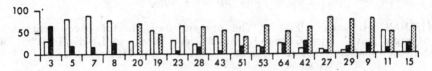

Combination of all glacier fluctuation records from the Himalayas

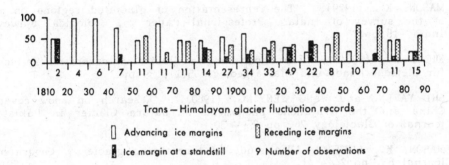

Trans—Himalayan glacier fluctuation records

☐ Advancing ice margins   ⬚ Receding ice margins

◼ Ice margin at a standstill   9 Number of observations

FIG. 33.   THE CHANGING STATE OF GLACIERS IN THE
HIMALAYAN REGION
(Modified and up-dated from Mayewski and Jeschke, (10))

ACKNOWLEDGEMENTS

This work was carried out by members of the International Karakoram Project, 1980.   We are indebted to the Royal Geographical Society,   the National Geographic Society, and the Universities of Oxford and London for financial support.   Dr. Ted Smith kindly assisted us with the Minapin Survey, Zafir Hashmet with the Ghulmet, Professor Ron Waters with the Pisan, and Dr. Ed. Derbyshire with Pasu.   The distance meter was provided by Hewlett Packard to whom we are most grateful.   In addition we are grateful to the cartographers, photographers, and secretarial staff of the School of Geography, Oxford, and the Departments of Geography at King's College, and the London School of Economics, London.   At base camp we were fortunate in the assistance we received from N. de N. Winser, Miss Shane Wesley-Smith and Jamal Khan.   Transport was supplied by British Leyland (Land Rover V8).   We are indebted to Dr. J. M. Grove and Dr. J. B. Auden for comments on a first draft of this paper.   We thank Mr. S. A. Bucknell, M.A., FRGS, Senior Geography Master, Monmouth School for Boys, for providing photographs by Melaram.

REFERENCES

1) SCHNEIDER, H. J., (1961). 'The German Karakoram Expedition, 1959'. In Berge der welt 1960 - 1961, pp 86 - 105.

2) WISSMAN, H. von, (1959). 'The present day glacier cover and snowline in high Asia' (in German), Akad. d. Wiss. und. d. Litt im Mainz Abh. der math.-naturw. Klasse. Heft 14, pp 1101 - 1434.

3) SCHNEIDER, H. J., (1969). 'The Minapin Glacier and Men; NW Karakoram'(in German), Die Erde 100, pp 266 - 286.

4) MASON, K., (1929). 'The representation of glaciated regions on maps of the survey of India', Professional Paper 25, Geological Survey of India, 18 pp.

5) KICK, W., (1964). 'The Chogo-Lungma Glacier: Karakoram', Zeitschrift für Gletscherkunde und Glazialgeologie 5(1), 59 pp.

6) SHI YAFENG and WANG WENYING, (1980). 'Research on snow cover in China and the avalanche phenomena of Batura Glacier in Pakistan', Journal of Glaciology 26, pp 25 - 30.

7) MASON, K., (1935). 'The study of threatening glaciers', Geographical Journal 85, pp 24 - 41.

8) MASON, K., (1930). 'The glaciers of the Karakoram and neighbourhood', Records Geological Survey of India 63, pp 214 - 278.

9) MERCER, J. H., (1975). In W. O. Field (ed.) Mountain Glaciers of the Northern Hemisphere, pp 371 - 409. (US Cold Regions Research and Engineering Laboratory, Hanover, NY, Vol. 1).

10) MAYEWSKI, P. A. and JESCHKE, P. A., (1979). 'Himalayan and trans-Himalayan glacier fluctuations since AD 1812', Arctic and Alpine Research 11, pp 267 - 287.

11) HAYDEN, H. H., (1907). 'Notes on certain glaciers in Northwest Kashmir', Records Geological Survey of India 35, pp 127 - 137.

12) VISSER, P. C. and VISSER-HOOFT, J., (1938). 'Karakoram; Scientific Results from the Netherlands Expedition to the Karakoram and adjacent areas in 1922 - 25, 1929 - 30 and 1935'(3 vols.) Brill:Leiden. (in German)

13) ZHANG XIANGSONG, (1980). 'Recent variations in the glacial termini along the Karakorum Highway', Acta Geographica Sinica 35, pp 147 - 160.

14) BATURA GLACIER INVESTIGATION GROUP, (1976). Investigation report of the Batura Glacier in the Karakoram Mountains, the Islamic Republic of Pakistan (1974 - 5). 123 pp.

15) PAFFEN, K. H., PILLEWIZER, W. and SCHNEIDER, H. J., (1956), 'Forschungen im Hunza-Karakoram', Erdkunde. 10, pp 1-33.

16) MacBRYDE, D. H., (1964). Scientific report of studies carried out on the Minapin Glacier by members of the Cambridge Expedition to Nagir, Karakoram, 1961.

17) CONWAY, W. M., (1894). Climbing and exploration in the Karakoram Himalaya, 3 vols., Unwin, London.

18) MASON, K., (1914). 'Examination of certain glacier snouts of Hunza and Nagar', Records, Geological Survey of India 6, pp 49 - 51.

19) VISSER, P. C., (1928). 'Glaciers of the higher Indus' (in German), Zeitschrift für Gletscherkunde 16, pp 169 - 229.

20) WORKMAN, W. H. and WORKMAN, F. B., (1910). 'The tongue of the Hasanabad Glacier in 1908', Geographical Journal 36, pp 194 - 196.

21) MAYEWSKI, P. A., PREGENT, G. P., JESCHKE, P. A. and AHMED, N., (1980). 'Himalaya and trans-Himalayan glacier fluctuations and the south Asian monsoon record', Arctic and Alpine Research 12, pp 171 - 182.

Quaternary glacial history of the Hunza Valley, Karakoram mountains, Pakistan

E. Derbyshire – Keele University
Li Jijun – Lanzhou University, P.R.C.;
F.A. Perrott – Oxford University; **Xu Shuying** – Lanzhou University, P.R.C.;
R.S. Waters – Sheffield University

ABSTRACT

The glacial landforms and glacial and paraglacial sediments, of the upper Hunza valley were mapped and resolved into a chronological sequence on the basis of morphostratigraphy, superposition, and degree of surface weathering. Sediments recognised include fluvial, glaciofluvial, debris-flow and lacustrine types together with tills of predominantly meltout origin. Sediments were analysed for particle size, projection roundness, surface texture, clast and microfabric and clay mineralogy.

Weathering of surface boulders, degree of subsurface weathering, point compressive strength of boulder surfaces, desert-varnish development and percentage lichen cover were used to establish a relative chronology, the desert varnish index providing results consistent with other tests.

Eight glacial phases of varying magnitude were recognised, as follows: (1) A widespread glaciation, leaving isolated and deeply weathered erratics on summit surfaces above 4150 m altitude. This occurred when valleys were much less incised and is considered to be the earliest Pleistocene glaciation represented in the area and termed Shanoz. (2) and (3) A second series of valley glaciations (c. 2500 m) was recognised from weathered till remnants on benches in the main Hunza valley. These are attributed to the Yunz and Borit Jheel glacial stages, the former being older than 139,000 yr BP (TL dating of lake silts) and correlated with the penultimate Pleistocene glaciation. The Borit Jheel glaciation filled the high diffluence cols and is dated (TL) as early last Pleistocene glaciation. (4) A stage of 'expanded foot' and minor valley glaciation (Ghulkin I);this is younger than 47,000 yr (TL). (5) A more limited glacial advance of several km, blocking the Hunza River, occurred in the Ghulkin II stage; considered to be a late phase of the last Pleistocene glaciation. (6) A glacial extension of 1-2 km left larger, well defined moraines with well varnished and weathered surfaces (Batura stage: considered mid-Holocene). (7) A minor advance restricted to tributary valleys is indicated by mildly varnished moraines (Pasu I). It is bracketed by ^{14}C dates of 830 ± 80 and 325 ± 60 yr BP. (8) Minor oscillations of the 19th and 20th centuries which left sharp, unweathered metastable moraine forms and termed the Pasu II stage.

INTRODUCTION

Notwithstanding the long-established scientific attraction of the Himalaya, the study of its geomorphic evolution and Quaternary glacial stratigraphy is still in a rudimentary stage. To a large extent, this is a result of the scale and formidable difficulties posed by the terrain itself and, in the past, by the restrictions arising from political considerations. Nowhere is this truer than in the Karakoram. The considerable engineering achievement of the rebuilt Karakoram Highway in the 1970's, together with the increasing encouragement of scientific research and international co-operation by the governments of Pakistan and China, has now made areas such as the Hunza Karakoram much more accessible.

The following parts of the Hunza valley were investigated: the area between the Batura and Gulmit glaciers, the gorge section of the upper Hunza River around Saret, the stretch from Baltit to Aliabad, and the middle and lower Minapin, Pisan and Ghulmet glaciers on the northern slopes of Rakaposhi (7788 m) near Hindi (Fig. 1). The study areas, therefore, span the northern sedimentary, axial granitic, central-southern metasedimentary and Chalt green schist zones of Desio (1): (see also Figs. 4 and 5 in (2)). The arid climatic environment of the middle and lower slopes and the resulting sparse vegetation cover, while facilitating field observation and sampling, pose absolute dating problems. Very little wood or charcoal occurs, there is no peat, few lakes exist and lichens are sparse. The skeletal soils contain very little organic matter and palaeosols do not provide the same potential for dating as in the European Alps (3).

REVIEW OF EARLIER WORK ON THE LANDFORM EVOLUTION
AND GLACIAL HISTORY OF THE KARAKORAM AND ADJACENT REGIONS

Detailed study of the glacial chronology of the Himalaya including the Karakoram was initiated early in this century by the pioneering work of De Filippi's Italian Expedition to the Himalaya, Karakoram and Chinese Turkestan (4). On the basis of mapping of the Kashmir Basin and the upper Indus, especially around Skardu, a four-fold glacial sequence was proposed by analogy with Penck and Bruckner's sequence for the European Alps (5). However, the first stage recognised was ascribed to the Mindel and the fourth to a post-Würm (possibly Bühl) advance. A map of the extent of the third glaciation (Würm correlate) in the upper Indus shows a large glacial lake in the main Indus valley, dammed by ice from the Astor valley and stretching from below Gilgit into the upper Hunza valley. This four-fold sequence was retained by de Terra and Paterson (6), with the difference that they equated the fourth glaciation with the Würm.

Norin (7) believed the ice of the last Pleistocene glacial maximum (termed the Shigar glacial subepoch) filled the whole of the Indus drainage system, descending to altitudes of between c. 1675 and 1830 m on the Kashmir slope. This was deduced from dissected outcrops of 'bottom moraine' within old river terraces. The Shigar stage was considered to date from 14,000 - 13,000 yr BP (based on the Swedish varve chronology) and to be followed by the 'Khapalu glacial subepoch' between 11,000 and 9,800 yrs BP, characterised by local but impressive loess-covered terminal moraines at about 2440 m with glaciolacustrine sediments on the upstream side. Norin regarded the Indus valley between Skardu and the junction with the Suru River as having had no through-flowing glacier at this stage, and to have been affected only by local glacier incursions. Finally,

458

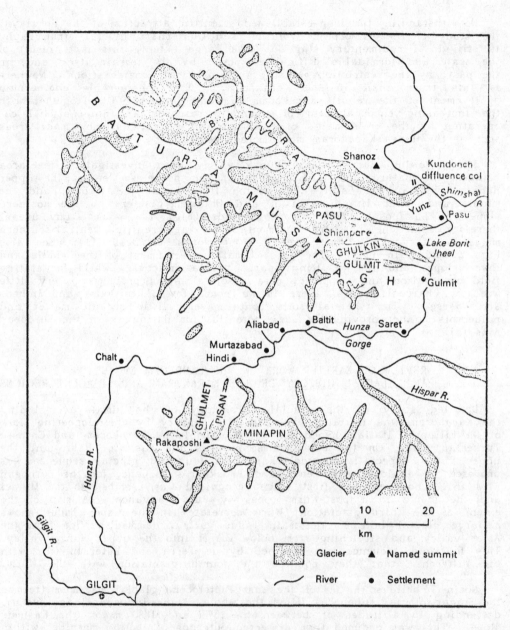

FIG. 1. LOCATION MAP OF THE UPPER HUNZA RIVER REGION.

he classified the moraines 1 – 5 km below the ice fronts as sub-recent (3,000 – 2,000 yr old), formed during the 'post-glacial subepoch'.

The terrace sequences of the Indus and its tributaries, the Eocene and later tectonic history, and the associated planation surfaces of the western Himalaya and southern Qinghai-Xizang (Tibet) provide data on the geomorphological evolution of the region. Woldstedt (8) following Movius (9), used the Alpine terminology for the terraces of the River Soan (Fig. 2). He regarded the main boulder bed of Morris (10) as indicative

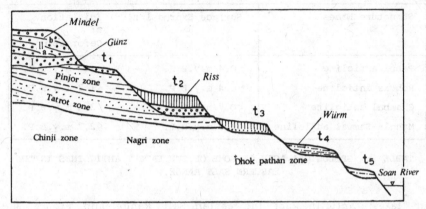

FIG. 2. CROSS-SECTION OF THE INDUS TERRACES
NEAR THE SOAN RIVER; REFERENCES (8) AND (9).

of the first glaciation of South Asia, followed by the first interglacial (Cromerian) and then the Mindel glaciation, this sequence being combined in the well known Boulder Conglomerate Formation. In contrast, de Terra and Paterson (6) placed the first glaciation in the Tatrot Zone, the maximum extent of Pleistocene ice occurring in the second glaciation and its till merging with the Boulder Conglomerate in the piedmont. The work of Gill (11) and the recent palaeomagnetic dating reported by Johnson et al (12) for the Siwaliks make it clear that both of these earlier interpretations are in error. It is now clear that the Tatrot Zone is much older than the first glaciation and no unconformity can be found separating the Tatrot and Dhok Pathan Zones. Moreover, there are two conglomerates. The Boulder Conglomerate (last member of the Siwalik sequence) is underlain by a clear disconformity, slightly younger than the Olduvai event. Clearly, a major change in sedimentary regime is indicated, reflecting vigorous upheaval of the Himalaya. Post-dating the Boulder Conglomerate is the Lei Conglomerate of Gill (11), a valley-fill mostly consisting of local lithologies. This is a major post-Siwalik formation (13). Since the Upper Siwalik is Pliocene to early Pleistocene in age (magnetic epoch 5 to early Brunhes), the Lei Conglomerate must be of Middle Pleistocene age. Accordingly all the Indus-Soan terraces postdate this period.

The Potwar Plateau, Peshawar region and southern Qinghai-Xizang extend the Himalayan sequence back in time. The Potwar Plateau – Peshawar region of Pakistan is a portion of the north-west Himalayan syntaxis (14), which was subjected to successive uplifts in the late Cenozoic. It is thus of considerable importance in the study of the morphological development of the Himalaya as a whole. The successive uplifts of the Himalaya were associated with expansion of the mountainous region and the southward migration of the piedmont depression. The Potwar

Plateau, 500 m high, was the latest portion to be involved in this mountain expansion. The Salt Range with the Main Front Thrust fringes the plateau on the south, and the famous Main Boundary Thrust on the north. Data obtained by recent palaeomagnetic dating suggest an age of less than 0.7 m.y. for the Chambal ridge, and about 0.4 m.y. for the Rohtas and Pabbi anticlines on the banks of the Jhelum River in the eastern Salt Range. It seems, therefore, that the Main Front Thrust was activated at the same time and that the Potwar Plateau began to rise, resulting in the rejuvenation of the rivers across the surface. Table 1 gives the dates of the

Structure Names	Surface Expression	Initiation of Deformation
Pabbi anticline	~0.4 m.y.B.P.	~1.2 m.y.B.P.
Rohtas anticline	~0.4 m.y.B.P.	~1.7 m.y.B.P.
Chambal anticline	<0.7 m.y.B.P.	<2.4 m.y.B.P.
Mangla-Samwal anticline	<1.5 m.y.B.P.	<2.7 m.y.B.P.

TABLE 1: GEOLOGICAL AGES OF SOME OF THE LATEST ANTICLINES IN THE EASTERN SALT RANGE.

the four latest anticlines of the eastern Salt Range and Figure 3 shows the morphology of the Potwar Plateau and adjacent areas. Two main surfaces occur in the piedmont zones of the Himalaya. One is the Potwar Plateau and the other is a dissected planation surface at an elevation of about 1,600 m a.s.l. The dates of the surface expression of the four anticlines in the eastern Salt Range become younger from north to south (range 1.5 m.y. to 0.4 m.y.) reflecting the southward expansion of the rising belt. As an erosion surface, the Potwar Plateau must have been formed before that time and thus is of early Pleistocene age. The 1,600 m surface, situated north of the Main Boundary Fault may be of Pliocene age because it truncates the Murree Series and has been dissected by a broad valley system which plunges southward into the Potwar Plateau surface.

Although landform development of the inner part of the Himalayas has been discussed for a long time, the age of the surfaces is still to be determined (15). Recent Chinese work has shown that this region has been planated twice since its emergence from the sea in the late Eocene (16). The lower planation surface developed in a warm humid environment with Hipparion fauna present over the whole region, and it is now well preserved at an elevation of 4,500 - 5,000 m a.s.l., taking the form of low and gently-sloping ridges with tabular uplands in the interior of the plateau. Towering above this 'plateau surface' are the remnants of the upper planation surface, thought to be of Eocene age, which often take the form of flat-topped mountains, usually at an elevation of more than 6,000 m a.s.l. As regional equilibrium lines are generally below this height, these flat-topped residuals are often covered with ice, including some large ice caps reaching hundreds of km^2 in extent. Just below the 'plateau surface' is a widely developed third surface made up of broad valleys and lake basins and considered to be of early Pleistocene age. Incised into this surface is the modern valley system, with tier upon tier of terraces on the valley sides, indicating successive uplift of the crust in the late Quaternary. When this outline is compared with

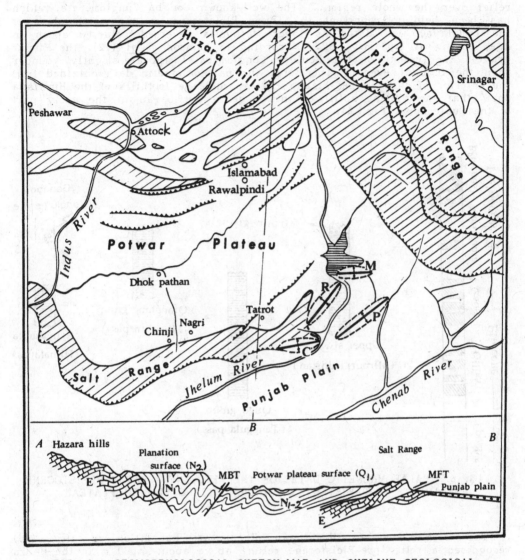

FIG. 3. GEOMORPHOLOGICAL SKETCH-MAP AND OUTLINE GEOLOGICAL
SECTION OF THE POTWAR PLATEAU AND ADJACENT AREAS.

the morphological elements of the Potwar Plateau and adjacent regions,
it is evident that similar morphological events have occurred in both
regions. Recent palaeomagnetic dating of lacustrine formations on the
Qinghai-Xizang Plateau gives the Qiangtang Group near the Kunlun Pass
an age of 2.7 - 1.4 m.y. BP and the Qugo Group, west of the
Tanggula Pass 4.32 - 2.2 m.y. BP (17). It appears that they are approximately the same age as the Tatrot and Pinjor zones in the Potwar Plateau.
Following deposition of the Qiangtang Group, strong tectonic movements
induced the formation and dislocation of the lacustrine beds. In addition,
a change in the sedimentary regime caused a shift from fine-grained material
to coarse clastics reflecting the vigorous uplift and an increase in relative

462

relief over the whole region. The well-known Gongba Conglomerate, which has been held to represent the Lower Pleistocene on the north slope of the Himalaya, was dated as lower Matuyama reversed polarity epoch in age, ending at 1.63 m.y. BP. According to Johnson et al (12), the Mirpur gravels (equivalent of the Boulder Conglomerate) appear slightly younger than the Olduvai event. Therefore, the palaeomagnetic data obtained from the interior of the Qinghai-Xizang Plateau and the foothills of the Himalaya show that the violent uplift of this huge region began in the early part of the Quaternary (Fig. 4).

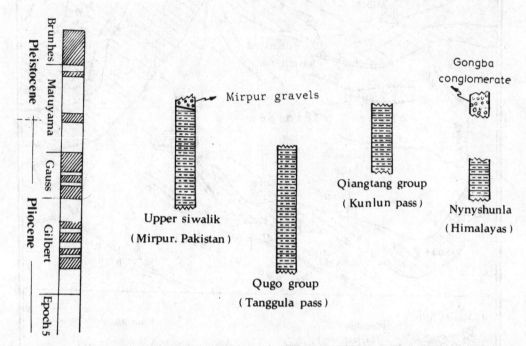

FIG. 4. PALAEOMAGNETIC DATES OF THE PLIOCENE - EARLY PLEISTOCENE SEDIMENTS IN PAKISTAN AND QINGHAI-XIZANG (TIBET) PLATEAU.

The German-Austrian Himalayan Karakoram Expedition of 1954 (18) recognised a relic 'pre-Pleistocene relief' above about 4,000 m in the Hunza region, with a shoulder below it at 3,000 m and then a deep gorge incised 1,000 m or more. The presence of a surface at 4,100 - 4,200 m a.s.l. was confirmed by the I.K.P. in the Pasu area. It is especially well-developed between the Batura and Pasu glaciers, where there is a 5 km long flat-topped ridge. This is referred to here as the 'Patundas surface' after the numerous shepherds' huts situated on it. Remnants of this surface are common along the Hunza valley, often making up the highest valley shoulder. They appear to represent a former valley floor with a still higher surface above it, e.g., the 5,200 m rock bench on Mirschikar (5445 m), opposite Aliabad. In view of the Rb/Sr age for the Karakoram granodiorite of 8.6 m.y. (19), both the upper (5,200 m) and lower (Patundas) surfaces must be much younger. It is reasonable to regard the 'Patundas surface' as early Pleistocene in age and the 5,200 m surface as Pliocene. The 'Patundas surface' may thus correlate broadly with the Potwar Plateau surface.

Apart from Trinkler (20), who refers to a snowline depression for the 'later glacial periods' of 600 - 800 m in the Kunlun, there is a hiatus of several decades in the work on Pleistocene chronology and snowlines in the upper Hunza region itself. A series of papers followed in the 1950's, notably those of Paffen et al. (18), Wiche (21) and Schneider (23). Paffen et al. suggest a late Pleistocene snowline lowering here of the order of 1,000 m. They argue, however, that, because of the very steep slopes, this downward shift did not increase the glacier accumulation areas sufficiently to maintain a trunk glacier in the main Hunza valley. Expansions of the tributary glaciers were recognised, including evidence for damming of the Hunza to form lakes near the mouth of the Hasanabad Glacier. They recognised a lowest limit of ice advance at about 2,000 m a.s.l. (of late or post-glacial age and said to be equivalent to Dainelli's fourth glaciation).

The Austrian Karakoram Expedition of 1958 (21) worked on the glacial landforms on the north and northwest sides of Haramosh (southeast of Rakaposhi) and in several valleys on the south side of the Gilgit Range, northwest of Chilas. North of Haramosh, Wiche recognised a maximum glacial advance during the last Pleistocene cold stage of only 4 - 5 km despite the severe snowline depression. In contrast, the 1,200 - 1,300 m snowline depression estimated by Wiche for the area northwest of Chilas gave rise to an extensive glaciation because there is a large area of terrain in the range 3,000 - 5,000 m a.s.l. This emphasis upon topographical control of ice extent during the late Pleistocene is also evident in the work of Heuberger (22) on the Everest region.

The paper by Schneider (23) is an important contribution as it was the first to consider in detail the morphology and weathering of Pleistocene sediments in the northwest Karakoram. In the Hunza valley, a threefold sequence was recognised, the oldest unit consisting of high indurated terrace remnants some 200 - 300 m above the younger terrace suite. Erratic blocks of granodiorite, devoid of matrix, resting at 4,000 - 4,150 m on the limestone ridge at Shanoz on the north side of the Batura Glacier also indicated an old and extensive glaciation of unknown age.

The second stage was thought to be represented by the 'young diluvial terrace of the Hunza valley', said to be of interglacial age, which is almost continuous between Khaibar and Gilgit and is c. 100 m thick in places, and the 'diluvial moraine ridges' extending down to 1,900 - 2,300 m in the Hunza valley. The terrace comprises thin, coarse to fine fluvial gravels at the base, followed by a thick diamict ('Blockschutt') with an unsorted loamy matrix. This material was considered to be of debris-flow rather than of glacial origin, and Schneider uses the term Fanglomeratstrôme to describe it. It is capped by thin fluvial beds ranging from gravels to clays. The 'diluvial moraine ridges' were described as lying on or against the terrace sediments and all were regarded as products of side glaciers, with no evidence of through-flowing glaciers in the trunk valley. The Batura system was considered to have extended as far as Saret (30 km), the Minapin only to Hindi (8 km) and the Hasanabad to Murtazabad (6 km). In the discussion on the age of the moraines of this stage (his 'Hunza valley stage') Schneider adopted Paffen et al.'s (18) view that the snowline depression approximated 1,000 m but, because of the steep terrain, no trunk glacier was maintained in the Hunza valley. He did not exclude the possibility that these moraines correlated with the fourth (i.e., first major post-glacial) stage of Dainelli(4), although a Würm maximum date was thought more likely. A third group of deposits consisting of moraines closely adjacent to the present glaciers was also recognised. The older of these was called the 'earth pyramid series' because of its severe dissection. They are somewhat indurated and carry an almost complete vegetation cover. They are best developed, according to Schneider, around the Batura Glacier, where their upper limits lie at over 600 m above the snout. Whether they correlated with the 'young diluvial moraines' was left cryptic and the suggestion was made that they might be older at Batura than elsewhere. The younger moraines close to the glaciers have a very sparse vegetation cover and were ascribed by Schneider to nineteenth century glacial advances (the 'historical advance' moraines).

A similar sequence of moraines near the Minapin Glacier was recognised in the unpublished Report of the Cambridge Expedition to Nagir, Karakoram (1961). The oldest remnants, some 200 m above the snout, were attributed to Schneider's earth pyramid series. Below these, five high 'Pleistocene' moraines were recognised on the valley wall and closer to the glacier, on the west side, a large young lateral moraine with a two-thirds vegetation cover was described. The youngest moraine was considered to be a lateral one close to the snout.

The first modern study applying relative age criteria, including morpho-stratigraphy of moraines and terraces, weathering properties of surface clasts and degree of soil development, was the work of Porter (24) in Swat Kohistan. Three 'drift sheets' were recognised. The Laikot Drift is the oldest. It is severely weathered, patchy in distribution and shows little or none of its primary depositional morphology. The moraines of the succeeding, or Gabral stage, are broad but smooth with few surface stones. A thick loess cover (c. 2 m) occurs on correlative terraces. The Kalam Drift is the most widely exposed and may correlate with the last Pleistocene glacial maximum. The equilibrium line altitude (ELA: altitude of the line dividing accumulation area from ablation area) was estimated as 914 m lower than today's for this stage. The Kalam Drift forms bouldery, sharp-crested moraines and is well forested. A thin (< 1 m) loess cover occurs on the correlative terraces. Little-weathered drift adjacent to modern glaciers was attributed to the 'Neoglacial', i.e., only 'a few thousand years old'. Two stages were recognised on the basis of side-slope angles and plant colonisation. Although this sequence is superficially similar to some suggested for the western Karakoram, Swat Kohistan is much wetter than the Hunza, loess is widely distributed and soil development is much more advanced, so that it cannot be used as a framework or model for the Hunza. This is also true of the findings of the Japanese Glaciological Expeditions to Nepal 1973 – 76 as presented by Iwata (25) and Fushimi (26, 27). The problem is compounded by the apparent inconsistencies between these three accounts. Two old glaciations (termed the Ghat and the Lughla) were inferred by Fushimi (27) from valley forms alone, in the absence of deposits. The oldest of the four stages of deposition succeeding the Ghat (the Thyangboche of Iwata (25)) represented by large but subdued and vegetated lateral moraines with a loess and buried soil cover, is thought to date from the maximum of the last glaciation, although its distribution is similar to that of Heuberger's (28) 'late glacial' moraines. The high (200 m) moraines of the succeeding Periche stage, covered with a buried soil and loess mantle and completely overgrown by grasses and shrubs, were considered to date from a 'late-glacial re-advance'. Available radiometric age determinations, however, conflict with this view (see below). The succeeding Thuklha stage is characterised by smooth moraine surfaces, partly covered by loess, soils and vegetation. The boulders are weathered and a date in the range 2,000 to 6,000 yr BP was suggested. The youngest moraines of all carry only a light vegetation cover, and are sharp crested, the most recent being lichen free. They are ascribed by Iwata (25) to the Lobuche stage.

A detailed study of the drift of the upper Hunza valley was undertaken in 1974 and 1975 by teams from the Chinese Academy of Sciences in response to the threat to the Karakoram Highway posed by the advances and variable water discharge from the Batura Glacier. Their first report (29) contains a map showing several stages of moraine deposition, from the last Pleistocene glaciation (termed the Dali Glaciation in western China (30)) to the present. The oldest remains recognised were tills found at over 3,500 m on the mountain sides both north and south of the Batura. The Batura Glacier, together with the Pasu, Ghulkin and Gulmit was thought to have formed a system which filled the Hunza valley and reached a point at least 30 km downstream from the present terminus ((31), following (23)). Using cirque threshold values and glaciation limit estimates, they suggested an equilibrium line during the Dali Glaciation of 4,200 m, that is about 800 m below the present equilibrium line which ranges from 4,700 to 5,300 m ((31) cf. the 4,800 - 5,000 m cited by von Wissman (32): Fig. 14). An extensive area of dark brown-coloured moraines some 60 - 70 m thick, extending over 2 km downstream from the Batura snout, was ascribed to the 'early Neoglacial' with an estimated age of about 3,000 yr BP (30). Inside these, and extending at most 400 - 500 m in front of the outermost ice-cored drift is the 'yellowish moraine' ascribed to the culmination of the Little Ice Age 'about 200 - 300 yr ago'. This advance cut the local irrigation canals which suggests an ice thickness 50 - 60 m greater than the present close to the Hunza River. Inside this zone, an area of 2 km^2 of 'greyish moraines' was recognised and ascribed to the period 1880 - 1974. Except for two small areas near the river, most of this grey till has a substantial ice-core.

Subsequently, Zhang and Shi (33) cited geomorphological and sedimentary evidence for at least three Pleistocene glacial stages in the Batura region, which they termed Shanoz (oldest), Yunz and Hunza. Weathered and isolated erratics are the only sedimentary signs of the Shanoz glaciation, although they rest on glacially-moulded, high level surfaces around the summit of Shanoz Mountain (4350 m, cf. Schneider (23)). More extensive, but also markedly weathered glacial sediments were described around the north entrance to a large diffluence col running southward from Yunz to the Pasu Glacier. Finally, thick accumulations of weathered tills, glaciofluvial and glaciolacustrine deposits occurring as a plateau overlooking the Hunza River 1.5 km north of Pasu village were ascribed to the third, or Hunza stage. Relics of this age were also mapped on the north side of the Batura Glacier in the diffluence col near Kundorich, on the north side of the Shimshal River, and in several outcrops to the west of the Pasu plain.

Deposits classified as 'post-Neoglacial' include debris-flow and alluvial fan sediments which are extensive on the northern side of the Hunza River and which have partially buried the 'Neoglacial' moraines. Substantial thicknesses of young lacustrine silts produced by blocking of the Hunza by the Batura Glacier occur in several places, the largest outcrop being to the west of the glacier. The terraced outwash plain north of Pasu was also mapped as 'post-Neoglacial' and as containing meltwater sediments of the present phase. Debris-flow activity continues today especially on the fans north of the snout of the Batura (34) amounting to perhaps four major fan-forming events in the past century.

In sum, a combination of morphostratigraphy and degree of weathering of tills suggests three levels of ice inundation during the Pleistocene, though in the absence of superposed tills, it is difficult to ascertain whether this indicates three distinct glaciations or three phases in a declining series of a single glaciation. The postglacial stratigraphic framework is clearer, with dates for the past 200 years being derived from local sources including the truncation of irrigation canals of known age and other oral records (31).

No radiometric age determinations on Pleistocene deposits have been available for the Hunza region, although a limited number have been recorded in Kashmir and the Himalaya. Singh and Agrawal (35) report two assays on peat and lake deposits at 3120 m a.s.l. at Toshmaidan in Kashmir. The older of the two dates (15,250 ± 760/820 yr BP) is for a clay mud, 56 cm above the basal date on a lake clay of 13,850 + 900/785 yr BP. These unsatisfactory dates are taken to indicate that deglaciation was in progress in the period 15,000 - 14,000 yr ago. Radiocarbon dating of buried wood and charcoal on the south side of the Everest Range in east Nepal indicate that the Periche stage is as young as 1200 ± 100 yr old (27), the Khumbu II moraine 1150 ± 80 yr old (collected by Muller, Radiocarbon, 1961), and terrace II, overlain by aeolian sand, 1155 ± 160 yr old (collected by Benedict, Radiocarbon, 1976). A much younger series of dates for the oldest Thuklha moraine, Khumbu II moraine and outwash terrace III is 410 ± 110, 480 ± 80, 530 ± 165 and 550 ± 85 respectively. Again, there is the systematic problem of deriving meaningful inferences for the Hunza Karakoram from these dates, given that they are derived from the much more humid Everest Region. There is also some uncertainty in ascribing stages to particular moraines, e.g., Iwata (25) expresses the view that the Periche moraines are of late-glacial age, because of the likelihood of the buried soil dating from the Hypsithermal, and that the Thukhla moraines are probably of 'early Neoglacial' age.

At least four Pleistocene glaciations have been recognised in western China, much of the evidence coming from the northern slopes of the Himalaya, Karakoram and Kunlun Shan. The estimated snowline depression during the last (Dali) glaciation ranges from 300 to 1200 m, the lower figure applying to the north slope of the central Himalaya (30), although the effects of neotectonics have not been fully evaluated. A number of glacier expansions indicating cool phases have been recognised in the postglacial record. An advance of over 2 km by the Arza Glacier (southeast Xizang) has been dated by one of the authors (LJ) at 2980 ± 150 yr BP and a snowline depression of 100 - 300 m inferred. A further cold phase occurred in southeastern Xizang 1920 ± 40 and 1500 ± 85 yr BP, followed by a warm interval dated at 915 ± 80 and 700 ± 90 yr BP, a sequence which appears to agree well with the Nepal dates. Three morphologically fresh moraines surround many glaciers in southern Qinghai-Xizang, and lichenometric dating by Li (36) suggests that those of the Konbukou Glacier formed in 1818, 1871 and 1895. It appears that the advances of the Little Ice Age were much smaller than those of c. 3,000 yr ago (30).

TECHNIQUES EMPLOYED IN THIS STUDY

The glacial landforms and sediments were mapped using standard methods (37). Morphological and surface sediment maps were constructed from good quality stereoscopic aerial photographs (approximate scale of 1:30,000) kindly loaned by the Government of Pakistan. The photographs were an invaluable aid and work was concentrated in the areas covered by them, mainly a strip along the Karakoram Highway from the Batura Glacier to

Sarat and in the Hindi embayment – Pisan Glacier – Minapin Glacier area 10 km south-southwest of the base camp at Aliabad. These areas offered the most complete sequences of materials, but intermediate points were also sampled as opportunity arose. The study areas were used to trace sedimentation from currently active sites outward to areas of older Pleistocene deposition.

The combination of variations in rock resistance between the different geological zones, with very high erosion rates and high sedimentation rates by the glaciers, make the region well-suited to the use of morpho-stratigraphy for studying the Pleistocene landform history. The interplay of debris flow, glacial sedimentation and fluvial incision has also provided some very large and informative sections on the valley floor and lower valley slopes, to which have been added smaller but higher altitude sections along the Karakoram Highway. These provided detailed information on superposition of sediments and on certain facies relationships.

Some substantial exposures of till were available for study on the ice-proximal side of the lateral moraines. Many of the older and higher moraines, however, yielded no exposures and examination was limited to shallow excavations.

Sediments were classified on the basis of hand-specimen evaluation of particle size and shape, together with examination of bedforms, bedding, jointing, faulting, fabric, and superposition. Particular attention was paid to sediments in the process of deposition upon, and adjacent to, selected glaciers. Measured sections of the deposits were drawn in the field and constructed using stereophotography of vertical faces. High level outcrops were located on the air photographs and their height established by field altimeter.

The sediments were classified in the field as tills, glaciofluvial sands and gravels, lacustrine silts, colluvial diamicts (debris-flow deposits: the "mudflows" of Goudie et al. (2)) and fluvial deposits. The tills appear to be predominantly of meltout origin, although some lodgement tills were recognised. This field classification was tested by laboratory analysis of samples from over 100 documented sites with respect to particle size, particle shape, clast fabric, microfabric and mineralogy of the clay-grade fraction (38).

Sample collection from exposed sediments was combined with documentation of moraine morphology, weathering of surface boulders, subsurface weathering and soil properties at a total of 25 sites in an attempt to differentiate the moraines into age groupings (38)(39)(40); see Table 2. In addition, the point compressive strength of boulder surfaces as indicated by the Schmidt hammer test, desert varnish development and percentage lichen cover were measured. The latter technique worked only in the Rakaposhi area. Even there, no lichens were found on the till of the current stage. Lichens were also absent in the much drier Batura-Pasu area. Microclimate, with shade favouring lichen growth, was a further factor which hampered the use of this technique. The most consistent and successful field techniques proved to be: a desert varnish index on an ordinal scale from negligible (1) to very strong (5), with a sixth category in which the maximal varnish had flaked off or been otherwise almost destroyed; soil colour in the upper 10 cm (yielding increased redness with age though with considerable 'overlap'); presence/absence of $CaCO_3$ films or crusts beneath stones in the upper 20 cm (which proved useful in the older tills

468

only), point compressive strength of boulder surfaces, and boulder
weathering ratios, which proved consistent in the granodiorite-dominated
lithologies of the Batura-Pasu area; and the morphology and disposition
of moraines in relation to present glacier snouts. Although there is some
evidence that the inner slope angle of moraines decreases with time, this
property is too much affected by the original form and trimming by later

CRITERIA USED	LOCATION
Surface boulders:-	
a. Desert Varnish Index (D.V.I.)	B, R
1. Negligible (fresh rock)	
2. Slight (yellowish)	
3. Moderate (pale brown)	
4. Strong (brown)	
5. Very strong (dark brown)	
6. Post-maximal (being destroyed by weathering)	
b. Lichen Cover %	R
c. Point Compressive Strength (Schmidt Hammer)	B
d. Weathering Ratios	
% flaked and split boulders	B
% pitted boulders	B
Subsurface Weathering:-	
e. Soil Munsell Colour (10cm)	B, R
f. Boulder Grussification	B
g. Soil Carbonate	B, R
Not useful: Surface boulder frequency, slope angles, other soil properties	

B: Batura-Pasu area

R: Rakaposhi area

TABLE 2: RELATIVE DATING OF MORAINES - USEFUL MEASURES

ice advances to be reliable. Similarly, the percentage of carbonate in
the <2 mm soil fraction is locally variable and too much affected by limestone
and dolomite content. Outer moraine slope angles, surface boulder
frequency, subsurface stone weathering (appreciable only in the case of
t_3 and older tills; see below) and all soil properties except colour and
carbonate content provided few useful results.

Laboratory analysis of weathered materials from 10 cm and 20 cm depth
included pH (using a B.D.H. soil test kit), electrical conductivity (1:5
at 25 °C), percentage carbonate (determined manometrically using a Bernard
calcimeter), particle size and clay mineralogy. Measurements of free iron

content and cation-exchange capacity were not attempted because of insufficient variation in the simpler measures.

THE TILLS AND RELATED SEDIMENTS

All the glaciers in the Hunza valley carry supraglacial debris derived
from mass wasting of the very steep and actively-degrading valley sides,
and some glaciers, including the Batura, Ghulkin and Hispar are mantled
by such debris for many km above their snouts. Dominant modes of
deposition are by meltout and sliding as the glacier surface melts down.
Till is also being deposited by subglacial lodgement but deposition of
flow till by slurrying of debris down ice slopes was not observed and
no flow till was recognised in till assemblages of Pleistocene age. This
dominance of meltout till is common in low latitude montane glaciers and
agreed with predictions (41).

The mode of debris entrainment in the glaciers, the importance of
granitic rock types as sources of granular debris, and the large volumes
of meltwater produced as the glaciers descend to valley floors with hot
desert summer climates, has resulted in widespread meltwater modification
of both modern and Pleistocene tills. This process, occurs at all scales
from slow mechanical eluviation of silt and clay in the matrix of tills
to fluvial incision and sorting in the proglacial plains. As a result,
all tills in this region are relatively coarse-grained, with particle size
curves showing positive skewness and varying degrees of depletion of their
finer grades. The grading envelopes of sediments classified in the field
as tills, eluviated tills and glacial outwash debris show appreciable overlap. Both the Batura-Pasu-Ghulkin and the Minapin-Pisan sediment groups
show this transition from extremely poorly sorted subglacial tills to eluviated
meltout tills and weathered tills and finally to better sorted meltwater-
stream sediment (38). The tills in the high diffluence cols and bedrock
benches up to at least 3,400 m, together with tills from the gorge section
of the Hunza, have a rather finer mean grain size (2.48ϕ) than the Batura
(1.37ϕ), the Ghulkin (1.33ϕ) and the Pasu (1.24ϕ) glaciers, and much
finer than those of the Minapin and Pisan glaciers (0.93ϕ). Although some
of the high level tills have an appreciable amount of limestone source
material in the matrix, this does not entirely explain their generally
fine grades which are more likely to derive from the very much thicker,
vigorously flowing glaciers which deposited them. The greater frequency
of compacted and jointed tills of lodgement type in protected sites on the
main Hunza valley floor (Fig. 5) and on benches and in cols at the higher
altitudes, is consistent with this interpretation.

The glaciofluvial gravelly sands vary from poorly sorted proximal
facies to well sorted, well rounded distal facies. Silt and clay rarely
exceed 10 per cent by weight. Other sediments deposited in glacial meltwaters include proximal glacial lacustrine beds with dropstones and a
finely-bedded to laminated distal glaciolacustrine group, all containing
at least 70 per cent in the silt and finer grades. The debris-flow sediments
are finer grained than the tills of the region. The mean sorting is higher
than in the tills, all being extremely poorly sorted. The material contains
much glacial debris redeposited by flow, the matrix being supported during
emplacement by positive porewater pressures. As a result, these sediments
have not suffered the depletion of the fine fraction noted in the tills, and
constitute a much more uniform population.

FIG. 5. LODGEMENT TILL (PALE) BENEATH STONY MELTOUT TILL AND GLACIOFLUVIAL AND FLUVIAL GRAVELS IN THE LEE OF A ROCHE MOUTONEE (LEFT, VERY PALE) IN AN ISLAND IN THE HUNZA RIVER BELOW THE PASU GLACIER SNOUT.

THE SEQUENCE OF FORMS AND SEDIMENTS

Shanoz stage

Weathered tills occur as residual clusters of boulders at high levels. The highest erratics of granodiorite and granite gneiss were found resting on slate and limestone bedrock at approximately 4150 – 4300 m on the 'Patundas surface', west of Lake Borit Jheel, between the Pasu and Ghulkin glaciers. Occurrences have also been confirmed at 3900 m north of the Batura Glacier. These are the oldest till remnants (t_1) of the Shanoz phase of Zhang and Shi (33).

Yunz stage

High patches of compact tills in various stages of dissection, but with an undisturbed matrix, occur on two benches (3200 and 3650 m) and in patches up to 3900 m on the upper, western slopes of the Pasu-Ghulkin diffluence col. This t_2 stage is probably the equivalent of the Yunz of Zhang and Shi (33). Similar material was also found on Mount Kundorich (north of the Batura Glacier) and on the high level (3,200 – 3,500 m) glaciated limestone bedrock surface of low relief between the lower Batura and Pasu glaciers. Till remnants at high levels elsewhere are probably correlatives of this material. For example, the strongly cemented fluvial gravels capped by eroded till 800 m above the valley-floor near Sarat (Fig. 6) rest on a bench representing a former valley-floor remnant at about 3,000 m. Similar material is found at an estimated altitudinal range of 3,300 - 3,800 m on a high rock bench east of the Hunza River facing the Pasu Glacier (Fig. 7). High remnants also occur in the Batura-Pasu diffluence col at over 3,200 m near Yunz and at over 3,100 m at both ends of the Pasu-Ghulkin col.

Borit Jheel stage

Near Sarat, a valley-side bench at approximately 2,700 m (Fig. 6) is mantled with till which is weathered but much less dissected than the t_2 remnants described above. This t_3 till occurs much more extensively than the high level remnants and is rarely found above 3,000 m a.s.l. It occurs in the Batura-Pasu diffluence col and also on the slopes of the Pasu-Ghulkin diffluence col around Lake Borit Jheel where it shows subdued morainic morphology. Tills of this stage are predominantly of lodgement and subglacial meltout origin. Cemented conglomeratic variants with duricrusts reaching 2 m or more in thickness are found at several places where the t_3 till has a high limestone content. They occur at just over 2,500 m at the base of the Batura 'drift plateau' (see below), at about 2,900 m south of the Pasu snout and at a similar altitude northeast of Lake Borit Jheel. They also occur at river level beneath 20 m of debris-flow deposits in the large fans south of Gulmit village (Fig. 8). Northeast of Lake Borit Jheel, where the till is richer in granodiorite, there is typically a lag of grus and small slate fragments on its weathered surface and the underlying calcrete is being exposed over quite a large area. Here, the desert varnish is very strong to post maximal on about 84 per cent of all boulders sampled, although only one site was examined. A shiny and almost black coating covers the sides of some tafoni indicating a 'second generation' of desert varnish (Fig. 9). Weathering pits are very large, about 84 per cent of granodiorite boulders being pitted. Limestone shows strong vermicular weathering and slate is shattered. The soils have calcrete in the upper 20 cm with pendant growths beneath stones. Soil colours are all in the 10 YR range. Some subsurface granodiorite boulders are significantly grussified.

The glaciation which produced the t_3 tills clearly filled the trunk valleys and the glacial diffluence cols (Fig. 10). Compact but jointed lodgement tills of the Minapin Glacier, overlying glaciolacustrine silts, can be traced on to a till-covered bench at 2,500 - 2,600 m a.s.l., some 500 m above the Hunza River. This till is dissected but can be traced as far as a till plateau remnant in the upper trough of the Hunza beneath which is an intermediate trough with fluvial sediments indicating interglacial conditions (Fig. 11). In view of its extensive development in the Pasu-Ghulkin diffluence col, the glaciation which laid down the t_3 till is here referred to as the Borit Jheel stage. It may be a correlative of the Hunza stage of Zhang and Shi (33). Beneath the t_3 tills, 0.5 km southeast of the Ghulkin Glacier snout, glaciolacustrine silts have yielded a thermoluminescence (TL) age of 139,000 ± 12,500 yr BP (personal communication, Pei Jingxian,. Institute of Geology, Academia Sinica), indicating that the two high level glaciations (t_1 and t_2) antedate the last full glaciation in the global Pleistocene record. In some places, there are well preserved series of retreat moraines indicating by their morphology an orderly downwasting of ice within the diffluence cols and on the Hunza valley-side benches at the end of the t_3 stage. This is well shown by the series northeast of Lake Borit Jheel where the ice was retreating back into the Pasu valley (Figs. 7, 10)

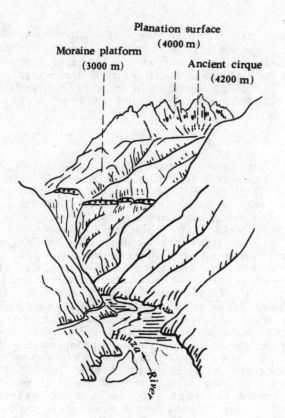

Planation surface
(4000 m)

Moraine platform
(3000 m)

Ancient cirque
(4200 m)

Hunza River

FIG. 6. SKETCH OF THE
UPPER HUNZA GORGE,

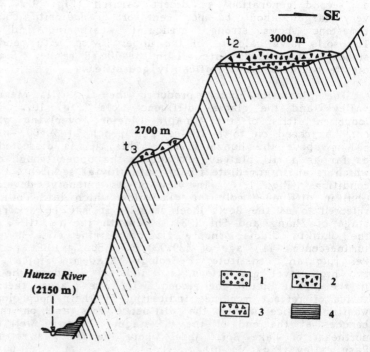

SE

3000 m

t_2

2700 m

t_3

AND A SECTION
ALONG THE LEFT
BANK OF THE
HUNZA RIVER
NEAR SARAT:

Key

1. Conglomerate;
2. Deposits of Yunz
   glacial stage
3. Deposits of the
   Borit Jheel
   glacial stage
4. Lacustrine beds.

Hunza River
(2150 m)

1   2

3   4

FIG. 7.   THE UPPER HUNZA VALLEY FROM GHULKIN GLACIER
(LOWER RIGHT) TO THE SHIMSHAL RIVER, SHOWING LAKE BORIT
JHEEL IN $t_3 - t_4$ MORAINE – MANTLED DIFFLUENCE COL.
DISSECTED HIGH–LEVEL TILL OF THE YUNZ STAGE (t_2) IS
VISIBLE IN RIGHT MIDDLE DISTANCE.

474

FIG. 8. LODGEMENT TILL (TO RIGHT OF ARROW) BENEATH THICK DEBRIS FLOW AND ALLUVIAL FAN SEDIMENTS IN HUNZA VALLEY FLOOR SOUTH OF GULMIT VILLAGE.

FIG. 9. TILL OF THE BORIT JHEEL (t_3) STAGE IN THE DIFFLUENCE
COL BETWEEN THE PASU GLACIER AND LAKE BORIT JHEEL, SHOWING
"POST-MAXIMAL" WEATHERED BOULDER SURFACES, TAFONI, ETC.

Ghulkin I stage

Both Batura and Pasu glaciers developed large 'expanded foot' termini
in the subsequent Ghulkin I (= t_4) stage, and also extended across minor
watersheds and into the diffluence cols for short distances. Here they
are represented by rounded but well-formed arcuate moraines. On the
south side of the Batura, the t_4 till of this stage has been preserved
near the surface of a triangular-shaped drift plateau remnant at
2,600 - 2,700 m a.s.l. It underlies glaciofluvial gravels and scree and
overlies thick glaciolacustrine silts which, in turn, rest on another till
assumed to be of t_3 age. Tills with weathering characteristics similar
to those of the upper tills in the Batura drift plateau occur near the
Hunza River on the north side of the Batura Glacier, at the northern ends
of the Batura-Pasu and Pasu-Ghulkin diffluence cols, and in the col
1.75 km east-south-east of the Pasu Glacier snout, indicating a much more
restricted glacial extent than in the preceding Borit Jheel stage
(Fig. 10). The desert varnish on the t_4 deposits is very strong to post-
maximal with splitting affecting about 52 per cent of the boulders and
cavernous weathering occurring on about 76 per cent. Soil colours are
in the 10 YR range and oxidation is evident below 20 cm depth. There
is incipient calcrete and pendant growth of carbonate beneath stones.
Granodiorite cobbles with incipient grussification occur at 10 - 20 cm depth.

476

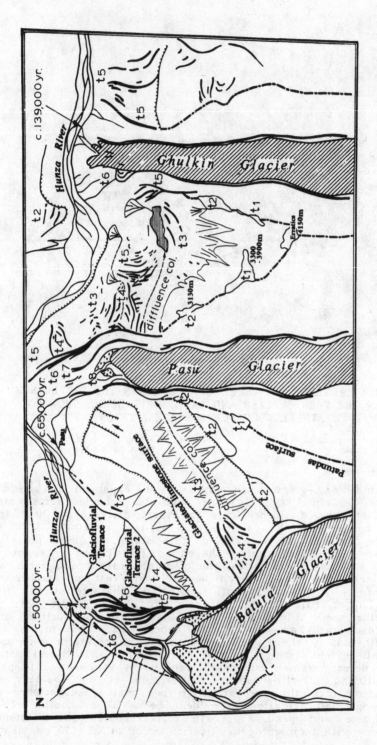

FIG. 10. SKETCH MAP OF THE BATURA–GHULKIN AREA SHOWING PRINCIPAL MORAINES, EXTENT OF MAIN GLACIAL STAGES, AND SITES OF TL DATES.

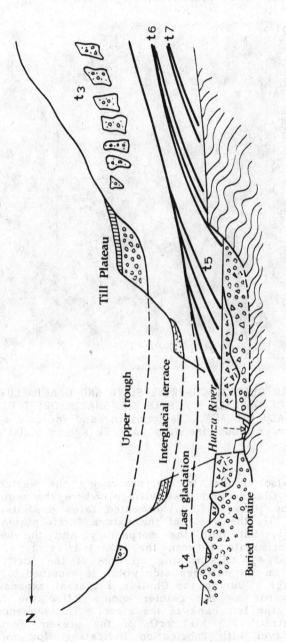

FIG. 11. CROSS-SECTION OF THE HUNZA VALLEY NEAR HINDI VILLAGE.

The Batura 'drift plateau' and part of the smaller one north of the Pasu snout, overlooking Pasu village, probably both date from the Ghulkin I (t3) stage. In both cases, tills are interbedded with thick glaciolacustrine and glaciofluvial sediments (Fig. 12). Similar

FIG. 12. UPPER GLACIOLACUSTRINE SILTS AND GLACIOFLUVIAL SANDS BENEATH TILL OF GHULKIN I AGE IN THE SMALL 'DRIFT PLATEAU' OVERLOOKING PASU VILLAGE. Bedforms in sand indicate meltwater flow to N E. Lower lacustrine silts have TL age of 65,000 ± 300 yr BP. View NNW.

relationships are also evident in outcrops along the western valley side between the Batura Glacier and Pasu village, where the sequence is almost completely hidden by thick (c. 10 m) cemented talus deposits. In the map of Zhang and Shi (33), the whole of the Batura 'drift plateau' is ascribed to the Hunza stage, but both the morphology and the weathering data presented here discriminate between the cemented t3 tills, the weathered but weakly cemented t4 tills making up most of the 'drift plateau', and the retreat moraine on its northern side which is ascribed to the succeeding Ghulkin II (t5) stage. During the Ghulkin I glacial expansion, the Hunza was diverted eastwards as the glacier snouts filled the large erosional rock embayments on the left bank of the river. The sequence in the small residual 'drift plateau' 1.5 km north of the present Pasu snout shows a lower fluvial gravel with imbrication indicating flow to the southeast below a lower lacustrine fine sandy silt with dropstones, a limestone breccia with granodiorite erratics, an upper series of lacustrine silts with dropstones and sands and gravels with bedforms (crossbeds and imbrication) indicating flow to the northeast, all overlain by a coarse meltout till (partially covered by limestone scree) which grades northwards into a glaciofluvial gravel. A TL date for the lower lacustrine silts, obtained by the Institute of Geology of Academia Sinica, is 65,000 ± 300 yr BP. This, of course, may refer to a phase of sedimentation during the Borit Jheel or between the Borit Jheel and Ghulkin I stages, as the lower lacustrine silts are some tens of metres below the surface of this plateau which appears to be of Ghulkin I age. The evidence of northeastward flowing meltwaters rules out a major trunk valley glaciation in the upper Pleistocene in favour of a major 'expanded foot' tributary glaciation which, however, delivered considerable thicknesses of ice (c. 200 m) to parts of the main valley and caused major damming of the Hunza River. Similar major 'expanded foot' advances occurred during this stage in the Hindi embayment (Minapin and Pisan Glaciers) and elsewhere.

About 2.5 km south of the Batura snout, sheared and highly distorted dense tills, interbedded with glaciolacustrine silts with dropstones and glaciofluvial gravels, stratigraphically underlie (beneath a major shear plane) younger subglacial and supraglacial tills of the Batura stage. The lacustrine silts in this sequence have a TL age of 50,000 ± 2,500 yr (Institute of Geology, Academia Sinica). In the Hindi embayment (Figs. 11, 13), moraines of c. 100 m thickness are preserved as inliers in a terrace made up largely of debris-flow material laid down only from the northern valley side after the last glaciation. These moraines were deposited at a time when expanded foot lobes of both Pisan and Minapin glaciers extended up to 1.45 km north of the present Hunza River channel, diverting and ponding the drainage. Lacustrine silts intercalated with the till of moraines from the Pisan Glacier yielded a TL age of 47,000 ± 2,350 yr BP (Institute of Geology, Academia Sinica).

Ghulkin II stage

The moraines of this stage occur as multiple series of large but rounded ridges, well seen on the south side of the Pasu Glacier and the north side of the Ghulkin. Similar moraines are found on the right bank of the Minapin Glacier (Fig. 11).

Between 68 and 76 per cent of the surface boulders of this t_5 till display very strong or post-maximal varnish conditions (desert varnish index 4.86-5.08), between 42 and 74 per cent of the granodiorite boulders being pitted. Granular disaggregation of boulder surfaces is also evident. The Schmidt hammer readings were all fairly low (22 - 29 kN/m^2). The differences between this Ghulkin II stage and the earlier Ghulkin I stage are minor in terms of surface weathering, resting on a slightly weaker development of boulder pitting (with much smaller tafoni), a lower proportion of fragmented and split boulders and slightly weaker $CaCO_3$ cementation apart from crusts under stones. In addition, no subsurface grussification was noted on till of t_5 age. The rather similar weathering state of the surface material suggests that this Ghulkin II stage may be a late still-stand in the retreat from the Ghulkin I expansion: both appear to date from the second half of the last Pleistocene glaciation. It seems clear that the Ghulkin stage was characterised, in both early and late phases, by only limited advances down the trunk valley of the Hunza River.

Batura stage

Large, well preserved moraines confined to the tributary glacial troughs of the Hunza (t_6) include the outer ramparts of the large half-cone end

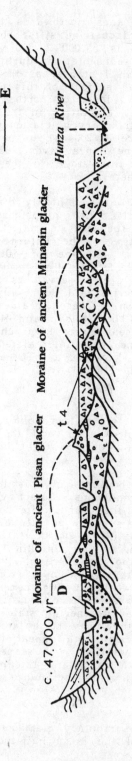

FIG. 13. LONGITUDINAL SECTION ALONG RIGHT BANK OF THE HUNZA RIVER NEAR HINDI.

A – till of last glaciation (t_4);   B – fluvial sediments of last glaciation;
C – debris flow deposits post–   D – torrent deposits of Holocene age.
 dating last glaciation;

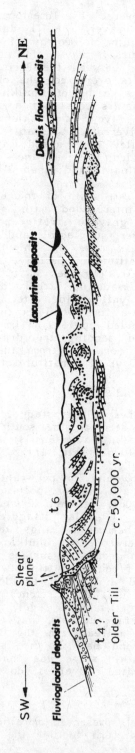

FIG. 14. SECTION OF THE SHEARED AND DEFORMED TILLS NEAR THE BATURA GLACIER.

moraine complexes of the Ghulkin and Gulmit glaciers, a group of prominent laterals on the south side of the Pasu Glacier, and the extensive deposits stretching almost 2.5 km downstream of the Batura snout on both sides of the Hunza River. Here, between 20 and 40 m of sediments are exposed. Rather loose meltout tills intercalate with lodgement tills, supraglacial slide debris and thrust (shear–plane) tills. There are local gradations to glaciofluvial gravels and sands and intercalated lacustrine silts with dropstones, suggesting impedance of the Hunza drainage by the Batura and Pasu glaciers. The whole sequence shows signs of disturbance in the form of thrusting by overriding ice and local loading structures and normal faults (Fig. 14). The upper part of the till is also intercalated with debris–fan sediments which substantially bury it on the left bank of the Hunza, leaving a few moraine crests as inliers (Fig. 15). The

FIG. 15. MORAINE OF THE BATURA (t$_6$) STAGE (MIDDLE DISTANCE) PARTLY BURIED BENEATH THICK DEBRIS FLOW AND ALLUVIAL FAN SEDIMENTS. IMBRICATED GRAVELS OF THE UPPER GLACIO-FLUVIAL TERRACE OF THE BATURA IN LEFT FORE-GROUND. HUNZA-SHIMSHAL CONFLUENCE. VIEW EAST.

sedimentary structures suggest that the till is the product of a relatively prolonged stillstand of the ice and its intercalation with debris–flow deposits indicates that the bulk of the fans to the east were penecontemporaneous with the later phases of moraine deposition. The desert varnish is very strong to post-maximal on 6 - 74 per cent of the boulder surfaces (D.V.I. 3.78 - 4.74), and there is clear evidence of carbonate translocation. Schmidt hammer readings are clearly higher than for moraines of the Ghulkin II stage (means in the range 24 - 38 kN/m^2) suggesting a significant interval between stages $t_4 - t_5$ and t_6 (Fig. 16). Although there

FIG. 16. MORAINES OF THE BATURA STAGE ON LEFT SUPERIMPOSED ON THOSE OF THE GHULKIN II STAGE (RIGHT) IN COL ON SOUTH SIDE OF PASU GLACIER.

are no unweathered boulders, 54 - 82 per cent show flaking, and 18 - 46 per cent are split (18 - 26 per cent having pits). Soil oxidation is noticeable at 10 cm depth, but soil colours are commonly in the 2.5 Y range. These are the 'brown moraines' of the Batura Glacier Investigation Group (29) for which a 'Neoglacial' age (c. 3,000 yr or greater) has been suggested. They represent the largest ice extension during the post-glacial period. Similar remnants occur in dissected form on both sides of the Pisan and Minapin glaciers, near Hindi, where the average lichen cover is 49.8 - 65.4 per cent and the cover of grasses, herbs, lichens and mosses is clearly much more extensive than on younger moraines closer to these glaciers. This ice expansion was confined, however, to the southern side of the Hunza and did not divert the river. In view of the well-developed nature of the t_6 moraines around the Batura Glacier, this ice advance is called the Batura stage.

Pasu I stage

Inside the brown moraines of the Batura stage is to be found a series of high, sharp-crested moraines with a yellow surface colour (t_7). They were considered to date from the earlier part of the Little Ice Age advance by the Batura Glacier Investigation Group (29, 31). They cut across the t_6 moraines on the south side of the Batura Glacier and occur as a series of very high, straight lateral ridges on both sides of the Pasu Glacier and as fluted till on small plateaux beyond its terminus (Fig. 17). Equiv-

FIG. 17. LATERAL MORAINE OF PASU I STAGE (RIGHT OF CENTRE) IN FRONT OF PASU II END MORAINE (LOWER RIGHT) ADJACENT TO PASU GLACIER (FAR RIGHT). MORAINES TO LEFT ARE OF THE BATURA STAGE. NOTE TRUNCATED, ABANDONED IRRIGATION CANAL IN LOWER LEFT OF PHOTO. VIEW SOUTH.

alent lateral moraines also occur around the Ghulkin Glacier where they are superimposed on, but do not breach those of the t_6 stage (Fig. 18). The many topographical discordances between these moraines and those of the Batura stage moraines, together with the evidence for truncation of irrigation canals by this later ice advance near the Batura terminus, led to the suggestion by the Batura Investigation Group (29, 31) that

484

FIG. 18. BATURA STAGE LATERAL MORAINES OF THE GHULKIN GLACIER (LOOPED, FOREGROUND) WITH PASU I STAGE LATERAL MORAINE ON THE ICE-PROXIMAL SIDE (STRAIGHT RIDGE). DEBRIS MANTLED LOWER GHULKIN GLACIER IN BACKGROUND. VIEW SOUTH-EAST.

these yellowish t_7 moraines are about 200 – 300 yr old. Over half of pebble-sized clasts in the Batura t_7 moraines of this advance show significant edge rounding, indicating substantial reworking of older glaciofluvial and fluvial sediments. Although more limited than the Batura stage advance, this was a major event. The ice-front moraines of the Ghulkin Glacier dating from this stage have been washed out, but the Gulmit Glacier front was 0.6 km further down valley than at present and the Batura terminus lay about 1 km further east. The greatest advance occurred in the Pasu Glacier, however, the ice extending about 2.5 km beyond its present margin. The surface boulders on the t_7 moraines have only incipient desert varnish (D.V.I. 1.83 – 3.12) and high point compressive strengths (Schmidt hammer readings 40 – 45 kN/m^2 in the Batura – Pasu area). Flaking affects only 18 – 40 per cent of boulders: no split or pitted boulders were found. Occasional, presumably older transported boulders have a strong desert varnish. Yet others show no varnish at all. The typical yellowish varnish appears to consist of a pale, clastic coating made up of silt-grade rock flour. Although soil development is negligible in the Batura-Pasu area, it is perceptible in the moraines of the Rakaposhi glaciers which also have a slight lichen cover (0-10 per cent) and woody vegetation, notably juniper shrubs. A log of Himalayan pencil cedar (Juniperus) buried in the outer flank of a right lateral moraine by the Minapin Glacier has yielded a radiometric age of 830 $\pm$ 80 yr BP (HAR – 4345). Wood fragments from the ice-proximal face of the same moraine yielded a radiocarbon age of 325 $\pm$ 60 yr BP (Lanzhou University). A Little Ice Age date for this glacial advance thus appears certain.

Pasu II stage

The youngest tills (t_8) are contained in sharp-crested, unstable and sometimes ice-cored moraines of the nineteenth and twentieth century advances (37). They sometimes occur merely as ledges on the ice proximal side of the Pasu I moraines, e.g., parts of the Pisan, Minapin and Lower Pasu glaciers, or as a splay of boulders overtopping the Pasu I stage moraines, as in the Ghulmet Glacier. They occur with tills of the Batura and Pasu I stages in a single high lateral moraine around much of the lower Ghulkin Glacier, indicating a remarkably consistent lateral margin of this glacier over several thousand years. On the lower Batura Glacier they form a substantial hummocky moraine, much of it ice-cored, and in one location, meltout till of this stage overlies lodgement till of Pasu I age separated by a thin (1 – 2 m) glaciofluvial gravel. These are the grey tills of the Batura Glacier Investigation Group (29, 31). They carry no vegetation (even lichen), there are no split or pitted boulders and point compressive strengths are the highest recorded (Schmidt hammer readings 52 – 56 kN/m^2 in the Batura-Pasu area). Strongly developed varnish does not occur and 28 – 58 per cent of boulders have negligible varnish. Soil development is negligible and there is no appreciable oxidation of these Pasu II (t_8) stage deposits.

The data on which the relative dating of the moraines of the upper Hunza region is based are summarised in Table 3. It is worth noting that none of the clay-grade material subjected to mineralogical analysis by X-ray diffraction showed any evidence of secondary clay minerals. There is no evidence of mixed layer clays. Moreover, there is no significant difference in the mineral population between the sub-surface (>1 m) samples of tills of t_4 – t_5, t_6, t_7 and ice-cored moraines (t_8) derived from the same glacier (41).

Age	Mean D.V.I.	% D.V.I. >4	Lichen cover %	Schmidt hammer kN/m²	% Split	% pitted	Soil colour (10cm)	Subsurface boulder weathering	Soil CaCO₃
Area	B, R	B, R	R	B	B	B	B	B	B, R
t_8	1.48-2.14	0	0	52-56	0	0	5Y	-	present
t_7	1.83-3.12	0-4	0-10	40-45	0	0	5Y	-	present
t_6	3.78-4.74	6-74	49.8-65.4	30-38	18-46	18-26	2.5Y	-	crusts under stones
t_5	4.86-5.08	68-76	*	22-29	*	42-74	2.5Y-10YR	-	crusts under stones
t_4	*	76	*	n.d.	*	76	10YR	slight	incipient calcrete
t_3	*	84	*	n.d.	*	84	10YR	significant	calcreted

*   Not useful - too weathered

n.d.   No data

B   Batura-Pasu area

R   Rakaposhi area

TABLE 3: RELATIVE DATING RESULTS: SUMMARY

DISCUSSION AND CONCLUSIONS

There is some divergence of view about the maximal extent of ice in the Karakoram' during the Pleistocene, even leaving aside the extreme view, as represented by early reports of erratics as far south as Potwar. Desio and Orombelli (42) favoured an extensive reticular glacier system in the Indus drainage, leaving three huge moraine complexes in a waning series at Shatial (800 m), Dudishal (840 m) and Gunar (1220 m). Desio (1) states that Pleistocene glaciers certainly reached Shatial and, very recently, Ambraseys et al. (43) proposed three distinct glacial advances in the Besham-Kamila section of the Indus on the basis merely of matching abraded surfaces on valley shoulders. A more conservative view is represented by that of Paffen et al. (18), Wiche (21) and Schneider (23) who found no evidence of through-flowing trunk valley glaciers.

Reconnaissance of the valley sides from the Batura Glacier down the Hunza, Gilgit and Indus rivers as far as the junction with the Shatial River at about 800 m revealed abundant dissected till remnants perched 200 - 800 m above the valley floors. Many of these remnants, however, lie in erosional embayments such as those along the Hunza opposite the snouts of glaciers like the Batura, Pasu, Ghulkin and Minapin-Pisan. They clearly record 'expanded foot' style extensions of tributary valley ice which repeatedly diverted the river and so caused the enlargement of these rock embayments by fluvial erosion. The record of retreat from the expanded foot stage includes not only high level tills but a well ordered sequence of retreat moraines exemplified by the Hindi embayment. High level till remnants also occur, however, between the embayments along straight stretches of the trunk valley. One such is the remnant of a huge lateral moraine at about 500 m above the valley floor near Aliabad. Other evidence of a glacial phase of through-flowing valley ice made up of confluent tributary glaciers includes ice polished and striated surfaces (mean striae direction N 010° E) at about 3,100 m on the summit of the mountain between Lake Borit Jheel and the Hunza and moutonnée surfaces on the flat, limestone ridge crest between the Pasu terminus and the Pasu-Batura outwash plain. The tills along the Karakoram Highway at about 2,000 m altitude west of the Hunza between the Pasu and Ghulkin glaciers and those around Sarat are of subglacial meltout and lodgement type and, from their position, suggest a minimum ice thickness in the trunk valley of 500 m. The argument that Pleistocene snowline depression of 800 - 1,000 m would not appreciably enlarge the glacial extent beyond the tributary valleys (cf. Paffen et al. (18), Schneider (23), Heuberger (28)) is difficult to sustain for the Hunza above the Batura terminus in view of the high elevations of the trunk valley floors and their manifestly glacially-eroded form and residual till deposits. There can be no doubt that substantial valley ice was generated by the growth of the larger transfluent glaciers. The Hispar Glacier, for example, overrode the 4,000 m high ridge separating it from the Hunza valley, leaving high lateral moraines along the summit crest (Fig. 19). Ice over 500 m thick invaded the Hunza valley and its course can be traced beyond the complex deposits at the Hasanabad confluence down to the Hindi embayment. Following the regional (Shanoz) glaciation of the surface indicated by the high level erosional remnants, at least two trunk valley stages (Yunz and Borit Jheel) occurred. The discernible record is complicated, but the actual sequence was probably much more complex than the record now seems because of the interaction of glacial, fluvial, and mass wasting processes in this tectonically active region. With respect to the detailed glacial sequence in the Hunza-Indus region, most authors have recognised

FIG. 19. LATERAL MORAINES ON SUMMIT OF INTERFLUVE
NEAR HISPAR-HUNZA CONFLUENCE. Deformed tills of last
stage of trunk valley glaciation overlying glaciofluvial
beds resting, in turn, on fluvial gravels and bedrock.
View SSE; composite photograph.

the following series of stages:

(i) One or more older stages with little or no moraine depositional
morphology, strongly weathered and often dissected tills and, in many
cases, a thick loess cover on the moraines.

(ii) One or more intermediate stages with subdued, but discernible
moraine morphology related to the existing tributary valley systems.
The moraines show considerable weathering and soil development and some
loess cover.

(iii) One or more younger stages with sharp morainic forms and
weathering and soil development only slight to moderate. The youngest
of these are the advances in the historical period.

The morphological, sedimentary and weathering data presented here appear to indicate the following glacial stratigraphy for the upper Hunza and adjacent regions (Table 4).

TILL UNIT	STAGE NAME	TENTATIVE DATES (yr. B.P.)
t_1	SHANOZ	nil
t_2	YUNZ	nil
		$139,000 \pm 12,500$
t_3	BORIT JHEEL (? Hunza correlate)	$65,000 \pm 3,300$ $50,000 \pm 2,500$
t_4	GHULKIN I	$47,000 \pm 2,350$
t_5	GHULKIN II	nil
t_6	BATURA	nil
t_7	PASU I	830 ± 80 325 ± 60
t_8	PASU II	'Historical'

TABLE 4:   PROPOSED QUATERNARY GLACIAL STAGES IN THE UPPER HUNZA RIVER VALLEY.

(I) a distinct phase of regional glaciation of a higher level Hunza valley system with ice levels above 4,150 m as indicated by isolated erratics. The extent of ice at this stage in the area bounded by latitudes $35^\circ - 37^\circ$N and longitudes $74^\circ - 76^\circ$E approximated 23,000 km^2 compared to 11,900 km^2 at present. There is no material yet available to fix the age of this valley stage, but an age much older than the last Pleistocene pleniglacial is very probable, to judge from the degree of dissection of the glaciated surfaces. This is the Shanoz stage.

(II) a phase of ice extension in which ice filled and overtopped the high diffluence cols and entered the trunk valley of the Hunza in considerable volume. Major glacial disruption of the Hunza River occurred, and substantial glaciolacustrine sediments were deposited at several levels. Ice occupied a landscape in which the Hunza valley floor was about 800 m above the present as indicated by bench remnants at 3,000 m a.s.l. and above. These benches were created by subsequent vigorous river incision resulting from uplift. An age greater than 139,000 yr appears

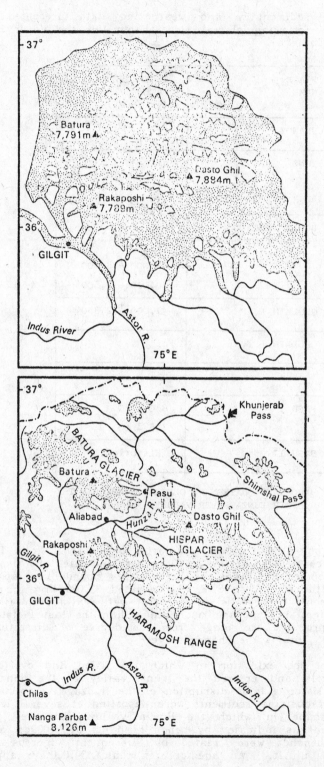

FIG. 20.
GLACIER COVER IN
THE HUNZA VALLEY
(A) PLEISTOCENE-
BORIT JHEEL STAGE;
(B) PRESENT.

likely for this Yunz stage, i.e., older than the last major Pleistocene glaciation.

(III) a phase of glaciation which, although it left subglacial tills at altitudes up to about 3,000 m or slightly above, was characterised by greater topographical control of ice extent. Till of this major advance can be traced along the Karakoram Highway between the Pasu and Ghulkin glaciers, for example. Ice occupied diffluence cols, and tributary glaciers coalesced into extensive Hunza valley ice bodies, extending beyond the Indus junction to the Astor confluence (Fig. 20). Pending further absolute dating of intercalated silts, it is not possible to give an age range to this major Borit Jheel ice expansion, although available TL dates seem to indicate an age greater than 50,000 - 65,000 yr, and correlation with the Hunza stage of Zhang and Shi (33) is uncertain. However, well-preserved retreat moraines high on rock shoulders up to 3,100 m indicate a protracted withdrawal.

(IV) following the above was a phase of more limited glacial extent (Ghulkin I). Ice was predominantly restricted to the tributary valleys although there was limited extension into the diffluence cols and some coalescence within the Hunza valley (e.g., the Batura and Pasu glaciers coalesced and dammed the Hunza River on several occasions). The prominent but rounded and weathered moraines typify this Ghulkin I stage which appears to have an age of 47,000 yr or younger, i.e., second half of the last glaciation of the Pleistocene.

(V) the succeeding Ghulkin II stage appears to date from the latter part of the last Pleistocene glaciation. It is not clear whether this was a distinct advance following marked withdrawal of the Ghulkin I stage or a major ice stillstand at the end of the Ghulkin II stage.

(VI) in the following Batura stage, the Batura-Pasu and Ghulkin glaciers again blocked the Hunza valley but were confined to their troughs and built large moraine complexes, those of the Ghulkin and Gulmit resting on rock benches above the Hunza. The Batura ice extended barely 2 km beyond the present ice-cored moraines, but an extensive glaciofluvial valley train developed between the ice front and the Pasu Glacier. The moraines of this stage have a well developed, varnished and weathered surface but the glacial depositional morphology is very well preserved. An early to middle Holocene age is suggested for this Batura stage (t_6).

(VII) a more restricted advance succeeded the Batura stage and no expanded foot forms developed. There is a marked change in the surface weathering characteristics between the Batura stage and this (t_7) stage. The yellow to pale brown, mildly varnished moraines are restricted to the tributary valleys and faithfully parallel the present ice margin. This Pasu I advance was relatively more pronounced in the Pasu valley than elsewhere in the areas investigated. It appears to be at least as old as the Little Ice Age, since it is bracketed by dates of 830 and 325 yr BP. The radiocarbon ages of Holocene moraines elsewhere in the Great Himalaya (reviewed above) are consistent with those suggested by this study. However, the Himalayan dates obtained so far appear to be too young for the degree of surface weathering observed. Descriptions of the Periche stage moraines (for example Fushimi (27)) are reminiscent of the Batura or the Ghulkin II moraines in the sequence adopted here. This apparent inconsistency may arise from site identification or strati-graphic problems causing an incorrect attribution of ^{14}C dates to moraines. Moreover, the humid conditions of the Great Himalaya compared with the

Karakoram may be an element in the apparent inconsistency.

(VIII) an even more limited series of minor advances in the nineteenth and twentieth centuries left sharp, unstable, unweathered and commonly ice-cored moraines, albeit very locally within the limits of the Pasu I advance (37). Considerable downwasting of ice surfaces has occurred since the outer moraines of this Pasu II (t_8) advance were deposited, commonly 30 m and as much as 50 m on the larger glaciers (Batura and Pasu).

To conclude, it appears from the limited new evidence presented here that topographical control of ice form and extent has increased progressively through the Quaternary in this part of the Karakoram. It has long been argued that multiple glaciation of alpine mountains leaves an incomplete depositional record because of the increasing efficiency of glacial accumulation basins through time (44). In the case of the Hunza Karakoram, the glacial depositional record of three extensive glaciations, the third of which (with an age range of 139,000 to 50,000 yr BP) appears to coincide with the first half of the last (Dali) glaciation in montane Xinjiang and Qinghai-Xizang (30), has been preserved because of the rapidity of uplift in the late Pleistocene and Holocene. As a result, the glacial stratigraphic record of these recent periods is complex and cryptic: a detailed stratigraphy for the last 40,000 yr will require much more detailed research, not only on glacial sediments but in those of paraglacial and lacustrine origin.

ACKNOWLEDGEMENTS

This work was carried out as part of the Geomorphology Programme of the International Karakoram Project organized by the Royal Geographical Society in 1980, in collaboration with Academia Sinica and the Pakistan Ministry of Science and Technology. We are greatly indebted to the National Geographic Society, to the Royal Society and to the Royal Geographical Society for providing funding. Travel costs for three of the authors were borne by the Government of the People's Republic of China. The Pakistan Army is also thanked for invaluable helicopter photo-reconnaissance transport.

REFERENCES

1) DESIO, A., (1974). 'Karakoram Mountains'. In Spencer, A. M., (ed.) Mesozoic-Cenozoic Orogenic Belts: Data for Orogenic Studies. Geological Society of London, Spec. Pub. No. 4, pp 255 - 256.

2) GOUDIE, A., et al., (this volume). 'The geomorphology of the Hunza valley, Karakoram Mountains, Pakistan'.

3) ROETHLISBERGER, F. and SCHNEEBELI, W., (1979). 'Genesis of lateral moraine complexes, demonstrated by fossil soils and trunks: indicators of postglacial climatic fluctuations'. In Schluchter, Ch. (ed.) Moraines and Varves, Balkema, Rotterdam, pp 387 - 419.

4)   DAINELLI, G., (1922).   Studi sul glaciale: Spedizione Italiana de
     Filippi nell'Himalaia, Caracorum e Turchestan/Cinese (1913-1914),
     Ser. II, Vol. 3, 658 pp.

5)   PENCK, A. and BRUCKNER,   (1909).   Die Alpen im Eiszeitalter.
     Tauchnitz, Leipzig.

6)   TERRA, H. de and PATERSON, T. T., (1939).   'Studies on the Ice
     Age in India associated with Human Cultures', Carnegie Inst.
     Washington Publ. 493, 354 pp.

7)   NORIN, E., (1925). Preliminary notes on the Late Quaternary glaciation
     of the North Western Himalaya.   Geografiska Annaler, 7, pp 165 -
     194.

8)   WOLDSTEDT, P.,   (1965).   Das   Eiszeitalter.   Verlag, Stuttgart,
     374 pp.

9)   MOVIUS, H. L., (1949). 'Lower Paleolithic archaeology in southern
     Asia and the Far East', Studies in Physical Anthropology, No. 1,
     pp 17 - 82.

10) MORRIS, T. O., (1938).   'The Bain Boulder Bed: a glacial episode
     in the Siwalik Series of the Marwat Kundi Range and Sheik Budin,
     North-west Frontier Province, India'.   Quart. J. Geol. Soc., London,
     94, pp 385 - 421.

11) GILL, W. D., (1952).   'The stratigraphy of the Siwalik Series in
     the northern Potwar, Punjab, Pakistan. Quart. J. Geol. Soc. London,
     107, pp 375 - 394.

12) JOHNSON, G. D., JOHNSON, N. M., OPDYKE, N. D. and TAHIRKHELI,
     R. A. K., (1979).   'Magnetic reversal stratigraphy and sedimentary
     tectonic history of the Upper Siwalik Group, Eastern Salt Range and
     Southwestern Kashmir'.   In Farah, A. and DeJong, K. A., (eds.)
     Geodynamics of Pakistan, Geological Survey of Pakistan, Quetta,
     pp 149 - 165.

13) RENDELL, H. M. and THOMAS, K. D., (1980).   'Environmental change
     during the Pleistocene and Holocene in the Peshawar Basin, Pakistan',
     J. Archaeological Science, 7 (1), pp 69 - 79.

14) WADIA, D. N., (1953). Geology of India. London.

15) MACHATSCHEK, F., (1955). Das Relief der Erde. Berlin.

16) LI JIJUN, WEN SHIXUAN, ZHANG QINGSONG, WANG FUBAO, ZHANG BENXING,
     and LI BINGYUAN, (1979).   'A discussion on the period, amplitude
     and type of the uplift of the Qinghai-Xizang Plateau', Scientia Sinica,
     22, pp 1314 - 1328.

17) QIANG FANG and MA XING-HUA, (1979).   A preliminary discussion
     on some problems about the Quaternary magnetic stratigraphy of
     China. Kexue Tongbac, 24.

18) PAFFEN, K. H., PILLEWIZER, W. and SCHNEIDER, H.-J., (1956).
     'Forschungen im Hunza-Karakoram', Erdkunde 10, pp 1 - 33.

494

19) DESIO, A., TONGIORGI, E. and FERRARA, G., (1964). On the geological age of some granites of the Karakorum, Hindu Kush, and Badakhshan, Central Asia', Proceedings 22nd International Geological Congress, New Delhi, 11, pp 479 - 496.

20) TRINKLER, E., (1930). 'The Ice-Age on the Tibetan Plateau and in the adjacent regions', Geogr. J. 75, pp 225 - 232.

21) WICHE, K., (1959). 'Klimamorphologische Untersuchungen im westlichen Karakorum', Verhandl. dtsch. Geographentages 32, pp 190 - 203.

22) HEUBERGER, H., (1974). 'Alpine Quaternary Glaciation'. In Ives, J. D. and Barry, R. G., (eds.) Arctic and Alpine Environments, pp 319 - 338. Methuen, London.

23) SCHNEIDER, H.-J., (1959). 'Zur diluvialen Geschichte des NW-Karakorum', Mitt. geogr. Ges. München 44, pp 201 - 216.

24) PORTER, S. C., (1970). 'Quaternary glacial record in Swat Kohistan, West Pakistan', Geological Society of America, Bulletin 81, pp 1421 - 1446.

25) IWATA, S., (1976). 'Late Pleistocene and Holocene moraines in the Sagarmatha (Everest) region, Khumbu Himal, Seppyo 38, pp 109 - 114.

26) FUSHIMI, H., (1977). 'Glaciations in the Khumbu Himal (1), Seppyo 39, pp 60 - 67.

27) FUSHIMI, H., (1978). 'Glaciations in the Khumbu Himal (2), Seppyo 40, pp 71 - 77.

28) HEUBERGER, H., (1956). 'Beobachtungen über die heutige und eiszeitliche Vergletscherung in Ost-Nepal', Zeits. für Gletscherkunde, 3, pp 349 - 364.

29) BATURA GLACIER INVESTIGATION GROUP, (1976). Investigation Report on the Batura Glacier in the Karakoram Mountains, the Islamic Republic of Pakistan (1974-1975). Batura Investigation Group, Engineering Headquarters, Peking, 123 pp.

30) SHI YAFENG AND WANG JINGTAI, (1981). 'The fluctuations of climate, glaciers and sea level since late Pleistocene in China', In Sea Level, Ice and Climatic Change., I.A.H.S. Pub. No. 131, pp 281 - 293.

31) BATURA GLACIER INVESTIGATION GROUP, (1979). 'The Batura Glacier in the Karakoram Mountains', Scientia Sinica, 22, pp 958 - 974.

32) WISSMAN, H. von., (1960). 'Die heutige Vergletscherung und Schneegrenze in Hochasien mit Hinweisen auf die Vergletscherung der letzten Eiszeit', Abhandl. Akad. Wiss. Mainz, math. naturwiss. Klasse 1959, 14, pp 1101 - 1407.

33) ZHANG XIANGSONG and SHI YAFENG, (1980). 'Changes of Batura Glacier in the Quaternary and recent times', Professional Papers on the Batura Glacier, Karakoram Mountains, Science Press, Beijing, pp 173 - 190.

34) CAI XIANGXING, LI NIENJIE, and LI JIAN, (1980). 'The mudrock flows in the vicinity of the Batura Glacier', Professional Papers on the Batura Glacier, Karakoram Mountains, Science Press, Beijing, pp 146 - 152.

35) SINGH, G. and AGRAWAL, D. P., (1976). 'Radiocarbon evidence for deglaciation in north-western Himalaya, India', Nature, 260, p 232.

36) LI JIJUN, (1975). New data on glaciers in southeastern Xizang. J. Lanzhou University (Natural Sciences), 2, pp 115 - 123. (In Chinese).

37) GOUDIE, A. S., JONES, D. K. C. and BRUNSDEN, D., (1983). 'Recent fluctuations in some glaciers of the western Karakoram Mountains, Hunza, Pakistan', this volume, pp 411 - 455.

38) LI JIJUN, DERBYSHIRE, E. and XU SHUYING, (1983). 'Glacial and paraglacial sediments of the Hunza valley, north-west Karakoram, Pakistan: a preliminary analysis', this volume, pp 496 -535.

39) PORTER, S. C., (1970). 'Weathering rinds as a relative age criterion: Application to subdivision of glacial deposits in the Cascade Range', Geology 3, pp 101 - 104.

40) BIRKELAND, P. W., COLMAN, S. M., BURKE, R. M., SHROBA, R. R., and MEIERDING, T. C., (1979). 'Nomenclature of alpine glacial deposits, or What's in a name?', Geology, 7, pp 532 - 536.

41) DERBYSHIRE, E., (1982). 'Sedimentological analysis of glacial and pro-glacial debris: a framework for the study of Karakoram glaciers', Procs. Inter. Karakoram Project, Vol. 1, pp 347 - 364.

42) DESIO, A. and OROMBELLI, G., (1971). 'Notizie preliminari sulla presenza di un grande ghiacciario vallivo nella media valle dell' Indu (Pakistan) durante il Pleistocene', Atti della Accademia Nazionale de Lincei 51, pp 387 - 392.

43) AMBRASEYS, N., LENSEN, G., MOINFAR, A. and PENNINGTON, W., (1981). 'The Pattan (Pakistan) earthquake of 28 December 1974: field observations', Quarterly Journal of Engineering Geology, 14, pp 1 - 16.

44) GAGE, M., (1965). 'Accordant and discordant glacial sequences', Geological Society of America, Spec. Pub. 84, pp 393 - 414.

Glacial and paraglacial sediments of the Hunza Valley North-West Karakoram, Pakistan: A preliminary analysis

Li Jijun – University of Lanzhou, P.R.C.
E. Derbyshire – Keele University, U.K.
Xu Shuying – University of Lanzhou, P.R.C.

ABSTRACT

The tills of the upper Hunza region include both lodgement and meltout types. Particle size analysis shows them to be relatively coarse-grained and somewhat positively skewed. There is evidence of modification of the particle size distribution by meltwater eluviation, especially in the case of the Ghulkin and Pasu glacier deposits but the role of bedrock lithology and glacier size is also discernible. Tills currently being deposited are predominantly of meltout type: these have diffuse fabrics, rather granular matrixes and contain evidence of eluviation and collapse. Re-deposition of the abundant coarse supraglacial debris by sliding under gravity is common. The thickest exposures of compact and sheared lodgement till, on the other hand, are of Pleistocene age, especially those of the main valley (t_2), diffluence col (t_3) and expanded foot (t_4) stages. There is no evidence of direct deposition of till by viscous flow either during or since the Pleistocene although the abundant debris flows do contain glacial material. It appears likely that the present combination of high-activity glacial regimes and arid to semi-arid valley floor climatic conditions have prevailed since at least the beginning of the late Pleistocene. The distribution of the dense, jointed lodgement tills of Pleistocene age is consistent with their deposition in a generally free-draining environment beneath thick glaciers. The genetic classification of tills based on field relationships is generally confirmed by laboratory-derived data on particle size, particle shape, stone orientation and fabric.

INTRODUCTION

Scientific studies of glaciers and glaciation of the Himalayan region go back over 60 years [1]. The dominant approach throughout this time has been morphostratigraphical, the main emphasis being on moraine and terrace correlations. Very little attention has been given to the sedimentary properties of these and associated landforms. This is hardly surprising in view of the recency of the emergence of sedimentology as a scientific discipline whose prime aim is to 'narrow the gap between modern process and past product' [2]. This development included growth in glacial sedimentology, although this field remains relatively retarded, especially in the area of facies discrimination in deposits of Pleistocene age.

Nevertheless, there has been a rapid increase in interest in the genetic classification of glacial sediments (3). This approach is proving fruitful in the study of recent glacial sedimentation (e.g., 4, 5, 6) and the development of general models for application to Pleistocene deposits is promising (7, 8).

The principal sedimentary properties (mineralogy, petrology, particle size, particle shape and fabric) reflect the interaction between the source materials and the energy of the process system. In the case of glacial sediments, the source rocks of the glacier bed and slopes above the glacier influence the susceptibility to weathering and glacial comminution which, in turn, influence the grain size of the glacial sediments produced. The energy of the glacial process system is strongly controlled by climate which is reflected in ice thickness and mass flux. These, in turn, affect erosion potential, and transportation and deposition modes (9).

This study presents preliminary results of the mapping and analysis of glacial and paraglacial sediments in the upper Hunza River valley, which emerged as part of a larger, more comprehensive study of the glacial morphology, sedimentology and stratigraphy of the region. The area studied (see Fig. 1 of reference 1; this volume) extends from the Batura Glacier to the northern slopes of Rakaposhi south of Hindi and includes parts of the northern sedimentary, axial granitic, central-southern sedimentary and Chalt green schist zones (10).

In the field, sediments were mapped using altimeter, plane-table, and aerial photographic cover at a nominal scale of 1:30,000. Classification of sediments was based on particle size and shape and components of the macrofabric (bedding, bedforms, jointing, faulting and stone orientation). Laboratory analysis of over 100 samples included particle size, particle shape, clast fabric, microfabric, petrology and mineralogy.

Particle size distribution was determined by a combination of wet sieving and sedimentation (Andreasen pipette) methods according to B.S. 1377 (British Standards Institution (11)). Edge rounding of particles was estimated by comparing their projection shape with the roundness charts of Rittenhouse (12) and Shepard and Young (13). This was done for the pebble grade in the field and for the medium sand grade in the laboratory using low power stereomicroscopy. Clast fabric was measured in the field and, with greater accuracy, in the laboratory using oriented and levelled blocks of till following the technique of McGown and Derbyshire (14). Microfabric was examined using small orientated and levelled till samples in a scanning electron microscope (cf. 15). Sand grain surface textures were also studied using this technique. Mineralogy of selected sand grade samples was determined by optical microscopy, and mineralogy of the clay grades was established by X-ray diffractometry.

PROCESSES, FORMS AND SEDIMENT ASSOCIATIONS

The glaciers of the Hunza Karakoram have developed in a landscape in which tectonically-stimulated fluvial erosion has produced some of the deepest valleys on earth. The relative relief of the region of up to 5600 m is of fundamental importance in understanding both past and present glacial sediment systems and the associated geomorphological processes.

The glaciers of the region extend from the high snowfall and perennially frozen soil zone down to the aridity of the valley floors, mean annual

498

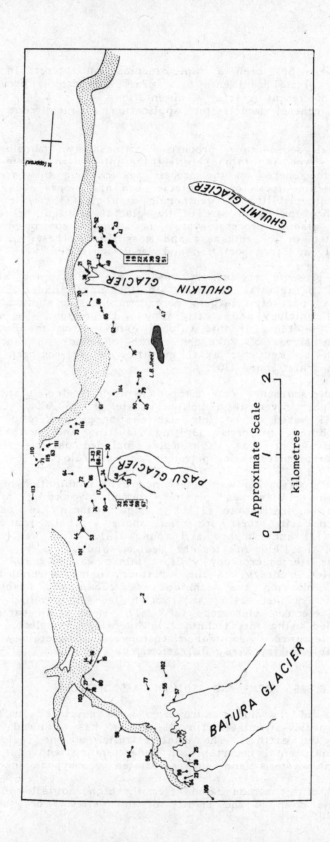

FIG. 1. SEDIMENTS MAPS.

A. GENERAL MAP OF THE SAMPLING AREA.

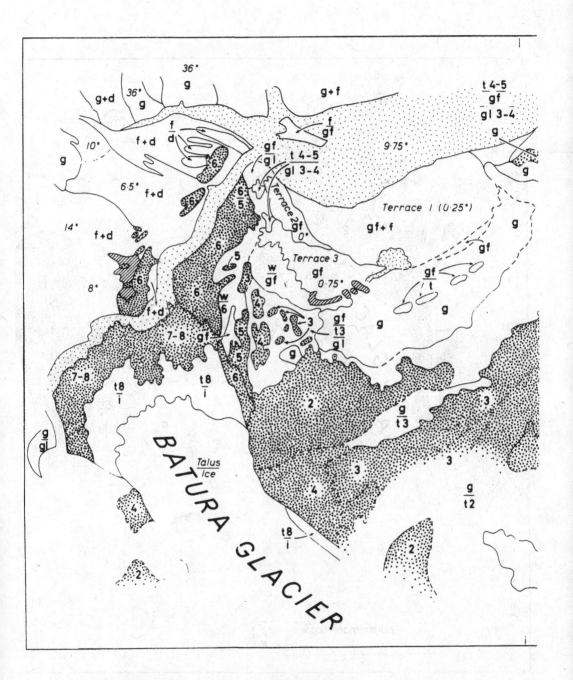

FIG. 1. SEDIMENT MAPS.

A. Surficial sediment map in three zones between the Batura Glacier
   and Gulmit fan.   (cont'd...)

500

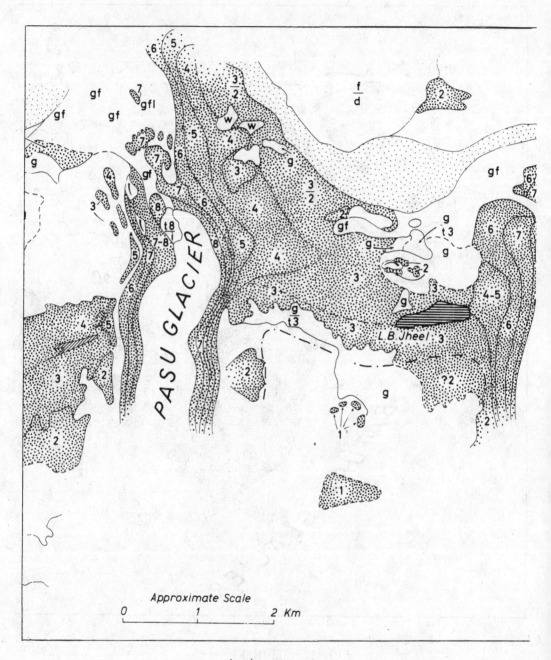

A. (cont'd...)

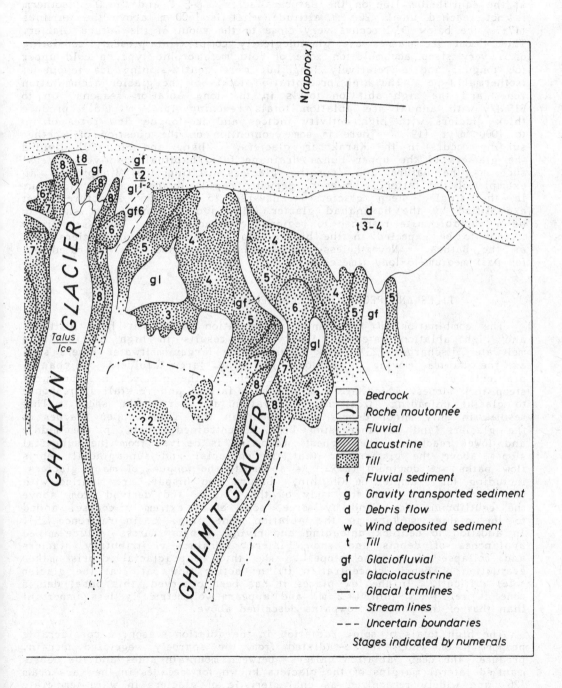

A.

precipitation for source and snout areas of the Batura Glacier being
2000 mm and 100 mm respectively (16). The mean annual air temperature
at the equilibrium line on the Batura Glacier is -5°C and the 0°C isotherm
is not reached until 4200 m altitude, that is 1600 m above the terminus
(17). Ice below 0°C occurs very close to the snout of the Batura Glacier.
Thus, most ice masses are glaciologically complex at present, consisting
of a very steep accumulation zone of cold metamorphic type, a cold upper
ice tongue, and a relatively long but more gently-sloping ice tongue of
isothermal type. The high precipitation totals of the glacier accumulation
zones and the high ablation totals in the long ablation seasons (up to
315 d/y with annual ice ablation totals reaching 18.41 m (18)) produce
thick glaciers with high activity indices and ice-tongue flow rates of up
to 1000 m/yr (19). There is some contention on the question of whether
surging occurs in the Karakoram glaciers. Zhang and Shi (20) divided
the glaciers of the upper Hunza drainage into 'habitually active glaciers'
such as the Batura and 'surging glaciers', citing the Minapin as an
example of the latter. However, periodic rapid advances appear to occur
in the single, steep glaciers transverse to the rock structures best
exemplified by the Hasanabad glacier (21), and this has been shown to
be a variable rate response to regional climatic change. Such sensitivity
is not to be expected in the bigger, flatter and compound glaciers such
as the Batura. Nevertheless, surge behaviour cannot yet be ruled out
for past medium-to-long timescales (22).

TILLS AND PROGLACIAL SEDIMENTS

The combination of high snow accumulation values at high altitudes
and high ablation rates in the valleys results in high but variable
meltwater discharges (23), peak flows in the larger meltwater rivers such
as the Batura moving a bedload including large boulders on channel
gradients up to 18°. The deep incision by the glaciers has produced
steep rock slopes, which are susceptible to failure and rockfall in response
to glacial unloading, frost-riving, nivation and earthquake shock. This
results in only minor debris entrainment in the 'cold' upper reaches of
the glaciers and an increasing bulk of entrained debris in the middle
and lower reaches. The highest proportion is derived from the subaerial
slopes above the glaciers so that the englacial and supraglacial debris
flow paths are dominant (24). As a result, the tongues of many glaciers,
including for example the Ghulkin, Batura and Hispar, are mantled with
debris melted out from the body of the glacier and derived from above
the equilibrium line, and by scree, some of it extremely coarse, added
to the glacier surface in the ablation zone (Fig. 2 in reference 25).
In addition to bedrock unloading and riving, debris sources include mixed
avalanches of debris and snow, lateral moraines of tributary glaciers
and collapse of moraine slopes. The thick supraglacial debris makes
evaluation of debris concentration in the active flow path (at the glacier
sole) difficult. In the few places it has been observed, this basal debris
zone is relatively thin (< 2 m) and appears volumetrically less important
than that of the passive flowpaths described above.

The high totals of solar radiation in the ablation season, a considerable
proportion of which is re-radiated from the sparsely vegetated terrain,
produce the deep 'ablation valleys' between mountain sides and the debris
mantled lateral margins of the glaciers known for decades in the Karakoram
(26) and widely recognized as characteristic of glaciers in warm-temperate
to tropical regions. The importance of radiation asymmetry in this
connection has recently been demonstrated by Hastenrath (27). One result

of this process in the Hunza region is the concentration of debris into high, initially ice-cored moraines. Abundant debris supply and high rates of sediment throughput in the glacial system are thus conducive to large moraines and thick, generally granular tills intimately related to outwash sediment complexes, particularly close to the valley floors.

Tills lying at altitudes below 3000 m, although weathered and dissected to varying degrees (1), have retained their primary morphological expression in the form of moraines. They are distributed in the glacial diffluence cols and the tributary valleys as well as on benches in the main Hunza valley, providing a clear record of the declining series of a deglacial hemicycle (Fig. 1). The relatively minor differences in glacier extents from the Ghulkin II (t_5) stage through the Batura (t_6), and the Pasu I (t_7) stages (Table 1) has resulted in the stacking of till and the construction of some huge lateral moraines. This process of remoulding, deformation and local superposition can be seen in tills of all ages. Intercalation with glaciolacustrine silts, glaciofluvial sands and gravels, talus and debris flow deposits is characteristic.

time stage	stage name	time stage	stage name
t_1	SHANOZ	t_5	GHULKIN II
t_2	YUNZ	t_6	BATURA
t_3	BORIT JHEEL (? Hunza correlate)	t_7	PASU I
t_4	GHULKIN I	t_8	PASU II

TABLE 1: PROPOSED QUATERNARY GLACIAL STAGES IN THE UPPER HUNZA RIVER VALLEY.

The most dense tills in the region are those with lithologies predominantly of limestone, schist and slate. They occupy a wide altitudinal range from the Hunza valley floor to about 3000 m on high valley benches and the lower summits (e.g., east of Lake Borit Jheel). Compact matrixes (Fig. 2) and an abundance (over 20%) of faceted and striated clasts with a strongly anisotropic pebble fabric indicate that many of these tills were laid down by subglacial lodgement in the last major trunk valley stage of glaciation (Borit Jheel – t_3 stage). The limestone tills, in particular, are subject to calcreting. A series of exposures in limestone tillite attributed to the Yunz (t_2) stage and calcreted to at least 3.5 m depth, occurs 2 km north of the Ghulkin Glacier and 0.5 km N.E. of Lake Borit Jheel (Fig. 1). Similar remnants of calcareous till occur at about 3000 m west of this lake. A series of exposures in these tills can be seen along the Karakoram Highway between the Ghulkin and Pasu glaciers. On slate bedrock about 6 km north of the Ghulkin a melange of slaty,

FIG. 2. COMPACT LODGEMENT TILL WITH FACETED AND STRIATED CLASTS AT C. 2680 M IN MORAINE ON EAST SIDE OF MINAPIN GLACIER VALLEY.

rather loose meltout tills, glaciofluvial gravels and deformed glaciolacustrine silts is overlain by an extremely poorly sorted, dense limestone-rich lodgement till with over 60 per cent by weight of silt and clay (Fig. 3). There is a well developed series of shear planes just below the interface. The sequence is completed by an upper meltout till with rather coarse, angular limestone clasts and only 20% fine fraction (Fig. 3). This sequence represents a change from tributary valley ice deposition of meltout type from an expanded Pasu Glacier, with local ponding of meltwaters, to subglacial deposition by through-flowing Hunza valley ice with a change to limestone-rich tills derived from the north.

The highest lateral moraine remnants of the Minapin Glacier dating from the latter part of the trunk valley glaciation (Borit Jheel (t_3) stage: Fig. 4; see also Fig. 7 in (1)) contain compact, very poorly sorted lodgement tills resting on striated schist pavements at c. 2680 m (altimeter). Even at this altitude, however, there are abundant intercalated lacustrine silts laid down in high-level, ice marginal ponds, the highest recorded laminated silts lying at 2774 m.

Lodgement tills occur at lower levels in moraines of all ages but there is some evidence in the form of particle size and shape changes that meltwater and the melting out process became more important as ice volumes progressively decreased in the late Pleistocene. Moraines of the Ghulkin stages (t_4 and t_5) are locally gullied and degraded (to gradients of about $28°$). Exposures up to 1000 m in height reveal subglacial lodgement tills and shearing zones, especially on the ice-proximal moraine slopes,

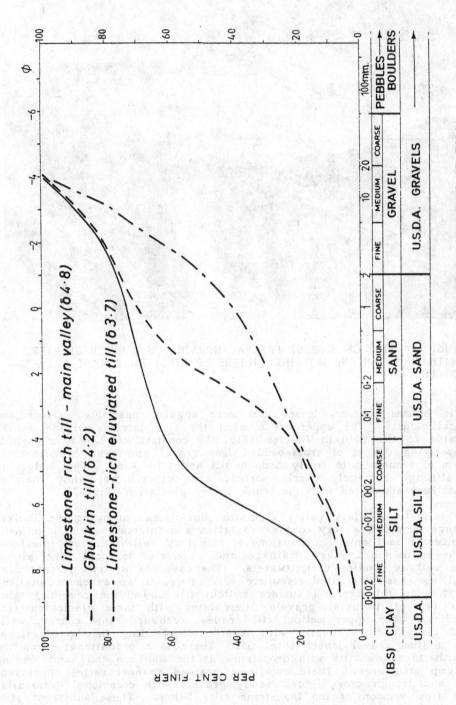

FIG. 3. CUMULATIVE FREQUENCY CURVES OF THREE TYPES OF TILL.

FIG. 4. TILL OF HIGHEST LATERAL MORAINE OF MINAPIN GLACIER, WITH HUGE FAN AND MORAINE INLIERS OF HINDI EMBAYMENT IN BACK-GROUND.

but tills become coarser, looser and more angular near the surface and on distal slopes. The uppermost 6 m of the $t_{4/5}$ lateral moraine on the south side of the Ghulkin Glacier (Fig. 1) consists of 2.5 m of meltout till, overlying 0.5 m of cross-bedded fine gravel and sand beneath which is 3.5 m of laminated to finely bedded silt and silty sand. The lodgement tills, although extremely poorly sorted, are rather more sandy than in many of those associated with the trunk valley glaciation (Fig. 3).

The shift to tributary valley glaciation during and following the Ghulkin (t_4) stage produced a very complex variation and intercalation of glacigenic and meltwater sediments. Expansions of tributary valley glaciers interfered with the main Hunza River drainage and gave rise to local reversals of flow as well as ponding of meltwaters. The case of the drift terrace near Pasu village has been cited elsewhere (1). Here, in an exposure totalling about 30 m in thickness, a surface meltout till underlying cemented talus overlies two glaciofluvial gravels intercalated with three glaciolacustrine silt bodies. The upper meltout till grades northwards into coarse, well-rounded glaciofluvial gravels, with some cross-bedding in the occasional fine to medium gravel lenses (Fig. 5). There is a gradational transition downwards to sandy silts with dropstones at the southern end, and loading and slump structures. These overlie a braided channel series of current bedded and trough cross-bedded sandy gravels with occasional 'mudballs', derived from erosion of the lacustrine silts below. These subjacent silts are much less sandy than the upper silt unit and load structures are

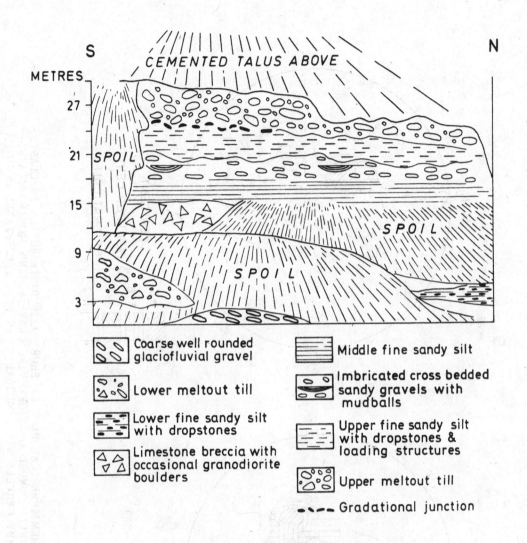

FIG. 5. GENERALIZED SUCCESSION IN DRIFT TERRACE
OVERLOOKING PASU VILLAGE. V.E. x 3

less common.   At the south end, this silt horizon overlies a limestone
breccia with occasional granodiorite erratics resting, in turn, on a meltout
till.   The basal unit is a coarse, well-rounded glaciofluvial gravel.
A third unit of glaciolacustrine silts with dropstones is even more uniform
with up to 20 units made up of fine to very fine laminae.   A sample of
this lowermost silt yielded a TL age of 65,000 ± 3,300 yr B.P.   It is evident
from the gravel fabric that the lowermost glaciofluvial member was deposited
by the Hunza River, but both fabric (Fig. 6) and bedforms indicate
deposition of the upper gravels by northeasterly flowing meltwaters.   The
high slate and schist content and rather low limestone content of this
unit confirms provenance from the Pasu Glacier to the southwest of the
exposure (Table 2).   Moreover, the intercalated lacustrine silt units can
be traced northwards at similar altitudes as far as the larger drift plateau

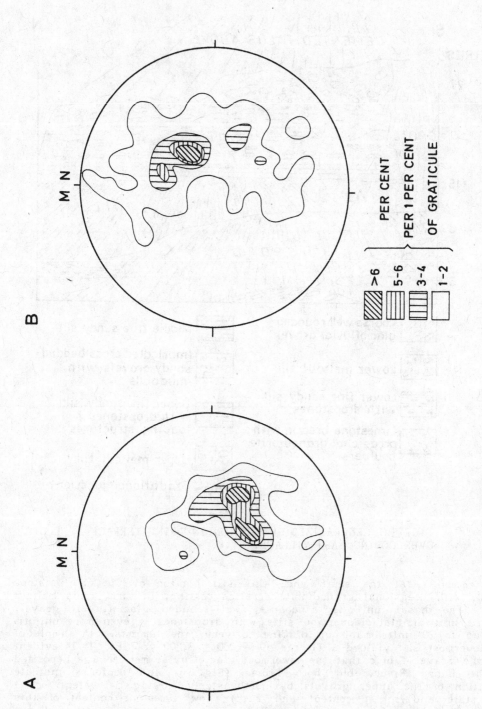

FIG. 6. LOWER HEMISPHERICAL PLOT OF SHORT AXIS DISTRIBUTION IN CLASTS IN (A) COARSE, WELL-ROUNDED GLACIOFLUVIAL GRAVELS, AND (B) IMBRICATED CROSS-BEDDED SANDY GRAVELS WITH 'MUDBALLS' IN SITE ILLUSTRATED IN FIG. 5.

Stratum	Roundness % (Shepard and Young 1961)						Lithology of clasts %				
	A	B	C	D	E	F	Lime-stone	Granite	Schist	Sand-stone	Other
Cross-bedded sandy gravels with 'mudballs'	2	18	24	20	32	4	22	4	50	14	10
Lowermost glaciofluvial gravel	8	28	20	6	34	4	40	20	0	4	36

TABLE 2: COMPARATIVE DATA ON ROUNDNESS AND LITHOLOGY FOR TWO GLACIOFLUVIAL HORIZONS IN THE DRIFT PLATEAU REMNANT WEST OF PASU VILLAGE.

south of the Batura Glacier (Fig. 1), clearly indicating two or three periods since c. 65,000 yr B.P. in which the lower Pasu Glacier effectively dammed the Hunza River, creating a lake between its terminus and that of the Batura Glacier. Remnants of these glaciolacustrine episodes in the form of 30 cm of silts overlying well-rounded glaciofluvial gravels (imbricated 300 - 320° magnetic) have been preserved beneath at least 12 m of cemented talus between the Pasu and Batura glaciers about 30 - 40 m above Hunza terrace 1. These relationships are replicated at different scales in many other localities. Distal progression from a thick series of meltout tills to poorly sorted and crudely bedded outwash deformed during ice retreat by thrusting and buried ice collapse can be seen at the junction of the Hispar and Hunza rivers (Fig. 22 in (1)). The sequence, dating from the latter part of the trunk valley ice stage rests on bedrock and also on glaciofluvial sands and gravels with dips of 2 - 10°. Lacustrine silts 10 - 20 m thick are important units in the sedimentary succession. Glacio-lacustrine sandy silts with dropstones reach 15 m in thickness in the lateral moraines of Ghulkin II (t_5) age on the south side of the Pasu Glacier.

The most complex sedimentary sequences occur in places where ice fronts have oscillated back and forth over a distance of only 1 or 2 km during and since the last Pleistocene glaciation. This is illustrated in the exposures along the Hunza River between the Batura Glacier terminus and the confluence with the Shimshal River. A moraine of the end phase of the last Pleistocene glaciation (Ghulkin II (t_5) stage) consisting largely of intercalated coarse, angular meltout tills, proximal glaciofluvial facies and abundant glaciolacustrine silts with dropstones was overthrust by ice of the post-Pleistocene Batura (t_6) stage, leaving a thick sequence of coarse, rather granular tills. The two sequences are separated by a well-developed shear zone dipping at 40° towards 330° magnetic, in which clast anisotrophy is very high and silt coatings and smears occur throughout. The effects of thrusting extend at least 50 m into the t_5 moraine and are clearly shown by the plastic deformation of the glaciolacustrine silts. High angle thrust planes can be measured in many locations in this end moraine zone and a well developed shear fissility occurs in the Ghulkin ($t_{4/5}$) stages on the south side of the Pasu Glacier.

The Batura stage tills and glaciofluvial gravels, with a total thickness of at least 18 m, consist, in the type area, of a lower series separated from the upper sequence (11 - 12 m thick) by a planar erosion surface. Large-scale structures observed include shear zones indicating successive accretion of till in the upper sequence, steeply dipping ($\sim 40^\circ$) coarse gravels deposited by sliding on buried ice and a variety of deformation structures. The latter include glaciofluvial bedded sand with vertical dips and concentric structures in glacigenic gravels around large boulders. Some of the clasts show signs of crushing and fracture. Loading and shearing appear indicated here, especially as thrust planes rise up on the distal sides of the concentric features, but the precise mechanism remains cryptic.

These depositional and re-moulding processes are still occurring in the terminal zone of the Batura Glacier. On the edge of the present ice-cored moraine zone (location 75, Fig. 1) tills about 4 m thick deposited by meltout and sliding overlie unconformably a thin glaciofluvial gravel which rests, also unconformably, on a severely deformed complex consisting of a sheared, rather granular lodgement till, meltout tills, eluviated meltout and slide-deposited tills and glaciolacustrine silts with dropstones having dips of up to 67° (Fig. 7). This may represent the latest 'historical' advance over proglacial sediments deposited prior to about 200 yr ago. Certainly, glaciofluvial gravels laid down by meltwaters during retreat have been incorporated in tills of the advance of the past century, as can be seen from the high proportion of fluvially rounded clasts (up to E and F degrees in the six point scale of Shepard and Young (13)) in the supraglacial meltout till of the Batura and Pasu glaciers (maximum percentages of 35 and 55 respectively). The lower till has the microfabric properties characteristic of the lodgement type.

Glaciofluvial sediments occur at all topographical levels, usually inter-bedded with tills. The relationship is particularly complex in areas where supraglacial accumulation has given rise to buried ice. The proglacial tract of the Pasu Glacier in which ice wastage and retreat has occurred over the past 1000 yr or so shows a gradation from fluted tills and small push moraines, variously dissected and replaced by meltwater gravel trains, to the present dead-ice area of 'hummocky moraine' and supraglacial melt-water gravels (Fig. 8).

Extensive spreads of glaciofluvial sands and gravels are best developed near the floor of the Hunza valley and its larger tributaries. These are preserved as terraces such as those on the west side of the Hunza River between the Batura and Pasu glaciers (Fig. 1). The lower terrace is about 5 m thick with a surface gradient of no more than 0.25°. It is dominated by slate and schist clasts which are moderately to well rounded (Table 2). The medium sand grade is also more rounded than in most till samples examined (Table 3). There is a gentle imbrication up-valley and trending slightly west of north (Fig. 9) but the material is very poorly sorted (sorting coefficient 2.46) and leptokurtic, in contrast to the fluvial gravels of the Hunza which are poorly sorted (s.c. 1.72-1.92) and mesokurtic. The desert varnish on the surface of this terrace is similar to that on the postglacial Batura (t_6) stage moraine. The gravels of the second (upper) terrace are much thicker (up to 20 m), coarser, and somewhat more angular (Table 4). The lithology is dominated by limestone and granite and is clearly derived from the Batura Glacier debris. Also there is no significant difference in the mineralogy of the clay grade of glaciofluvial terrace sediments and Batura tills, quartz, mica, Fe-chlorite, feldspars and calcite occurring in all samples tested.

FIG. 7. EXPOSURE ADJACENT TO ICE-CORED MORAINE AT BATURA GLACIER TERMINUS. SCALE BARS (METRES).

512

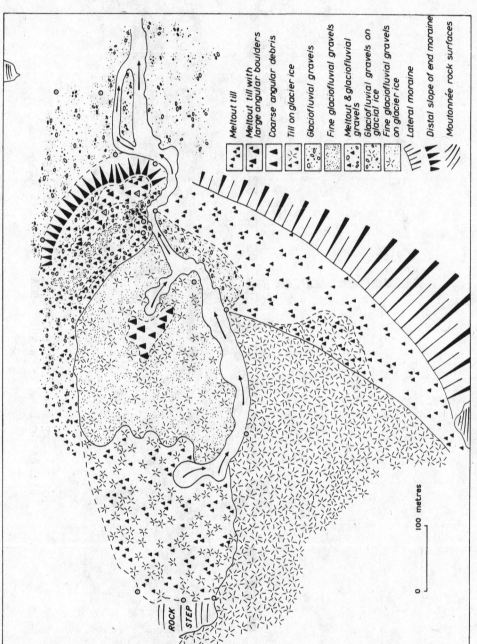

FIG. 8. FIELD MAP OF SEDIMENT COMPLEX AROUND 1980 TERMINUS OF PASU GLACIER.

	PASU	BATURA	GHULKIN	MINAPIN AND PISAN	MAIN VALLEY	FLOODPLAIN
Roundness of coarse sand grade mean:-	0.2	0.2 - 0.3	0.1	0.2	0.2	0.5
range:-	0.1 - 0.4	0.1 - 0.7	0.1 - 0.3	0.1 - 0.7	0.2 - 0.5	0.3 - 0.9
Minerals (coarse sand grade)	Mainly QUARTZ. Feldspars. 10% biotite, shale and opaque	Mainly QUARTZ with fine-grained GRANITE feldspars. 12-20% biotite	Mainly QUARTZ Feldspars 8% biotite shale, and opaques	>90% black and green SCHIST fragments with Fe stain.	LIMESTONE, DOLOMITE & CALCITE 30-40% QUARTZ 25-35% Feldspars. Remainder schist and opaques	Up to 45% OPAQUES with abundant biotite and slate
Minerals (clay grade)	Quartz Micas Feldspars Chlorite (Fe)	Quartz Micas Chlorite (Fe) Feldspars Calcite	Quartz Micas Chlorite (Fe) Feldspars	Micas Quartz Chlorite (Fe) Feldspars Additional in Flow Debris Calcite Halite	Quartz Calcite Feldspars Micas Chlorite (Fe)	

TABLE 3: ROUNDNESS AND PRINCIPAL MINERAL GROUPS FROM SIX LOCATIONS IN THE UPPER HUNZA VALLEY.

514

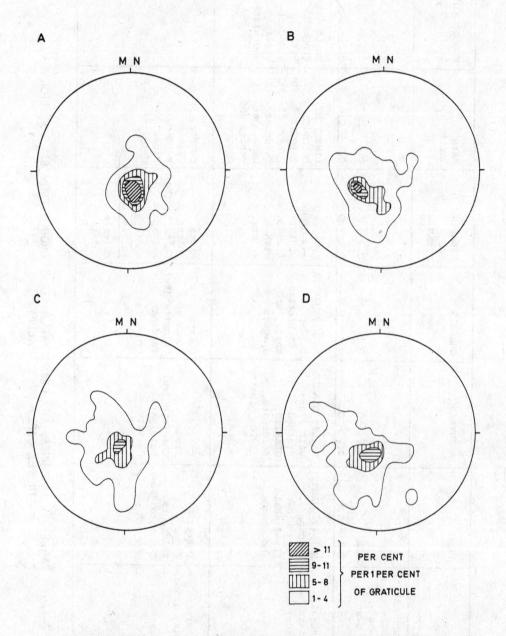

FIG. 9.   LOWER HEMISPHERICAL PLOTS OF SHORT AXIS DISTRIBUTION
OF PEBBLES AND COBBLES IN LARGE TERRACES BETWEEN BATURA AND
PASU GLACIERS; A – LOWER TERRACE; B – UPPER TERRACE, TOP LAYER;
C – UPPER TERRACE, MIDDLE LAYER; D – UPPER TERRACE, BOTTOM
LAYER.

Terrace	Thickness (m)	Pebble Roundness % (Shepard and Young 1961)						Lithology %					Genetic Type
		A	B	C	D	E	F	Limestone	Granite	Schist	Sandstone	Other	
Lower	5	O	10	26	20	36	8	2	6	46	28	18	glacio-fluvial
Top	3	O	10	18	30	16	26	26	20	28	6	20	alluvial
Upper middle	15-20	O	10	34	18	32	6	64	26	0	4	6	glacio-fluvial
Bottom	1	O	4	26	22	20	28	48	2	30	16	4	glacio-fluvial

TABLE 4: COMPARATIVE DATA ON ROUNDNESS AND LITHOLOGY FOR HORIZONS IN THE TWO LARGE GLACIOFLUVIAL TERRACES BETWEEN PASU VILLAGE AND THE BATURA GLACIER.

It is extremely poorly sorted (s.c. 4.95) and leptokurtic and passes upstream into meltout till and glaciolacustrine silts (Fig. 10). At both upper and lower boundaries it passes into fluvial gravels, the fabric of all three units being imbricated to the north (Fig. 9).

COLLUVIAL AND ALLUVIAL SEDIMENTS

A substantial proportion of the lower valley sides and valley floors in the Hunza is surfaced by debris flow and talus fans. They vary in their size and surface gradients according to their age, mean sediment grain size and dominant formative process. Fluvial outwash fans develop rapidly with shifts in the point of emergence of meltwater streams such as those of the Batura and Ghulkin glaciers and seasonal meltwater plays an important role in debris flow sedimentation (28). Periodically high sliding rates in some of the higher and steeper glaciers have also been regarded as instigators of debris flow events (29).

The largest fans in the area, e.g., on the east bank of the Hunza River southeast of Pasu and east of the Batura terminus, and in the Hindi embayment (see below), have mean gradients of less than 15° and have accumulated in the mountain-foot angle at the point of emergence of small snow- and glacier-fed streams. Their detailed stratigraphy has not yet been elucidated but Chinese work suggests two principal periods of growth, namely Hypsithermal and recent. They are made up predominantly of re-deposited glacial and fluvial sediments.

A group of much steeper fans with gradients of the order $36 - 37^{\circ}$ mantle the lower mountain slopes and are locally interbedded with the debris flow and alluvial sediments of the more gently sloping fans. The steeper talus fans are fed by fall and slide of rock fragments produced

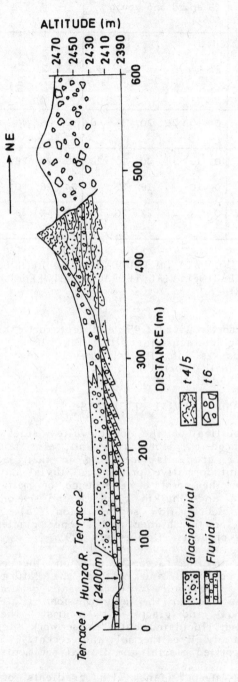

FIG. 10. SECTION SHOWING TRANSITION FROM POSTGLACIAL BATURA (t_6) STAGE TILL (RIGHT) TO GLACIOFLUVIAL TERRACE.

by frost and salt weathering, unloading and spalling. Some of this taluvium is cemented, as in the cliffs to the west of Pasu village where it is overlain by tills and so clearly ante-dates them. Many examples of such alternation of subaerial mass movement and alluvial and debris fan sedimentation with till deposition were noted, recording shifts from glacial to non-glacial conditions in the valleys. In addition, there is clear evidence that glacial and debris flow sedimentation occur simultaneously and have always been intimately related. Since the last invasion of the main Hunza valley by glacial ice, there has been massive replacement of glacial depositional surfaces by those of alluvial and debris flow origin along the Hunza valley floor and lower valley sides. Tills have been removed by river action and also buried beneath thick fanglomerates. In an area of 12 km^2 along the Hunza valley between the Ghulkin Glacier and the western edge of the Batura Glacier terminus, no less than 5.23 km^2 (43.5%) is surfaced by fan debris and 3.71 km^2 (31%) by fluvial and glaciofluvial sediments, only 1.69 km^2 (14%) being mantled by glacigenic sediments, the remaining 11.5% being bedrock.

The oldest and highest fluvial sediments discovered lie at approximately 3000 m, some 800 m above the present Hunza River channel close to Sarat. They are strongly cemented, well rounded and sorted fine (<5 mm) sandy gravels and underlie a dissected till of the last major glaciation of the main Hunza valley system (Yunz or t$_2$: Table 1).

The finest examples of the interaction of glacial, debris flow and alluvial sedimentation, however, have been preserved in a series of embayments cut into bedrock by the repeated diversion of the Hunza River by the advance of large glaciers in some tributary valleys. The simplest case occurs in the embayment on the south bank of the Hunza and opposite the unnamed glacier between Baltit and Sarat (Fig. 1 of reference (1), this volume). Here hummocky remnants of recessional moraines have been gently dissected and mantled with up to 1.5 m of sandy, silty loam with gravel layers. Locally, an upper yellow stoneless silt up to 20 cm thick mantles the hummocks and has the appearance of loess redeposited by colluviation and slopewash.

The Batura Glacier terminus area provides more complex sequences. On the north side of the glacier, over 10 m of cemented debris flow deposits with a silty matrix and local pockets of openwork gravel are overlain by 2.5 to 4 m of lacustrine grey to buff sandy silts, finely laminated with cross-bedded units and fine ripple-bedded sets (Fig. 11). Further to the west it is very poorly sorted (sorting coefficient 2.95) increasingly faulted and contorted owing to its proximity to the glacier and loading by overlying sediments. These include up to 5 m of rounded, coarse to medium fluvial gravels below at least 6 m of debris flow deposits and a capping of 2.5 m of lacustrine silts. The debris flow units differ little from the Batura Glacier tills and probably represent ice distal flow and slide debris.

The fanglomerates of the embayment opposite the Batura terminus are locally interbedded with, and have buried to a depth of 30 – 40 m, the coarse meltout tills of the Ghulkin II (t$_5$) and Batura (t$_6$) stages, only the larger moraines protruding as inliers (see Fig. 15 in (1)). The upper parts of the fans show parallel and convoluted bedding dipping at c. 20°, but the mean gradients of the main fan surfaces average 6.5°. The clasts are generally angular to subangular (B-D in the scale of Shepard and Young (13)), there being perceptible increase in edge-rounding with increased age and distance down-fan. All but the most recent fan sediments

FIG. 11. ABOUT 4 M OF FINELY BEDDED LACUSTRINE SILTS INTER-CALATED WITH FLOW DEBRIS AND FLUVIAL GRAVELS JUST NORTH OF BATURA TERMINUS.

are calcreted to a depth of c. 0.5 m, although the calcrete appears to be degenerating in many areas.

In the Hindi embayment, debris flow sediments have accumulated to thicknesses of up to 90 m in an area of almost 4 km^2 (Fig. 12, also Fig. 4). The sequence of deposits includes coarse (>10 mm) fluvial sandy gravels mainly of granodiorite and sandstone resting on a perched valley floor remnant c. 300 m above the present channel of the Hunza River. The half-buried moraines of the Minapin and Pasu glaciers north of the river contain lenses of glaciolacustrine sandy silts (mean grain size c. 0.005 mm) with a T.L. age of 47,000 ± 2,350 yr B.P., i.e., within the last Pleistocene glaciation. The upper slopes of these inliers are somewhat degraded and strongly calcreted. The debris flow sediments making up the extensive fan surface about 100 m above the Hunza River clearly post-date not only the moraines but the thick fluvial sediments deposited when 'expanded foot' enlargement of the Minapin and Pisan glaciers diverted the Hunza around the northern perimeter of the embayment.

These thick debris flow deposits are remarkably uniform in their particle size characteristics (see below), being matrix-supported diamicts with clay (<0.002 mm) contents of less than 8%. They are made up for the most part of a series of relatively thin (<4 m) silty sandy gravels with occasional waterlain sets of plane-bedded sands and fine gravels. The debris flow material shows in the size, shape and lithology of both clasts and matrix its derivation from the deposits of the adjacent glaciers (Table 3).

The glacial, debris flow, alluvial and lacustrine sediment systems are thus intimately related in the sedimentary record of the upper Hunza River region. All were present throughout the Quaternary, although varying in relative and local importance.

OTHER PROCESSES

The periglacial processes, notably frost-shattering and nivation, are important at high levels but, under the prevailing arid conditions at lower altitudes, degradation of glacial and other sedimentary landforms does not include solifluction on valley sides and floors. Instead, salt weathering, deflation and wind abrasion have produced some distinctive surface forms and sediments.

Silts have accumulated from time to time in small ephemeral lakes on moraines, fans and glaciofluvial terraces, e.g., on the Batura Glacier and east of Baltit, and on the outwash gravel terraces between the Batura and Pasu glaciers. The mineralogy of the clay-grade material in these lacustrine silts closely reflects the mineralogy of the glacier system adjacent to them. Wind-blown sand has accumulated in small dunes, sheets and sand shadows.

The evidence of moraine weathering characteristics has been referred to elsewhere (1, 22), but attention is drawn to erosional etching by the wind affecting moraine boulders of granite and limestone in particular. Unlike the granite tafoni and the networks of fine solutional grooves on the limestone, wind etching produces thorn-like projections on granite and large radiating troughs on limestone, one trough measuring 28 cm long by 10 cm wide and 6 cm deep.

It is striking that the only area in the Hunza valley in which this sand-blasting of till stones was observed to such a marked degree is the southern side of the Batura Glacier terminus. This localization is clearly a product of the strike of the Hunza valley here (S 20° - 30° E) which funnels the near-surface winds. The southeastern flow is particularly important in winter and spring when the lower boundary layer of stable air over the Qinghai-Xizang (Tibet) Plateau induces such a 'return flow' in the near-surface circulation. This produces peak wind speeds from October to April with a maximum recorded gust of 22.6 m/s, sufficient to cause failure of the steel cable of the Bailey bridge over the Hunza (30). The lower reaches of a lateral gully on the south side of the 'drift plateau' south of the terminus and opposite the Shimshal River confluence have this same S.E. trend and display some fine examples of aeolian grooving.

The silt coatings of the moraines of the Pasu I (t_7) glacial stage appear to be subject to deflation. Certainly, the grain size composition of surface samples tends to become poorer in fines with age although it is not clear whether this is a result of deflation or illuviation. The distinctive wind-cut grooves seen near the Batura Glacier have not developed on the boulders of the t_7 stage. A period of 200 - 300 years (the estimated age of the t_7 moraines) is evidently not long enough for this degree of etching to develop. As wind etching of till stones is common on moraines of Batura (t_6) age, a period of at least 3000 years may be needed to produce forms on the scale observed. Relict wind-worn micromorphology was also noted on moraines dating from the last Pleistocene glaciation. It appears probable, therefore, that at least two phases of markedly hot and dry

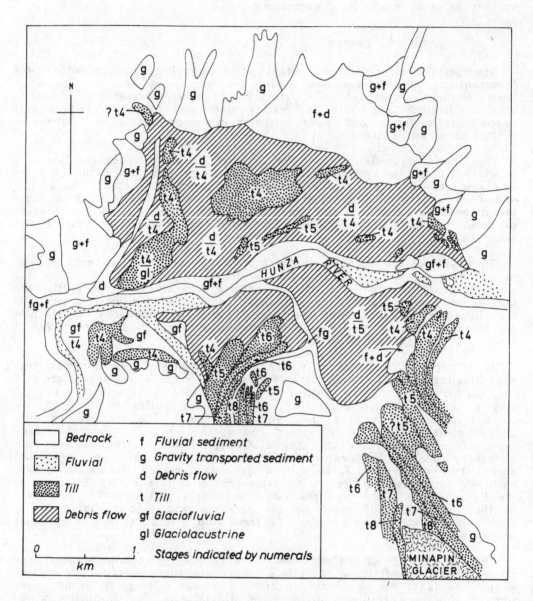

FIG. 12. SURFICIAL SEDIMENTS OF THE HINDI EMBAYMENT.

climate have produced wind-etched boulders in the Hunza valley since the last glaciation. It is suggested that the Hypsithermal and the time since about 3000 yr ago are the most likely periods.

LABORATORY ANALYSIS: SYNTHESIS

The classification of the Hunza valley sediments arrived at in the field was tested in the laboratory with special reference to particle size characteristics.

The tills of the region are relatively coarse-grained, very poorly sorted, somewhat positively skewed and generally mesokurtic. Many show in their particle size characteristics some evidence of depletion of fines. The particle size envelopes of tills, eluviated tills and glacial outwash sediments overlap (Fig. 13). This is best shown by the deposits of the terminal zone of the Pasu Glacier. Samples in group A in Figure 14A were classified as subglacial tills in the field, while those in group B were regarded as of meltout origin. The assemblage of supraglacial debris on the dead ice in the terminus zone of the Pasu Glacier classified in the field map (Fig. 8) as coarse angular talus, meltout till, fine and coarse glaciofluvial gravels and various mixtures of these, displays a broad envelope along the gravel-sand axis of the ternary plot. In only one of these samples (number 34) does silt and clay exceed 20% by weight. This material consists of ripple-bedded sands and fine gravels occurring as a lense in meltout till adjacent to the meltwater stream and in contact with the ice front. It was classified as proximal glaciofluvial but its high fines content (34%), very poor sorting (s.c. 3.82) and platykurtic distribution match the properties of the subglacial till with which it is so intimately associated. The poorly sorted to very poorly sorted sandy silt (sample 53) is the lowermost lacustrine bed making up the 'drift plateáu' near Pasu village. The co-plot of mean grain size and sorting coefficient shows a similar transition from subglacial tills, to eluviated tills and the supraglacial meltwater assemblage (Fig. 15A). The co-plot of sorting coefficient versus skewness shows all the tills to be positively skewed and very to extremely poorly sorted, only the sandur material being negatively skewed (Fig. 15B). For comparison, Figure 15 includes an indication of the approximate zones occupied by tractional debris (near the glacier bed) and high level debris (englacial and supraglacial) as identified on glaciers in Iceland and Spitzbergen by Boulton (24). The Pasu subglacial tills plot well within the former zone and the eluviated tills are distributed close to the transition between the two zones. The supraglacial and glaciofluvial samples all plot outside the 'traction' zone. Their classification appears consistent with the distribution of subglacial till, meltout till and proximal outwash on co-plots of glacial sediments from several Pleistocene and recent glaciers (9).

Similar trends can be seen in the summary statistics for the Ghulkin and Batura deposits. The Batura ternary plot (Fig. 14B) shows four eluviated meltout tills occupying a transitional area between a coarse, distal outwash sample (number 2) and a group consisting of eight subglacial tills and one subglacial ice-contact waterlain deposit (number 27). Also shown are two sandy glaciolacustrine silts with dropstones (80, 103) and four samples of a much finer facies which originated in distal parts of the glaciolacustrine system. The same groupings emerge on the co-plot of sorting against mean size (Fig. 15C). The Ghulkin samples (Fig. 14C) include a cluster of four subglacial tills, one coarse subglacial till (40) and a group of five eluviated meltout tills separating subglacial tills

522

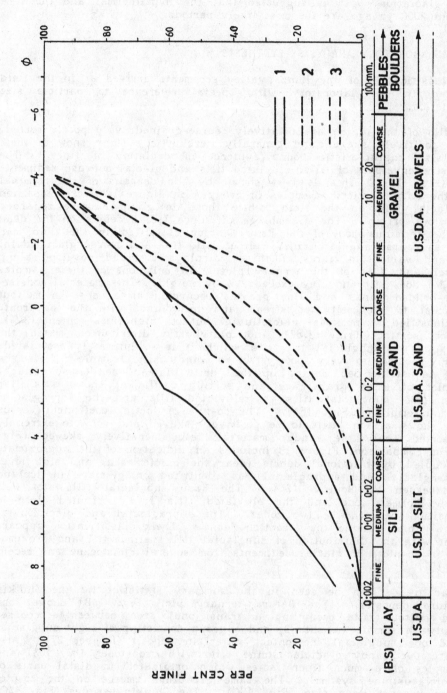

FIG. 13. PARTICLE SIZE ENVELOPES FOR MATERIALS CLASSIFIED IN THE FIELD AS (1) TILLS, (2) ELUVIATED TILLS, AND (3) GLACIAL OUTWASH.

from five glaciofluvial samples. One of the three glaciolacustrine silts (52) underlies compact till of the Yunz (t$_2$) or main trunk valley glaciation and its laminations are severely deformed. Sample 20 is a near-surface sample of weathered and varnished coarse glaciofluvial debris of the Ghulkin II (t$_5$) stage. The mean size-sorting co-plot provides good separation (Fig. 15D) the exceptions being the weathered glaciofluvial sample (20) and a glaciofluvial sample (21) intercalated with a meltout till complex close to the ice front. The distribution is completed by two finer-grained distal glaciolacustrine samples.

The range of variation in mean grain size of the tills (Table 5) is greatest in the Pasu Glacier (s.d. 1.11: coefficient of variation 79.47) followed by the Batura (s.d. 0.83: c.v. 57.93) and the Ghulkin (s.d. 0.43: c.v. 30.14). Despite the presence of a large area of the relatively easily comminuted Pasu slates beneath the lower reaches of the Pasu and Ghulkin glaciers, the mean size of the Batura tills is rather finer and, with a mean sorting coefficient of 4.03, the tills tend to be extremely poorly sorted rather than very poorly sorted as in the case of the Pasu (s.c. 3.73) and Ghulkin population (s.c. 3.78). The larger size and greater throughput of the Batura produces a larger proportion of comminution products in the matrix of the subglacial tills, although the influence of the Pasu slates and limestones and dolomites beneath the lower Batura is also a factor, as indicated by the mineralogy of the clay grades (see above). The lower range of values of positive skewness in the Ghulkin till samples, compared to those of the Pasu and Batura tills, reinforces the interpretation based on field classification and the range of mean sizes and sorting coefficients, namely that deposits around the Ghulkin have suffered more marked depletion of fines than have the tills of any other glacier investigated in the upper Hunza. Better sorting of the sand component results in these tills being rather more leptokurtic than the average for the Upper Hunza as a whole (Table 5).

The tills of the Minapin and Pisan glaciers include a somewhat dispersed group of lodgement origin and a coarser, loose-structured meltout group (Figs. 14D, 15E). It can be seen that five samples of debris flow sediments from the thick deposits mantling these tills in the Hindi embayment plot within the envelope for the Minapin and Pisan subglacial tills. The debris flow samples are relatively finer grained than tills from all glaciers examined, extremely poorly sorted, positively skewed and platykurtic. The mean sorting coefficient (4.57) is higher than in any of the Hunza tills and the envelope very narrow (Fig. 16). The similarity in the petrology and mineralogy of these debris flow deposits and the Minapin and Pisan tills and the presence in them of glacially-crushed sand grains suggests mixing with and redeposition of glacial debris by flow with the slurry matrix supported by positive porewater pressures. This explains the maintenance of the extremely poor sorting and the fine-grained matrix of glacially-comminuted debris, characteristics noted in some so-called flow tills. Size fractionation is clearly discernible in field exposures only at contacts between successive flows.

The tills of the high diffluence cols and bedrock benches up to at least 3 400 m, together with a few tills taken from the Hunza Gorge reach between Sarat and Baltit (Fig. 14E) represent an 'extended valley phase' of much greater ice cover during the Pleistocene (1), with very much larger ice bodies than the present. They appear to have been characterised by higher throughput of ice and sediment and high confining pressures (high effective stresses), with the result that most of these tills have a rather finer mean grain size (2.24∅) than the tills of the tributary

AREA	Sample No.	Mean	Sorting	Skewness	Kurtosis
	30	-0.02	3.44	0.38	0.85
	33	0.01	2.98	0.09	0.75
	36	0.33	3.22	0.13	0.75
	46	1.00	3.90	0.24	0.89
PASU	48	1.97	3.62	0.11	1.04
	60	0.51	3.71	0.31	1.28
	64	2.43	3.93	0.03	1.34
	66	2.36	4.09	-0.05	0.84
	72	3.13	4.27	0.05	1.53
	74	1.15	4.16	0.30	1.13
$\bar{x}$		1.287	3.732	0.159	1.040
s.d.		1.119	0.421	0.141	0.271
	25	2.31	4.97	0.23	1.01
	28	0.05	3.45	0.34	0.86
	54	0.40	2.55	0.09	0.98
	55	0.23	3.10	0.18	0.85
	56	1.15	4.18	0.29	0.79
BATURA	57	2.31	4.13	-0.03	0.83
	58	1.57	4.09	1.19	0.89
	75	1.97	4.55	0.16	0.94
	77	2.22	4.39	0.03	0.96
	78	1.37	4.22	0.42	1.15
	99	1.50	4.67	0.30 ·	0.92
$\bar{x}$		1.371	4.027	0.291	0.925
s.d.		0.833	0.718	0.327	0.100
	19	0.88	3.17	0.14	1.22
	37	1.15	3.50	0.06	1.29
	38	1.05	3.76	0.06	1.07
	40	1.27	5.04	0.46	0.81
	42	2.17	3.61	0.07	1.30
GHULKIN	47	1.45	3.89	0.04	1.06
	49	1.03	3.44	0.02	1.02
	50	1.73	3.74	-0.02	1.08
	51	1.88	4.29	0.14	1.43
	62	1.02	3.44	0.09	1.07
$\bar{x}$		1.363	3.788	0.106	1.135
s.d.		0.431	0.535	0.134	0.177

AREA	Sample No.	Mean	Sorting	Skewness	Kurtosis
MINAPIN AND PISAN	29	-0.02	3.67	0.54	1.08
	81	-0.12	3.35	0.56	1.14
	82	1.05	4.23	0.42	0.67
	83	-0.19	3.22	0.48	0.95
	84	0.45	3.58	0.32	0.77
	85	3.00	4.76	-0.02	0.83
	88	2.58	4.50	0.14	0.99
	95	0.78	4.69	0.49	0.97
	97	0.53	3.99	0.41	1.00
	98	2.33	3.99	0.08	1.08
	104	-0.15	3.23	0.49	1.11
	$\bar{x}$	0.931	3.928	0.355	0.963
	s.d.	1.176	0.566	0.200	0.149
HINDI FLOW DEBRIS	26	1.69	4.32	0.10	1.07
	89	2.51	4.88	0.16	0.81
	91	2.46	4.64	0.19	0.83
	93	2.13	4.35	0.17	0.71
	96	2.33	4.65	0.17	0.65
	$\bar{x}$	2.224	4.568	0.158	0.814
	s.d.	0.333	0.234	0.034	0.161
MAIN VALLEY	31	2.15	3.87	-0.13	1.04
	45	2.07	3.98	0.07	0.81
	63	0.83	4.21	0.39	0.83
	71	0.25	3.79	0.52	0.79
	73	3.62	4.80	-0.34	0.89
	76	1.80	4.69	0.40	0.88
	87	3.68	4.13	0.02	1.59
	92	3.56	4.94	-0.14	0.84
	$\bar{x}$	2.245	4.301	0.0987	0.959
	s.d.	1.304	0.447	0.307	0.266

TABLE 5: PARTICLE STATISTICS FOR TILLS FROM THE UPPER HUNZA VALLEY.

526

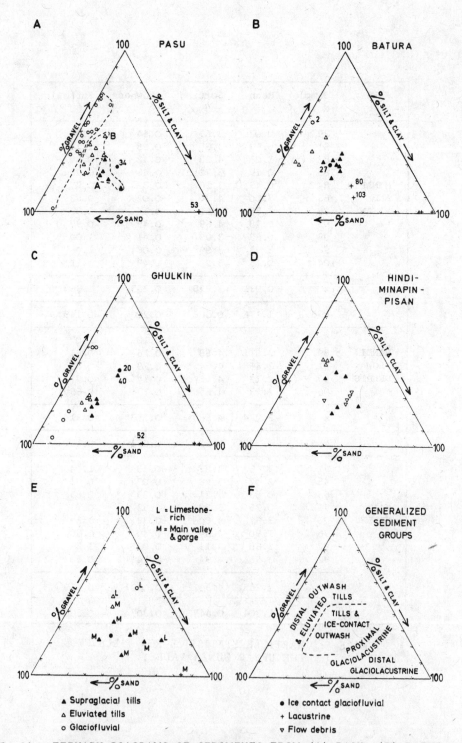

FIG. 14. TERNARY DIAGRAMS OF SEDIMENTS FROM (A) PASU, (B) BATURA, (C) GHULKIN, (D) HINDI-MINAPIN-PISAN, (E) HUNZA GORGE AND COLS, AND (F) GENERALIZED DISTRIBUTION FOR REGION.

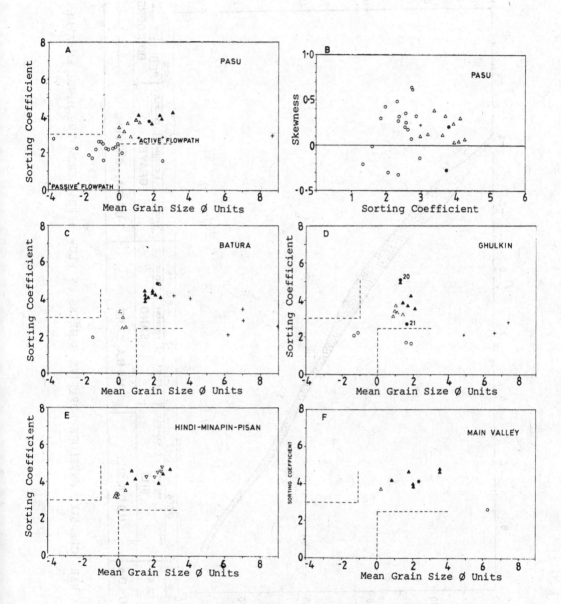

FIG. 15.   CO-PLOTS OF STATISTICAL MEASURES OF PARTICLE SIZE.

528

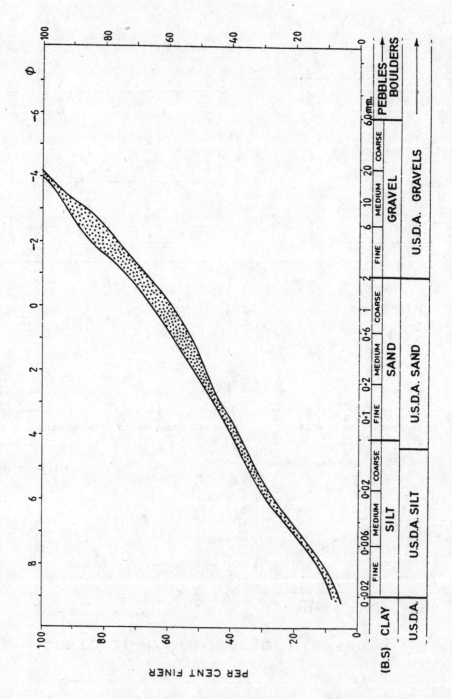

FIG. 16. PARTICLE SIZE ENVELOPE FOR FIVE SAMPLES OF FLOW DEBRIS FROM THE HINDI EMBAYMENT.

valley glacial stages. They also have a mean sorting coefficient higher even than the Batura Glacier tills (4.30 clearly within the extremely poorly sorted range), and all but one plot well within the 'active' flowpath zone confirming the field classification (Fig. 15F). Both limestone-rich lodgement tills and granitic and gneissic tills deposited in the Sarat-Baltit embayment are included. The limestone-rich tills on the valley-side shoulders between the Ghulkin and Batura glaciers are dominated by limestone and dolomite fragments in the sand grade (reaching 30 – 40%) with a substantial component of dark minerals and schist fragments (Table 3). Their clay grades also include calcite as well as the quartz, mica, feldspar and chlorite typical of the Hunza tills.

Tills of lodgement and meltout origin discriminated on the basis of field evaluation and particle size analysis show clear differences in the degree of anisotropy in both clast and matrix populations. Particle orientation and dip are more diffuse in the meltout tills which also have a higher apparent voids ratio in the matrix (Fig. 17A). In contrast, the matrix of many lodgement tills is compact, anisotropic (Fig. 17C) and frequently pierced by sets of microshears. The matrix of much of the main valley till is compact, fine and rich in dolomite and calcite (Fig. 17B). Diamicts produced by weathering and debris flow have a microfabric which is quite different from the tills, consisting of predominantly angular, mechanically weathered (by salt) silts including many flakes of chlorite and mica with fragile, irregular edges and corners (Fig. 17D). Clast anisotropy in some lodgement tills is comparable to values measured in the regelation ice of active glaciers (31). It is important to record, however, that this distinction is clearest in the oldest tills and those dominated by limestone lithology: the distinction becomes much less clear in younger, granodiorite-rich tills.

It has been recognized that sediments produced by glacial comminution and deposited under a confining pressure at the glacier sole plot as two relatively straight lines representing clast and matrix populations, separated by an inflexion. McGown and Derbyshire (32), using data from tills from several parts of the world, have shown that this characteristic inflexion occurs in the sand range in most subglacial tills. In the Hunza subglacial tills, an inflexion occurs in the range $+ 1\emptyset$ to $+ 3.3\emptyset$ for the granitic materials (cf. $- 1\emptyset$ to $+ 3.5\emptyset$ in McGown and Derbyshire), but as fine as $+ 5.3\emptyset$ in the tills dominated by other lithologies.

Where meltwater eluviation has occurred, as in supraglacial and sub-glacial meltout tills deposited in regions with warm summers, the curve undergoes varying degrees of modification and increasingly approximates the form typical of very poorly sorted proximal outwash (Fig. 13). The inflexion is clearly discernible in some tills, notably those from the Batura Glacier and those of the main valley glaciation, but all show some degree of divergence from the form typical of Pleistocene tills from higher latitudes. It is suggested that some meltwater eluviation during deposition has always been a characteristic of the glaciers of this region, including the very much larger ice bodies of the Pleistocene glaciations. To a degree, this view is supported by the roundness values of a 10% sample of the coarse sand from the tills. Although quartz grains with glacially-crushed edges can be found in all samples, the proportion of subangular to rounded grains is usually higher in the tills of meltout type.

Subglacial tills which have not been eluviated and are little altered by weathering tend towards a symmetrical to strongly positive skewness

530

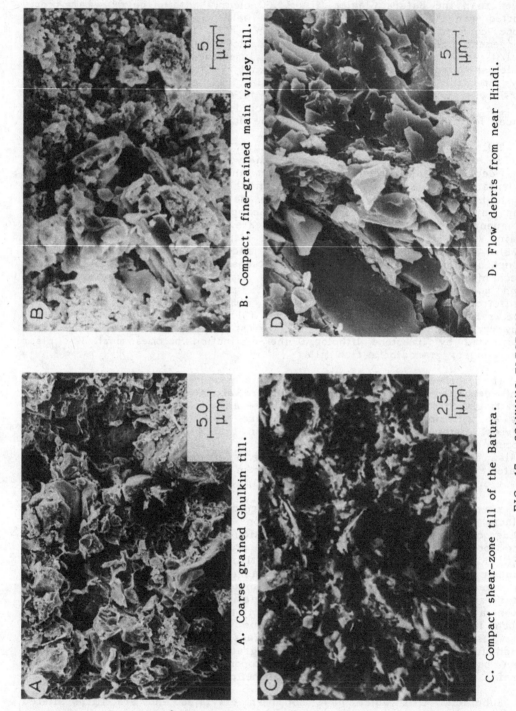

A. Coarse grained Ghulkin till.

B. Compact, fine-grained main valley till.

C. Compact shear-zone till of the Batura.

D. Flow debris from near Hindi.

FIG. 17.   SCANNING ELECTRON MICROGRAPHS.

(31) and to range from mesokurtic to platykurtic. The tills of the upper Hunza (together with the ice-proximal outwash and glaciolacustrine samples noted above) generally show such a distribution. The granodiorite-rich lodgement tills of the Batura Glacier plot within a fairly narrow envelope, similar to that of Pleistocene tills from the North Sea basin (Fig. 18A). All other Hunza tills show varying degrees of divergence from this distribution. The schist- and phyllite-rich tills of the Minapin and Pisan glaciers (Fig. 18B) tend to be mildly to very positively skewed probably reflecting the influence of fine-grained bedrock lithology on the matrix. The tills of the main valley glaciation from the gorge and high valley-side shoulders (Fig. 18C) have a similar range, but the limestone-rich tills show the widest dispersion in skewness. Two end members are shown, one being a compact lodgement till and the other a loose meltout till with granular clasts reminiscent of supraglacial scree. The tills of the Pasu Glacier (Fig. 18D) have the widest kurtosis range with 4 being leptokurtic, while those of the Ghulkin Glacier (Fig. 18E) are distinctly leptokurtic in trend, reflecting a high to very high (up to 60%) proportion of sand in the population of the laboratory-tested samples (+ 9 to - 4∅). Both of these glaciers flow across quite a large area of the 'Pasu slates' which include large outcrops of quartzite and marble. However, most of their courses (like that of the Batura) lie on the granodiorite. The coarser mean grain size of the till matrix of the much smaller Pasu and Ghulkin glaciers (20 and 16 km long respectively, compared to the 58 km length of the Batura) is interpreted as a function of less effective comminution over relatively short transport distances as well as, in many cases, more marked translocation of silt and clay grades by interstitial meltwater during and immediately following deposition.

CONCLUSIONS

The glacial sediments of the Hunza Karakoram make up a gradational series from subglacial tills of lodgement type to subglacial and supraglacial meltout tills, very poorly sorted and rather angular ice-contact glaciofluvial sandy gravels, well-rounded distal glaciofluvials, and proximal and distal glaciolacustrine silts, as shown in their gross particle size characteristics (Fig. 14F: cf. 33, 6). This sequence is also evident in co-plots of the principal grain size statistical parameters, and in the particle size distribution curves. Both of these plotting techniques provide a good first approximation, after due consideration is given to type of bedrock source, of depositional type, which is broadly consistent with observed flowpath distribution of the debris, and the thermal, hydrological, and dynamic character of the glaciers.

Very high summer air temperatures in the lower reaches of the glaciers produce abundant meltwaters which modify most of the glacigenic deposits by slow eluviation of silts and clays in the matrix and fluvial incision and sorting in proglacial fans and plains. The combination of these processes with extensive areas of granodiorite bedrock and an abundance of supraglacially transported debris results in a predominance of relatively coarse-grained tills. Tills currently being deposited around glacier margins are predominantly of meltout type with diffuse fabrics, and rather loose matrixes. Deposition of supraglacial debris by sliding is important but the high infiltration rates and granular nature of much of the debris results in little or no deposition of flow till as commonly understood. Glacial debris remoulded and redeposited as debris flows, on the other hand, is widespread in the sedimentary record. This material does not

532

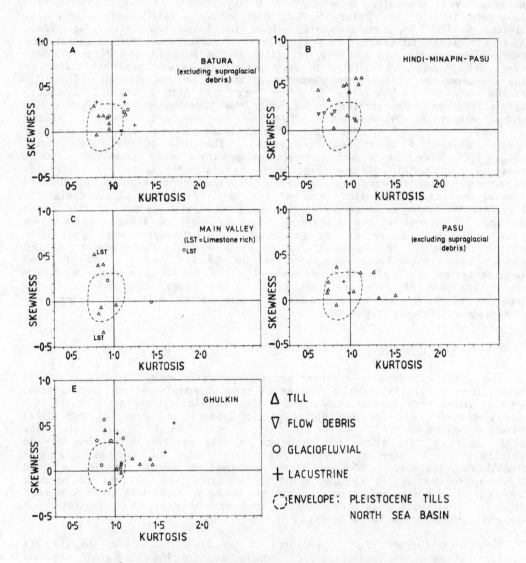

FIG. 18.   SKEWNESS V. KURTOSIS CO-PLOTS FOR (A) BATURA,
(B) HINDI-MINAPIN-PISAN, (C) HUNZA GORGE AND COLS, (D)
PASU, AND (E) GHULKIN AREAS.   SUPRAGLACIAL DEBRIS OF
PRESENT ICE-CORED STAGE EXCLUDED.

exhibit depletion of the fine fraction because the matrix was supported during emplacement by positive porewater pressures, only minor relocation of fines occurring towards distal margins during the dewatering phase.

Complex interbedding and facies variation over short distances is characteristic, and serves to obscure distinctions between conventionally recognized sediment types. This complexity is characteristic of the depositional record of the last eight glacial phases of the Quaternary extending back to at least 130,000 yr B.P. The current conjunction of semi-arid valley floors and high activity glaciers of compound thermal type thus appears to have prevailed since the beginning of the Upper Pleistocene. However, there is some evidence that the older tills, and those derived from finer-grained bedrock such as limestone and schist, have higher bulk densities, finer mean grain sizes and more markedly anisotropic fabrics. These characteristics are also a reflection of the fact that the glaciers of the main valley and diffluence col phases were larger than their modern equivalents.

Modification of the particle size distribution of the surfaces of Quaternary sediments by deflation is widespread but rarely severe. With the exception of the oldest glacial remnants in the form of isolated erratics at high altitudes, change resulting from weathering, notably in the form of surficial oxidation, desert varnish coatings and grussification, has had a significant effect on grain size and shape only in the upper half metre or so. Calcrete development, however, may extend to depths of 2 m or more in some of the older tills and outwash gravels and to c. 40 - 50 cm in debris fan sediments of mid-Holocene and younger age.

The genetic classification of glacial and paraglacial sediments on the basis of field data is generally confirmed by analysis of the laboratory-derived data on particle size, particle shape, lithology, mineralogy, stone orientation, and fabric.

ACKNOWLEDGEMENTS

This work was carried out as part of the Geomorphology Programme of the International Karakoram Project organized by the Royal Geographical Society in 1980 in collaboration with Academia Sinica and the Pakistan Ministry of Science and Technology. We are greatly indebted to the National Geographic Society, to the Royal Society and to the Royal Geographical Society for providing funding. Travel costs for the authors were borne by the Government of the People's Republic of China. The Pakistan Army is also thanked for invaluable helicopter photo-reconnaissance transport.

REFERENCES

1) DERBYSHIRE, E., LI JIJUN, STREET-PERROT, F. A., XU SHUYING and WATERS, R. S. 'Quaternary glacial history of the Hunza valley, Karakoram Mountains, Pakistan'. This volume.

2) READING, H. G., (ed.) (1978). Sedimentary environment and facies. Blackwell, Oxford pp 557.

3) DREIMANIS, A., (1976). 'Tills: their origin and properties'. In Legget, R. F. (ed.). Glacial Till Roy. Soc. Canada Spec. Publ., No. 12, pp 11 - 49.

534

4) BOULTON, G. S., (1968). 'Flow tills and related deposits on some Vestspitsbergen glaciers', J. Glaciol., 7, pp 391 - 412.

5) BOULTON, G. S., (1970). 'On the deposition of subglacial and meltout tills on the margin of certain Svalbard glaciers', J. Glaciology, 9, pp 231 - 245.

6) MILLS, H. H., (1977). 'Differentiation of glacier environments by sediment characteristics: Athabasca Glacier, Alberta, Canada', J. Sed. Petrology, 47, pp 728 - 737.

7) BOULTON, G. S., (1972). 'Modern Arctic glaciers as depositional models for former ice sheets', J. Geol. Soc. London, 128, pp 361 - 393.

8) BOULTON, G. S. and EYLES, N., (1979). 'Sedimentation by valley glaciers; a model and genetic classification'. in Schluchter, Ch. (ed.) Moraines and Varves, Balkema (Rotterdam), pp 11 - 23.

9) DERBYSHIRE, E., (1982). 'Sedimentological analysis of glacial and proglacial debris: a framework for the study of Karakoram glaciers', Volume 1, these proceedings. Int. Karakoram Project,

10) DESIO, A., (1974). 'Karakoram Mountains'. In Spencer, A. M., (ed.) Mesozoic-Cenozoic Orogenic Belts: data for orogenic studies. Geological Society of London, Spec. Pub. No. 4, pp 255 - 256.

11) BRITISH STANDARDS INSTITUTION, (1967). Methods of testing soils for civil engineering purposes. British Standards Institution (London).

12) RITTENHOUSE, E. G., (1943). 'A visual method of establishing two-dimensional sphericity', J. Sed. Petrology 13, pp 79 - 81.

13) SHEPARD, F. P. and YOUNG, R., (1961). 'Distinguishing between beach and dune sands', J. Sed. Petrology, 31, pp 196 - 214.

14) McGOWN, A. and DERBYSHIRE, E., (1974). 'Technical developments in the study of particulate matter in glacial tills', J. Geology, 82, pp 225 - 235.

15) DERBYSHIRE, E., (1978). 'A pilot study of till microfabrics using the scanning electron microscope'. In Whalley, W. B., (ed.) Scanning Electron Microscopy in the Study of Sediments, Geo. Books (Norwich), pp 41 - 59.

16) BATURA GLACIER INVESTIGATION GROUP, (1979). 'The Batura Glacier in the Karakoram Mountains', Scientia Sinica, 22, pp 958 - 974.

17) ZHANG XIANGSONG, CHEN JIANMING, XIE ZICHU and ZHANG JINHUA, (1980). 'General features of the Batura Glacier'. Professional Papers on the Batura Glacier, Karakoram Mountains, Science Press (Beijing) pp 8 - 27 (In Chinese).

18) ZHANG JINHUA and BAI ZHONGYUAN, (1980). 'The surface ablation and its variation of the Batura Glacier'. Professional Papers on the Batura Glacier, Karakoram Mountains, Science Press (Beijing), pp 83 - 97. (In Chinese).

19) SHI YAFENG and WANG JINGTAI, (1980). 'Research on snow cores in China and the avalanche phenomena of Batura Glacier in Pakistan', J. Glaciology, 26, pp 25 - 30.

20) ZHANG XIANGSONG and SHI YAFENG, (1980). 'Changes of the Batura Glacier in the Quaternary and recent times'. Professional Papers on the Batura Glacier, Karakoram Mountains, Science Press (Beijing), pp 173 - 190. (In Chinese).

21) GOUDIE, A. S., JONES, D. K. C. and BRUNSDEN, D. 'Recent fluctuations in some glaciers of the western Karakoram Mountains, Hunza, Pakistan. This volume, pp 411 - 455.

22) GOUDIE, A. S. et al. 'The geomorphology of the Hunza valley, Karakoram Mountains, Pakistan'. Vol. 2, these proceedings, pp 359 - 410.

23) COLLINS, D. N. 'Hydrology and hydrochemistry of glacial meltwaters'. To be published.

24) BOULTON, G. S., (1978). 'Boulder shapes and grain-size distribution of debris as indicators of transport paths through a glacier and till genesis', Sedimentology, 25, pp 773 - 799.

25) FRANCIS, M., MILLER, K. J. and DONG ZHIBIN. 'Impulse radar, ice-depth sounding of the Ghulkin Glacier'. This volume, pp 111 - 123.

26) MASON, K., (1930). 'The glaciers of the Karakoram and neighbourhood', Records of the Geological Survey of India, LXIII.

27) HASTENRATH, S. and PATNAIK, J. K., (1980). 'Radiation measurements at Lewis Glacier, Mount Kenya, Kenya', J. Glaciology, 25, pp 439 - 444.

28) BATURA GLACIER INVESTIGATION GROUP, (1976). Investigation report on the Batura Glacier in the Karakoram Mountains, the Islamic Republic of Pakistan (1974 - 1975). Batura Investigation Group, Engineering Headquarters, Peking, 123 pp.

29) CAI XIANGXING, LI NIENJIE and LI JIAN, (1980). 'The mud-rock flows in the vicinity of the Batura Glacier'. Professional Papers on the Batura Glacier, Karakoram Mountains, Science Press (Beijing), pp 146 - 152. (In Chinese).

30) LIU GUANGYUAN, (1980). 'The climate of the Batura Glacier and its adjacent areas'. Professional Papers on the Batura Glacier, Karakoram Mountains, Science Press (Beijing), pp 99 - 110. (In Chinese).

31) DERBYSHIRE, E., (1980). 'The relationship between depositional mode and fabric strength in tills: schema and test from two temperate glaciers'. In Stankowski, W., (ed.) Tills and Glacigene Deposits, Universytet im. Adama Mickiewicza W Poznaniu, Seria Geographia, Nr. 20, pp 41 - 48.

32) McGOWN, A. and DERBYSHIRE, E., (1977). 'Genetic influences on the properties of tills', Q. J. Engng. Geol. 10, pp 389 - 410.

33) LAWSON, E. E., (1979). 'Sedimentological analysis of the western terminus region of the Matanuska Glacier, Alaska', U. S. Corps of Engineers, CRREL Report 79 - 9, 122 pp.

Particle size distribution on the debris slopes of the Hunza Valley

D. Brunsden – King's College, University of London; **D. K. C. Jones** – London School of Economics; **A. S. Goudie** – University of Oxford

ABSTRACT

The rock fall scree slope model in which there is a size difference down-slope on screes formed by rock fall processes was shown to apply to Karakoram screes. These simple relationships were shown to be disturbed by subsequent mass movement processes to produce distinctive patterns of stripes, festoons and scallops. The patterns are described for large and small screes and various processes. Considerable differences in grain size are caused by dry grain flow, shallow sliding and mudflow or debris avalanche.

INTRODUCTION

During the course of the geomorphological programme (1) undertaken by The Royal Geographical Society, International Karakoram Project (IKP), 1980, it was observed that the numerous, extensive and often spectacularly large debris slopes of the Hunza valley consisted of a continuum of features that could be subdivided morphogenetically into the following groups:-

(1) Rockfall screes

This term is here restricted to simple and usually relatively small accumulations of rock debris formed at the base of unstable rock faces by the dominant process of discrete rock fall. They occur as simple cones, Figure 1, coalesced cones and rectilinear talus slopes, with supply areas that are normally broken rock faces without any substantial drainage inputs.

(2) Patterned screes

The most numerous debris slopes in the valley occur below rock faces which are incised by steep shallow ravines or chutes (Figure 2) that act as small catchments for water and debris as well as providing pathways for avalanches. The resultant

FIG. 1. ROCKFALL SCREE CONE NEAR PASU WITH MOD-IFICATIONS DUE TO SLIDING AND MUDFLOW ACTIVITY.

FIG. 2. PATTERNED SCREE WITH INCISED CATCHMENT AREAS, AVALANCHE OR DEBRIS CHUTES.

talus accumulations are generally much larger than those produced by simple rockfall processes alone, ranging up to base widths of 1,000 m and heights of 500 - 600 m. Form varies from cone, through compound cone, to irregular sheets, and although substantial portions are rectilinear in form, they generally tend to develop a concave run-out at the base. The formative processes involved are dominated by rock fall from gully sides, pinnnacles and steep headwalls above, but there is also substantial redistribution of debris by subsequent and contemporaneous mass movement processes, acting on the debris accumulations, most particularly dry grain flow, small mudflows generated on the upper parts of the scree, shallow sliding and, very occasionally, gully incision where stability has been increased by weak cementation of finer-grained matrix-supported layers in well stratified deposits. The result is a variety of surface patterns including: (a) stripes of alternating coarse and fine material (Figure 3); (b) shallow scallops, festoons and lobes of coarse debris separated by arcuate and diamond-shaped patches of fine material (Figure 4); and (c) a complex of lobes (Figure 5) and shallow levée-flanked channels (Figure 6).

(3)   Scree-fans

Where catchment size in the valley-side rockslopes above is of sufficient size to generate true runoff, then rock fall debris, valley confined mudflows and overland flow, all combine 'contemporaneously' to produce complex debris slopes. Such slopes (Figure 7) may be of a form and steepness (15° - 30°) that is intermediate between true screes (32° - 37°) (Types 1 and 2) and fans (Type 4) produced by mudflow activity or running water (5° - 10°).

(4)   Fans

This category contains a wide range of features which can be broadly sub-divided into mudflow fans, catastrophic mudflow fans and alluvial fans, according to the dominance of the formative processes. Alluvial fans are found in valley bottom locations at the mouths of meltwater streams. They have generally smooth surfaces, slightly concave downslope, sloping at less than 7°, into which the drainage line is usually incised in a clearly defined channel. Mudflow fans normally have no clear channel produced by running water, although they may be gullied around their distal margins. The gently concave surfaces stand at 7° - 20° and have an irregular micro-topography due to a complex veining of shallow, levée-flanked surface channels and the presence of lobes of coarser debris (Figure 8). Local relative relief may be up to 5 m and many of the narrower channels (2 - 5 m wide) have vertical sides up to 3 m high. They have a characteristic mottled appearance due to the presence of lobes and tracks of variable age and, therefore, variable degree of desert varnish development. Although the upper parts of the fans may be fed by rockfall, the slopes are generally moulded by repeated mudflows which emanate from gullies and small valleys (Figure 9). Catastrophic mudflow fans lie intermediate between alluvial and mudflow fans. Produced by occasional enormous discharges of sediment down pre-existing gullies and valleys, due to major storms or meltwater surges, the resultant fans tend to be similar to alluvial fans in that their surfaces appear uniform and contain single

FIG. 3. STRIPED SCREE OF ALTERNATING FINE AND COARSE MATERIAL.

FIG. 4. SCREE AT ALIABAD WITH SHALLOW SCALLOPS, FESTOONS
AND LOBES OF COARSE DEBRIS.

FIG. 5. DOWNSLOPE VIEW OF SCALLOPING DUE TO DRY PLANAR SLIDING.
The photographer is standing on one slide surface immediately in front
of a coarse lobe. Downslope is the next slide scar and lobe.

FIG. 6. LEVEED MUDFLOW CHANNELS ON LOWER DEBRIS SLOPE AT HINDI.

FIG. 7. SCREE-FAN INTERMEDIATE BETWEEN A TRUE ROCKFALL SCREE AND A MUDFLOW-RUNOFF FAN. PASU, HUNZA.

FIG. 8. MUDFLOW FAN NEAR CHALT, HUNZA.

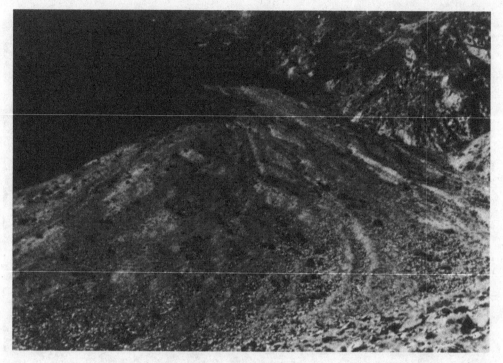

FIG. 9.   LOBATE SCREE SURFACES AT HINDI, HUNZA.

incised   channels   where   the   drainage-lines   have   reestablished
themselves.   The   main   differences   are   (1)   a   generally   lobate
form,   (2)   truncated   distal   margin   due   to   erosion   by   main   valley
drainage,   (3)   the   presence   of   thick   sequences   of   matrix-supported
sediments   rather   than   stratified   deposits,   (4)   an   inverse   grading
of   clasts   and   huge   boulders   distributed   on   the   surface   margins,
(5)   steeper   surface   slopes,   (6) super-elevated   or   depressed   margins
as   the   mud   'rode'   up   and   down   the   confining   slopes.   A   wide
range   of   intermediate   forms   exists   whose   surface   roughness   varies
according   to   which   process   was   last   dominant,   and   in   certain
instances   all   three   forms   are   to   be   found   intimately   associated
in   the   same   major   sedimentary   accumulation.

Some   of   the   characteristic   features   of   these   debris   accumulation   types
are   shown   diagrammatically   in   Figure   10.   Their   volume,   variety   and
conspicuousness   merited   some   investigation   by   members   of   IKP   1980,   although
time   constraints   limited   the   size   and   scope   of   the   studies   that   could   be
undertaken.   This   paper   presents   the   results   of   the   studies   into   rockfall
screes   and   patterned   screes.   Those   relating   to   the   examination   of   mudflows
will   be   reported   elsewhere   (2).   Before   discussing   the   nature   of   the   screes
in   the   Hunza   Valley,   it   is   first   necessary   to   examine   briefly   the   relevant
background   literature   concerning   scree   development.

PREVIOUS STUDIES

Many   authors   (3)   -   (11)   have   shown   that   there   is   often   a   well   developed
sorting   of   particles   from   small   to   large   in   a   downslope   direction   on   screes.
This   distribution   is   attributed   by   most   writers   to   the   rockfall   process.

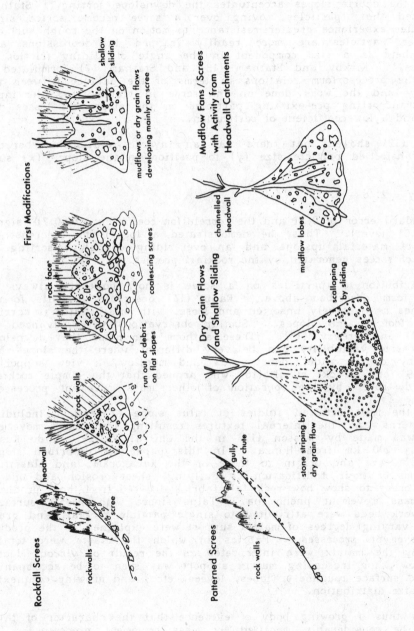

FIG. 10. DIAGRAM OF SCREE TYPES.

Gardner (12) states that 'Assuming an equal velocity, a large rock will fall further than a smaller rock.' and 'increase in roughness, further accentuates the downslope sorting in that it acts as a "friction" element in the rockfall process. In the same manner decreasing slope angle toward the foot of the debris slopes accentuates the downslope sorting.' Statham (7) explained that 'particles moving over a scree become sorted since small particles experience greater resistance to motion on the rough surface. Thus smaller particles are more readily trapped in depressions and effectively there is a size component in the angle of sliding friction of the particles.' Kirkby and Statham (13) and Statham (7) formulated a model for the process/form relations in terms of the balance between the input energy and the work done on the scree in transporting the input particles, transporting pre-existing particles on the slope and losses due to friction and a low coefficient of restitution.

Gardner (12) showed that there was an inverse relationship between the $\log_{10}$ transformed particle size (y) to position on the slopes (x) .such that:-

$$y = 2.46 - 0.61x$$

with a standard error of 0.21 and the correlation coefficient of -0.71 (significant at .01 level). Thus, he demonstrated a logarithmic decrease in mean size of material upslope and an over-riding downslope sorting on the surface of screes generated by the rockfall process.

The distribution of particles on a scree is not, however, always in the simple form described above. Rapp (14) described deposits formed by avalanches as extremely unsorted and loose, with small material carried toward the foot of the slopes. Similar observations were advanced by Gardner (12) and Drewry (15). These authors also noted that downslope sorting and size variability can be very different where the slopes are influenced by debris slide, debris flow and mudflow, a view supported by Bones (9) and Bruckl et al (16) who argued that the simple rockfall pattern is destroyed by the operation of other mass movement processes.

One of the most detailed studies of talus slope properties, including surface patterns and the internal textures resulting from mass movement processes, was made by Wasson (17) in the Chitral Valley, Hindu Kush, approximately 200 km from Hunza. In this paper he described slopes of comparable size and origin to those of the Karakoram, and classified them in terms of their modification by gullying, sheet erosion and undercutting, in order to show how rock fall fabrics differed from those produced by other mass movement phenomena. Talus slopes with smooth surfaces and open work beds were attributed to single particle rockfall and grain flow, while varying degrees of matrix support were explained as the product of mass movement processes. Fabrics in which the clasts were totally supported by the matrix were interpreted as the result of viscous debris flow (mudflow), for increasing matrix support was seen to be accompanied by increased surface roughness (lobes, levées, etc.) and growing complexity of particle size distribution.

There is thus a growing body of evidence that the character of talus slopes can be considerably modified by mass movement processes. The main deficiency at the present time, is the lack of results from detailed field investigations undertaken in areas where a wide variety of processes operate together. Arid high mountains, such as the Hindu Kush and the Karakoram, provide ideal settings for such studies because of high debris production rates, due to salt weathering, freeze-thaw and tectonic instability, and the presence of a variety of sliding and flowing mechanisms, including those associated with snow cover. The study by Wasson (17) concentrated on the differing internal structures of the resultant deposits. The Karakoram investigation described here, was designed to be complementary by focussing attention on the surface sorting characteristics.

THE HUNZA VALLEY STUDY

Introduction

The widespread occurrence and variety of patterned screes in the Hunza Valley, suggested that talus accumulations are subject to a range of sorting mechanisms. Therefore, the objectives of the study were to (i) show how particle size distributions varied between 'representative' types of scree (simple rockfall, striped, patterned) and (ii) identify the mechanisms responsible for these variations in surface characteristics.

Screes are distributed throughout the length of the Hunza Valley, the largest examples being located near Pasu (in the north), Aliabad, Hindi, Gulmet and Khanabad (in the Middle Reach), and along the western side of the Lower Reach, see Figure 11. As no maps were available and time

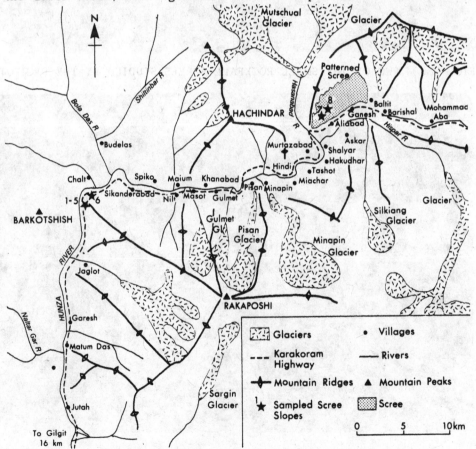

FIG. 11. LOCATION MAP OF STUDY AREA.

FIG. 12a. TWO OF THE SMALL ROCKFALL SCREES STUDIED BY THE PROJECT.

FIG. 12b. SIMPLE ROCKFALL SCREE NEAR CHALT, HUNZA.

constraints precluded the preparation of maps from air-photographs, no attempt was made to sample scree slopes on a statistically representative basis.   Instead, eight slopes were purposefully chosen so as to illustrate the variable nature of debris slopes in the Hunza Valley.   The choice was based on visual inspection of screes over the entire 129 km from Gilgit to Gulkin, with a view to examining "representative" screes from a range of categories.   Selection was largely influenced by problems of access and the enormous physical dimensions of many of the slopes, further, competing demands for man-power by other project tasks imposed severe restrictions.   Despite these limitations, the results are considered of both interest and value.   They should be regarded as preliminary, descriptive, and hypothesis-forming and it is hoped that they will stimulate further work on the area as well as make some inferences of relevance to the general debate on the origin of variable particle size distribution on screes.

The eight slopes actually studied were:-

(a)   five small rockfall generated scree cones (Type 1) (Figure 12) located close to the Karakoram Highway (Figure 11; 1 - 5), which were studied so as to determine whether the rockfall sorting model was applicable to the study region.

(b)   one striped rectilinear scree (Type 2a) (Figure 11, No. 6), 125 m long (Figure 13), to show the early stages of rockfall scree modification by the generation of small mudflows.

(c)   two very large patterned screes (Type 2b) (Figure 4) at Aliabad (Figure 11; 7 - 8) to show the nature and origin of mature patterns.

FIG. 13.   STRIPED RECTILINEAR SCREE STUDIED BY THE PROJECT.

The following discussion has three main components. First, an account is given of the field observations undertaken. These consisted of (a) slope form measurements using Suunto inclinometer and 30 m tape and, (b) particle size measurements of either $\underline{a}$, $\underline{b}$ and $\underline{c}$ axes or $\underline{b}$ axis alone. Second, the particle size distributions on the simple rockfall screes are described and compared with previous work and third, an attempt is made to illustrate the nature of some of the distributions which arise from the operation of small mudflows and from dry sliding processes operating on the scree surface. It was not possible to measure the effect of talus creep due to the short time period available for field work. The characteristics of scree-fans and mudflow fans are also omitted since they form a group of landforms which, although related, possess distinctive morphologies, surface roughnesses and matrices deserving of separate treatment.

Rockfall screes

The fine screes which showed visual evidence of rockfall activity only were small (28, 30, 42, 74 and 100 m slope length) debris accumulations beneath vertical rock chutes and rock faces 50 - 250 m high at 50 - 90°. The rockfalls were partly generated by construction activity on a jeep track higher up the face. Scree composition essentially consisted of blocky fragments of chlorite schist with some quartzite, and mean Zingg shape categories for a total of 560 clasts measured along the centre lines (see below) revealed 43.9% blades, 32.1% discs, 17.5% rods and 6.4% spheres.

The measurement programme was carried out as follows. A centre line was measured and subdivided for sampling purposes into five equally spaced points - together with an additional point at the apex. At each point the $\underline{a}$, $\underline{b}$ and $\underline{c}$ axes of twenty clasts were recorded using a 1 m steel tape. Two lines, radial from the apex, and equally spaced between the centre line and the side margins of the scree, were also measured, subdivided, and sampled at points equivalent to those on the centre line, to check for any variation in size across the scree, and 20 $\underline{b}$ axes were measured at each point. In some cases less than five equally spaced points were used due to the asymmetrical shape of the scree. The slope profiles were measured between the points on all three lines (see Figure 14).

The following computations were undertaken:-

(1)   Particle shape, using the Zingg classification (18), for samples along the centre lines.

(2)   The $\underline{b}$ axis was used as a measure of particle size and means, standard deviations and coefficients of variation calculated. The coefficient of variation was used as a measure of dispersion because of the wide range of mean particle size.

(3)   Since particle sizes vary considerably, and the distributions are of lognormal form skewed, the data were transformed using a $\log_{10}$ transformation.

(4)   Analysis of variance was used to isolate variation between samples and within samples both down and across the slopes.

(5)   Mean $\underline{b}$ axis size of the central line samples were plotted against mean $\underline{b}$ axis sizes for left-hand and right-hand samples so as to identify any variation across the scree.

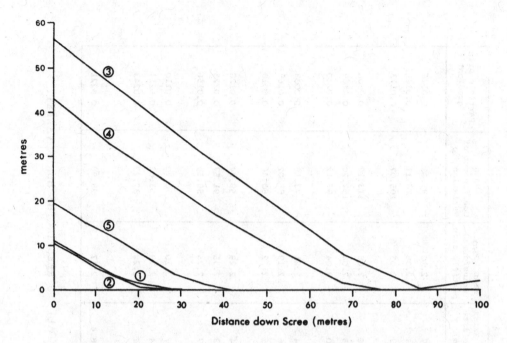

FIG. 14.   CENTRE LINE PROFILES FOR ROCKFALL SCREES.

(6)   The systematic variation in a downslope direction was examined using a regression of transformed mean particle size (y) at each of the points to position on the slope (x), position on the slope being defined by Gardner (12) as the ratio of the distance of the sample point from the foot of the slope to the length of the slope.

(7)   Particle size gradients (12), ($\log_{10}$ mean particle size at lowest sample point $-$ $\log_{10}$ mean particle size at highest sample point/ slope length) was calculated so as to examine the rate of change of size down the screes.

(8)   Particle size $-$ slope angle relations were examined using a regression of mean particle size transformed by $\log_{10}$ (y) and mean slope angle for the unit (x), measured downslope between sample points.

Particle size distribution

The b axis of pebbles was used as a measure of particle size, and coefficients of variation, standard deviation, and mean particle size were calculated for each sample point on the five slopes.   These data are summarised in Table 1.   The material varies greatly in size, with b axes ranging overall from less than 1 cm to 432 cm (Table 2), and particle size distribution at the sample points are of log normal form.

The material tends to be unsorted at any sample point, as demonstrated by the large coefficients of variation, and extreme variations occur because of the presence of very large blocks (Table 2) which continue to influence size distributions until they become totally buried.   Although these boulders may rest at any point on the slope, they are most common in the middle

	Length (m)	Mean slope (°)	Maximum slope (°)	Minimum slope (°)	Mean b axis particle size (mm)	Mean Log_{10} transformed particle size	Mean Coefficient of variation (%)	Particle size gradient
Scree 1 Left	25.2	23.6	33.0	15.5	127	2.10	91.01	0.0526
Centre	28.0	20.9	35.5	8.0	114	2.06	78.17	0.0341
Right	28.0	20.9	35.5	1.0	113	2.05	105.07	0.0173
Scree 2 Left	25.0	15.2	31.0	1.0	120	2.08	113.12	0.0306
Centre	25.0	18.2	29.5	4.0	115	2.06	159.20	0.0469
Right	25.0	17.5	30.0	4.5	110	2.04	97.12	0.0240
Scree 3 Left	85.0	25.7	37.0	3.0	164	2.21	78.18	0.0008
Centre	85.0	28.2	36.5	5.0	156	2.19	82.02	0.0168
Right	85.0	27.5	36.5	8.0	165	2.22	60.46	0.0010
Scree 4 Left	62.5	25.5	37.0	7.5	194	2.29	59.24	0.0128
Centre	62.5	27.5	36.5	9.5	223	2.35	91.57	0.0225
Right	62.5	27.4	35.5	7.0	177	2.25	58.54	0.0156
Scree 5 Left	35.0	24.9	33.0	6.5	127	2.10	70.70	0.0110
Centre	35.0	23.1	31.0	12.0	154	2.19	103.02	0.0214
Right	35.0	24.8	31.0	14.0	167	2.22	87.22	0.0243
Mean	-	23.39	33.9	7.1	148.4	2.17	88.98	0.0221

TABLE 1: SOME DESCRIPTIVE STATISTICS FOR ROCKFALL SCREES IN THE HUNZA VALLEY.

SCREE	1 28 m			2 30 m			3 100 m			4 74 m			5 42 m		
LENGTH LINE	L	C	R	L	C	R	L	C	R	L	C	R	L	C	R
TOP	-	1	-	-	1.8	-	-	1.6	-	-	2.7	-	-	1	-
1	6.2	8.9	46.5	12.8	22.5	18.6	28.0	36.3	15.0	15.5	23.2	12.0	10.5	15.5	43.5
2	31.5	17.0	26.6	10.1	25.5	27.5	41.0	42.2	20.0	10.3	32.9	6.0	31.1	24.0	21.0
3	88.0	31.4	58.0	30.0	432.0	34.0	35.0	78.8	16.5	62.6	182.0	44.7	37.3	81.0	26.5
4	83.4	54.5	38.3	239.0	39.5	29.5	63.0	38.3	53.0	93.0	123.0	99.1	49.6	80.0	32.5
5	-	86.0	61.5	82.5	34.5	100.1	95.0	96.0	81.5	51.0	133.0	88.0	80.0	209.0	180.0

— = Largest clast on profile

═ = Largest clast on scree

L, C and R = Left, centre and right hand lines as viewed facing the scree.

TABLE 2: SIZE OF LARGEST CLAST MEASURED AT SAMPLING POINTS ON THE FIVE ROCKFALL SCREES.

552

and lower portions (Table 2). However, in three out of the five screes the largest measurements were recorded in a mid-slope position. This is considered an accidental over-emphasis resulting from the sampling scheme adopted, for visual inspection (Figure 12) indicates that the very largest boulders actually occur in the lowest one-third of the slope. The distribution of large boulders is wholly consistent with the view (12) that larger blocks travel further, although this investigation does suggest that two additional points may be of importance. First, that in rockfalls consisting of clasts of very different sizes, the largest boulders actually reach the debris cone first and before a rain of increasingly fine debris. Therefore, those blocks coming to rest on the higher parts of the slope will become buried by fine debris, thereby contributing to the overall downslope increase in particle size. Second, large blocks falling on to steeply-sloping and unstable scree surfaces, will slide rather than 'bounce', and be brought to rest in a mid-slope position by friction. The latter is considered especially true of very large and long scree slopes.

Particle size across and down the slope

The particle size distribution across the slope at any elevation was investigated by using analysis of variance and is also shown graphically in Figure 15. The variance within samples is generally greater than

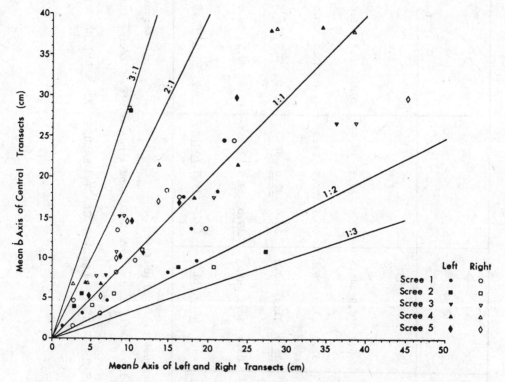

FIG. 15.   GRAPHICAL PLOT OF VARIATION OF PARTICLE SIZE ACROSS
THE SCREE SLOPES 1 - 5.

variance between samples and in only six out of twenty-nine cases (Table 3) does there appear to be a significant difference in size across the slope. This is emphasised in Figure 16 which shows no consistent

Scree sample	Sample points downslope	Probability of Null hypothesis being correct	
Scree 1	1	34.5615	
	(2)	(27.3436)	
	3	3.8142	(Significant)
	(4)	(4.1586)	(Significant)
	5	41.4569	
	(6)	(29.7400)	
	7	3.7045	(Significant)
	(8)	(23.6748)	
	9	10.1413	
Scree 2	1	18.7421	
	2	3.8884	(Significant)
	3	48.1837	
	4	22.8648	
	5	21.1616	
Scree 3	1	47.9198	
	2	44.1559	
	3	10.0493	
	4	23.6027	
	5	19.7572	
Scree 4	1	4.9268	(Significant)
	2	0.6332	(Significant)
	3	40.8192	
	4	49.7923	
	5	31.6347	
Scree 5	1	33.6022	
	2	44.6848	
	3	36.2076	
	4	29.5211	
	5	20.1514	

NB The figures in brackets represent additional data that were collected
for Scree 1.

TABLE 3:   RESULTS OF ONE-WAY ANALYSIS OF VARIANCE TEST TO ESTABLISH
WHETHER OR NOT DIFFERENCES IN PARTICLE SIZE EXIST BETWEEN
THE CENTRE AND THE TWO FLANKS OF THE DEBRIS CONES PROD-
UCED BY ROCKFALL PROCESSES.

554

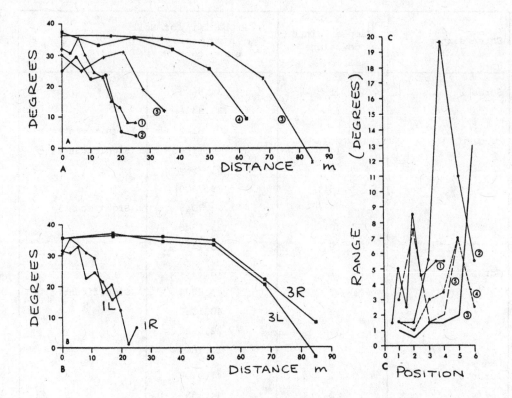

FIG. 16.   (A) SLOPE ANGLE PLOTTED AGAINST DISTANCE FOR CENTRE LINES.
           (B) SLOPE ANGLE PLOTTED AGAINST DISTANCE FOR FLANK LINES.
           (C) RANGE OF SLOPE ANGLE PLOTTED AGAINST SAMPLE   POSITION
               FOR ALL FIVE SCREES.

differences in particle size between the centre and flank sampling points.
It can be assumed, therefore, that central line data are representative
of the scree as a whole and that lateral sorting is insignificant.

There is, however, a marked change in size downslope, due to the
rock fall process supplying debris of variable size.   This characteristic
is often visually apparent, and analysis of variance showed a statistically
highly significant difference between samples in a downslope direction
in all but one case (Table 4), with a very low probability that the null
hypothesis should be accepted.   The one aberrant case is caused by the
presence of a particularly large boulder (b axis of 432 cm) in a midslope
position on Scree 2 (Table 2).

The systematic variation in a downslope direction was examined using
a regression of $\log_{10}$ transformed mean particle size (y) at each of the
sample points, (a, b and c axes on the centre lines and b axis only on
the flank lines) to position on the slope (x) as defined by Gardner (12)
(ie the ratio of the distance of the sample point from the foot of the slope
to the length of the slope).   In all cases an inverse relationship is
described (Figure 17).   The best fit cases are frequently linear (16 out
of 25 transects), although in seven cases a hyperbolic curve gives a
slightly better level of explanation, while in a further two cases a negative

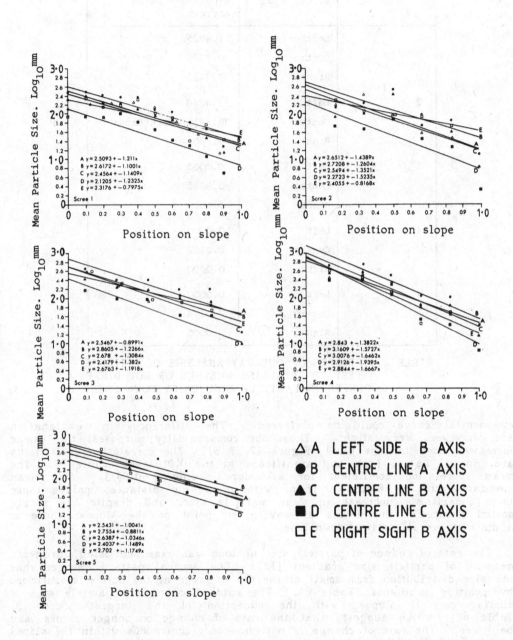

Legend:

△ A LEFT SIDE   B AXIS
● B CENTRE LINE A AXIS
▲ C CENTRE LINE B AXIS
■ D CENTRE LINE C AXIS
□ E RIGHT SIGHT B AXIS

Scree 1
A $y = 2.5093 + -1.211x$
B $y = 2.6172 + -1.1001x$
C $y = 2.4564 + -1.1409x$
D $y = 2.1205 + -1.2325x$
E $y = 2.3176 + -0.7975x$

Scree 2
A $y = 2.6512 + -1.4389x$
B $y = 2.7208 + -1.2604x$
C $y = 2.5494 + -1.3521x$
D $y = 2.2723 + -1.5235x$
E $y = 2.4055 + -0.8168x$

Scree 3
A $y = 2.5467 + -0.8991x$
B $y = 2.8605 + -1.2266x$
C $y = 2.678 + -1.3084x$
D $y = 2.4179 + -1.382x$
E $y = 2.6763 + -1.1918x$

Scree 4
A $y = 2.843 + -1.3822x$
B $y = 3.1609 + -1.5727x$
C $y = 3.0076 + -1.6462x$
D $y = 2.9126 + -1.9395x$
E $y = 2.8844 + -1.6667x$

Scree 5
A $y = 2.5431 + -1.0041x$
B $y = 2.7554 + -0.8811x$
C $y = 2.6387 + -1.0346x$
D $y = 2.4037 + -1.1489x$
E $y = 2.702 + -1.1749x$

FIG. 17. LINEAR REGRESSION OF MEAN PARTICLE SIZE OF b AXIS IN TERMS OF LOG_{10}MM PLOTTED AGAINST POSITION ON SLOPE DOWNSLOPE BETWEEN SAMPLE POINTS.

Scree number	Sample points across slope	Probability of null hypothesis being correct
1	Left	1.9525
	Centre	0.2375
	Right	0.3493
2	Left	3.0048
	Centre	39.3419 (Significant)
	Right	1.9903
3	Left	0.0000
	Centre	0.2382
	Right	0.0000
4	Left	0.0000
	Centre	0.0122
	Right	0.0000
5	Left	0.0000
	Centre	2.0224
	Right	0.0000

TABLE 4: RESULTS OF ONE-WAY ANALYSIS OF VARIANCE TEST TO ESTABLISH WHETHER OR NOT DIFFERENCES IN PARTICLE SIZE EXIST DOWN SCREE.

exponential curve could be preferred. The differences in explanation are, however, very slight. Thus, for comparability purposes, the linear form was used for the plots (Figure 17, A-E). The correlation coefficients are, in all cases, high and significant at the 0.01 level (Table 5). The mean correlation coefficient for all slopes being -0.8973. The mean standard error is 0.8253. Thus, with increasing distance upslope there is a logarithmic decrease in mean particle size, and despite the poorly sorted nature of the debris at any given point on the slopes, there is a distinct overall sorting downslope.

The rate of change of particle size upslope was examined using Gardner's measure of particle size gradient (12). The general pattern emerged that the size distribution from small at the top to large at the bottom is confirmed by positive gradients (Table 6). The rate of change of particle size is similar on all slopes, with the exception of the largest (Scree 3; Table 6C), which suggests that the rate of change on longer slopes may be lower. The rate of change is not, however, continuous within the slopes (Table 6, A-E). Occasionally the gradients are reversed, especially towards the bottom of the slopes, where mean particle size tends to decrease on the concave portion near to the junction with the valley floor. There are also some reversals from positive to negative gradients in the centre of the slopes due to the influence of very large particles.

Scree Number		Correlation coefficient	Standard error
1	Left b axis	-0.8815	0.7353
	Centre a axis	-0.9259	0.6296
	Centre b axis	-0.9029	0.6518
	Centre c axis	-0.8657	0.7000
	Right b axis	-0.7399	0.5492
2	Left b axis	-0.9037	0.9206
	Centre a axis	-0.7777	0.8874
	Centre b axis	-0.7120	0.9479
	Centre c axis	-0.8256	0.9899
	Right b axis	-0.9770	0.6423
3	Left b axis	-0.9029	0.7108
	Centre a axis	-0.8532	0.8821
	Centre b axis	-0.8794	0.9139
	Centre c axis	-0.8965	0.9392
	Right b axis	-0.9687	0.8187
4	Left b axis	-0.9077	0.9015
	Centre a axis	-0.9506	0.9649
	Centre b axis	-0.9505	0.9957
	Centre c axis	-0.9680	1.1080
	Right b axis	-0.9096	1.0165
5	Left b axis	-0.9884	0.7233
	Centre a axis	-0.9192	0.6784
	Centre b axis	-0.9802	0.7374
	Centre c axis	-0.9736	0.7862
	Right b axis	-0.9115	0.8013
		$\bar{X}=-0.8973$	$\bar{X}=0.8253$

TABLE 5: CORRELATION COEFFICIENTS AND STANDARD ERRORS FOR LINEAR REGRESSIONS OF TRANSFORMED LOG $_{10}$ PARTICLE SIZE VERSUS SLOPE POSITION (X) (IE. RATIO OF THE DISTANCE OF THE SAMPLE POINT FROM THE FOOT OF THE SLOPE TO THE LENGTH OF THE SLOPE FOR ROCK FALL SCREES).

A. SCREE 1

Scree 1 Right line

	0	2.8	5.6	8.4	11.2	14.0	16.8	19.6	22.4	25.2	28.0
0	–	–	–	–	–	–	–	–	–	–	–
2.8	–	–	.12329	-.00282	.05618	.05235	.06036	.09554	.03674	.04092	.01914
5.6	–	.15889	–	-.12893	.02263	.02870	.04463	.02999	.02232	.02915	.00612
8.4	–	.12591	.09293	–	.17418	.10752	.10249	.06971	.05257	.05549	.02541
11.2	–	.12261	.10448	.11600	–	.04086	.06664	.03489	.02217	.03176	.00062
14.0	–	.10030	.08077	.07470	.03339	–	.09243	.03191	.01594	.02948	-.00743
16.8	–	.07905	.05908	.04781	.01374	-.00596	–	-.02861	-.02230	.00850	-.03239
19.6	–	.06424	.04531	.03340	.00587	-.00789	-.00982	–	-.01600	.02705	-.03365
22.4	–	.06011	.04364	.03379	.01323	.00651	.01275	.03532	–	.07011	-.04248
25.2	–	–	–	–	–	–	–	–	–	–	-.15507
28.0	–	–	–	–	–	–	–	–	–	–	–

Scree 1 Left line

TABLE 6: PARTICLE SIZE GRADIENTS ($\bar{x}b$ axis) BETWEEN SUCCESSIVE SAMPLE POINTS FOR ROCKFALL SCREES.

	0	2.8	5.6	8.4	11.2	14.0	16.8	19.6	22.4	25.2	28.0
0	-	-	-	-	-	-	-	-	-	-	-
2.8	-	-	.09243 .11754 .13579	.07498 .09021 .11130	.07669 .08783 .08719	.06298 .07279 .08521	.06436 .06862 .08186	.05825 .06366 .07870	.05110 .05543 .06304	.04872 .05416 .05625	.03685 .03787 .04036
5.6	-	-	-	.05754 .06289 .08682	.06882 .07298 .06289	.05317 .05787 .06835	.05735 .05639 .06838	.05141 .05289 .06729	.04421 .04508 .05091	.04144 .04510 .04489	.02990 .02791 .02843
8.4	-	-	-	-	.08011 .08307 .03896	.05098 .05536 .05911	.05729 .05423 .06224	.04988 .05038 .06240	.04155 .04151 .04373	.03876 .04214 .03790	.02595 .02291 .02009
11.2	-	-	-	-	-	.02186 .02764 .07925	.04588 .03980 .07388	.03981 .03949 .07021	.03191 .03113 .04492	.03049 .03395 .03769	.01692 .01289 .01695
14.0	-	-	-	-	-	-	.06989 .05196 .06850	.04879 .04541 .06570	.03526 .03229 .03348	.03265 .03554 .02730	.01594 .00936 .00449
16.8	-	-	-	-	-	-	-	.02768 .03884 .06284	.01795 .02245 .01596	.02024 .03005 .01357	.00245 -.00057 -.01152
19.6	-	-	-	-	-	-	-	-	.00814 .06604 -.03096	.01652 .02564 -.01109	-.00596 -.01371 -.03632
22.4	-	-	-	-	-	-	-	-	-	.02482 .04525 .00879	-.01305 -.02359 -.03900
25.2	-	-	-	-	-	-	-	-	-	-	-.05093 -.09243 -.08679
28.0	-	-	-	-	-	-	-	-	-	-	-

a	
b	axes
c	

TABLE 6 (CONT'D): SCREE 1 MIDDLE LINE

560

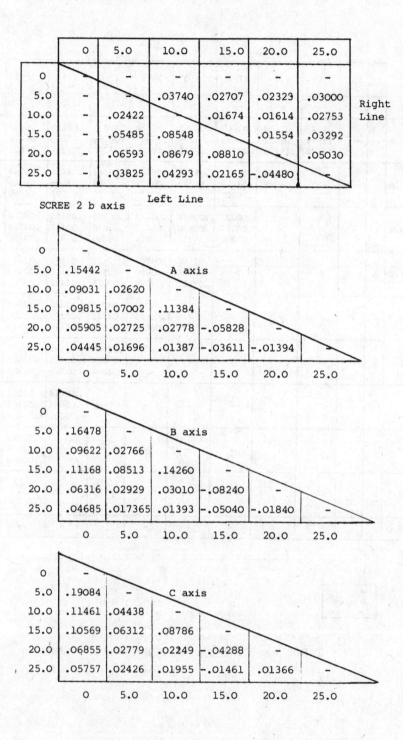

	0	5.0	10.0	15.0	20.0	25.0
0	-	-	-	-	-	-
5.0	-	-	.03740	.02707	.02323	.03000
10.0	-	.02422		.01674	.01614	.02753
15.0	-	.05485	.08548		.01554	.03292
20.0	-	.06593	.08679	.08810		.05030
25.0	-	.03825	.04293	.02165	-.04480	

Right Line

SCREE 2 b axis   Left Line

A axis

	0	5.0	10.0	15.0	20.0	25.0
0	-					
5.0	.15442	-				
10.0	.09031	.02620	-			
15.0	.09815	.07002	.11384	-		
20.0	.05905	.02725	.02778	-.05828	-	
25.0	.04445	.01696	.01387	-.03611	-.01394	-

B axis

	0	5.0	10.0	15.0	20.0	25.0
0	-					
5.0	.16478	-				
10.0	.09622	.02766	-			
15.0	.11168	.08513	.14260	-		
20.0	.06316	.02929	.03010	-.08240	-	
25.0	.04685	.017365	.01393	-.05040	-.01840	-

C axis

	0	5.0	10.0	15.0	20.0	25.0
0	-					
5.0	.19084	-				
10.0	.11461	.04438	-			
15.0	.10569	.06312	.08786	-		
20.0	.06855	.02779	.02249	-.04288	-	
25.0	.05757	.02426	.01955	-.01461	.01366	-

TABLE 6 (CONT'D): B. SCREE 2.

	0	17.0	34.0	51.0	68.0	85.0	
0	-	-	-	-	-	-	
17.0	-	-	.01005	.00555	.01113	.01233	Right
34.0	-	.01312	-	.00263	.01167	.01309	Line
51.0	-	.00278	-.00756	-	.02348	.01832	
68.0	-	.00818	.00571	.01899	-	.01592	
85.0	-	.01051	.09643	.01825	.01750	-	

Left Line

b axis

	0	17.0	34.0	51.0	68.0	85.0
0	-					
17.0	.05614	-	A axis			
34.0	.02380	.00046	-			
51.0	.02264	.00589	.01132	-		
68.0	.01703	.00399	.00576	.00019	-	
85.0	.01637	.00625	.00841	.00696	.01372	-

	0	17.0	34.0	51.0	68.0	85.0
0	-					
17.0	.05280	-	B axis			
34.0	.03039	.00799	-			
51.0	.02317	.00836	.00187	-		
68.0	.01828	.00677	.00168	.00360	-	
85.0	.01678	.00778	.00771	.00720	.01079	-

	0	17.0	34.0	51.0	68.0	85.0
0	-					
17.0	.05312	-	C axis			
34.0	.02749	.00185	-			
51.0	.02317	.00820	.01455	-		
68.0	.01907	.00772	.00977	.00675	-	
85.0	.01731	.00836	.01053	.00851	.01027	-

TABLE 6 (CONT'D): C. SCREE 3.

	0	12.5	25.0	37.5	50.0	62.5	
0	-	-	-	-	-	-	
12.5	-	-	-.01873	.01847	.02067	.01806	Right line
25.0	-	.01302	-	.05567	.04037	.03033	
37.5	-	.02925	.04549	-	.02506	.01765	
50.0	-	.02382	.02922	.01294	-	.01024	
62.5	-	.01604	.01705	.00284	-.00727	-	

Left line

b axis

	0	12.5	37.5	50.0	62.5	
0	-					
12.5	.04948	-				A axis
25.0	.02506	.00063	-			
37.5	.02943	.01940	.03818	-		
50.0	.02707	.01961	.02909	.02001	-	
62.5	.02117	.01409	.01874	.00878	.00246	-
	0	12.5	37.5	50.0	62.5	

	0	12.5	25.0	37.5	50.0	
0	-					
12.5	.05352	-				B axis
25.0	.02600	-.00152	-			
37.5	.03084	.01950	.04051	-		
50.0	.02810	.01962	.03020	.01988	-	
62.5	.02246	.01469	.02010	.00989	-.00010	-
	0	12.5	25.0	37.5	50.0	62.5

	0	12.5	25.0	37.5	50.0	
0	-					
12.5	.04559	-				C axis
25.0	.02746	.00932	-			
37.5	.03410	.02836	.04739	-		
50.0	.03156	.02689	.03567	.02395	-	
62.5	.02512	.02000	.02356	.01164	-.00066	-
	0	12.5	25.0	37.5	50.0	62.5

TABLE 6 (CONT'D): D. SCREE 4.

	0	7	14	21	28	35
0	-	-	-	-	-	-
7	-	-	.01613	.01322	.01559	.03032
14	-	.03821	-	.01031	.01532	.03505
21	-	.02404	.00987	-	.02033	.05258
28	-	.02585	.01966	.02946	-	.07451
35	-	.02496	.02055	.02589	.02231	-

Right line

Left line

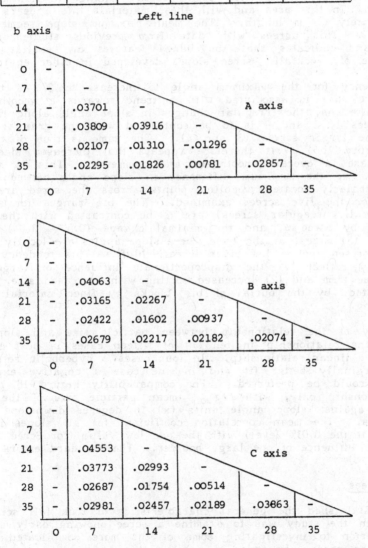

b axis

A axis

	0	7	14	21	28	35
0	-					
7	-	-				
14	-	.03701	-			
21	-	.03809	.03916			
28	-	.02107	.01310	-.01296	-	
35	-	.02295	.01826	.00781	.02857	-

B axis

	0	7	14	21	28	35
0	-					
7	-	-				
14	-	.04063	-			
21	-	.03165	.02267	-		
28	-	.02422	.01602	.00937	-	
35	-	.02679	.02217	.02182	.02074	-

C axis

	0	7	14	21	28	35
0	-					
7	-	-				
14	-	.04553	-			
21	-	.03773	.02993	-		
28	-	.02687	.01754	.00514	-	
35	-	.02981	.02457	.02189	.03863	-

TABLE 6 (CONT'D): E. SCREE 5.

Particle Size and Slope Angle

The centre line profiles are shown in Figure 14 and reveal typical generalized concave forms. Slope angles ranged from 37° (Screes 3 and 4) to -5° at the base of Scree 3 where large blocks had banked against a low terrace. Average inclinations ranged from 15.2^o to 28.2^o (Table 1), the overall mean for 15 transects being 23.9^o.

Maximum recorded slope angles generally increased with scree size (2 = 31°; 1 = 35.5^o; 5 = 33°; 4 = 37°; 3 = 37°), as did mean slope (Table 1). There is an indication that once a scree reaches a certain length there appears to be little or no further change in maximum angle (Figure 16a). In this area and with these materials this appears to occur at approximately 60 m length. The mean maximum slope segment angle was 34.3^o. This agrees with data from previous studies elsewhere (Table 7) and indicates that the Hunza features are similar to, and representative of, rockfall scree slopes developed in other environments.

The tendency for the maximum angle to increase to 37° as the scree grows (Figure 16a) is associated with a trend towards a smoother form. Thus in Figure 16a, the irregular changes in slope angle along the centre lines of Screes 1, 2 and 5 is to be contrasted with the gradual changes shown by the larger screes (3 and 4). The same is true of the flank profiles (Figure 16b). It therefore appears that increased size results in an increase in overall smoothness and symmetry. This is shown in Figure 16c where the maximum difference in slope angle between each of the three similarly located sampling points across the scree are plotted as graphs for the five screes examined. The big ranges for screes 1, 2 and 5 (small, irregular screes) are to be contrasted with the reduced range shown by Scree 4, and the minimal change (0.5^o - 2.0^o) for the largest scree (3) except at the base where slope angles are locally reversed due to the presence of a low river terrace bluff. This tendency towards symmetry may reflect (1) the disproportionate influence of large blocks on small scree form and (2) increased sorting with increased size, largely, it is suggested, by the burial of big blocks by finer material and by surface sliding.

The study of the relationship between particle size and slope angle shows that the relationship in ten out of fifteen cases is best explained by a negative linear relationship. In four cases a hyperbolic relationship gives a marginally better fit and in one case a negative exponential relationship could be preferred. For comparability Figure 18 plots the linear relationship only, with $\log_{10}$ mean particle size of the b axis (y) plotted against slope angle units (x) in degrees downslope between sample points. The mean correlation coefficient for all slopes is 0.7777 (significant at the 0.01% level) with the one low value (for Scree 2) again reflecting the influence of one large boulder. The standard errors average 2.0459 (Table 8)

Patterned Screes

Having established the general characteristics of rock fall screes, the next stage in the study was to examine a scree showing early stages of patterning prior to investigating some of the more complicated features present in the area. In order to achieve this progression, a scree was chosen which clearly displayed vertical (downslope) stripes, a patterning due to small surface mudflows and characteristic of many scree slopes in Hunza. The scree (Figure 11, No. 6, Figure 13) was both accessible to the

Location	Source	Process Operating	Angle (°)	Type of angle	Geometry of long profile
Rocky Mountains	Gardner (1972)	Rockfall, Avalanche, Debris flow	11.5–28.6 23.5 25.7 34.7	Range Mean Median Maximum	Concave
Austria	Brückl et al (1974)	Miniature overlapping landslides	34.9	Maximum	Rectilinear
Devon Island	Howarth & Bones (1972)	Rockfall Meltwater Basal Erosion	32.5–36.3 20.9–37.6 36.9–38.8	Mean Mean Mean	Concave Concave Concave
North Wales	Tinkler (1966)	Rockfall	35	Mode	Concave-rectilinear
IKP 1980	(this paper)	Rockfall	15.2–28.2 37 34.3	Mean Maximum Mean maximum	Concave-rectilinear
		Patterned scree shallow sliding	33	Mean	Rectilinear
		Mudflow	36	Maximum	Concave
Hindu Kush, Chitral	Wasson (1979)	Rockfall, Mudflows, Grain flow, Sheet wash	9–30	Mean	Rectilinear to Concave

TABLE 7: SLOPE ANGLE DATA FOR SCREE SLOPES IN VARIOUS MOUNTAIN ENVIRONMENTS.

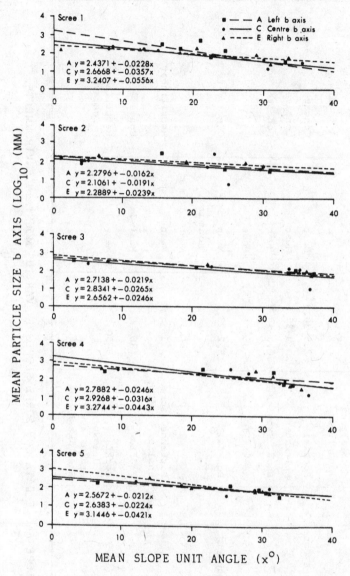

FIG. 18. LINEAR
REGRESSION OF MEAN
PARTICLE SIZE OF b
AXIS (LOG$_{10}$) PLOTTED
AGAINST SLOPE ANGLE
UNITS (x)$^{\circ}$ DOWNSLOPE
BETWEEN SAMPLE POINTS

MEAN SLOPE UNIT ANGLE (x$^{\circ}$)

Karakoram Highway and of a size (124.8 m slope length) which permitted relatively easy measurement. The scree was divided into nine longitudinal units (A – I) representing alternate strips of coarser and finer material which occurred reasonably regularly across the slope (Figure 13). A profile was measured down the central line (Figure 19) and subdivided into six units (as with the rockfall study) so as to yield six equally spaced sample traverses, including one at the crest. The b axes of twenty particles were measured within each stripe encountered on these traverses.

Because of the shape of the headwall and the presence of minor chutes, only five stripes were sampled at the crest. Downslope, the number of stripes encountered increased to eight on the first traverse (Table 9) and nine on the subsequent three prior to decreasing to five on the last because

Scree number	Transect Line	Correlation Coefficients (Linear)	Standard Error of the Estimate
1	Left	−0.8231	1.7086
	Centre	−0.9045	2.0922
	Right	−0.8078	2.0160
2	Left	−0.7211	2.3317
	Centre	−0.3676	1.7935
	Right	−0.8579	1.7380
3	Left	−0.9786	2.4553
	Centre	−0.6086	2.2665
	Right	−0.9731	2.3725
4	Left	−0.6790	2.0888
	Centre	−0.7798	2.3511
	Right	−0.7524	2.4613
5	Left	−0.8364	1.8071
	Centre	−0.6299	1.3926
	Right	−0.9468	1.8127
	$\overline{x}$ =	−0.7777	2.0459
	(significant at 0.01 Level)		

TABLE 8: PARTICLE SIZE (y) VERSUS MEAN SLOPE ANGLES (x) FOR BASIC DATA ON THE FIVE ROCKFALL SCREES.

of the termination of two mudflow lobes (D and H on Table 9). A tape was laid orthogonally across the slope on each traverse and the width of each stripe recorded (Table 9). Twenty $\underline{b}$ axis measurements were obtained for each stripe from twenty equally spaced sampling points across the unit. These were analysed to provide the data tabulated in Table 10. There is, therefore, a length bias to the data since the belts of finer material (mudflow tracks) were narrower than the intervening unmodified areas (Table 9), and yet the same number of pebbles were measured in each. In other words, equal status was given to units of different width. This subjective element was introduced in order to display the particle distribution across the whole slope so that mean particle distributions could be derived and compared (Figures 20 and 21). The procedure is not considered to hinder seriously the essentially descriptive nature of the exercise.

The particles measured were essentially composed of blocky and elongate pieces of chloritic schist, but there was also a reasonable portion of silt, sand and fine gravel sized particles which give a fine matrix to the openwork of the upper parts of the scree. It is the presence of this fine material at the base of the 80° backface which is a primary cause of mudflow

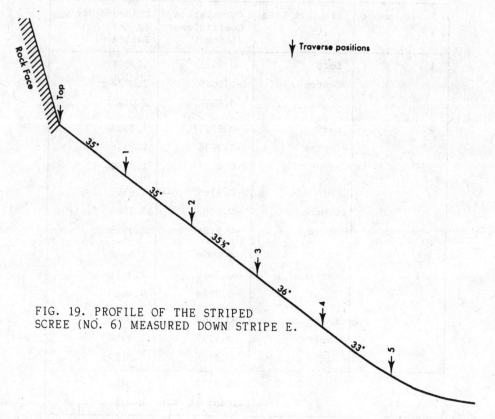

FIG. 19. PROFILE OF THE STRIPED
SCREE (NO. 6) MEASURED DOWN STRIPE E.

generation by reducing permeability and providing the necessary matrix for large particle buoyancy.

In the analysis which follows the graphical plots are based on the raw data since the size distributions are very obviously different. The data were not transformed to Log_{10} since no statistical analyses were intended. Furthermore, transformed data were not as useful as raw data for graphical presentation and analysis.

Results

The size data (Table 10) for this scree (maximum slope $36°$) clearly display the initial effects of mudflows on a talus slope. The observed process was that mudflows generated on the scree at the base of the head-wall brought lobes of fine material downslope so as to bury the underlying coarser debris. The effect of this is clearly shown in Figure 20 where the mean b axis data have been plotted as graphs for each stripe. Two distinct populations emerge, an upper group (coarse stripes) showing the natural rockfall scree tendency for particle size to increase downslope (see previous section), and a second group, reflecting the finer stripes, where mean b axis particle size is constant or slightly decreasing downslope (Table 10). An additional, and unexpected, effect was that mudflow generation caused the removal of fines from the upper part of the scree slope but left a lag of coarser debris. Thus the 'finer' stripes were, in fact, slightly coarser in their upper parts than the unaffected belts

Cross Profile	Mean Stripe Width	I (C)	H (F)	G (C)	F (F)	E (C)	D (F)	C (C)	B (F)	A (C)
(C) = Coarse stripe					(F) = Fine stripe					
TOP	3.5	-	-	-	Abs	-	-	Abs	Abs	Abs
1	2.9	4.0	2.8	6.0	0.9	4.4	2.9	1.4	0.7	Abs
2	3.3	5.3	1.1	7.3	2.4	2.4	1.2	6.6	2.4	1.3
3	4.3	4.4	4.1	6.4	3.2	3.9	1.7	6.1	4.3	4.4
4	5.2	10.2	3.6	6.4	4.3	4.4	1.8	4.5	4.9	6.4
5	8.0		13.0		0.7		18.0		5.0	3.3

TABLE 9: STRIPE WIDTH, SCREE 6.

of scree surface on either side (Table 10, Figure 20).

The extent of modification is even more dramatically depicted in Figure 21 which shows variation in mean b axis along each of the six cross-profiles. The 'saw-tooth' pattern shows: (1) how the difference in mean particle size on the fine and coarse stripes increases downslope (Nos. 1 - 5) and (2) the regularity of the change in a downslope direction. It must be borne in mind that this slope was deliberately chosen as a clear example of a rockfall scree that was suffering from its first mudflow redistribution of debris. As this process develops it should be expected that the original regular downslope size distribution will be progressively destroyed and that subsequent mudflows will not only bury or disturb original debris but will further redistribute material already involved in previous mudflows. The pattern will thus become increasingly complex and be still further disturbed by levées and lobes. It should be noted that the movements on this scree tended to bury the underlying debris. On other slopes where the mudflows are generated in the higher source areas and not on the screes themselves, the mudflows tend to be more erosive and to leave a channelled form. In these cases the size distribution will reflect excavated layers of underlying material rather than simple surface redistribution. This problem was not, however, studied in this project. Finally, it was observed that scree material characteristics had some influence in determining the impact of small shallow mudflows. The blocky nature of the striped scree studied (Scree 6) meant that mudflows could pass easily over the surface with little disturbance to the underlying fabric. On slate screes, by contrast, shallow mudflows generate sliding so that pronounced terminal lobes are formed of imbricate particles.

While vertical striping was seen to be the product of minor mudflow activity, many screes carried complex patterning, including stripes, mottles and scallops, which cannot be explained by this processs. The most obvious generating mechanisms for intricate patterns include dry grain flow and planar sliding (or avalanching). In order to investigate this possibility, two very large adjacent screes near Aliabad (Figure 11) were chosen for study (Figure 4). These had a computed relative relief of 533 m, overall slopes close to 33°, and displayed fine vertical stripes of alternate coarse and fine sediment units in their upper parts, below which they became scalloped with festoons of coarse material (Figure 5) which arced around the base of steeper lighter-coloured surfaces. Surface particle distributions

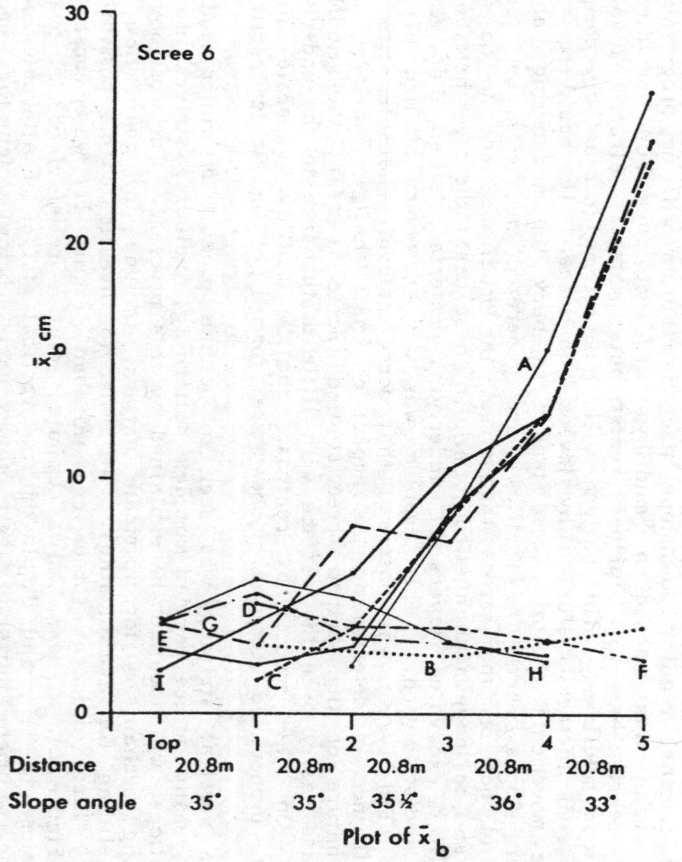

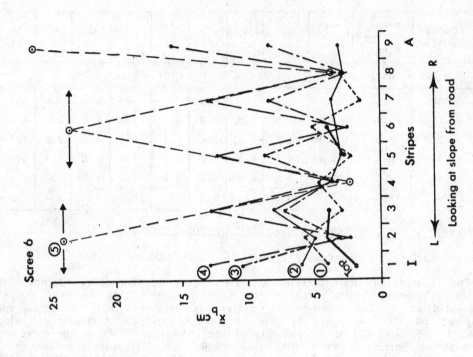

FIG. 21. GRAPH TO SHOW VARIATION IN MEAN b AXIS FOR EACH STRIPE ALONG THE SIX TRAVERSES.

FIG. 20. PLOT OF MEAN b AXIS AGAINST POSITION ON SLOPE FOR EACH STRIPE OF SCREE 6.

	A COARSE	B FINE	C COARSE	D FINE	E COARSE	F FINE	G COARSE	H FINE	I COARSE
TOP	—	—	—	7.1 3.95 1.5 37.5 5.6	15.8 2.71 0.9 117.3 14.9	—	12.4 3.80 0.9 82.9 11.5	11.4 3.96 1.0 64.1 10.4	5.0 1.93 1.0 49.2 4.0
1	—	5.3 2.98 1.7 40.6 3.6	4.3 1.44 0.3 66.7 4.0	53.5 5.17 1.0 222.6 52.5	5.0 2.06 0.5 58.2 4.5	20.3 4.72 1.1 116.9 19.2	12.0 2.89 0.7 96.5 11.3	23.0 5.69 1.5 108.3 21.5	8.0 3.87 1.7 38.0 6.3
2	8.2 3.09 1.5 56.9 6.7	10.2 2.72 0.9 86.4 9.3	9.9 3.65 1.0 77.5 8.9	9.6 3.20 0.3 78.4 9.3	14.2 2.90 1.0 117.2 13.2	22.5 3.73 1.0 133.8 21.5	63.2 8.0 1.5 175.6 61.7	23.0 4.93 1.0 117.0 22.0	15.5 5.99 1.0 80.0 14.5
3	27.3 8.21 0.8 80.6 26.5	7.5 2.51 1.0 71.7 6.5	16.2 8.25 3.2 41.3 13.0	10.6 3.00 1.2 73.0 9.4	15.0 8.68 5.0 34.3 10.0	11.0 3.74 1.3 63.6 9.7	16.3 7.34 3.1 48.6 13.2	11.0 3.07 0.9 88.9 10.1	24.6 10.44 1.0 51.0 23.6
4	35.0 15.69 8.2 40.1 26.8	13.5 3.09 1.0 86.7 12.5	22.7 12.94 7.2 12.94 15.5	10.6 2.50 0.9 98.4 9.7	23.6 12.21 7.4 35.9 16.2	14.7 3.18 0.4 97.8 14.3	31.1 12.85 6.5 47.2 24.6	4.7 2.28 0.8 50.9 3.9	29.0 12.87 6.0 44.5 23.0
5	56.0 26.2 5.8 51.1 50.2	12.0 3.69 0.8 75.9 11.2		50.0 23.41 12.0 41.4 38.0		9.0 2.33 0.5 103.4 8.5		49.0 23.99 10.0 50.6 39.0	

KEY

Max	Mean
Min	Coefficient of variation
Range	

TABLE 10: PARTICLE SIZE CHARACTERISTICS ON STRIPED SCREE (SCREE 6).

572

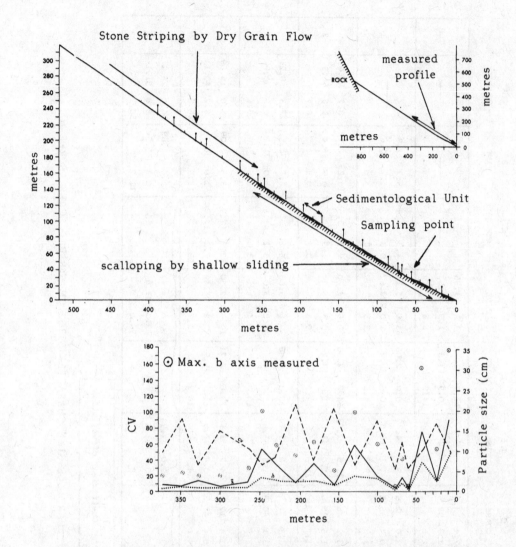

FIG. 22. PROFILE AND DETAILS OF THE IMMENSE SCALLOPED SCREE AT ALIABAD. In the lower diagram the dashed line (---) is the coefficient of variation, the full line (——) mean particle size and the dotted line (...) standard deviation. The change in process from dry grain flow to sliding has a clear effect on mean size distribution at 260 m.

appeared to be modified by dry grain flow in the higher areas and by the shallow sliding of coarse debris at lower elevations to reveal underlying finer material below the slip surfaces. The lower slopes were therefore irregular in morphology with steep portions in the failure areas and shallow benches or lobes in the slide deposition areas. In order to describe this pattern a profile was surveyed up the centre of one scree (Scree 7) from the base to a relative elevation of 227 m, sufficient to cross the scalloped zone and pass on to the vertical striping. It became physically too difficult to measure further on this immense slope. A total slope length of 450 m was sampled and measured in this exercise, which is believed to give a representative section of the patterns present. The profile (Figure 22) was measured with a 30 m tape and Suunto inclinometer.

The slopes were measured with respect to breaks of slope and the sedimentological units. The $\underline{b}$ axes of a sample of twenty pebbles were measured in the centre of each sedimentary unit and computed to yield the range, mean and variation of particle size (Figure 22).

The adjacent scree (No. 8), which also showed very clear vertical stripes of coarse and fine material, was examined by means of a horizontal transect across the slope at an elevation of c. 220 m, to provide a comparison with the distributions on the much smaller scree No. 6. Ten stripes (A-J in Figure 23) were covered over a measured distance of 51 m. The

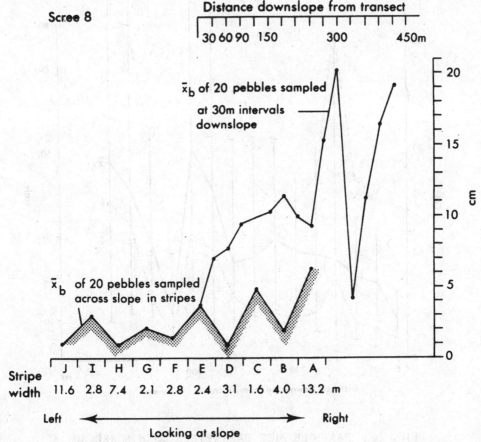

FIG. 23. MEAN PARTICLE SIZE DISTRIBUTION ACROSS AND DOWN SCREE NO. 8.

width of each stripe was recorded and the b axes of twenty pebbles measured at regular spacing across the unit. A profile was then measured down the middle stripe from this point to the base (420 m long) and twenty b axes measured on particles selected at 30 m intervals. The purpose of this last exercise was to provide a different picture of downslope distribution from that obtained for Scree No. 7 (Figure 22). The data were analysed to yield the range, mean b axis size and variance of the sample at each site (Figures 23 and 24).

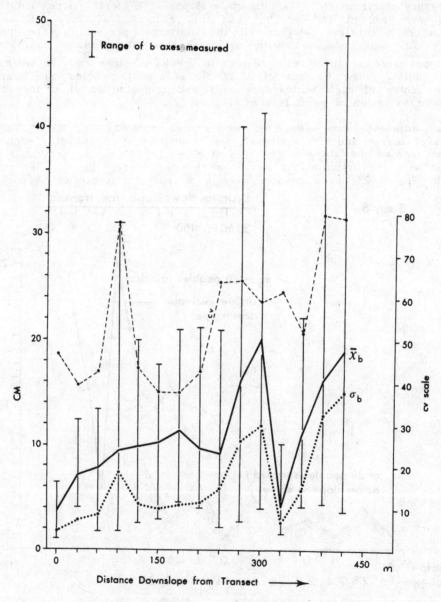

FIG. 24. PARTICLE SIZE PARAMETERS DOWN SCREE NO. 8.

RESULTS

The particle size distribution on Scree 7 (Figure 22) demonstrates that on the upper portion of the scree, where the dominant process is dry grain-flow, there is little change in maximum and mean particle size in a down-slope direction, although the coefficient of variation fluctuates. This is reflected in slope form which is smooth and at a constant 33°. Immediately that shallow planar sliding becomes the dominant process, however, a very distinctive downslope pattern is created. The light-coloured failure surfaces stand at 24 - 33° (generally steepening with elevation) and are developed in fine-grained material. There is no apparent downslope increase in particle size in these failure surface units. Much of the coarse material which once entirely covered the scree surface has now been redistributed downslope by shallow sliding, tending to accumulate in scallop-shaped lobes (Figure 22). Slope angles increase upslope from 26 - 30° (riser) and 18 - 21° (tread) for the lowest four lobes, to 27 - 33° for the higher coarse patches (Figure 22). The mean grain size for these lobes shows a tendency to increase downslope for each successive lobe (Figure 22), as does maximum recorded particle size.

This probably reflects the original grain-size distribution displayed by natural rockfall screes (see Figures 17 and 20). In this instance, however, it is important to note how a normally fairly regular downslope increase in particle size has been made highly irregular by the impact of shallow surface sliding. In addition, the effect of grainflow appears to be progressively burying coarser debris, thereby creating internal stratification. This stratification is no doubt of great relevance in the creation of patterning because it results in internal instability, and, in consequence, facilitates surface sliding in response to differential loading by debris, interference by animals and man, variations in the pattern of snow thaw or earthquake shocks. The occurrence, on the lower slopes, of shallow coarser layers overlying finer materials, suggests that this stratification extends well down the scree and may reflect alternating episodes of coarse and fine debris supply (very little coarse debris appears to be arriving on the scree at present). Finally, the relatively low slope angle of this scree (33° as compared with 35 - 37° for Screes 1, 3, 4 and 6) indicates the influence of smaller mean particle size, a conclusion supported by slope measurements on the finer material of Scree 8 which was similarly standing at 32 - 33°.

The transect across part of Scree 8 revealed the lateral variation in particle size caused by dry grain-flow (Figure 23). The pattern is one of fluctuating particle size, with lower values on the grain-flows than on the intervening older scree surfaces, and is similar to that recorded for mudflow modification (Scree 6, Figure 21) although the surface appearance is visually rather different when viewed from close range. The fixed sampling internal profile down the scree (Figure 23) revealed a regular increase in mean particle size over the first 180 m (31.5 - 33°), after which size fluctuated with variations in slope angle (32° to 20°) due to slumping. The pattern is, therefore, similar to that demonstrated for the lower part of Scree 7 (Figure 22), the sudden peaking at the base of the profile being produced by a large slide with well developed reversed backslopes. These data are replotted in Figure 24 to reveal how changes in mean size are really a response to the presence of large blocks on the slope.

CONCLUSION

The main purposes of the exercise described in this paper were:

(a) to investigate whether the rock fall model described by other authors, in which there is a size distribution downslope on screes formed by rock fall processes, was applicable to the screes in the Hunza valley, and

(b) to attempt to describe the way in which the size distribution pattern resulting from rock fall is modified by the operation of other mass movement processes.

The sample of five screes that were measured to establish the characteristics of the simple Type 1 rockfall screes, produced results that were compatible with previous studies, namely:

(1) The scree slopes possessed similar slope angles and rectilinear to concave slope form to those slopes reported from other areas of the world.

(2) The screes, which varied in size and maturity from small and fresh to relatively large and mature, showed an increasing regularity in slope form and symmetry with increasing size, and the development of a rectilinear slope segment at a characteristic angle for the material which seemed to remain constant after a certain critical length (c. 60 m long) had been attained.

(3) The material was unsorted at any sample point, with large coefficients of variation; there was no significant difference in size across the slope, but there was a significant linear increase in size downslope. The rate of change of particle size was similar on all slopes except the largest slope where a slower rate of change was evident. The change in size downslope was not, however, completely regular and there were occasional reversals caused by the presence of large boulders or by changing gradients at the base of the slope.

(4) Particle size also showed a negative linear relationship with slope angle and this relationship had a high correlation coefficient.

However, the examination of patterned screes revealed that these simple relationships are considerably modified by mass movement processes, as suggested by Gardner (12), Bones (9) and Drewry (15). Although the downslope size sorting pattern could still be distinguished on striped screes, in the undisturbed debris surfaces between mudflows (Figure 20) and cohesionless grain-flows, the general tendency is for increasingly complex particle size patterns with increasing mass movement activity. Mudflows and grain-flows introduce stripes of finer material into zones of generally coarser surface debris, thereby causing rapid and dramatic lateral fluctuations in particle size (Figures 21 and 23), as well as reducing the overall downslope rate of particle size increase. Slipping and slumping, by contrast, produce downslope changes due to the production of lobes and scallops of unsorted coarser particles where the debris comes to rest, separated by patches of finer material. The size distribution downslope tends to fluctuate dramatically, particularly in the case of shallow sliding on stratified screes (Figure 22), although traces of the rockfall size distribution can still be distinguished in the overall pattern of the coarser lobes. Particle size distributions across the slope are also greatly complicated by sliding, and the resultant patterns are more irregular and complex than those produced by dry or wet flowage.

One danger of this interpretation is the assumption that the largest screes originally developed as rockfall screes and were subsequently modified by mass movement processes. It is possible that they could be the product of complex evolutionary histories involving a wide range of processes acting either successively or more or less contemporaneously. Since no observations could be made on a temporal basis, this question remains unresolved. Nevertheless, it is felt that the observations presented here do reflect changes caused by recent mass movement and not the surface expression of complex internal structures.

Generally, mudflows and dry grain flows produce patterns in which there is an irregular decrease or evening out of grain size downslope, with some sorting of larger sizes into levées and lobes. Repeated mudflow activity, in particular, results in the creation of extremely complex morphological and sedimentological patterns and the formation of scree-fans and mudflow fans (Types 3 and 4 in the introduction). Repeated grain-flow can result in marked visual patterning (rays, stripes) but generally leads to progressive uniformity of particle size (Figure 22, upper part). Shallow sliding tends to occur on these slopes as spatially discrete events and only affects the individual patches of material in which they occur. They sort the material into a fine slide area and a coarse deposition area without necessarily affecting the distribution on adjacent areas. However, the change to an irregular slope form results in the creation of catchments for rolling clasts. The sorting process is therefore continued so that slides probably increase in size as coarse material accumulates on the upper parts of lobes in effect as smaller screes climbing up the slide scar. Snow will also tend to be thicker and lie longer at such locations, thereby causing further movement.

Finally, it is worth repeating that the patterned screes chosen for study were relatively simple cases in which mainly first-time alterations of the scree were measured. The patterns become increasingly complex when the mass movement processes occur in debris that has already been moved by superficial sorting mechanisms. The effect of this was not studied in this project. Similarly, the effects of undercutting on scree slopes were not examined although several screes showed surface sliding due to basal erosion by the Hunza River or following the creation of the Karakoram Highway, and in certain instances entire screes of large dimensions were observed to be slipping (Figure 25).

The large volume and complex patterning of the Hunza screes indicate that talus accumulation and redistribution is relatively rapid in this area. The coexistence of arid conditions and winter snowfall make for an interesting and unusual mix of modifying processes. Although little is yet known regarding the pattern of late Pleistocene and Holocene climatic change in the area, it is possible that scree formation and modification have undergone rapid changes in the recent past. It is hoped that this preliminary investigation into patterning will stimulate further interest and work on this problem and in this area.

578

FIG. 25. LARGE-SCALE MASS MOVEMENT OF AN ENTIRE SCREE SLOPE
CAUSED BY UNDERCUTTING, NEAR GHULMET, HUNZA.

ACKNOWLEDGEMENTS

We are deeply indebted to Martin Edge for assistance with data analysis, and to the cartographic, secretarial and photographic staffs at the University of London, King's College, the London School of Economics and the University of Oxford for invaluable technical assistance.

This work was undertaken during the International Karakoram Project, 1980 and received financial assistance from the National Geographic Society, the Central Research Fund of the University of London, the University of Oxford, the Royal Society and the Royal Geographical Society.

REFERENCES

1) GOUDIE, A. S., et al. This volume. 'Geomorphology of the Hunza Valley, Karakoram Mountains, Pakistan'.

2) BRUNSDEN, D., JONES, D. K. C., WHALLEY, W. B. and GOUDIE, A. S. 'Debris flows of the Karakoram Mountains', (in preparation).

3) ANDREWS, J. T., (1961). 'The development of scree slopes in the English Lake District and Central Quebec-Labrador', Cahiers de Géographie de Québec 10, pp 219 - 231.

4) GARDNER, J., (1968). 'Debris slope form and processes in the Lake Louise district: a high mountain area'. Unpublished Ph.D. Thesis, McGill University, 263 pp.

5) STATHAM, I., (1973). 'Scree slope development under conditions of surface particle movement', Transactions, Institute of British Geographers 59, pp 41 - 53.

6) STATHAM, I., (1973). 'Process form relationships in a scree system developing under rockfall'. Unpublished Ph.D. Thesis, University of Bristol, 263 pp.

7) STATHAM, I., (1976). 'A scree slope rockfall model', Earth Surface Processes 1, pp 43 - 62.

8) RAPP, A., (1960). 'Talus slopes and mountains walls at Templefiorden, Spitzbergen', Norsk Polar Institutt Sk. 119, 96 pp.

9) BONES, J. G., (1973). 'Process and sediment size arrangement on high arctic talus slopes, South west Devon Island, North west Territories, Canada', Arctic and Alpine Research 5, pp 29 - 40.

10) CARSON, M. A., (1977). 'Angles of repose, angles of shearing resistance and angles of talus slopes', Earth Surface Processes 2, pp 363 - 380.

11) STATHAM, I., (1977). 'Angle of repose, angles of shearing resistance and angles of talus slopes - a reply', Earth Surface Processes 2, pp 437 - 440.

12) GARDNER, J., (1972). 'Morphology and sediment characteristics of mountain debris slopes in the Lake Louise district', Zeitschrift für Geomorphologie NF 15, pp 390 - 402.

13) KIRKBY, K. J. and STATHAM, I., (1975). 'Surface stone movement and scree formation', Journal of Geology 83, pp 349 - 362.

14) RAPP, A., (1959). 'Avalanche boulder tongues in Lappland', Geografiska Annaler 41, pp 34 - 48.

15) DREWRY, D. J., (1973). 'Rockwaste sliding - a surface transportation mechanism of screes', Institute of British Geographers Special Publication.

16) BRUCKL, E., BRUNNER, F. K., GERBER, E. and SCHEIDEGGER, A. E., (1974). 'Morphometrie einer Schutthalde', Mitt. Ost. Geogr. Gesell. 116, pp 79 - 96.

580

17) WASSON, R. J., (1979). 'Stratified debris slope deposits in the Hindu Kush, Pakistan', Zeitschrift für Geomorphologie 23, pp 301 - 320.

18) PETTIJOHN, F. J., (1957). Sedimentary Rocks, Harper Bros. New York, 718 pp.

Sediment load of the Hunza River

R. I. Ferguson – University of Stirling

ABSTRACT

The Hunza River drains 13200 km^2 of the western Karakoram mountains, in which heavily-glacierised mountains rise 5 - 6000 m above arid incised valleys. Measurements by the Pakistan Water and Power Development Authority show the Hunza contributes 22% of the runoff and 39% of the suspended sediment load of the upper 143,000 km^2 of the Indus basin. Its sediment yield of 4,800 t km^{-2} a^{-1}, equivalent to a denudation rate of 1.8 mm a^{-1}, is amongst the highest in the world but the precise rate is uncertain because of infrequent sampling, considerable short-term and seasonal scatter about sediment rating curves, and big year-to-year variations in load. Ninety-nine percent of annual suspended load is carried in June-September when melt-water discharge is maximum and concentrations exceed 10 kg m^{-3}. IKP measurements in summer 1980 showed sediment concentrations in a major proglacial tributary averaged only 12% of those at the basin mouth. This downstream increase, and the pronounced seasonal hysteresis in sediment rating curves, suggest fresh sediment stores are flushed and then depleted not only as melting extends to higher elevations but subsequently as swollen rivers overtop unvegetated valley floors and undercut terraces. Solute concentrations estimated from electrical conductivity also increase downstream but less so. At the downstream site annual solute yield is estimated as 90 t km^{-2} a^{-1}, three times the world average but only 2% of the solid load. Hydraulic properties calculated from gauging data suggest bedload transport could exceed 100,000 t/day in summer and forms 2 - 10% of total annual load.

INTRODUCTION

Young mountain ranges, of which the Greater Himalaya are the supreme example, undergo exceptionally rapid erosion by a combination of mass movement, glacial transport, and fluvial transport. According to data compiled by Meybeck (1) the three great Himalayan rivers (Indus, Ganges, Brahmaputra) carry almost 10% of the 20 x 10^9 tonnes of particulate sediment delivered from continents to oceans each year. Their combined load is forty times that of the three main N. Asian rivers (Lena, Ob, Yenisei) even though the latter drain an area three times larger.

Rapid erosion in young mountains is a consequence of high relief due to continuing tectonic uplift. High relief is associated with steep slopes on which large and rapid mass movements can occur; it leads to higher annual precipitation and runoff with a consequent increase in fluvial erosion; and the upper slopes are sufficiently cold for snowfall to nourish glaciers that cause additional erosion and improve the coupling between valley-side and valley-floor sediment transport systems. The results of rapid erosion include spectacular landforms, geomorphic hazards within the mountains, and sedimentation problems downriver. Quantification of erosion rates in young mountains is important in assessing both the inter-play between tectonic and denudational processes over geologic time, and the practical consequences of erosion and sedimentation at a human and engineering timescale.

In Pakistan this practical importance has led to an official programme of discharge and sediment transport measurement in the upper Indus River and its headwaters in the Karakoram Mountains, for estimation of the likely rate of infill of the 84 km long Tarbela reservoir. Fluvial measurements during the International Karakoram Project (IKP) were designed to complement this routine data collection by investigating the sources of sediment and the short-term fluctuations in transport rate. This paper summarises both long-term and short-term patterns of fluctuation in sediment transport, makes the first estimates of dissolved load and bed load transport in the Karakoram, and assesses the reliability of the estimated average erosion rate.

STUDY AREA AND RIVER REGIME

The Indus River rises in the Xijang (Tibet) region of China, flows westwards through Indian Kashmir north of the Himalayan Range, then turns south past the western end of the Himalaya to the Tarbela reservoir near the edge of the Punjab plains (Fig. 1). Below here it is joined by the monsoon-swollen rivers of Kashmir and the Punjab, but the upper Indus is fed largely by snow and ice meltwaters from the Karakoram mountains in the catchments of four major right-bank tributaries. From east to west these are the Shyok, Braldu, Hunza and Gilgit Rivers, the Hunza being itself a tributary of the Gilgit. Fluvial research during the IKP was centred on the Hunza basin in the western Karakoram.

The flow of the upper Indus has been gauged since the early 1960's by the Pakistan Water and Power Development Authority, WAPDA, at a string of flood-warning stations upstream of Tarbela. The furthest upstream of these is at Partab Bridge just below the confluence of the Gilgit River. The 143,000 km^2 catchment above this station (marked in Fig. 1) includes the whole of the Karakoram mountains except for their NE slopes draining to interior basins in western China. WAPDA also operate gauging stations on the upper Indus at Kachura near Skardu south of the central Karakoram; on the Shyok River which drains the eastern Karakoram; on the Hunza River at Dainyor Bridge near Gilgit in the western Karakoram; and on the Gilgit River just downstream of its confluence with the Hunza. This last station replaces that shown on Figure 1 above the Hunza confluence, which was discontinued in 1972. At each station water level is measured daily in winter, hourly in summer, and converted to discharge by a periodi-cally updated rating curve based on velocity-area measurements from road bridges at intervals of 1 to 6 weeks. The rating curves are subject to frequent shifts because of bed mobility and some are poorly defined at high stages. WAPDA (2) considers published discharge figures for the

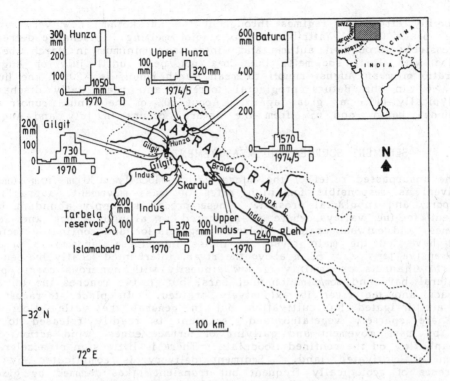

FIG.1.   KARAKORAM RIVER GAUGING STATIONS: LOCATION AND SEASONAL
REGIME (WAPDA DATA).

Hunza and Gilgit Rivers at Gilgit to be accurate to 10%, those for other
stations accurate to 15%.

The hydrology of the Karakoram mountains is summarised by the author
elsewhere in this volume (3).   In brief, the valleys and eastern plateaux
are arid with annual precipitation averaging 130 mm at Gilgit and still
less at Skardu.   Since annual runoff is considerably greater than this
(Fig. 1) precipitation must increase greatly with altitude, as was confirmed
by Chinese investigations of the Batura Glacier in the northwest Karakoram
(4).   Annual precipitation at an elevation of 2560 m near the glacier
snout was only 100 mm but three snowpits at 4800 – 5000 m, near the equilibrium line altitude, indicated a net accumulation of 1030 – 1250 mm water
equivalent.   Gross snowfall within the accumulation zone probably exceeds
2000 mm because annual runoff from the 690 km^2, 48% glacierised basin
exceeds 1500 mm.

The Batura is the first of a series of large glacier-fed tributaries
of the Hunza River which increase its annual runoff from 320 mm above
the Batura confluence (catchment 5000 km^2) to a 1966 – 79 mean of 910 mm
near Gilgit (catchment 13200 km^2).   The other gauged catchments in the
Karakoram region have lesser percentages of permanent snow and ice cover
and correspondingly lower annual runoff depths (Fig. 1) so that the
Hunza River contributes 22% of the discharge of the Indus at Partab from
only 9% of the total basin area.   Much of the remainder is probably derived
from comparably small areas of the high central and eastern Karakoram
within the Braldu and Shyok basins, but the former is not separately
gauged and the latter only well downstream from its main glacier sources.
The Hunza, and within it the Batura, are thus of all the Karakoram rivers
the ones with the best hydrometric database for studies of sediment transport.

Monthly streamflow regimes throughout the Karakoram show very strong summer peaks (Fig. 1) attributable to glacier melting. Discharge decreases progressively throughout autumn and winter to a minimum in March, begins to rise with April snowmelt, but does not peak until July or August. The ratio of mean August runoff to mean March runoff is > 30 in the Hunza River, > 60 in the Batura proglacial torrent, and peak daily discharges are typically twice as great again. About 60% of the annual runoff from the Hunza basin, and 65% from the Batura, occurs in July and August.

SEDIMENT SOURCES AND MEASUREMENTS

The unsurpassed relief of the Hunza valley (locally 6000 m from summits to river) is responsible for high rates of mass movement, supraglacial transport, and subglacial erosion. These processes supply abundant loose sediment to the valleys where it is stored in alluvial cones and fans, moraines, and outwash trains. Holocene deglaciation and continuing tectonic uplift have led the main rivers to incise into these deposits to produce an extensive terrace c 50 m above the river, interrupted locally by bedrock gorges. Channels are mainly of low sinuosity with numerous coarse point or lateral bars and some midchannel bars, but in two reaches the terraces lie back and the river is extensively braided. In places terraces and fans are irrigated for cultivation but in general the valley floor and sides are sparsely vegetated and sediment is readily released to the river by mass movement and gullying of terrace edges, wind action, and overbank flow on the confined floodplain. There is little sign of accelerated erosion of cultivated land. Sediment delivery is complicated by the occurrence of geologically frequent but transient lakes dammed by glacier advances, rockfalls or mudflows. A 10 km long lake formed in 1973 on the Hunza River downstream of the Batura confluence is aggrading rapidly while further downstream the river is eroding a lakefill formed in 1858.

Although mechanical weathering and entrainment are conventionally regarded as dominant in high mountain areas there is indirect and direct evidence that chemical weathering is also important in the Karakoram (5)(6)(7).   Weathering rinds and sulphate efflorescences are found up to and above glacier snouts, and can be attributed to intense solar heating of surfaces wetted by snowmelt. Desert varnish and honeycomb weathering are widespread in the arid valleys and springs in calciferous bedrock have electrical conductivities exceeding 300 μS cm^{-1}.

Sediment transport measurement at WAPDA gauging stations is based on collection of depth-integrated water samples at the same time as gauging of discharge (Q), i.e., at generally weekly intervals in summer and less frequently at other times of year. Concentrations of suspended sediment (c) are determined by filtration and weighing using standard techniques. Sediment loads ($\int Qc\ dt$) are estimated using flow duration curves and sediment rating curves. The accuracy of such estimates is discussed later in this paper. Published WAPDA data in American units (Q in ft^3/s, c in ppm by weight, load in short tons) have been converted by the author to SI units. Water chemistry is not routinely measured, although a few analyses were made in the Hunza and Gilgit Rivers in the 1960s, and no bedload measurements have been made.

Investigations during the IKP in July-August 1980 were designed to complement WAPDA's routine measurements at the Dainyor gauging station at the mouth of the Hunza basin. In order to assess short-term fluctuations during the peak flow season, and to investigate source areas, samples of 300 - 800 ml were collected automatically by vacuum sampler (ALS Mk 4 and Mk 4B Super) from the Hunza River near Dainyor and from the Batura proglacial torrent some 140 km upstream. At Dainyor sampling was at intervals of 6h or 8h between 19 July and 12 August, then daily to 24 August. At Batura hourly samples were obtained between 14 July and 10 August. Some samples were lost through mechanical failure, especially following duststorms, and some during unexpectedly big recessions when the river fell below the sampler intake nozzle. Samples were pressure-filtered in the field through preweighed Whatman 542 papers of nominal pore size 2.7 µm and concentrations determined gravimetrically on return to the UK.

Results at the two sites can be compared directly with each other but not necessarily with WAPDA data, since pump samplers collect point samples from the river's edge where concentrations are typically lower than in a depth-integrated sample from the centre. To minimise this problem both sites were positioned where the main flow impinged on the steep outer bank at the outside of a bend. The mean concentration at Dainyor was actually higher than the long-term July and August mean for the nearby WAPDA gauging station, probably because of the failure to sample low flows. A replication experiment in which concentrations were determined from six samples at one-minute intervals showed a 14% standard deviation: short-term fluctuations in concentration associated with turbulent eddies are thus small compared to the almost sixfold overall range in concentration over the study period, though large enough to warrant caution in interpreting minor changes.

The characteristics of suspended sediment at both sites are being examined by scanning electron microscopy and size distributions are being determined by Coulter counter. This work is still in progress and results will be reported elsewhere.

Solute transport was investigated at both sites by continuous in situ recording of electrical conductivity using battery-powered WPA CM25 conductivity meters with EIL dipcells and Rustrak chart recorders. In river water of constant temperature and ionic balance, predominantly inorganic chemistry, and pH>4, the total concentration of dissolved solids (TDS, mg/l) is linearly proportional to electrical conductivity (EC, µs/cm.$25\,^{\circ}$C). Calibration of this relationship by evaporative determination of sample TDS was not feasible but WAPDA data for Dainyor in June 1969 show a constant TDS/EC ratio of 0.58; this conversion factor has been assumed to apply also in July-August 1980. The same data show fairly constant pH ($\approx$8) and ionic composition (dominated by Ca^{2+}, SO_4^{2-} in winter, and Ca^{2+}, HCO_3^- in summer) and this was confirmed by analyses of samples, collected in 1980, although neither can be relied on because analyses were made after prolonged storage without filtration or refrigeration. Water temperature was not continuously recorded but frequent measurements show a restricted range , 0 - $2\,^{\circ}$C at Batura and 9 - $11\,^{\circ}$C at Dainyor; recorded conductivities have been increased by the conventional correction of 2% per $^{\circ}$C from the averages of these ranges to the standard reference temperature of $25\,^{\circ}$C. TDS concentrations in 1980 estimated with these assumptions are thought to be accurate to within 10 - 20%.

MEAN EROSION RATES

Estimates of the average suspended sediment load of Karakoram rivers have been made by WAPDA (8) and are summarised in Table 1. In an average year the entire Karakoram and trans-Karakoram region above

River and site	years of record	area km^2	runoff m^3s^{-1}	runoff $mm\ a^{-1}$	Sediment load per annum Mt	Sediment load per annum $t\ km^{-2}$
1 Hunza (Dainyor Bridge)	1966-75	13200	380	910 (640-1170)	63 (28-101)	4800
2 Gilgit (Gilgit Town)	1963-72	12100	280	740 (690-830)	14 (4-28)	1100
3 Gilgit (Alam Bridge)	1966/7/9-75	26200	700	840 (620-1020)	70 (33-104)	2700
4 Shyok (Yugo)	1973-5	33700	310	290 (230-390)	34 (21-55)	1000
5 Indus (Kachura)	1970-5	112700	960	270 (240-370)	87 (57-133)	770
6 Indus (Partab Bridge)	1963-75	142700	1760	390 (290-530)	160 (81-201)	1100

Source: WAPDA 1976, converted to SI units. Figures in brackets are ranges of values. Site numbers refer to map of Figure 1. Basins 1 and 2 lie within 3, 4 within 5, 3 and 5 within 6.

TABLE 1: ANNUAL RUNOFF AND SUSPENDED SEDIMENT LOAD FROM KARAKORAM RIVER BASINS.

the Partab gauging station yields 390 mm or 57 km^3 of runoff, and 160 Mt (million tonnes) of sediment. This load is equivalent to 0.1 - 0.15 km^3 of unconsolidated deposit. Since the Tarbela reservoir (which receives an estimated extra 110 Mt a^{-1} of sediment from tributaries below Partab) had an initial capacity of 11.5 km^3 its lifespan could be as short as 50 years, though a sedimentation survey after the first six years of operation showed less infill than expected even allowing for the persistently low discharges of those years (9).

The data of Table 1 show a marked spatial variation in annual runoff and sediment yield. The trans-Karakoram headwaters of the Indus in Xijang and Ladakh, upstream of the Shyok confluence, appear to provide relatively little runoff and sediment whereas the main Karakoram range extending across the Shyok, Braldu and Hunza basins into the edge of the Gilgit basin contributes disproportionately much more. The Hunza in particular yields 39% of the total sediment load at Partab from only 9% of the area.

These differences in basin output appear to correlate with valley-to-summit relief and % glacier cover, both of which are at a maximum in the Hunza region (10). However, although both places and time of maximum runoff suggest a glacier-melt origin it need not follow that sediment loads are derived mainly from subglacial or proglacial source areas. The rivers of the western Karakoram flow mostly on unvegetated glacial, colluvial, or alluvial valley fills rather than on bedrock, and thus have the opportunity to entrain sediment from these stores as well as flushing it through from the vicinity of glacier snouts.

In the absence of routine streamflow monitoring and sampling in major proglacial tributaries it is impossible to make precise calculations of their sediment loads for comparison with downstream sites, but the sediment concentrations measured in the Batura torrent in summer 1980 can be compared with those found at Dainyor over the same period. The proglacial concentrations averaged 1.1 kg m^{-3} (1100 mg/1 or ppm by weight) with a standard deviation of only 9%. Such concentrations are similar to those found in higher-latitude proglacial rivers (11) but almost an order of magnitude lower than those measured downstream at Dainyor, where the mean over the same period was 8.9 kg m^{-3} with a standard deviation of of 29%. A first estimate of the load of the Batura stream can be made by multiplying the summer 1980 mean sediment concentration by the July-August 1974/5 mean discharge of 132 m^3s^{-1} (4), which gives a load averaging 0.012 Mt/day compared to 0.99 Mt/day estimated in the same manner for Dainyor. This suggests that the Batura catchment, occupying 5% of the entire Hunza basin, supplies about 10% of its runoff but little over 1% of its sediment load.

This order of magnitude difference indicates either that the Batura is atypical and other proglacial rivers yield far more sediment, or that most of the load of the Hunza River is entrained from valley fills rather than glacier snouts. The latter seems more likely since the Batura is not obviously different in kind from other Karakoram glaciers or deficient in sediment, and since preliminary examination of sediment size, distributions and surface textures shows clear differences between proglacial and down-stream samples. It may be concluded that most of the load of the Hunza River is picked up from valley floors even if it is ultimately of glacial origin and is transported by glacier meltwater. A variety of sediment inputs may be involved including contributions from meltwater streams on alluvial fans, entrainment of alluvial and aeolian sediment from braid bars and floodplain margins at peak discharges, and mass failure of under-cut terrace margins; all these processes were observed to occur in July-August 1980.

Whatever its source, the average of 63 Mt of suspended sediment carried each year by the Hunza River is a very high load for a river draining only 13200 km^2. The mean yield on a per unit area basis is 4800 t km^{-2} a^{-1}, about twenty-five times the world average. Comparison with other areas is complicated by the small but definite scale dependence of sediment load per unit area, as a result of increased depositional opportunities in larger basins, so in Figure 2 load is plotted against basin area to allow comparison of relative erosion rates in terms of vertical departures from the general, less than 45°, trend. Only two other mountain regions plot as high as the western Karakoram: the Nepal Himalayas and the Southern Alps of New Zealand (14), both similar to the Karakoram in tectonic activity and extent of glacier cover but each experiencing much greater annual rain and snowfall. The only river system with an even higher erosion rate for its size is the middle Yellow River and its tributaries in the semiarid plateau northeast of the Himalayas, where thick loess deposits are being dissected by gullying.

The sediment yield from the Hunza basin can also be expressed in volumetric terms. Assuming a rock density of 2.65 t m^{-3} the mean denudation rate is 1800 $m^3km^{-2}a^{-1}$ ('Bubnoffs'), equivalent to lowering of the landscape at an average rate of 1.8 mm per year. Lowering is of course not spatially uniform but concentrated along major valleys, where it is probably more rapid than tectonic uplift since most of the 45 mm/year convergence of the Indian and Eurasian plates is thought to be taken up in underthrusting and strike-slip faulting (15). Sediment transport data are thus compatible with the antecedent appearance of the river gorges

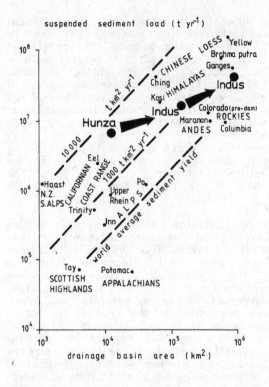

FIG. 2. SEDIMENT LOADS OF THE
HUNZA AND UPPER INDUS RIVERS
COMPARED WITH OTHER MOUNTAIN
RIVERS (1)(8)(12)(13)(14).

that run across the WNW-ESE structural grain of the Karakoram Range.

FLUCTUATIONS IN SEDIMENT LOAD

The mean annual denudation rate in a basin can validly be used for long-term comparisons, but it must be recognised that erosion proceeds at a fluctuating rate. The degree of variability, and the magnitude and timing of maximum erosion, may have important practical consequences and give clues to the internal workings of the sediment transport system.

Fluctuations in sediment transport in the Hunza River occur at three timescales over which measurements exist: year to year, seasonal, and daily. These frequencies are almost certainly only part of a wider spectrum. Over $10^2 - 10^3$ years one would expect to find abnormal years, of both low and high total load, as a result of the formation and outburst of lakes dammed by glacier advances or major mass movements (see Goudie et al, pages 411-455, for historical occurrences). Over $10^4 - 10^5$ years fluctuations associated with glacial-deglacial cycles would be apparent. At the other extreme, continuous recording of sediment concentrations estimated by photoextinction turbidity meter would reveal very short-term $(10^{-4} - 10^{-6}$ year) fluctuations associated with the passage of turbulent eddies, the collapse of short stretches of river bank, and other essentially random events. In the absence of appropriate data, however, it is unprofitable to discuss either very short-term or very long-term fluctuations and attention is restricted here to annual, seasonal and daily behaviour at timescales of $10^0 - 10^1$, 10^{-1}, and $10^{-2} - 10^{-3}$ years.

The range of annual runoff totals and sediment loads in different

Karakoram rivers over periods of 3 - 13 years is indicated in Table 1 below the corresponding mean values. There is an almost twofold year to year range in runoff at five of the six sites, and a three- to seven-fold range in annual sediment load. Variations in load at different stations are synchronous (Fig. 3)which suggests a common cause. At each site annual

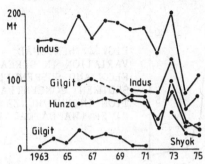

FIG. 3. ANNUAL VARIATION IN SUSPENDED SEDIMENT LOADS OF KARAKORAM RIVERS (DATA OF WAPDA (8)).

load and runoff are found to be strongly and almost linearly correlated: higher than average runoff is associated with a predictable increase in load. This relationship is to some extent inherent in the method of estimation of sediment load (see next section) but a real causal link probably does exist since more runoff implies larger runoff source areas and higher river levels, thus tapping more extensive sediment stores. To the extent that runoff in the Karakoram is predominantly glacial its annual amount must depend on glacier mass balance, which in mid-latitude mountains depends mainly on winter snowfall depths and summer tempera-tures. These in turn depend on the duration each year of different air-mass types over the whole region, so it is not surprising that annual runoff (and thus sediment load) varies in a similar way from year to year throughout the Karakoram. It should be noted that years of high runoff are not necessarily 'wet' years in the sense of higher than normal precipitation: glacier melting is on the contrary enhanced by a high snowline at the start of the summer, i.e., little winter snowfall, and sunny weather during the summer (16).

The implications of annual variability for the estimation of long-term average rates of erosion are considered later.

By far the biggest fluctuations of sediment transport in Karakoram rivers occur at a seasonal timescale (Fig. 4). WAPDA data show that a roughly twenty-fold increase in discharge from March to July in the Hunza River is associated with a still greater rise in suspended sediment concentration, from 0.01 - 0.1 kg m^{-3} in winter to peaks of over 10 kg m^{-3} in summer. Since load is the product of discharge and concentra-tion it varies even more, from a few hundred tonnes per day to upwards of two million tonnes, and WAPDA (8) estimated that 97 - 99% of the annual load in each year from 1966 to 1975 was carried between 1 June and 30 September. The other Karakoram gauging stations show similar seasonal regimes with at least 95% of the annual load transported in the four summer months in every year at every station. Since peak runoff and sediment transport occur during spells of sunny weather rather than after storms

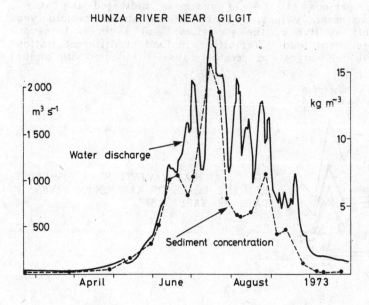

HUNZA RIVER NEAR GILGIT

FIG. 4. SEASONAL VARIATION IN STREAM-FLOW AND SUSPENDED SEDIMENT CONCENTRATION IN THE HUNZA RIVER (WAPDA DATA).

there is less day to day variation in load than in wetter climates: estimates for the Hunza River suggest no day contributes more than about 3% of the total annual load.

Winter concentrations are almost certainly limited not by the transporting capacity of the main rivers but by the restricted availability of fine sediment to winter runoff which is probably derived almost entirely from groundwater and geothermal melting beneath thick glaciers. The huge increase in sediment concentrations from March to July reflects the tapping of fresh sediment sources as snow and ice melt occur at progressively higher elevations and subsequently as swollen rivers spill over previously dry parts of the valley floors.

This explanation accounts also for a more subtle feature of the seasonal fluctuation shown in Figure 4. A given discharge tends to be associated with a higher sediment concentration in spring or early summer than in later summer or autumn, so that a clockwise hysteresis loop is present when concentration is plotted sequentially against discharge (Fig. 5). This is a consistent feature every year for which data exist for the Hunza River, and is also apparent at the other western Karakoram gauging stations and in the Shyok River in the east though not so clearly in the Indus at Kachura. Sediment hysteresis in individual floods lasting days rather than months, and between successive floods, is common in small temperate-zone rivers where it is usually attributed to initial flushing and subsequent depletion of a limited supply of available sediment (e.g., (17)). Those parts of the Karakoram below the snowline are fairly completely mantled by loose sediment but significant depletion of the finer grain sizes may nevertheless occur because of the restricted runoff source areas and the extremely rapid rate of removal of sediment. Much redeposition of silt and sand also occurs at channel margins during falling stages to replenish the store for next year.

The seasonal regime of sediment concentration, and especially its

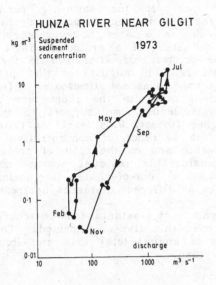

FIG. 5. SEASONAL HYSTERESIS OF
SUSPENDED SEDIMENT IN THE
HUNZA RIVER (DATA OF FIG. 4).

hysteresis, have consequences for the reliability of estimates of annual loads as discussed below.

It is apparent from Figure 4 that considerable discharge fluctuations occur in summer within the weekly or longer intervals at which sampling is carried out by WAPDA for determination of sediment concentrations. Discharge may be halved or doubled in a few days of cloudy or sunny weather respectively and the results of frequent sampling near the Dainyor gauging station during the IKP in July–August 1980 show that associated fluctuations in sediment concentration are considerable (Fig. 6). A total

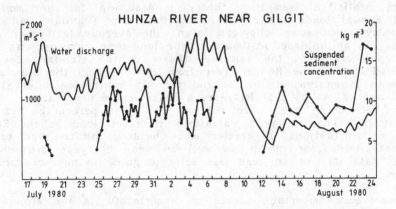

FIG. 6. SHORT-TERM VARIATION IN STREAMFLOW AND
SUSPENDED SEDIMENT CONCENTRATION IN THE HUNZA RIVER.

of sixty-four samples over five weeks during which discharge varied from 450 to 1890 $m^3 s^{-1}$ showed a six-fold range in suspended sediment concentration, from 3.1 to 17.3 kg m^{-3}. Although the streamflow and concentration curves in Figure 6 are broadly parallel the correlation between Q and c is not significant (0.05 level), partly because no samples were obtained during the recessions of 20 – 21 July and 10 – 12 August. Concentrations at 6h or 8h intervals in the first part of the sampling period do not show any clear relationship to the definite but subdued diurnal oscillation of streamflow. Inspection of Figure 6 does however suggest a tendency for concentrations to rise at the start of a period of day-on-day increase in discharge, then stabilise or tail off after two to three days (e.g., 25 - 28 July, 2 - 4 August, 12 - 15 August) so that peak concentrations lead peak discharges as at the seasonal timescale. This suggests flushing and depletion of sediment occur in the short term as well as seasonally and statistical analysis lends some support to this idea. Concentrations are significantly higher (t test, 0.05 level) on rising than on falling discharges, and 50% of the variance of concentration is explained by regression on its previous value and on the ratio of present to previous discharge. Some of the considerable residual scatter may be due to the superposition at Dainyor of out-of-phase fluctuations transmitted from major proglacial tributaries at different distances upstream.

Sediment loads estimated by the product of instantaneous discharge and concentration show a ten-fold range over this five-week period. This again has implications for the estimation of annual total load and long-term average erosion rate, to which we now turn.

RELIABILITY OF ESTIMATED LOADS

Sediment load estimates based on infrequent sampling are notoriously unreliable (18)(19). From a statistical point of view errors in the estimation of annual loads are inevitable given the scatter about the relationships used to estimate loads on days when sediment concentration is not measured; and short-term estimates of long-term mean rates of erosion are additionally uncertain because of the substantial year to year fluctuations already discussed.

The uncertainty about the long-term mean can be quantified using elementary statistical sampling theory. Assuming for the moment that estimated annual loads are correct, and that they fluctuate randomly and independently about some long-term mean, the average load over a sample of n years is an unbiased estimate of the long-term average with standard error $s\sqrt{n}$ where s is the sample estimate of the standard deviation of annual loads. For the Hunza River from 1966 to 1975 the standard error of the mean is equivalent to 12% and the 95% confidence interval for the long-term average is 63 ± 17 Mt/year, a considerable uncertainty. Estimates for other Karakoram rivers are subject to similar percentage errors. To the extent that year to year variations in river load in the Karakoram are due to fluctuations in glacier mass balance, which tend to include decade-scale trends or cycles as well as year by year differences, the 1966 - 75 data may be an even less reliable guide to past or future rates of erosion.

An even more important source of uncertainty is the standard flow-duration and rating-curve procedure used to estimate annual sediment loads in the absence of daily or more frequent sampling. The true load $L = k\int Qc\ dt$ (where Q is discharge, c is sediment concentration, and k is scale conversion factor) is approximated by $\hat{L} = k\Sigma Q\ f(Q)\ \hat{c}$ where $f(Q)$ is the frequency of different discharges Q and $\hat{c}$ is the estimated concentration at each discharge, predicted by a rating curve $\hat{c} = \hat{a}Q^b$ which is the antilog of the least-squares regression of log c on log Q for sample data such as those plotted in Figure 5 above.

There are three potential sources of error in this procedure:
(1) Summation over discrete time intervals and discharge bands under-
estimates the integrated load by missing transient peaks of Q and c.
(2) $\hat{c}$ differs quasi-randomly from the true c because of scatter about the
rating curve, leading to uncertainty in predicted concentrations and thus
loads.   (3) $\hat{c}$ may differ systematically from the true c if the latter is
not a simple power-law function of Q over the full range of discharge,
or it is but the coefficients are wrongly estimated because of sampling
variability or unrepresentative non-random sampling; this leads to systematic
bias in $\hat{c}$ and thus over or under-estimation of load.

Error (1) increases with Δt and with the short-term variability of Q;
it is unlikely to be serious in the large, fairly slowly-fluctuating rivers
of the Karakoram given the time and discharge intervals used by WAPDA.

Uncertainty (2) is substantial: the least-squares rating curve fitted
to the Hunza River for 1966 - 75 predicts more than twice, or less than
half, the measured concentration in 50% of the 221 samples to which the
regression was fitted and is presumably at least as unreliable for unsampled
days.   It is commonly assumed that this error tends to cancel out in the
summation for load but the empirical results of Walling and Webb (19)
show systematic underestimation and this can be shown to be statistically
expectable given the assumptions made in fitting the log-log rating curve.
Any such bias is really another form of error (3), additional to any bias
due to nonlinearity, unrepresentative sampling (e.g., too few high-flow
samples to fix the vital upper end of the rating curve), or the sampling
error inevitable when the rating is estimated from a small sample.   The
magnitude of these errors is difficult to judge.   Karakoram rating plots
are reasonably log-linear (although the seasonal hysteresis shows that
the scatter is non-random), and WAPDA's sampling routine ensures some
sampling of high flows.   However, the infrequency of sampling leads to
substantial uncertainty in the regression coefficients; for example, ratings
for the individual years 1966 - 1975 have slopes ranging from 1.12 to
1.56.   Even though none of these is significantly different (0.05 level)
from the overall 1966 - 75 rating, predictions based on the steepest or
flattest curves would differ considerably from those using the pooled
relationship.   On the other hand, it may be argued that yearly rather
than cumulative ratings should be used because the availability of sediment
may vary from year to year.

Some increase in reliability should be gained by using rating curves
that take account of effects other than that of Q alone on sediment
concentration, and thus have less and more nearly random scatter leading
to smaller errors in $\hat{c}$.   One approach is to take account of seasonal
hysteresis by fitting separate ratings to 'rising' and 'falling' seasons,
each year being divided at the days of minimum and maximum discharge
(or into predefined seasons); another is to predict c from not only Q but
also ΣQ, the cumulative discharge from the start of the year, as an index
of the quantity of sediment already transported and thus the depletion
of the supply.   A third way of reducing scatter is to neglect the winter
months when concentrations are quite variable in logarithmic terms but
so low that they constitute a negligible part of the annual load.

Computations summarised in Table 2 show that the scatter about the
simple rating curve for the Hunza River is substantially reduced by these
stratagems.   The residual standard deviation (in units of log c) is almost
halved by neglecting the winter months and taking account of seasonal
flushing and depletion, and the proportion of measured concentrations

Data	Sample size	exponent of Q ΣQ	residual st. dev. (log)	predictions accurate to factor of 2
All	221	1.28	0.40	50%
June-Sept.	122	1.10	0.32	66%
Rising season	98	0.94	0.30	76%
Falling season	123	1.38	0.26	
June-peak	63	0.53	0.26	77%
Peak-Sept.	59	1.51	0.22	
June-Sept.	122	1.25 -0.65	0.24	82%

Source: data from WAPDA Annual Reports, computations by author.

TABLE 2:   ALTERNATIVE RATING CURVES FOR SUSPENDED SEDIMENT. HUNZA RIVER
1966 - 75.

predicted to within a factor of two rises from 50% to 82%.

Different load estimates may be obtained using different ratings and it is reasonable to have most confidence in the estimate based on the rating with least scatter.   Recalculations summarised in Table 3 show that for

Method of estimation	suspended sediment load (10^6 tonnes)		
	1975	1969	1973
WAPDA Q rating	28	69	100
Jun-Sept. Q rating	37	66	93
Jun-peak, peak-Sept. Q ratings	43	67	90
Jun-Sept. Q, ΣQ rating	39	64	106

Loads are for June-September only (c. 99% of annual total).
WAPDA estimates are made using flow duration curves, others
using daily mean discharges. Alternative rating curves as
in Table 2.

TABLE 3:   ALTERNATIVE ESTIMATES OF SUSPENDED SEDIMENT LOAD
OF HUNZA RIVER IN YEARS OF LOW, AVERAGE, AND HIGH
RUNOFF.

the Hunza River differences due to method of calculation are smaller than those from year to year, and there is no evidence of any clear bias due to the choice of estimation method.   However, this is no guarantee of accuracy and one would have more confidence in load estimates for Karakoram rivers if the sampling frequency was higher.   Best of all would be to record continuously the signal of a photoextinction turbidity meter calibrated against frequent sampling: rating-curve errors are thus bypassed as the load integral can be evaluated directly as $\hat{L} = k \Sigma \bar{Q} \, \bar{c}$ for a suitably short time interval, say one hour.

DISSOLVED LOAD

Landscape denudation embraces loss of rock products in solution as well as in particulate form. The total dissolved load of the world's rivers is only about 1/5 the suspended load, and some solutes are of atmospheric not denudational origin, but dissolved load should clearly not be disregarded in any discussion of erosion rates.

The annual solute load of a river is defined, like the suspended load, as $L = k \int Qc \, dt$ where Q is discharge and c is concentration, in this case of total dissolved solids (TDS). This integral is again generally approximated as $\hat{L} = k \Sigma Q \, \hat{c}$ where $\hat{c}$ is the rating-curve estimate of TDS at the mean discharge Q over an interval Δt. Unlike suspended sediment, however, the TDS concentration tends to decrease with increasing discharge because of the dilution of chemically enriched baseflow by purer storm or meltwater runoff. This means the dissolved load is more uniformly distributed over time and reduces greatly the statistical errors involved in its estimation from sample data. An adequate estimate may be possible using a coarse time interval (e.g., monthly) for the summation and even the crude approximation $\hat{L} = k \hat{Q} \hat{C}$ may be no more in error than a more sophisticated estimate of the suspension load.

The electrical conductivity of the proglacial Batura torrent in July and August 1980 fluctuated diurnally and increased during recessions when glacier ablation was reduced or stopped in cold cloudy weather. Its overall range was equivalent on assumptions previously explained to TDS concentrations from 0.013 to 0.061 kg m^{-3} (13-61 mg/1). Electrical conductivity at the Dainyor gauging station on the Hunza River over the same period showed more subdued fluctuations of the same character with a range corresponding to 0.069 - 0.083 kg m^{-3} TDS. The higher mean at the downstream site indicates that the Hunza River is chemically enriched by nonglacial runoff, probably groundwater in particular.

Solute concentrations at different discharges at Dainyor in summer 1980 continue the inverse trend of the few available WAPDA data for the same site in February - June 1969. By fitting a rating curve to the combined 1969 and 1980 data and applying it to the 1966 - 75 mean monthly discharges an estimate of 1.2 Mt/year is obtained for the dissolved load of the Hunza River. The dilution effect means that the 87% of annual runoff during the four summer months carries only 46% of the dissolved load, compared to 99% of the suspended load. The annual load is equivalent to a gross yield of 90 t km^{-2} a^{-1}, about three times the world average (Meybeck (1)). Since most Karakoram runoff is meltwater with a very low concentration of atmospheric solutes, and evaporative magnification of runoff chemistry is inhibited by low temperatures in winter and scanty vegetation, it is likely that the chemical denudation rate is only slightly less than the gross solute yield. This supports the view that chemical weathering, traditionally regarded as slow in high mountains, is active up to the snowline in the Karakoram. Chemical denudation is however responsible for only about 2% of the total river load.

BEDLOAD

Transport of solid particles by rolling or bouncing along the river bed is conventionally regarded as unimportant compared to transport in suspension or solution. A bedload to suspension-load ratio of only 5% is commonly assumed and was adopted in the World Bank's feasibility study for the Tarbela reservoir according to WAPDA (9). This assumption is however based on very limited data, since bedload transport is difficult and costly to measure, and it may be seriously in error in the steep powerful rivers of mountain regions.

The velocity and depth measurements made during discharge gauging by WAPDA at Dainyor allow calculation of hydraulic properties of the Hunza River at different times of year and these can be substituted in engineering formulae for estimating river bedload. There are many such formulae but most are calibrated with data for sand transport in canals and laboratory flumes rather than pebble and boulder transport in mountain rivers.

A recent formula (20) with a plausible physical basis and good empirical backing from gravel-river measurements relates bedload transport i (as immersed weight per unit width per unit time) to stream power w, flow depth Y, and modal bedload size D (when $Y_0 = 0.1$ m and $D_0 = 1.1$ mm) as follows;

$$\frac{i}{0.1 \text{ kg m}^{-1}\text{s}^{-1}} = \left(\frac{w - w_0}{0.5 \text{ kg m}^{-1}\text{s}^{-1}}\right)^{3/2} \left[\frac{Y}{Y_0}\right]^{-2/3} \left[\frac{D}{D_0}\right]^{-1/2}$$

where w_0 is the critical power for incipient motion, estimated from Shields' criterion and the logarithmic velocity profile for rough-boundary turbulent flow. This specific rate can be converted to a daily load by multiplying by river width. Daily loads in dry weight have been estimated in this way using hydraulic data from thirty-three gaugings at Dainyor in 1979 - 80 (unpublished WAPDA data) and a river gradient estimated from levelling near, but not at the site.

The chief uncertainty in these calculations is the choice of modal bedload diameter D, which affects w_0 as well as appearing directly in Bagnold's equation. The river bed at Dainyor was not accessible or visible in July-August 1980 and sediment exposed on nearby point bars ranged from sand to boulders of about 0.5 m b-axis diameter. Flow resistance over a pebbly bed is controlled by particles coarser than the median but most transport is of the finer particles, down to the threshold size for suspension. To obtain upper and lower bounds on the bedload of the Hunza River, calculations have been made separately for D = 16 mm and D = 128 mm between which the modal bedload size is thought to lie. Estimates of daily transport for D = 16 mm range from 1.2 kt at minimum gauged winter flow to 64 - 83 kt at gauged flows in July and August, and about 130 kt in peak conditions. The annual load in an average year would be about 7 Mt, which is over 10% of the suspended load. If D = 128 mm the bedload transport rate is estimated as zero in winter low flow, 27 kt/day at summer maximum, and about 1 Mt over the whole year. This is less than 2% of the suspension load but still a very high absolute bedload.

CONCLUSIONS

The sediment load of the Hunza River is amongst the greatest in the world for a drainage basin of its size, predominantly because of the very high suspended sediment concentrations in summer meltwater discharges. Dissolved load and bedload are both high by absolute standards but are estimated to make up only 2% and 2 - 10% respectively of total load. Fluvial denudation in the western Karakoram is driven by, and broadly proportional to, glacier-melt runoff but the importance of sediment storage on valley floors is shown by the big downstream increase in suspended sediment concentrations and the seasonal and short-term behaviour of concentration in relation to discharge. More reliable estimates of annual load require more frequent sampling supplemented by continuous recording of turbidity.

ACKNOWLEDGEMENTS

I thank Stirling University and the Carnegie Trust for financial assistance, and David Collins, Brian Whalley, Andrew Goudie and Denys Brunsden who helped with data collection . I am grateful to Syed Amjad Hussain (Director, Surface Water Project) and Mohd. Baksh (Engineer, Gilgit Office) for access to unpublished WAPDA data.

REFERENCES

1) MEYBECK, M., (1976). 'Total mineral dissolved transport by world major rivers', Hydrol Sci. Bull. 21, pp 265 - 284.

2) WAPDA (1975), Pakistan Water and Power Development Authority Ann. Rept. v. 1, Lahore.

3) GOUDIE, A. S. and ten others, this volume, 'Geomorphology of the Hunza Valley', pp 359 - 410.

4) BATURA GLACIER INVESTIGATION GROUP, (1979). 'The Batura Glacier in the Karakoram Mountains and its variations', Scientia Sinica 22, pp 958 - 974.

5) HEWITT, K., (1968). 'The freeze-thaw environment of the Karakoram Himalaya', Canad. Geogr. 12, pp 85 - 98.

6) GOUDIE, A. S., this volume, 'Salt efflorescences and salt weathering in the Hunza Valley' , pp 607 - 615.

7) WHALLEY, W. B., MCGREEVY, J. P. and FERGUSON, R. I., this volume, 'Rock temperature observations and chemical weathering in the Hunza region' , pp 616 - 633.

8) WAPDA (1976). 'Sediment appraisal of West Pakistan rivers 1966 - 75', Pakistan Water and Power Development Authority, Lahore.

9) WAPDA (1979). 'Tarbela dam project: sedimentation base survey September - October 1979', Pakistan Water and Power Development Authority, Lahore.

10) MERCER, J. H., (1975). 'Glaciers of the Karakoram'. In Field, W., (ed.) Inventory of world glaciers, pp 377 - 409.

11) COLLINS, D. N., (1979). 'Sediment concentration in melt waters as an indicator of erosion processes beneath an Alpine glacier', J. Glaciol. 23, pp 247 - 257.

12) CURTIS, W. F., CULBERTSON, J. K. and CHASE, E. B., (1973). 'Fluvial sediment discharge to the oceans from the conterminous United States', US Geol. Surv. Circular 670, 17 pp.

13) AL-JABBARI, M. H., MCMANUS, J. and AL-ANSARI, N. A., (1980), 'Sediment and solute discharge into the Tay Estuary from the river system', Proc. Roy. Soc. Edinburgh 78B, pp 15 - 32.

14) GRIFFITHS, G. A., (1979). 'High sediment yields from major rivers of the western Southern Alps, New Zealand', Nature 282, pp 61 - 63.

15) MOLNAR, P. and TAPPONNIER, P., (1975). 'Cenozoic tectonics of Asia: effects of a continental collision', Science 189, pp 419 - 426.

16) YOUNG, G. J., (1977). 'Relations between mass-balance and meteorological variables on Peyto Glacier, Alberta, 1967/74', Zeitschr. für Gletscherkunde und Glazialgeologie 13, pp 11 - 25.

17) GREGORY, K. J. and WALLING, D. E., (1973). 'Drainage basin form and process'. Arnold, London.

18) WALLING, D. E., (1977). 'Limitation of the rating curve technique for estimating suspended sediment loads, with particular reference to British rivers', Intern. Assoc. Sci. Hydrol. publ. 122, pp 34 - 48.

19) WALLING, D. E. and WEBB, B. W., (1981). 'The reliability of suspended sediment load data', Internat. Assoc. Hydrol. Sci. publ. 133, pp 177 - 194.

20) BAGNOLD, R. A., (1980). 'An empirical correlation of bedload transport rates in flumes and natural rivers', Proc. Roy. Soc. Lond. A372, pp 453 - 473.

A preliminary study of ancient trees in the
Hunza Valley
and their dendroclimatic potential

R. Bilham*
G. B. Pant
G. C. Jacoby

* Enquiries regarding the paper should be directed to R. Bilham, Lamont-Doherty Geological Observatory, U.S.A.

ABSTRACT

Trees of great age are found on the high mountain slopes bordering the Hunza Valley and tributary valleys in the Karakoram Mountains. The survival of ancient trees is partly attributable to the enlightened local practice of culling branches for fuel rather than felling entire trees. We have subjected two specimens of juniper found at elevations of 3,800 m and 4,150 m to ring analysis and determined maximum ring counts of 1200 and 479, respectively. The detection of incomplete rings in different radial sections from both specimens suggests that some rings may be missing and that the age of the trees may be more than their ring counts indicate. The lower and older of the two specimens consisted of a dead stump of unknown absolute age and an attempt to crossdate the specimen with the higher, living specimen was unsuccessful.

The two specimens are unlikely to be the oldest or most climatically sensitive trees in the region. We conclude that there is a strong possibility for deriving climatic information for the last several centuries from the ancient trees of the region. A knowledge of past climate has important consequences in the study of glacial movements, erosion, and agriculture in the Hunza Valley.

INTRODUCTION

Villagers in the Hunza Valley use wood for cooking and heating. The wood comes occasionally from local trees growing in the valley floor but more usually the wood is dragged down the steep scree slopes from high elevations bordering the valley and adjacent valleys. To ensure that their future source of fuel is not jeopardised, wood gatherers never cut trees down unless they are dead. Instead a branch is severed from a healthy tree so that it continues to grow. To discourage extravagant use of this renewable fuel resource, the villagers pay a modest tax that is related to the weight of the wood they use. This enlightened practice results in the survival of trees more than a thousand years old.

The routes taken by villagers to drag branches from the high slopes can be recognised as whitened paths amid the dull grey patina of tumbled rocks forming the mountainside. The paths descend from elevations between 3,000 and 4,000 m (10,000 to 13,000 feet) in straight lines to the cultivated terraces at elevations of about 2,100 and 2,700 m (7,000 and 9,000 feet). For much of the descent, the paths are inclined at 35 - 40°.

The trees in the valley floor are mostly poplars (Populus) or trees bearing nuts or fruits, but the valley sides, where water is less plentiful, tend to be sparsely populated by species of juniper (e.g., Juniperus macropoda (Boissier)) and cedar (e.g., Cedrus deodara (Roxburgh) G. Don). The trees growing at high altitude are stunted and seldom attain great height or girth. Many of these trees depend on winter snowfall and occasional summer rains for their water supply since they may be perched on rocky slopes far from any perennial moisture source.

The major weather systems in the region are the small depressions known as 'western disturbances' travelling east from the Mediterranean along the polar front. These disturbances, more frequent during winter, produce snow and rains in the northwestern parts of the subcontinent of India and Pakistan, Afghanistan and Baluchistan. These disturbances are also the main source of snow on the mountains of the Hindukush and Karakoram areas. The orographic effects are dominant on the rainfall and snow accumulation rate. As in many semi-arid to arid regions, there is much local variation in precipitation. Retention of subsurface moisture is limited by poor soil development and precipitous slopes.

It is likely that the limiting factors for tree growth on the steep slopes are moisture and wind stress. At higher elevations, cold temperatures could also influence tree growth. It is of interest to study the growth rings of these trees in order to establish whether variations in growth rate have occurred, since these may reveal long-term variations in climate.

LOCATIONS OF TREES STUDIED

Two samples were obtained of a species that we tentatively identify as a juniper (Juniperus macropoda (Boissier)). The identification is not absolutely certain because there is some question whether a specimen of twig used in our identification is from the same tree. Our tentative identification is also based on photographs and wood anatomy. The specimens are from elevations of 3,800 m and 4,150 m (12,500 and 13,000 feet). The trees were located on the side of a hill named Sir Sar above the village of Mur Khun about 25 km north of the town of Hunza (Figures 1, 2 and 3).

The samples were chosen to provide an approximate estimate of the maximum age of trees that might be suitable for future detailed study. The sample collected from 4,150 m (HJ-1) was one of the highest trees found on Sir Sar. The tree consisted of several trunks hugging the ground in the shelter of south-facing rocks. The tree was about 50 cm tall and the maximum girth of any of its trunks was less than 20 cm. The specimen (HJ-1) is a low branch from a living trunk. The tree was located approximately 10 m to the west of the N/S ridge leading to the Sir Sar summit.

The 3,800 m sample (HJ-2) was one of several stumps amid a healthy group of about 50 trees 2 - 3 m high found on a west-facing slope approximately 150 m below a saddle leading to the Sir Sar ridge. We would have sampled one of the larger living trees if a coring device had been available. The stump had approximately the same diameter as the living

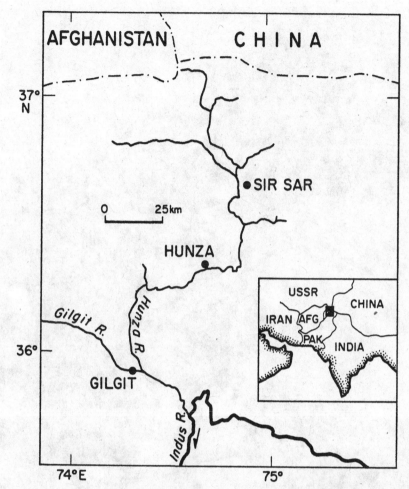

Fig. 1. The two specimens studied were found south
of the Sir Sar summit (4428 m)

members of the group. The minimum age of the tree was estimated on the
basis of counts of annual growth-rings in the laboratory and it is found
to be more than 1200 years old. At this site, the downslope creep of
surface slates and shales rotates trees downward during growth. As a
consequence, new growth occurs on the uphill side of the tree. In this
specimen (HJ-2), the original centre is on the lower edge and is not well
preserved, so the actual count of over 1200 is an underestimation of the
age in years. The two samples were sectioned and highly-polished prior
to microscopic examination.

TREE RING ANALYSIS

In conifers not subjected to tropical climates, the larger, thinly-walled
cells of spring growth and the narrow, thickly-walled cells of later growth
jointly constitute an annual growth-ring. As a tree grows older, the width
of each growth-ring generally diminishes as a function of age.

FIG. 2. A 1.2 M HIGH JUNIPER ON SOUTH FACING SLOPE.
Note lack of soil development. Altitude approximately 3,800 m.
Limbs severed by villagers for firewood have approximately 600
rings. Located near sample HJ2.

FIG. 3. A 70 CM HIGH MATURE JUNIPER ON SIR SAR RIDGE. Altitude approximately 4100 m. location close to sample HJ1.

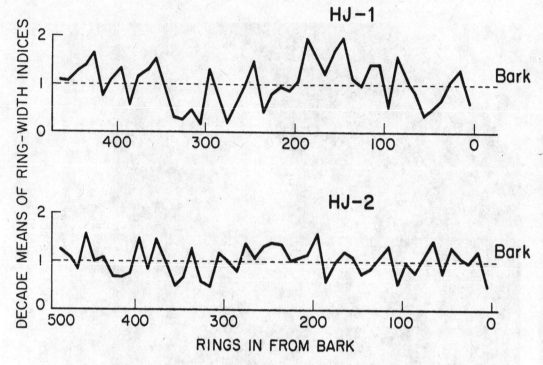

Fig. 4. Decade means of ring-width indices for the two specimens. The HJ-1 plot is one radius. HJ-2 is the plot of the mean value from three radii. Only the most recent 500 rings from HJ-2 are shown.

Superimposed on this smooth linear or power-law decay in ring-width, there are often significant variations in the widths of individual rings or groups of rings that reflect environmental conditions favourable or detrimental to growth. One of the most important variables determining growth rates is the effect of varying climatic conditions. Prolonged low temperatures, excessively high temperatures, or low annual rainfall will generally inhibit tree growth, whereas equitable temperatures and adequate moisture will stimulate growth. If the growth rings can be crossdated and the date of the outermost ring determined, the exact years of growth can be ascertained. Then the variations in width can be compared to growth-limiting factors.

Before deriving conclusions concerning climatic variations, it is important to establish that other environmental factors have not influenced variations in ring growth. It is generally assumed that the supply of mineral and organic nutrients to a tree does not vary significantly during tree life. Disease, insect-invasion, and fire can affect growth over a wide region. Locally, the competition for light from other trees and vegetation may restrict growth, although this does not appear to be a significant effect in the region described since the spacing of trees is several times their height. The selective harvesting of individual branches for fuel in the Hunza Valley may stimulate or suppress growth in following years. In some areas the effects of accelerated downslope creep may disturb the root structure of a group of trees adversely.

Local non-climatic effects can be removed from a tree-ring time series if a large population of trees from a wide area is studied. In this preliminary study, we discuss the potential for using less-accessible trees in future climatological studies. Of fundamental concern in any dendrochronological study is the relationship between ring number and year. Within a single radius or even an entire cross-section, there can be locally-absent rings, false rings or multiple rings for a year. The first is a problem if the trees are extremely physically stressed or have poor circuit uniformity and the second and last manifestations are commonly problems in tropical areas. If complete cross-sections of trees are available (as they are in this study), it is easier to identify locally absent rings; but if only cores are available, it is more difficult to cross-correlate (cross-date) the sequence of ring-widths from each radius in order to determine the continuity of growth-rings. A further property of the tree ring sequence that is important for developing and extending the chronology backwards in time is the ability to crossdate the ring sequence from a living tree with that from another living tree or perhaps a dead tree or piece of local timber that grew near it for part of its life.

In the samples studied, the rings appear to be annual in nature but there are problems with circuit uniformity. This is a common characteristic in juniper. Some rings, especially very narrow ones, pinch out as they are traced around the sections. Sample 1 (HJ-1) belongs to a living tree, whereas the sample 2 (HJ-2) appears to have been dead for many years. Both samples are with the bark intact, therefore the most recent ring could be counted, although the age of the last year of growth is unknown for HJ-2. Ring widths were measured to an accuracy of 0.01 mm. The ring-width data were standardized into ring-width indices using a cubic spline technique. For these two specimens, the cubic spline procedure was set as a 500-year filter, removing 50% of the variance with a period of 500 years or greater. Standardization divides the measured ring width by the standard curve value to produce a dimensionless ring-width index for each ring. It is commonly used in dendroclimatology in order to statistically compare ring-width data from trees of differing growth rates. The mean ring-width index of a standardized tree-ring data set is 1.0.

HJ-1 has a lobate growth and extremely poor circuit uniformity. The radius which appears most complete has a count of 479 rings and the outer ring is 1980. Because of the poor circuit uniformity and locally-absent rings, different radii from this branch do not crossdate. Branch growth is often less regular than growth of the main trunk. The average ring-width is 0.28 mm. The standard deviation of the ring-width indices of HJ-1 is 0.60. The serial correlation is 0.74.

HJ-2 has much better circuit uniformity and three radii could be cross-dated. Ring-widths were measured along the three radii for the most recent 500 years. The average ring-width is 0.23 mm and the standard deviation of the ring-width index chronology based on all three radii is 0.52. The serial correlation is 0.49. Through crossdating, we have determined that there are 4.6%. 2% and 2% missing rings on radii 1, 2 and 3 respectively. These percentages are large relative to most specimens used in dendroclimatic studies. Therefore missing or locally absent rings are a serious problem in specimens of this species from this area.

The decade values of ring-width indices (averaged for ten consecutive rings) are plotted in Figure 4 for each specimen. HJ-2 is a mean value for the three radii. It must be emphasized that these are decade values of rings and may not be decades of years, especially for HJ-1. Also, except for knowing the outer ring of HJ-1 is 1980, they are not dated absolutely nor with respect to each other.

The ring-width variation revealed by microscopic examination and the relatively high standard deviation have the same characteristics as climatically-sensitive conifers from the southwestern U.S. In the low precipitation and high altitude area of the Hunza Valley, one would expect to see trees whose growth is limited by climatic factors. The age of over 1,200 years shows the longevity is great enough to make dendroclimatological study very worthwhile. Another promising aspect is the possibility of long preservation of dead wood material. In a similar dry, high-altitude locale in the U.S., dead wood from Bristlecone Pine (Pinus longaeva (D. K. Bailey)) endures in a well-preserved state for thousands of years.

CONCLUSIONS

The study of the climate of the Hunza Valley in the last 1000 years has important consequences for the people of Hunza. Long-term fluctuations in climate have relevance to the agriculture of the region and to the relationships between precipitation, glacial advance and erosion rates in the valley floor. It is unlikely that in selecting the two samples described we have chanced upon the oldest trees in the region, or upon those most sensitive to climate.

There are problems, such as the lack of circuit uniformity in juniper and the practice of cutting branches from accessible trees, that result in unknown growth-rate disturbances. However, given the regional climate, good ring-width variation, the longevity of living trees, and the preservation of dead material, the dendroclimatic potential of this region must be considered very good. More samples from trees growing in locations that are inaccessible to wood gatherers could provide material to be absolutely dated and developed into a long, climatically sensitive chronology. Some technical climbing is indicated in order to reach undisturbed tree sites.

A combined study of dendrochronology and terrace sediment deposition upstream from each of the subsidiary glaciated valleys could yield a history of glacial advance and recession in the last several thousand years and its relationship to climatic variation. Tree studies in the region have potential use also in dating the recurrence times of major slumps on hill slopes.

ACKNOWLEDGEMENTS

We wish to thank Nigel Atkinson of the International Karakoram Project 1980 Survey team who reclimbed Sir Sar with a saw in order to collect the two samples discussed. Edward Cook, Chris Scholz, and David Simpson reviewed the manuscript. The work was made possible by an NSF travel award INT8016497 and the tree-ring analysis at the Tree Ring Laboratory of the Lamont-Doherty Geological Observatory was supported by the National Science Foundation under Grant No. ATM79-19223. Lamont Doherty Geological Observatory, Contribution No. 3184.

Salt efflorescences and salt weathering in the Hunza Valley, Karakoram mountains, Pakistan

A. S. Goudie – Department of Geology, University of Oxford

ABSTRACT

In the Karakoram Mountains, Pakistan, rainfall levels are low in the village, and under conditions of high evaporation salt efflorescences form, composed largely of magnesium sulphate (hexahydrite) and gypsum. Moraines and rock fall debris of the twentieth century have been severely weathered by salt attack.

PREVIOUS STUDIES ON SALT WEATHERING

Since the reviews of Wellmann and Wilson (1) and Evans (2) there has been renewed interest in the way in which mechanical disintegration of rock can be caused by the hydration and crystallisation of salts (3). While such activity may well be common in desert areas (4), it also occurs in such diverse environments as polluted cities (5), coastal locations (6) and polar regions (7). In addition to rocks in natural outcrops, building materials may also be adversely affected (see, for example, (8)(9)(10)). Some useful recent case studies of salt weathering are listed in Table 1.

Location	References	Location	References
Bahrain	(3) (11)	Indus Valley	(8)
Suez	(3) (10)	Saudi Arabia	(13)
South Australia	(12)	Baffin Island	(14)
Death Valley	(4)		

TABLE 1: RECENT CASE STUDIES OF SALT WEATHERING.

There are relatively few data on salt weathering and the chemistry of salt efflorescences in moderate altitude, dry mountainous areas such as the lower portions of the Karakorams. In the Karakorams themselves, Workman (15) noted sulphate impregnations and the physiological discomforts they caused up the nala at Hasanabad, while Hewitt (16) noted salt efflorescences in the subnival zone below around 4,500 m and stressed the role of salt weathering in creating tafonis and slope instability.

Salt weathering is an important geomorphological process, creating fine grained material suitable for deflation and loess formation (17), causing fan and lacustrine sediments to be fractured (4), and being involved in the formation of some tafoni (12).

THE AREA

The Hunza is one of the major tributaries of the Indus and passes through some of the steepest terrain on earth. It is fed by large glaciated catchments scoured into the ranges of the Karakoram Mountains which include Rakaposhi (7788 m).

The precipitation within the valley is low (see data for Gilgit in Table 2). This, combined with the high temperatures, creates arid

Month	Mean max.		Mean min.		Rainfall	
	$^{\circ}F$	$^{\circ}C$	$^{\circ}F$	$^{\circ}C$	in.	mm.
January	46.3	7.9	32.0	0.0	0.25	6.35
February	52.5	11.4	36.9	2.7	0.26	6.60
March	62.4	16.9	45.1	7.3	0.80	20.32
April	72.2	22.3	52.7	11.5	0.96	24.38
May	82.3	27.9	59.4	15.2	0.80	20.32
June	91.3	32.9	66.3	19.1	0.37	9.40
July	95.6	35.3	71.7	22.1	0.39	9.91
August	93.5	34.2	71.0	21.7	0.55	13.97
September	85.2	29.6	63.0	17.2	0.40	10.16
October	73.7	23.2	52.2	11.2	0.24	6.10
November	62.2	16.8	41.3	5.2	0.05	1.27
December	49.9	9.9	33.8	1.6	0.11	2.80
Year	72.2	22.3	52.1	11.2	5.18	131.57

TABLE 2: TEMPERATURE AND RAINFALL DATA FOR GILGIT
(altitude 1490 m)

conditions that favour salt accumulation. There is a very wide range of lithologies upon which the salt efflorescences occur, including granites, gneiss, schist, marble, limestone, shale, sandstone and volcanics.

EVIDENCE FOR SALT WEATHERING

White salt efflorescences are of frequent occurrence in the Hunza valley and its tributaries, and where they occur rock disintegration is evident in the form of flaking, splitting and tafoni development. For example, granite boulders and outcrops in the vicinity of a moraine-dammed lake Borrit Gil, (Jheel) show extensive cavernous weathering, while some man-made structures, including field walls (particularly near Hindi) and new concrete structures on the Karakoram Highway, are showing signs of efflorescence-related disintegration.

Some of the most interesting examples of the speed of disintegration are provided by some of the valleys that have been affected by twentieth century glacial retreat. For instance, in the Hasanabad valley some of the moraine and rock fall material postdating the retreat of the Hasanabad glacier as recorded between the 1920s and the 1950s by Visser (18) and Paffen et al. (19) has been rendered crumbly.

More impressively, in the Minapin valley, efflorescences derived from the evaporation of melt water seeping from the Glacier, have led to wide-spread disintegration of morainic boulders that have been deposited on a rock surface that was exposed at sometime between 1954 and 1961 (see Fig. 1 and ref (20)).

Analysis of rock hardness using the Schmidt Hammer (21) indicates a substantial difference in rock hardness between relatively unweathered till boulders on the present Minapin glacier surface and those laid down in the mid- to late- 1950s (Table 3) and subsequently softened by salt weathering. The weathered rocks have a mean value of around 36 (slates) and 33 (volcanics and schists) while the unweathered materials for the same rock categories both have values of 47.

THE NATURE OF THE SALT

Experimental work has shown that different salts have different capabilities of causing rock disintegration (22)(23)(24). In particular it has been found that sodium carbonate, sodium sulphate, and magnesium sulphate are highly effective (Table 4). Given the apparent rapidity with which this process appears to operate in the Karakorams it is important to know how much salt causes rock disintegration and the nature of the salt or salts involved.

An approach to the question of determining the quantity of salt in weathered rocks was to collect rock samples from areas where disintegration was prevalent and to determine the electrical conductivity of a 5:1 mix of deionised H_2O and the crushed rock. For comparative purposes rocks were collected from situations (such as quarries or newly blasted road sections) where little or no weathering was evident and the same method was employed for electrical conductivity determination. For each site at least five different rock samples were collected and the mean conductivity determined (Table 5).

The main result of this investigation is that in those areas where the rock is plainly highly weathered, salt contents (as represented by conductivity values) are on average between 4 and 5 times higher than in fresh, unweathered rock.

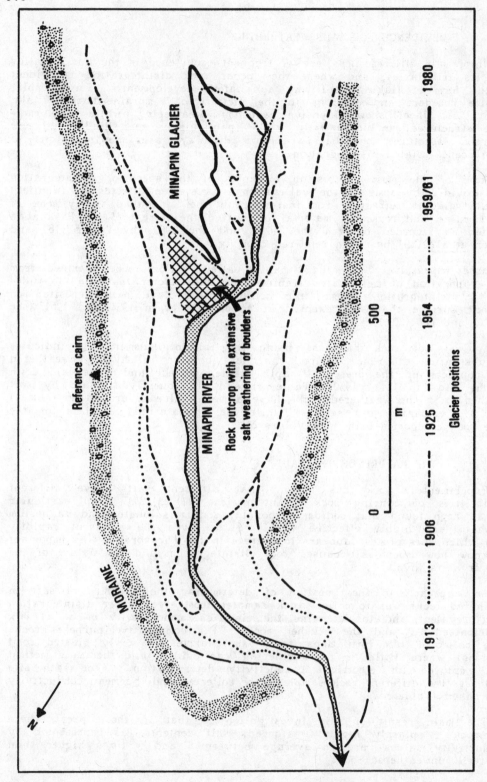

FIG. 1. THE LOCATION OF SEVERELY WEATHERED TILL ON THE MARGINS OF THE MINAPIN GLACIER.

TABLE 3: SCHMIDT HAMMER READINGS ON MINAPIN TILLS.

a) Unweathered till boulders on present day Minapin Glacier		b) Weathered till boulders from post 1954 moraine	
Slates	Volcanics and Schists	Slates	Volcanics and Schists
42.6	47.1	27.2	35.6
48.0	44.7	26.9	28.5
45.0	52.2	39.2	28.5
45.5	43.0	32.8	37.4
49.5	49.4	36.5	38.5
49.9	$\bar{x} = 47.28$	46.8	26.8
50.4		41.8	$\bar{x} = 32.55$
41.7		42.2	
49.5		41.1	
50.0		38.5	
48.7		26.2	
$\bar{x} = 47.35$		39.9	
		37.6	
		37.7	
		32.5	
		36.8	
		40.6	
		36.8	
		28.1	
		$\bar{x} = 36.27$	

TABLE 4: SALT WEATHERING EXPERIMENTS: EFFECTIVENESS OF DIFFERENT SALTS.

	Granite Kwaad (25)	Sandstone Goudie et al (22)	Sandstone Goudie (23)	Chalk Goudie (23)
1	Na_2SO_4	Na_2SO_4	Na_2SO_4	Na_2SO_4
2	Na_2CO_3	$MgSO_4$	$MgSO_4$	Na_2CO_3
3	$MgSO_4$	$CaCl_2$	$CaCl_2$	$NaHCO_3$
4	$NaCl$	Na_2CO_3	Na_sCO_3	$CaCl_2$
5	$CaSO_4$	$NaCl$	$NaNO_3$	$MgSO_4$
6				$CaSO_4$
7				$NaCl$

Numbers 1 – 7 in order of effectiveness.

(a) Weathered Rocks				(b) Unweathered Rocks			
Site	Rock Type	pH	K	Site	Rock Type	pH	K
AGKKW				AGKKW			
1	Quartzite	8.05	2450	11	Marble	8.05	188
2	Schist	7.85	620	12	Granodiorite	7.90	183
3	Slate	8.00	2890	13	Marble	8.40	240
4	Dacite	8.40	1310	14	Gneiss	8.15	146
5	Schist	7.95	1500	15	Gneiss	8.00	197
6	Slate	7.80	1720	16	Schist	7.95	206
7	Slate	7.80	1210	17	Slate	8.05	260
8	Slate	7.70	450	18	Marble	8.40	235
9	Schist	7.75	221	19	Granite	7.35	90
10	Granite	7.95	320	20	Granodiorite	7.60	124
Mean –	–	7.93	1269	Mean –	–	8.06	270

TABLE 5: ELECTRICAL CONDUCTIVITY K VALUES (μmho/cm/25°C) OF
ROCKS FROM THE HUNZA VALLEY.

The quality of the efflorescences was determined by scraping them carefully from rock exposures to minimise the amount of contamination with rock minerals. Ten samples, believed to be representative of the efflorescences found on different rock types between Dainyor Bridge (near Gilgit) and Pasu along the Hunza (Table 6) were subjected to analysis by X-Ray Diffraction. It is clear from these analyses that two main soluble salts are involved in the weathering process: Hexahydrite ($MgSO_4 . 6H_2O$) and gypsum ($Ca_2SO_4 . 2H_2O$). Of these two magnesium sulphate appears to have been generally dominant. This was also found to be the prime constituent of the waters in the enclosed lake at Borrit Gil.

That magnesium sulphate should have proved to be effective is not entirely surprising. It has a relatively high solubility in water (29.30% at 35 $^\circ$C) which means that when crystallisation takes place from solution (whether by evaporation or temperature change) a large volume of salt is available to crystallise out. Secondly, when $MgSO_4$ hydrates to $MgSO_4 . 7H_2O$, a marked volume change of 223% occurs which could cause rock fracture. Moreover experimental work shows that it is one of the more effective common salts (see Table 4) and its effects on the disintegration of certain structures - notably Cleopatra's Needles, the House of Commons and Durham Cathedral - have been noted (2). Indeed the ASTM sulphate soundness test (5) shows more attack by $MgSO_4$ than by Na_2SO_4, and Winkler attributes this to the former's faster rate of hydration.

The presence of hexahydrite as an efflorescence is by no means exceptional. It has previously been recorded from the Ross Sea area of Antarctica (26), the interior of Antarctica (27), underground chambers and elsewhere in Kansas, USA (28)(29), dolomite outcrops in Ohio (30) and shale exposures in California (31).

Sample	Location	Lithology on which Efflorescence Occurred	Efflorescence Mineralogy
AG1KP 1	Hunza Bridge, Hindi	Shale	Hexahydrite + gypsum (?)
AG1KP 3	13.7 km W of Aliabad on KKH	Shale	Hexahydrite + gypsum (?)
AG1KP 5	Minapin Glacier	Till (1950's)	Hexahydrite (dominant) + gypsum (?)
AG1KP 7	Minapin Glacier	Till (Present)	Hexahydrite + gypsum
AG1KP 9	Ghulmet Valley	Shale	Hexahydrite
AG1KP 11	Borrit Gil	Granodiorite	Hexahydrite + unknown
AG1KP 13	E of Sarrat on KKH	Granodiorite	Hexahydrite + gypsum
AG1KP 15	4 km E of Dainyor Bridge on KKH	Schist	Gypsum
AG1KP 17	Nilt on KKH	Till	Hexahydrite (dominant) + gypsum
AG1KP 19	25 km W of Pisan on KKH	Slate	Gypsum (dominant) + Hexahydrite

TABLE 6: LOCATIONS AND CHARACTERISTICS OF HUNZA EFFLORESCENCES.

ACKNOWLEDGEMENTS

This work was undertaken during the International Karakoram Project 1980 - a collaborative venture between the Royal Geographical Society, the Chinese Academy of Sciences, and the Pakistan Ministry of Science and Technology. Funding for this part of the Geomorphology programme was provided by the National Geographic Society, the Royal Society and the University of Oxford. The XRD analyses were kindly carried out by the Oxford Department of Geology and Mineralogy (Dr E J W Whittaker), and the analysis of the water of Borrit Gil by M Hogan-Guzkowska.

REFERENCES

1) WELLMANN, H. W. and WILSON, A. T., (1965). 'Salt weathering, a neglected geological erosive agent in coastal and arid environments', Nature 205, pp 1097 - 1098.

2) EVANS, I. S., (1970). 'Salt crystallization and rock weathering: a review', Revue Géomorphologie dynamique 19, pp 153 - 177.

3) COOKE, R. U., (1981). 'Salt Weathering in deserts', Proceedings Geologists' Association 92, pp 1 - 16.

4) GOUDIE, A. S. and DAY, M. J., (1981). 'The disintegration of fan sediments in Death Valley, California, by salt weathering', Physical Geography 1, pp 126 - 137.

5) WINKLER, F. M., (1975). Stone: properties, durability in man's environment. Springer Verlag, 2nd. ed. (Wien and New York), 230 pp.

6) TRICART, J., (1960). 'Expériences de désagrégation de roches granitiques par la cristallisation du sel marin', Zeitschrifte für Geomorphologie Supplementband 1, pp 239 - 240.

7) PREBBLE, M. M., (1967). 'Cavernous weathering in the Taylor dry valley, Victoria Land, Antarctica', Nature 216, pp 1194 - 1195.

8) GOUDIE, A. S., (1977). 'Sodium sulphate weathering and the disintegration of Mohenjo-daro, Pakistan', Earth Surface Processes 2, pp 75 - 86.

9) BLACKBURN, G. and HUTTON, J. T., (1980). 'Soil Conditions and the occurrence of salt damp in buildings of Metropolitan Adelaide', Australian Geographer 14, pp 360 - 365.

10) JONES, D. K. C., (1980). 'British applied geomorphology: an appraisal', Zeitschrift für Geomorphologie 36, pp 48 - 73.

11) DOORNKAMP, J. C., BRUNSDEN, D. and JONES, D. K. C., (1980). Geology, geomorphology and pedology of Bahrain. Geo. Abstracts, Norwich, 443 pp.

12) BRADLEY, W. C., HUTTON, J. T. and TWIDALE, C. R., (1978). 'Role of salts in development of granitic tafoni, South Australia', Journal of Geology 86, pp 647 - 654.

13) CHAPMAN, R. W., (1980). 'Salt weathering by sodium chloride in the Saudi Arabian Desert', American Journal of Science 280, pp 116 - 129.

14) WATTS, S. H., (1979). 'Some observations on rock weathering, Cumberland Peninsula, Baffin Island', Canadian Journal of Earth Sciences 16, pp 977 - 983.

15) WORKMAN, W. H., (1910). 'The tongue of the Hasanabad Glacier in 1908', Geographical Journal 36, pp 194 - 196.

16) HEWITT, K., (1969). 'Geomorphology of the mountain regions of the Upper Indus basin', University of London, Unpublished PhD dissertation, 2 vols., 310 and 121 pp.

17) GOUDIE, A. S., COOKE, R. U. and DOORNKAMP, J. C., (1979). 'The formation of silt from quartz dune sand by salt-weathering processes in deserts', Journal of Arid Environments 2, pp 105 - 112.

18) VISSER, P. C., (1928). 'Von den Gletschern am Obersten Indus', Zeitschrift für Gletscharkunde 16, pp 169 - 229.

19) PAFFEN, K. H., PILLEWIZER, W. and SCHNEIDER, H-J., (1956). 'Forschungen in Hunza-Karakorum', Erdkunde 10, pp 1 - 33.

20) GOUDIE, A. S., JONES, D. K. C. and BRUNSDEN, D., (1982). 'Recent fluctuations in some glaciers of the western Karakoram mountains Hunza, Pakistan', this volume, pp 411 - 455.

21) DAY, M. J. and GOUDIE, A. S., (1977). 'Field assessment of rock hardness using the Schmidt Test Hammer', Technical Bulletin, British Geomorphological Research Group 18, pp 19 - 29.

22) GOUDIE, A. S., COOKE, R. U. and EVANS, I. S., (1970). 'Experimental investigation of rock weathering by salts', Area 4, pp 42 - 48.

23) GOUDIE, A. S., (1974). 'Further experimental investigation of rock weathering by salt and other mechanical processes', Zeitschrift für Geomorphologie Supplementband 21, pp 1 - 12.

24) SPERLING, C. H. B. and COOKE, R. U., (1980). 'Salt weathering in Arid Environments. II. Laboratory Studies', Papers in Geography Bedford College, London, 9, 53 pp.

25) KWAAD, F. J. and P. M., (1970). 'Experiments on the granular disintegration of granite by salt action, University of Amersterdam', Fysisch Geografisch Laboratorium Publicatie 16, pp 67 - 80.

26) CLARIDGE, G. G. C. and CAMPBELL, I. B., (1968). 'Origin of nitrate deposits', Nature 217, pp 428 - 430.

27) TASCH, P. and ANGINO, E. E., (1968). 'Sulphate and carbonate salt efflorescences from the Antarctica interior', Antarctica Journal 3, pp 239 - 241.

28) TIEN, P. L. and WAUGH, T. C., (1970). 'Epsomite and hexahydrite from an underground storage area, Atchinson, Kansas', Bulletin, Kansas Geological Survey 199, pp 3 - 8.

29) TIEN, P. L., (1971). 'Ubiquity of hexahydrite in Kansas', Bulletin, Kansas State Geological Survey 202, pp 3 - 4.

30) FOSTER, W. R. and HOOVER, K. V., (1963). 'Hexahydrite ($MgSO_4 . 6H_2O$) as an efflorescence of some Ohio dolomites', Ohio Journal of Science 63, pp 152 - 158.

31) MURATA, K. J., (1977). 'Occurrence of bloedite and related minerals in marine shale of Diablo and Templor Ranges, California', Journal of Research United States Geological Survey 5, pp 637 - 642.

Rock temperature observations and chemical weathering in the Hunza region, Karakoram: Preliminary data

W.B. Whalley, J.P. McGreevy, R. I. Ferguson

Enquiries regarding this paper should be directed to Dr. W.B. Whalley, Dept. of Geography
Queen's University, Belfast

ABSTRACT

Air, rock surface, 5 cm deep and crack temperatures were recorded around a variety of rock types at various altitudes in the Hunza Valley in July/ August 1980. Clear sky and cloudy conditions were compared. Under clear skies rock surfaces can exceed air temperatures by as much as 24°C with actual rock temperatures around 40°C even at 4250 m. Cloudy conditions showed a smaller excess. Rock surface minima reached nearly 5°C below air temperatures under clear skies but were about the same under cloudy skies. Rock cooling rates vary considerably according to cloud cover and rock colour. At 4400 m the effects of chemical weathering are significant in rock breakdown. Desert varnish is rare above 3000 m.

INTRODUCTION

Certain geomorphologists, e.g. (1), have attempted to derive associations between weathering processes and intensity and climatic variables, essentially temperature and precipitation, to establish morphogenetic regions. Other workers, e.g. (2), have adopted a similar approach as regards variation in erosion (or denudation) rates in different global climatic regions. Such large scale schemes naturally contain local anomalies for a given region. This is especially true with mountainous areas. Not only are climatic controls on rock weathering altitudinally, rather than spatially, variable but there is a considerable lack of data on the exact nature of mountain climates. Climatic controls on rock weathering in mountains are particularly poorly understood. This paper presents some preliminary data on summer rock temperature regimes in the Hunza region of the Karakoram. Such data provide a useful basis for laboratory simulations of weathering processes.

Previous climatic data from the Karakoram and Himalayas is restricted to studies by De₃ Filippi (3) and Hewitt (4). More recently, Dronia (5) has obtained spot measurements on rock surfaces using an infra-red thermometer. However, no continuous measurements involving air, rock surface, and crack temperatures are available for this area. This paper is the first to attempt a detailed investigation of air, rock surface and crack temperatures in the Karakoram. Although data do exist for rock temperatures from elsewhere, these are confined to low altitude desert areas, see (6). The present investigations therefore represent an advance in the collection of basic data on which to assess the role of chemical weathering in the Karakoram.

METHODS

Temperature measurements obtained with two Grant Instrument Company Model D chart recorders are reported here Temperature sensors are bead, (FF) stainless steel rod probes (CM) and surface (EU) thermistor types.

Most surface and depth records were obtained using bead thermistors drilled in shallow surface depressions or to a depth of 5 cm. Thermal continuity was provided by sealing with a thermal transfer compound and, in the case of surface measurements, the probe was covered with powdered rock to give the same surface colouration as the original rock. In addition, 7 cm rock cubes were cut in the laboratory and pre-drilled for thermistor insertion. These cubes were then lagged on all sides except the exposed surface with expanded polystyrene to provide a standard for certain experiments. Basalt and sandstone from Northern Ireland were used in these experiments. Air temperatures are given as shade values.

Temperature precision is better than ±0.5°C: accuracy is no worse than ±1°C depending upon the recording scale used. Observations were taken at either 15 or 30 minute intervals over periods of variable length. Only selected time periods are presented in this paper.

DATA ANALYSIS

The data are presented according to the sites used. All of the sites were in the vicinity of the Aliabad Base Camp of the 1980 expedition (2200 m, 74° 37'E, 36° 18'N). Two sites were on the NNW ridge of Mirshikar (5445 m 74° 37'E, 36° 14'N); Mirshikar at 4250 m and Mirshikar Upper at 4400 m. The Nagar Road site was just off the road to Nagar village from the Karakoram Highway (approximately 2200 m, 74° 40.5'E, 36° 18.5'N). Names, elevations and positions just given are taken from Schneider (7) except for the Mirshikar sites which are given approximately in metres above sea level obtained by altimeter and checked against heights given by Schneider (7).

MIRSHIKAR RIDGE

The site was on a narrow section of the ridge at the very upper limit of grass growth.

The measurements from this site permit examination of temperature changes on differently oriented (south-east- and west-facing) rock surfaces, and within cracks of different widths (0.3, 0.5 and 10 cm), and relationships between these and air temperature variations.

The temperatures recorded during two out of the ten days of monitoring are depicted in Figure 1. Supplementary information abstracted from these graphs is presented in Table 1. The temperatures experienced at each thermistor location during the observation period fluctuated between these two extreme situations, but the relationships between the temperature variations at each location remained constant.

TABLE 1: MIRSHIKAR

	Maximum temperature	Minimum temperature	Diurnal temperature range	Cooling rate (^{o}C hr.$^{-1}$)
Air	17.4	2.0	15.4	1.231
South-east-facing rock surface	30.0	0.0	30.0	1.900
West-facing rock surface	41.0	1.1	39.9	2.467
0.3 cm wide crack	27.5	0.1	27.4	1.985
0.5 cm wide crack	19.9	1.1	18.8	1.464
10 cm cavity	17.5	0.3	17.2	1.342

(a) Date: 21.7.80 – 22.7.80 Clear conditions

	Maximum temperature	Minimum temperature	Diurnal temperature range	Cooling rate (^{o}C hr.$^{-1}$)
Air	8.9	2.0	6.9	0.394
South-east-facing rock surface	19.9	2.0	17.9	1.020
West-facing rock surface	14.9	2.5	12.4	0.689
0.3 cm wide crack	11.3	1.5	9.8	0.560
0.5 cm wide crack	10.2	1.5	8.7	0.483
10 cm cavity	8.9	1.0	7.9	0.465

(b) Date: 15.7.80 – 16.7.80 Cloudy conditions

619

FIG. 1. TEMPERATURE CURVES FROM MIRSHIKAR RIDGE.

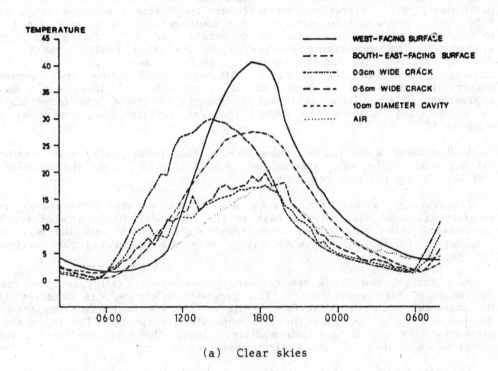

WEST-FACING SURFACE
SOUTH-EAST-FACING SURFACE
0·3cm WIDE CRACK
0·5cm WIDE CRACK
10cm DIAMETER CAVITY
AIR

(a) Clear skies

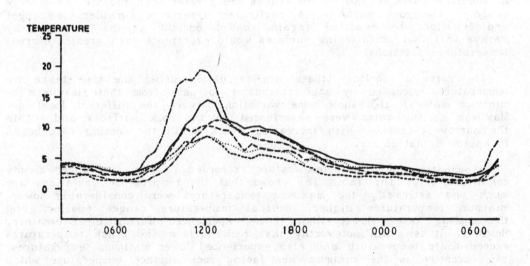

(b) Cloudy

The patterns recorded during clear weather conditions reveal some striking differences between the temperatures at the different locations, particularly with respect to the maximum temperatures attained (Fig. 1a; Table 1a). As would be expected, the south-east-facing surface was first to heat up, reaching its maximum temperature of 30 $^\circ$C during the early afternoon. The west-facing surface began its rapid heating phase some five hours after the south-east-facing surface, and attained a higher maximum temperature of 41 $^\circ$C. Although south-facing rock surface temperatures could not be monitored, it is possible that these would have approached 50 $^\circ$C: data published by Mathys (8) show that temperatures on south-facing surfaces can exceed those on surfaces facing west by as much as 10 $^\circ$C.

At the times when the maximum rock surface temperatures were attained the air temperature was exceeded by 18.7 $^\circ$C and 24.7 $^\circ$C on the south-east and west-facing surfaces respectively.

Temperatures within the cracks also exceeded air temperature, the greatest excesses being in the case of the narrowest (0.3 cm wide) crack. The wider cracks experienced lower temperature maxima with the maximum temperature in the 10 cm wide cavity only just exceeding the maximum air temperature.

Rock surface and crack temperature minima were in all cases less than the minimum air temperature. The greatest difference was observed for the south-east-facing rock surface, while the west-facing surface temperature dropped only slightly below the air temperature minimum. The temperature minima in the cracks are intermediate between the rock surfaces in the extent to which they fall below air temperature.

The net result of these variations is that diurnal temperature ranges at the rock surfaces and within cracks are greater than the air temperature range. The rock surfaces in particular experience considerable ranges and it might be expected (again, based on the graphs presented by Mathys (8)) that south-facing surfaces would experience even greater diurnal temperature variations.

The rates of cooling (these are calculated using the time taken for temperatures recorded by each thermistor to pass from their maximum to minimum values) also show some variation between the different locations. Maximum cooling rates were experienced on the rock surfaces and within the narrowest crack. With increasing crack width the cooling rate began to approach that of air.

Examination of the temperatures recorded at this site under cloudy conditions (Fig. 1b; Table 1b) shows that the temperature variations are much less extreme: the maximum temperatures were considerably lower, minimum temperatures higher, diurnal temperature ranges smaller, and the cooling rates slower than those experienced during clear weather. However, it is again noteworthy that rock surface and crack temperatures exceeded air temperature and also experienced lower minimum temperatures. (An exception is the minimum west-facing rock surface temperature which remained at 0.5 $^\circ$C above minimum air temperature.) The only difference in the actual patterns of temperature change was that under cloudy conditions the temperature at the south-east-facing rock surface exceeded that at the west-facing surface. An increase in cloud cover beginning at around midday would have prevented radiant heating of the west-facing surface during the afternoon.

MIRSHIKAR UPPER

This site is the highest (ca 4400 m) point easily accessible on the NNW ridge of Mirshikar. The ridge is very narrow at this point (Fig. 2) although broken rocks cover much of the surface. The gully to the

FIG. 2.   THE RIDGE AT THE MIRSHIKAR UPPER SITE.
NOTE THE BASALT BLOCK LAGGED BY POLYSTYRENE.

north contained snow and ice which nearly reached the ridge. The rocks may have been wetted by mist as well as rain two or three times during the record period although direct observation showed that they dried out rapidly when the cloud lifted.

A total of seven thermistors were used at this site to measure air, north and south facing rock surfaces and crack temperatures. In addition, surface and 5 cm depth temperatures were recorded from a basalt block embedded in polystyrene (Fig. 2). Temperature curves are shown in Figure 3 and summary data in Table 2.

The maximum temperatures at the north- and south-facing surfaces of the local (light grey) rock were 10.3°C and 9.2°C less than the maximum basalt surface temperature. These differences are of the same order or magnitude as those recorded by Kelly and Zumberge (9) and Moitke (10) for dark- and light-coloured rocks in Antarctica and illustrate the potential significance of colour as a control on rock temperature regimes. The importance of aspect in determining maximum rock surface temperatures is also highlighted.

All the maximum rock surface temperatures were considerably greater than air temperature. This also applies to the basalt temperature at 5 cm depth which exceeded the air temperature maximum by a considerable

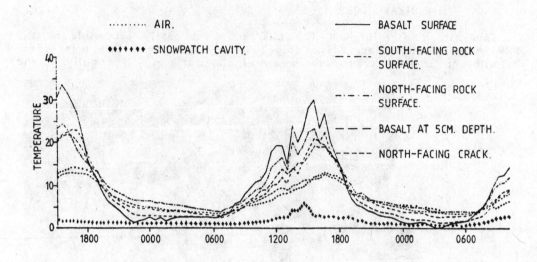

FIG. 3. SELECTED TEMPERATURE CURVES FOR MIRSHIKAR UPPER SITE

	Maximum temperature	Minimum temperature	Diurnal temperature range	Cooling rate ($^{o}C\ hr.^{-1}$)
Air	14.3	5.9	8.4	0.600
North-facing rock surface	23.0	3.0	20.0	1.481
South-facing rock surface	24.3	5.9	18.4	1.227
Basalt surface	33.5	1.1	32.4	4.629
Basalt : 5 cm depth	21.9	2.8	19.1	1.364
North-facing crack	12.9	3.3	9.6	0.711
Snowpatch cavity (5 cm wide)	5.9	1.1	4.8	-

TABLE 2: MIRSHIKAR UPPER STATION
Date: 30.7.80 - 1.8.80

7.6 °C. If it were assumed that the temperature gradient in the local rock was similar to that in the basalt, then 5 cm depth temperature maxima would be lower than in air. By controlling the extent of heating at the rock surface, rock colour would have some effect on the degree of heating experienced at depth.

Whilst the maximum temperatures in the crack and beneath the snowpatch did not exceed the air temperature maximum, the temperature changes at these locations nevertheless appear to be determined by rock temperature variations. The crack temperature is seen to rise more rapidly than air temperature during the early morning period when rock surfaces begin to respond to insolation. Also, the short term temperature fluctuation occurring below the snowpatch during the early afternoon (31/7/80) is seen to mirror the more pronounced variations experienced at the rock surfaces.

Apart from the south-facing rock surface, minimum temperatures at all locations dropped below minimum air temperature, the greatest difference being recorded for the basalt surface (4.8° C). The temperature at 5 cm depth in the basalt also dropped quite considerably below air temperature (3.1°C), and to a greater extent than the local rock surfaces.

The maximum temperature range of 32.4°C was experienced at the basalt surface, followed by that at the north-facing surface (20.0° C) and at 5 cm depth in the basalt (19.1°C). It is noteworthy that the range within the basalt block exceeded that recorded at the south-facing surface. Because of its lower night temperatures, the crack experienced a greater range of temperature change than air temperature which fluctuated minimally over a range of only 8.4°C.

Cooling rates are ranked in exactly the same order as for the temperature ranges at each thermistor location. The value of 4.629°C hr^{-1} is much greater than any values calculated from the previously described sites (the next most rapid value being 2.467°C hr^{-1} for the west-facing rock surface at the Mirshikar recording site).

NAGAR ROAD

This site was on desert varnished rocks by the site of the Chinese construction camp. The thermistors were used to measure air temperature and temperatures at two varnished rock surfaces (light and dark coloured), at a light-coloured, unvarnished rock surface and in the basalt block at 5 cm depth. Attention is focused on the time during which the greatest temperature variations occurred; see Figure 4 and Table 3.

The effects of rock colour on maximum surface temperatures attained are not as evident as at the Mirshikar Upper Site. The dark varnished surface shows an extremely rapid response to direct insolation, its temperature rising from 17°C to 46°C within three hours. After this initial rapid heating the temperature remained relatively constant and exceeded the maximum temperatures at the light-coloured rock and varnished surfaces by only 1.3°C and 2.0°C respectively.

Because they experienced lower night temperatures, the light-coloured surfaces had greater diurnal temperature ranges than the dark varnished surface. In all cases, surface temperatures exceeded air temperatures during the day and experienced considerably greater diurnal ranges.

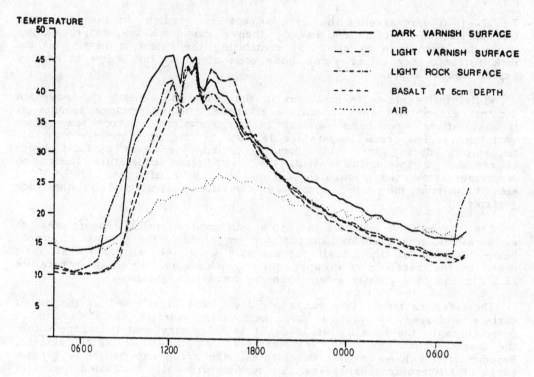

FIG. 4.   TEMPERATURE DATA FROM THE NAGAR ROAD SITE.

	Maximum temperature	Minimum temperature	Diurnal temperature range	Cooling rate (^{o}C hr.$^{-1}$)
Air	26.5	14.0	12.5	0.648
Dark-coloured varnish surface	46.3	14.0	32.3	1.724
Light-coloured varnish surface	44.3	10.5	33.8	1.745
Light-coloured rock surface	45.0	10.0	35.0	2.084
Basalt : 5 cm depth	39.5	10.2	29.3	1.519

TABLE 3.   NAGAR ROAD SITE: SUMMARY DATA
Date: 11.8.80 – 12.8.80

The minimum temperature of the dark varnished surface did not drop below that of air, but minimum temperatures at the light rock and varnished surfaces lay below minimum air temperature by $4.0^{\circ}C$ and $3.5^{\circ}C$ respectively.

The temperature variation at 5 cm depth in the basalt is also quite extreme as compared to the air temperature regime. Basalt surface temperatures were not measured. However, since basalt surface and 5 cm depth temperatures differed by $11.6^{\circ}C$ at the Mirshikar Upper Site, it is probable that basalt surface temperature at this location would have risen above $50^{\circ}C$. It is also probable that the diurnal temperature range at the basalt surface would have exceeded $40^{\circ}C$.

Cooling rates, even at 5 cm depth in the basalt, were considerably greater than in air. The most rapid rate was experienced at the light-coloured rock surface, over three times as fast as the rate of air temperature decrease.

ALIABAD CAMP SITE

The eastern boundary wall of the Expedition camp site, at Aliabad on the Hunza River Terrace, provided a useful location for some detailed experiments on heating rates and rock types. The wall was 1.6 m high and ran in an east-west direction.

Three thermistors were used to measure air temperature, and surface and depth (5 cm) temperatures of a sandstone block embedded in polystyrene. Temperatures were recorded every fifteen minutes during a period of twenty six hours (12/8/80 - 13/8/80). The data supply some further details on rock heating and cooling, at both surface and 5 cm depth, in relation to air temperature variations; see Figure 5; Table 4.

Both maximum temperatures and diurnal temperature ranges experienced are greater at this site than at any of the previously described locations. The most extreme temperature variations were at the sandstone surface where a maximum temperature of $54^{\circ}C$ and a diurnal range of $44.8^{\circ}C$ were recorded. Laboratory derived data show that maximum surface temperature of the dark coloured basalt used previously can exceed that of the sandstone by up to $4^{\circ}C$ thus suggesting that temperatures at dark rock surfaces at this location could approach $60^{\circ}C$.

The variation of temperature within the sandstone is also considerable, with maximum temperature exceeding that of air by $5.2^{\circ}C$, and the diurnal temperature range exceeding that of air by $5.5^{\circ}C$. When basalt surface and sub-surface temperatures were recorded simultaneously (at Mirshikar Upper Station), maximum temperatures differed by $11.6^{\circ}C$, and the sub-surface diurnal temperature range was 59% of the surface range. The corresponding values for the sandstone are $7.8^{\circ}C$ and 82%. These differences suggested that these two rocks may well have contrasting thermal properties which would control the extent to which surface temperature variations are experienced at depth. Subsequent measurement of thermal conductivities has shown this to be the case, the values for the basalt and sandstone being 0.96 $Wm^{-1}K^{-1}$ and 1.05 $Wm^{-1}K^{-1}$ respectively.

Rates of cooling were again most rapid at the rock surface and within the rock at 5 cm depth - even though the air temperature decreased at a rate of $2.145^{\circ}C$ hr^{-1}, the most rapid rate recorded for air during any period of observation of the sites discussed here.

626

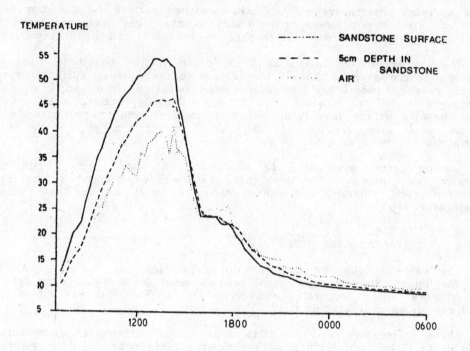

FIG. 5.   TEMPERATURE DATA FROM THE ALIABAD CAMPSITE,
WALL EXPERIMENT.

	Maximum temperature	Minimum temperature	Diurnal temperature range	Cooling rate (^{o}C hr.$^{-1}$)
Air	41.0	9.9	31.1	2.145
Sandstone surface	54.0	9.2	44.8	2.675
Sandstone : 5 cm depth	46.2	9.6	36.6	2.361

TABLE 4.   ALIABAD CAMPSITE
Date: 12.8.80 – 13.8.80

SUMMARY OF DATA FROM THE KARAKORAM RANGE

The relatively long time period (compared with other observers) during which temperatures were monitored in the Karakoram Range permits some general summary statements to be made:

a) The data support the contention that air temperature data provide a totally inadequate basis upon which to formulate ideas and hypotheses concerning rock weathering processes.

b) Discrepancies between air and rock temperatures are most marked under clear skies when radiant heating induces rock surface temperatures which are far higher than air temperatures. During cloudy conditions, such excesses are reduced, but can still be quite considerable, reaching 6.0°C on a west-facing rock surface at the Mirshikar North Ridge site. Maximum rock surface temperatures often exceeded 40°C and in one case (Aliabad Camp Site) exceeded 50°C. It is suggested, however, that dark-coloured rock surface temperatures can approach, and perhaps exceed, 60°C.

c) Under clear skies, rock surface temperature minima usually lie below minimum air temperatures: by as much as 4.8°C in the case of the basalt surface at Mirshikar Upper. During cloudy conditions, rock surface temperatures do not drop below air temperature to any great extent.

d) The net result of the higher temperature maxima and, quite often, lower temperature minima, is that rock surfaces experienced much greater diurnal temperature ranges than air. Under clear skies, diurnal rock surface temperature ranges varied between 18.4°C (south-facing rock surface at Mirshikar Upper) and 44.8°C (Aliabad Camp Site). Limits for air temperature ranges were 8.4°C (Mirshikar Upper) and 31.1°C (Aliabad Camp Site). Diurnal rock surface temperature ranges were normally between 30°C and 40°C, and those in air between 6°C and 16°C. Under cloudy skies, the diurnal temperature ranges were considerably reduced (to 12.4°C at the west-facing rock surface and 6.9°C for air at Mirshikar North Ridge).

e) Cooling rates were greatest at the rock surfaces during clear weather conditions, ranging from between 1.227°C hr^{-1} (south-facing rock surface at the Mirshikar Upper Site) to 4.629°C hr^{-1} (basalt surface at the Mirshikar Upper Site) as compared to values of between 0.6°C hr^{-1} (Mirshikar Upper Site) and 2.145°C hr^{-1} (Aliabad Camp Site), for air. Under cloudy skies, cooling rates decreased significantly to between 9.689°C hr^{-1} and 9.394°C hr^{-1} (west-facing surface and air at Mirshikar North Ridge).

f) The records from the Mirshikar Upper and Mirshikar North Ridge sites provide a limited amount of information on rock temperature variations in relation to aspect, but generalisations are not possible.

g) No general statements can be made concerning the effects of colour on rock temperature regimes beyond noting that dark-coloured rock surfaces attain higher temperatures than lighter-coloured rock surfaces during clear radiation weather.

h) Two rock types, basalt and sandstone, were used to measure temperature at 5 cm depth. The measurements show that maximum temperature, diurnal temperature ranges and cooling rates were all greater than those experienced in air.

i) Crack temperature regimes respond to rock temperature changes and are rarely affected by air temperatures. Maximum temperatures, diurnal temperature ranges and cooling rates were generally greater than in air.

j) Crack width seems to be an important control on the temperatures experienced within cracks. Narrow cracks are most likely to reflect rock temperature variations. With wider cracks, the product of circulation of air and increasing distances between opposite crack walls, would minimise the heating and cooling effects of the rock itself: temperature changes would therefore be less extreme.

ROCK BREAKDOWN IN HIGH MOUNTAINS

High mountain environments are apt to be thought of as essentially cold. Whilst this is undoubtedly true for winter and annual values, summer temperatures can be very high, as demonstrated in this paper. Furthermore, rock and crack temperatures can also be considerably higher than air temperature so that the latter alone provides no true guide to the temperature variations which rocks may experience.

The idea of high mountain regions (such as the Karakoram) being cold places gives the impression that frost shattering is the only process involved in rock breakdown. Although this may be true of the very highest summits (say above 5000 m) by far the greatest land area is at much lower altitudes than this. Only crude estimates can be made without good detailed maps but approximately one third of the Hunza area lies between 2500 and 5000 m in altitude. The temperature observations suggest that chemical weathering is likely to take place at least up to 4000+ metres and that very high rock temperatures can be achieved. Moisture and high positive temperatures are conducive to rock breakdown. Figure 6 was taken from below the Mirshikar Upper Site to the east and shows a typical, rather shattered mountainside with both bedrock and scree. The usual assumption would be that the scree has been produced by freeze-thaw action. This may be a contributor certainly but quantitatively its importance is not known.

OBSERVATIONS OF CHEMICAL WEATHERING ABOVE 4000 M

Figure 7 shows bedrock on the Mirshikar ridge just below the site of the Upper Station. The rocks are complex on this mountain and show diorite in this case. The cracks through the rock are picked out in a few places by lichen but mostly by colour differences as weathering takes place along them. Weathering rinds are prominent in hand section up to 10 mm thick. The small rocks seen in Figure 2 also have rinds of varying thickness. The importance of cracks and microcracks in weathering has been pointed out elsewhere (11) and close examination of the rock seen in Figure 7 (and elsewhere on the ridge) shows fine cracks 1 mm in width. Microcracks of the order of a few tens of micrometres wide would thus also be expected and it is along even these narrow cracks that weathering into small blocks can take place. On the whole, the weathering processes on the upper part of the Mirshikar ridge do not lend themselves to granular disintegration.

FIG. 6. VIEW ACROSS THE LOWER NORTH FACE OF MIRSHIKAR SHOWING
THE RATHER SHATTERED NATURE OF THE ROCK.

Other sites, near the main Mirshikar temperature site do show granular
disintegration of the "onion skin" or spalling type (Fig. 8). Mainly chemical
action is thought to be responsible for this process. Salt efflorescences
were nowhere seen on the rocks on Mirshikar.

DESERT VARNISH

Desert varnish is a concentration of iron and manganese oxides as a
thin coat on rocks. Its presence in the Hunza Valley is of particular interest
in relation to weathering processes. A general review of ideas about its
formation and significance is given by Whalley (13). It is common in the
roughly comparable (although somewhat drier) Hindu Kush Mountains of
Afghanistan (13) as well as the Karakoram. However, there are some
differences between desert varnish distribution in these areas. In the

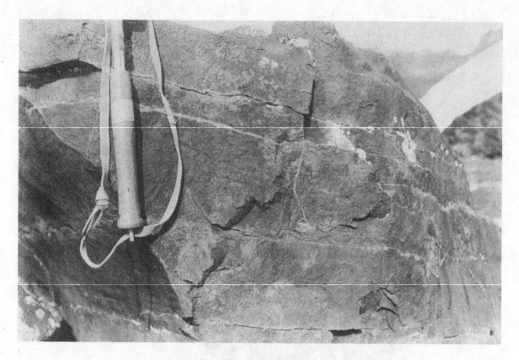

FIG. 7. WEATHERING ALONG CRACKS IN BEDROCK AT
MIRSHIKAR UPPER SITE.

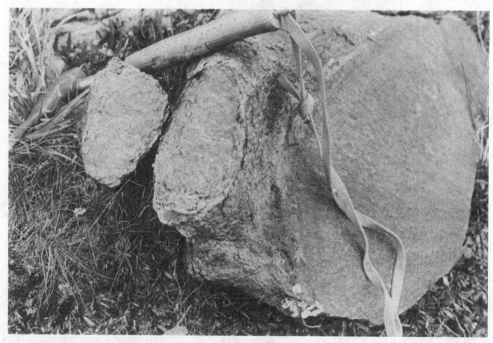

FIG. 8. ONION-SKIN TYPE WEATHERING TO FORM A CORESTONE
AT MIRSHIKAR.

Hindu Kush, varnish coats occur commonly on mountain ridges and summits well over 4000 m as well as in valleys at 2000 m (13). Some 300 km to the east, in the Hunza Valley of the Karakoram, varnishes are found commonly in the valley and on lower mountain slopes (Fig. 9). However,

FIG. 9.   DESERT VARNISH COVERING BOULDERS AT THE NAGAR ROAD SITE.

as altitude increases, varnishes are much thinner and above 3000 m are apparently non-existent.   Weathering rinds (12) do occur on most rocks however at 4000 m and above.   Weathering rinds differ from desert varnish in that ferric manganese concentrations on the outermost rock surfaces are not found but alteration of ferrous to ferric states of minerals in the rocks does occur.   The rinds may be several millimetres thick (dependent upon rock type).   Varnishes are of the order of tens of micrometres thick only.

Rock type differences do not account for the variation in varnish and rind formation.   The most likely reason is that of precipitation.   Rock temperatures are sufficiently great to enable weathering processes to take place at 4000 m but rainfall does differ widely between 4000 m and above and the valley bottoms (about 2000 m).

Although no precipitation data exist to enable comparison of valley and mountain precipitation, observations in the summer of 1980 suggest that clouds often surround certain mountain sides, even though the valleys may be sunny.   Light rain falling on rocks on mountain sides is probably an effective means of either removing desert varnish or preventing its formation. Elvidge (14) has shown that desert varnish on a rock from Arizona will be removed within months if it is moved to a humid climate.   The important climatic factor in varnish formation is thus aridity rather than temperature. We thus suggest that increases in precipitation (or humidity) vertically from valleys to mountain ridges are sufficient to prevent varnish formation.

Chemical weathering otherwise is not impeded (it may even be accentuated because of water presence) above the "limits" of varnish formation.

CONCLUSIONS

Rock surface temperatures can reach 10°C or more above air temperatures even at 4000 m and above. Such surface temperatures may easily reach 30°C under sunny conditions and 20°C under cloud. When coupled with moisture, either from cloud or precipitation, conditions very favourable to chemical weathering of rocks may be achieved in the summer. This is thought to be a significant part of rock breakdown on mountain flanks between 3500 and 4500 m.

Desert varnish is rare above 3000 m and is mainly confined to valley bottoms and lower sides. This shows the importance of the increased moisture at higher elevations as an inhibitor to desert varnish formation rather than a decrease in rock surface temperature or lessening of chemical weathering at high elevations.

REFERENCES

1) PELTIER, L., (1950). 'The geographic cycle in periglacial regions as it is related to climatic geomorphology', Ann. Assoc. Amer. Geogrs. 40, pp 214 - 236.

2) CORBEL, J., (1964). 'Erosion terrestre, étude quantitative (méthodes - techniques - résultats', Ann. Geogra., 73, pp 385 - 412.

3) DE FILIPPI, F., (ed.) (1929). 'Relazione scientifiche. Spedizione italiana de Filippi nell, Himalaia, Caracorum e Turchestan Cinese (1913 - 14)', Series 1, 3 vols., Bologna.

4) HEWITT, K., (1968). 'The freeze-thaw environment of the Karakoram Himalaya', Canadian Geographer, 12, pp 85 - 98.

5) DRONIA, H., (1978). 'Gesteinstemperaturmessungen im Himalaya mit einem Infrarot-Thermometer', Zeit. für Geomorphologie, 22, pp 101 - 114.

6) McGREEVY, J. P. and SMITH, B. J., (in press). 'Salt weathering in hot deserts: observations on the design of simulation experiments', Geog. Anner.

7) SCHNEIDER, H. J., (1969). 'Minapin - Gletscher und Menschen in N. W. - Karakorum', Die Erde, 100, pp 266 - 286.

8) MATHYS, H., (1974). 'Klimatische Aspekte zur Frostverwitterung in der Hochgebirgsregion', Naturforschenden Gesell, Bern, Mitt., 31, pp 46 - 62.

9) KELLY, W. C. and ZUMBERGE, J. H., (1961). 'Weathering of a quartz diorite at Marble Point, McMurdo Sound, Antarctica', J. Geol., 69, pp 433 - 446.

10) MOITKE, F. D., (1980). 'Microclimate and weathering processes in the area of Darwin Mountains and Bull Pass, Dry Valleys', Antarctic J. United States, 15, pp 14 - 16.

11) WHALLEY, W. B., DOUGLAS, G. R. and McGREEVY, J. P., (1982). 'Crack propagation and associated weathering in igneous rocks', Zeit. fur Geomorphologie, 26, pp 33 - 54.

12) WHALLEY, W. B., (1981). 'High altitude rock weathering processes', these proceedings, Volume 1, pp 365 - 373.

13) WHALLEY, W. B., (in press). 'Desert Varnish', in Goudie, A. S. and Pye, K., (eds.): 'Chemical Sedimentation and Geomorphology', Academic Press, London.

14) ELVIDGE, C. D., (1979). 'Distribution and formation of desert varnish in Arizona', Unpub. MS thesis, Arizona State University, Tempe, 112 pp.

The Karakoram research cell

S. A. R. Jafree

Director KKRC, Department of Earth Sciences, Quaid-i-Azam University, Islamabad, Pakistan

Northern Areas of Pakistan, including the Karakoram range are one of the most beautiful scenic areas of the world. Characterized by the greatest concentration of highest peaks on earth, shielded by the mountain systems of Afghanistan, U.S.S.R, China and India, topped by majestic glaciers and crisscrossed by countless streams, lakes and blank patches, it is a paradise and laboratory for social and natural scientists.

Countless numbers of international expeditions in the past few decades have been lured to this part of the world but their objectives and their results have been restricted; failing to produce any impact on the social and technological improvement of the region. One of the main reasons has been the lack of institutionalized, integrated and multidisciplinary research, pooling of expertise and knowledge of the Pakistani Universities and professional organisations. Paucity of funds is another significant deterrent. This has resulted in a lack of development and research programmes in this region despite its wonderful potential.

In 1980, as a follow-up of an 'International Conference on Recent Technological Advances in Earth Sciences' together with the 'Karakoram Project 1980', Karakoram research has been institutionalized by the Pakistan University Grants Commission with the establishment of the 'Karakoram Research Cell' (KKRC) with its office located at the Department of Earth Sciences, Quaid-i-Azam University. Its mandate has been broadened to include research in both the fields of Social and Natural Sciences involving Pakistan Universities and other professional organisations.

To date the KKRC has received 7 projects from three Universities of Pakistan, these are:

(1) Economic Geology Project (Punjab University, Lahore)

(2) Geophysical Project (Quaid-i-Azam University)

(3) Agricultural Research Project (Peshawar University)

(4) Hard Rock Geology Project (Peshawar University)

(5) Flora and Fauna Project (Quaid-i-Azam University)

(6) Cultural Issues and Sociological Perspective in the Planning and Development of Karakoram Region and Northern areas of Pakistan. (Quaid-i-Azam University)

(7) Establishment of a 'Karakoram Museum' at Gilgit to represent the natural and cultural heritage of the Northern areas. (Quaid-i-Azam University)

Each of these projects is headed by university professors who are experts in these fields and each project is likely to be completed in a 3 - 4 year period. The U.G.C. is financing some of these projects (1 to 5), while for project No. 6 the Ministry of Local Govt. and Rural Development of the Govt. of Pakistan is the sponsor. For project No. 7 the Agha Khan Foundation in Karachi is being approached for funding.

It is expected that these projects will be launched in the forthcoming field season starting May 1983. In order to initiate a process of multi-disciplinary, inter-university and inter-organisational involvement it is planned that each field season will be followed by seminars to adjudge, overview, plan and integrate the overall research effort and ensure a continually improving academic and practical output. Dissemination of expertise, methodologies and technology within Pakistani Universities and other professional organisations is being ensured by mandatory M.Sc, M.Phil., and Ph.D. student thesis inputs for each project.

The KKRC has planned to publish a scientific journal entitled 'PAKISTAN JOURNAL OF KARAKORAM RESEARCH' on an annual basis. The first issue (expected to be out by April 1983) is to be a bibliographical issue, highlighting the various disciplines and projects already completed by national and international scientists. Later issues will encompass all forthcoming research in the northern areas of Pakistan.

A reference library on Karakoram and northern areas research is already functioning in the Department of Earth Sciences, Quaid-i-Azam University, Islamabad and it has acquired a number of pertinent papers, books, maps and satellite imageries.

The Karakoram Research Cell is supervised by a high-powered national 'Project Identification and Implementation Committee' headed by the U.G.C. Chairman and is comprised of senior representatives from Ministries, Universities and other professional organisations. Its mandate is to identify, develop, execute and evaluate K.K. research projects, thereby functioning as a clearing house on all national and international research on the Karakoram and the Northern areas of Pakistan.

Every 4 or 5 years, the KKRC intends to hold an international conference similar to the one held in 1980 at the Quaid-i-Azam University, Islamabad. The next conference is scheduled to be held in 1985 or 1986.

Prof. Antonio Marussi, University of Trieste, Italy and Prof. Keith Miller, University of Sheffield, U.K. have kindly accepted to act as advisors to the KKRC.